Virus Diseases of Fruit Crops and Management

by

Y.S. Ahlawat

&

K.P. Srivastava

2016

Studium Press (India) Pvt. Ltd.

Virus Diseases of Fruit Crops and Management

ISBN: 978-93-80012-74-2

Published by:

Studium Press (India) Pvt. Ltd.
4735/22, 2nd Floor, Prakash Deep Building (Near PNB),
Ansari Road, Darya Ganj, New Delhi-110 002
Tel.: +91-11-4324200-15 (15 lines); Fax: +91-11-43240215;
jngovil@gmail.com; jngovil@hotmail.com

Printed at:
Global Printing Services, Delhi

FOREWORD

The total fruit production in the world is about 609,213,509 Metric tons. In India, fruit production without melons is about 84,791,100 metric tons. Most of the fruit plants have their origin in Asia, Australia, North America, and in Mediterranean regions. Of 3000 plant species only 12 species of grain crops, 23 vegetable crops and 35 species of fruits are grown all over the world. Important fruit plants grown all over the world are: Almond (*Prunus dulcis)*, apple ((*Malus domestica)*, Apricot (*Prunus armeniaca)*, Avacado (*Persea americana*), Banana, plantain (*Musa spp)*, Blackberries (*Rubus spp)*, Blueberries (*Vaccinium spp)*, Cacao or cocoa (*Theobroma cacao)*, Cashew (*Anacardium occidentale*), cherries ((*Prunus cerasus, P. avium)*, Chestnut (*Castanea spp.)*, Citrus (*Citrus species*), coconut (*Cocos nucifera*), Cranberry (*Vaccinium macrocarpon*), currents (*Ribes spp)*, Date (*Phoenix dactylifera)*, fig (*Ficus carica)*, gooseberry (*Ribes grossularia; R. hirtellum)*, grapes ((*Vitis vinifera*, other *Vitis spp)*, guava (*Psidium guajava* & related spp), Hazelnut or filbert (*Corylus avellana)*, Juneberry (*Amelanchier alnifolia)*, kiwifruit (*Actinidia deliciosa)*, Kumquat (*Fortunella spp*), Loquot (*Eriobotrya japonica)*, macademia (*Macadamia integrifolia)*, mango (*Mangifera indica)*, mayhaw (*Crataegus spp.)*, olive ((*Olea europaea)*, papaya (*Carica papaya)*, peach (*Prunus persica)*, pears (*Pyrus communis, P. pyrifolia)*, pecan (*Carya illinoensis)*, pineapple (*Ananas comosus*), pistachio (*Pistacia vera)*, plums (*Prunus domestica, P. salicina)*, pomegranate ((*Punica granatum)*, Quince ((*Cydonia oblonga)*, raspberries (*Rubus idaeus, R. occidentalis)*, strawberry ((*Fragaria X ananassa)*, and walnut (*Juglans regia)*.

Diseases particularly of virus nature posed serious threat to fruit industry as these pathogens are difficult to manage. Viruses of vegetatively propagated fruit crops such as apple, peach, citrus and banana etc. are much more important economically due to perpetuation of these pathogens in these crops. Inadvertent propagation from contaminated sources is responsible for wide spread of these diseases and pathogens to wider areas in different geographic regions. These diseases are managed by quarantine measures, eradication of infected plants and weed hosts, crop rotation and use of virus-free certified propagation material and use of pesticides to manage insects which are vector of viruses. Breeding for disease resistance provides a better protection for longer time. Reliable techniques of virus diagnosis have been developed all over the world and now it is a major component in certification programmes.

The book entitled "VIRUS DISEASES OF FRUIT CROPS AND THEIR MANAGEMENT" prepared by Professor Y.S. Ahlawat and Dr. K.P. Srivastava is aimed to describe fruit crop diseases caused by viruses and allied pathogens such as viroids and phytoplasmas will certainly be very helpful to grow virus-free fruit plantations in developing countries. The life long experience of the authors in this particular field and their original discoveries encouraged them to bring out this book.

The book has six chapters: introduction, virus and virus-like diseases of plantations of *temperate fruits* (85 diseases of 9 crops), *shell fruits* (15 diseases of six crops), *tropical and sub-tropical fruits* (69 diseases of 8 crops), *arid fruits* (42 diseases of 7 crops), *small fruits* (82 diseases of 9 crops). Most of these diseases have virus aetiology but 40 diseases are caused by Phytoplasmas, two diseases caused by Prokaryotes and 19 diseases are associated with viroids. The description of each disease and the pathogen associated with it have been described in relation to their geographical distribution, economic importance, disease symptoms, transmission, vector(s), host range, indicator and assay hosts, nature and structure of associated pathogen , their genome organization, methods of detection and management. The book contains 52 figures and 70 colour plates and over 2000 relevant references.

The book is designed to upgrade the knowledge of graduate and post graduate students, teachers, orchardists and farm Managers. It will also be a useful document in developing pest risk analysis in National plant quarantine system and management of fruit diseases in a holistic manner.

R.B. Singh

Chancellor, Central Agricultural University
Immediate Past President, NAAS
DI/1291, Vasant Kunj
New Delhi 110070

ABOUT THE AUTHORS

Yashvir S. Ahlawat

Yashvir S. Ahlawat was born in a village Kandera in Baghpat district of Uttar Pradesh in 1943. He obtained his M.Sc. and Ph.D degrees from Agra University. He carried out his research assignments at Indian Agricultural Research Institute, (IARI) New Delhi and its Regional Stations in Shimla and Kalimpong. During his research career he discovered several virus and phytoplasma diseases infecting agricultural and horticultural crops new to India and World. However, his major interest had been in virus and virus like diseases of fruit crops especially in citrus. At IARI, Prof. Ahlawat worked at the advanced centre for Plant Virology developed with the assistance of UNDP of FAO and headed Virology Unit of IARI. During his 40 years of research career, Dr. Ahlawat guided 10 Ph.D. and 4 M.Sc. students. He published over 150 research papers in reputed scientific journals. He handled both National and International projects on virus diagnostics, HRD, and National Facility for Virus diagnosis and quality control of tissue culture-raised plants. Under HRD project, Dr. Ahlawat conducted several group and hands- on trainings on various aspects of plant virology including one refresher course for senior professors from Indian Universities on molecular diagnosis of plant viruses.

The research contributions made by Prof. Ahlawat have been recognized nationally and internationally. Nationally, he received several honours and awards from Indian Council of Agricultural Research (ICAR), Indian Phytopathological Society (IPS), and Indian Society of Citriculture (ISC) for outstanding research contributions in plant virology. Internationally, he was awarded *James M. and Adeline Wallace* Research award 1998 by International Organization of Citrus Virologists, Universty of California for his outstanding contribution in citriculture. Dr. Ahlawat is a member and fellow of several national and international professional societies and served them in various capacities. He was elected as President, IPS in 2005 and served as Secretary General and Vice-President in Indian Society of Citriculture.

As FAO Fellow, Dr Ahlawat worked at the University of California, Riverside and at Instituto Biologico at Sao Paulo, Brazil. He visited University of California again as IBPGR Expert to develop guidelines for safe movement of citrus germplasm. He participated and chaired scientific sessions in several national and international symposia in India and abroad. Professor Ahlawat visited INRA at Bordeaux in France twice as visiting Professor. He was consultant to the Royal Govt. of Bhutan and visited University of Florida, CREC, Lake Alfred in 2010. He edited 8 books, contributed several book chapters and popular articles. After his superannuation in 2005, he had been working as Emeritus Scientist ICAR and a consultant to CABI and currently working as ES, Department of Science and Technology, Govt. Of India.

Dr. Krishna Prasad

Dr. Krishna Prasad, an eminent Entomologist was born on April 1, 1938 at Varanasi. He started his research career at Indian Agricultural Research Institute (IARI), New Delhi and continued there till his supernuation in 1998. He worked in this prestigious Institute in various capacities right from Entomologist in 1963 to Principal Scientist at the time of retirement. He worked under the PL 480 Project from 1965 to 1977 and completed his Ph.D. degree during the project period. His work on insect genetics and host plant resistance in Sorghum has been recognized globally. He upgraded his knowledge by visiting advanced laboratories in USA. Dr. Srivastava is not only an excellent researcher but also a very popular teacher. Lately in his research career, he shifted to Advanced centre for Plant Virology at IARI and did good work on virus-vector relationship.

Dr. Srivastava is a Fellow of Plant Protection Association of India PPAI), Entomological Society of India, Indian Phytopathological

Society and Indian Society of Genetics and Plant Breeding. He is life member of many Professional Societies and served them in various capacities. He served as Vice-President of PPAI and was in Editorial Board of several professional Journals. He published over 150 research papers, and authored two books on Plant Protection. He also published scientific dictionaries on agriculture Botany and Entomology. He keeps him busy in writing and reading books on various agricultural subjects.

ACKNOWLEDGEMENT

We wish to thank to the Science and Engineering Board (SERB), Department of Science and Technology, Govt. Of India who sanctioned this project for funding under USERS scheme. We are very much thankful to Dr. S.S. Kohli, Director and Scientist-F, Department of Science and Technology for all his help regarding finances and his valuable advice from time to time. We are indebted to our students, Drs. D.K. Ghosh, R.P.Pant and D.R. Das, Avijit Roy, V.N. Hoa for their help in various ways during this project. I wish to thank Dr. V.K. Baranwal who provided some material for the book. We are indeed grateful to our professional friends for their valuable suggestion and help. We are also thankful to Head, Division of Plant Pathology and Director, IARI, New Delhi and administrative staff of the Division and Institute involved in various activities of the project. At last but not the least we wish to put on record our sincere thanks to Mrs. Urmila Ahlawat who had been actively involved with this project.

PREFACE

I, Professor Yashvir S. Ahlawat have been actively working on virus and virus-like diseases of plants with major interest in viruses infecting fruit tees. Dr. K.P. Srivastava, co-author of this book is a well known Entomologist. Since viruses are spread by insect vectors, we both have worked together for several years. It had been long felt need by scientists and fruit growers in India and abroad to bring out a comprehensive publication on virus and virus-like pathogens infecting various fruit trees/ plants with special emphasis to the research work done in the Indian subcontinent which is considered to be the original home of several fruits. Therefore, we decided to write our experiences on virus diseases of fruit crops in India and abroad. While working at the University of California, Riverside (UCR), I was exposed to temperate fruit tree viruses and Phytoplasma by Dr. G.N. Oldfield working at USDA Lab in UCR campus. I earlier worked on temperate fruit tree viruses at the Indian Agricultural Research Institute (IARI) Regional Stations at Shimla and Kalimpong. I have very wide experience of viruses infecting citrus and discovered new viruses and viroids such as Indian citrus ringspot *Mandrivirus,* citrus yellow mosaic *Badnavirus* etc. while working at the Advanced Centre for Plant Virology at IARI, New Delhi. My exposer to citrus viruses at UCR, University of Florida, CREC, Lake Alfred, INRA Bordeaux and Corsica, France and at Instituto Biologico at Sao Paulo, Brazil helped me to understand National and International problems of virus diseases of trees. During my research career, it had been my privilege to handle several National and International projects on human resource development and virus diagnostics. To avoid introduction of viruses through the movement of citrus germplasm, IOCV, FAO and IBPGR jointly published *FAO/IBPGR Technical Guidelines for the safe Movement of Citrus Germplasm* (Frison and Taher, 1991) and I was a member of the team which prepared these guidelines.

The first chapter of the book provides brief introduction about fruit plants, their origin in Asia, Australia, North America, and in Mediterranean regions. Of 3000 plant species only 12 species of grain crops, 23 vegetable crops and 35 species of fruits are grown all over the world. Important fruit plants grown all over the world have been listed. Information has been provided on economic importance of fruits, fruit producing countries and world fruit production. Viruses of vegetatively propagated fruit crops such as apple, peach, citrus and banana etc. are much more important economically due to perpetuation of these pathogens in these crops. Diseases caused by viruses, phytoplasmas (mycoplasma-like organisms) and viroids are among the greatest threat to fruit industry all over the world. Diseases caused by these pathogens have a direct impact on export and import of horticultural plants and products. Basic information about viruses, phytoplasma and viroids and their diagnostics have been provided. Fruit crops are more susceptible to virus and virus-like infections. These diseases are managed by quarantine measures, eradication of infected plants and weed hosts, crop rotation and use of virus-free certified propagation material and use of pesticides to manage insects which are vector of viruses. Molecular breeding has been utilized to develop virus resistant plants in several fruit crops. Virus resistant transgenic crops such as Papaya in Hawaii have been developed for commercialization in several countries like USA, China and Canada.

In chapter-2 virus diseases of temperate (pome and stone fruits) fruits have been described in detail. Among pome fruits, 15 diseases of apple (11 virus diseases, 3 phytoplasma diseases and two viroid diseases), 5 diseases of pear (3 virus diseases, one each of phytoplasma and viroid disease), 5 diseases of quince have been described. Among stone fruits, 8 virus diseases of apricot, 5 diseases of almond (4 virus diseases and one phytoplasma disease), 22 diseases of cherry (18 virus diseases, 3 phytoplasma disease and one viroid disease), 12 diseases of peach and nectarine (5 virus diseases, 6 phytoplasma diseases and one viroid disease) and 7 diseases of plum (6 virus disease and one phytoplasma disease) have been described Descriptions have also been provided on their distribution and economic importance, disease symptoms, modes of spread by various methods including their vector(s), host range, indicator plant(s). The nature and structure of associated pathogen, their genomic organization have also been discussed. Methods of detection of associated pathogen and possible management strategies have been described in detail.

Chapter-3 describes diseases of shell fruits such as walnut (4 virus and one phytoplasma disease), Pecan (one virus disease), Chestnut (Two virus disease), Hazelnut (two virus and three phytoplasma disease, kivi and persimmon (one disease of each). Diseases of these crops have been described about their distribution, economic importance, symptoms, transmission, methods of diagnosis and management if known. In this group, either the disease has been described recently or no further information could be generated on already known diseases.

In chapter 4, virus and virus-like diseases of tropical and subtropical fruit crops totalling 71 diseases have been described. In avocado two virus disease, in banana six virus disease, In citrus 32 virus diseases, two diseases caused by Prokaryotes, two diseases caused by phytoplasmas and one by *Spiroplasma citri* and eight diseases caused by viroids, In papaya six virus diseases, in olive one phytoplasma disease, in coconut one virus, four phytoplasma and three viroid diseases, in date palm two phytoplasma diseases and in oil palm one virus disease have been described. These diseases have been described for their distribution, economic importance, symptoms, host range, associated pathogen(s), methods of pathogen detection and possible control measures.

Chapter-5 carries information on virus and virus-like diseases of arid fruits. In this chapter 30 diseases of seven crops have been described. In fig two virus and one phytoplasma disease have been described. In jujube (Ber) one phytoplsma and one viroid disease, in pomegranate one viroid disease, in mulberry four virus disease and one phytoplasma disease, in pineapple two virus disease, in grapevine 14 virus diseases, seven diseases of unknown aetiology, four phytoplasma diseases and two viroid diseases have been described with up to date information. In bael, no virus disease has been described but it is known as host of some viruses such as Citrus tristeza closterovirus.

In chapter-6, 82 virus and virus- diseases of eight small fruit crops (strawberry, raspberry, blackberry, blueberry, cranberry, gooseberry, black currant and red currant) have been described. In strawberry, 25 diseases associated with virus and virus-like pathogens (14 virus and five phytoplasma diseases) have been discussed. In raspberry, blackberry, boysenberry, thimbleberry, brambleberry and wineberry, 22 virus diseases have been described. In blueberry and cranberry, eight virus and one phytoplasma disease have been explained. These diseases have been described with detailed information on distribution, economic importance, disease symptoms, transmission, vector(s), host range, associated pathogen, methods of disease diagnosis and management.

At the end of each chapter relevant references are given. In this publication 52 photographs and about 70 coloured plates showing symptoms on various hosts have been put and most of them were collected by us and only a few were downloaded from internet.

We are fully conscious that in this publication, there may be shortcomings and defects. We therefore, request our colleagues working in plant virology/horticulture to communicate/suggest the defects, if any, to the senior author so that these could be rectified in future editions. However, this publication should be of value to the students and teachers of plant virology, plant pathololology and horticulture for diagnosis and management of virus diseases of fruit crops.

We wish to express our deep appreciation to my colleagues in India and professional friends all over the world who helped in various ways to bring out this publication. They will be duly acknowledged.

Y.S. Ahlawat
K.P. Srivastava

TABLE OF CONTENTS

LIST OF COLOURED PLATES

LIST OF FIGURES

1

INTRODUCTION

Most of the fruit plants have their origin in Asia, Australia, North America, and in Mediterranean regions (Annexure-1). Of 3000 plant species only 12 species of grain crops, 23 vegetable crops and 35 species of fruits are grown all over the world. Important fruit plants grown all over the world are: Almond (*Prunus dulcis),* apple (*(Malus domestica),* Apricot (*Prunus armeniaca),* Avacado (*Persea americana*), Banana, plantain (*Musa spp),* Blackberries (*Rubus spp),* Blueberries (*Vaccinium spp),* Cacao or cocoa (*Theobroma cacao),* Cashew (*Anacardium occidentale*), cherries (*(Prunus cerasus, P. avium),* Chestnut (*Castanea spp.),* Citrus (*Citrus species*), coconut (*Cocos nucifera*), Cranberry (*Vaccinium macrocarpon*), currents (*Ribes spp),* Date (*Phoenix dactylifera),* fig (*Ficus carica),* gooseberry (*Ribes grossularia; R. hirtellum),* grapes (*(Vitis vinifera,* other *Vitis spp),* guava (*Psidium guajava* & related spp), Hazelnut or filbert (*Corylus avellana),* Juneberry (*Amelanchier alnifolia),* kiwifruit (*Actinidia deliciosa),* Kumquat (*Fortunella spp*), Loquot (*Eriobotrya japonica),* macademia (*Macadamia integrifolia),* mango (*Mangifera indica),* mayhaw (*Crataegus spp.),* oilpalm (*Elaeis guineensis*), olive (*(Olea europaea),* papaya (*Carica papaya),* peach (*Prunus persica),* pears *(Pyrus communis, P. pyrifolia),* pecan (*Carya illinoensis),* pineapple (*Ananas comosus*), pistachio (*Pistacia vera),* plums (*Prunus domestica, P. salicina),* pomegranate (*(Punica granatum),* Quince (*(Cydonia oblonga),* raspberries (*Rubus idaeus, R. occidentalis),* strawberry (*(Fragaria X ananassa),* and walnut (*Juglans regia).*

India has varied agro-climate, which is highly favourable for growing large number of fruit and other horticultural crops. The total land under cultivation in India is about 175 million hectares and horticultural crops occupy around 13% of India's gross cropped area, producing 177.41 million metric tonnes. India is the second largest producer of fruits and vegetables. The total production of fruits has been estimated at 52.85 million metric tonnes from an area of 5.34 million ha during 2005-06. India is the second largest producer of fruits and vegetables in the world next only to Brazil and China. Around 10% of the world fruit production is accounted by India and leads the world in the production of mango, banana, sapota and acid lime and in productivity of grapes per unit land area.

In India, fruits are being grown for thousands of years but efforts have not been made to develop among modern lines as has been done in case of agriculture. Total area under fruit cultivation in India has increased from two and a half million acres in 1943 (Hayes, 1966) to 5346 million hectares during 2005-06. In production sector also the fruit industry is expanding rapidly. The important fruits grown in India are : mango, citrus, banana, guava, papaya, litchi, grapes, fig, custard apple and other tropical fruits like jackfruit, spota or sapodilla (chiku) etc.. Some subtropical fruits like Jujube (Ber), Loquot (*Eriobotrya japonica*), Pomegranate, Phalsa (*Grewia asiatica*), Avacado, Bel (*Aegle marmelos*), Cape Gooseberry (*Physalis peruviana*), the date (*Phoenix dactylifera*), mulberry, olive, passion fruit, persimmon (*Diospyros tomentosa*) and tree tomato (*Cyphomandra betacea*) are also grown in india. Temperate fruits in India are mainly grown along the Himalayan ranges. Among them important ones are: apple, pear, quince, plums, apricot, peach, almond, cherry, blackberries, strawberry, walnuts and pecan (*Carya illinoensis*).

Virus and virus-like diseases of fruit crops are attracting increasing attention because of the better understanding of the nature of viruses, mollicutes and viroids infecting fruit crops. During the last two decades researchers have intensified their research on virus and virus-like pathogens affecting fruit crops. During this period many of the unknown diseases and pathogens have been discovered and characterized (Chenulu and Ahlawat, 1993; Ahlawat, 2012). Mixed infection of viruses and viroids appeared with alarming frequency in fruit crops causing heavy losses. Viruses of vegetatively propagated fruit crops such as apple, peach, citrus and banana etc. are much more important economically due to perpetuation of these pathogens in these crops. Inadvertent propagation from contaminated sources is responsible for wide spread of these diseases and pathogens to wider areas in different geographic regions.

Diseases caused by viruses, phytoplasmas (mycoplasma-like organisms) and viroids are among the greatest threat to fruit industry all over the world. Diseases caused by these pathogens have a direct impact on export and import of horticultural plants and products. In earlier days, a farmer used to grow fruits mainly for his own consumption, but with the increase in population and urbanization, the demand for these commodities increased greatly resulting heavy commercialization of cultivation of fruit plants all over the world. Demand of fruits and their processed products encouraged farmers and business organisations to grow fruit plants on large scale since it has become more profitable. During this process elite species were introduced which inadvertently brought many new diseases in the country. Therefore, better commercialization facilitated the spread of parasitic organisms from one region to another throughout the globe.

Microbes especially fungi and bacteria were established as causal organisms of many diseases of horticultural plant species. But there were several contagious diseases where no microorganism could be established. For example, Colour breaking in tulips was known for a long time in The Netherlands, but the cause of the abnormality was not known. Some nurserymen were aware about the secret to the production of colour breaking in flowers of tulip plants (by way of tuber grafting) and by this secrecy they sold infected tulip bulbs with thousand times more price than healthy bulbs. Other historical examples are variegation in leaves of *Abutilon striatum* var. thompsonii and degeneration of potato which were known in Europe since 18th century. The cause of such maladies could not be established till Dr. Martinus Willem Beijerinck in 1895 established that the cause of mosaic disease of tobacco (already established a transmissible disease by Adolf Edward Mayer (1882 – 86) was not a microorganism. He described it as *Contagium Vivum Fluidum*. Henceforth, the term virus was used for any such infectious matter where no microorganism could be seen. Number of virus and virus-like diseases in plants greatly increased after 1920.

However, proper identification of viruses remained uncertain till a landmark discovery made by Wendell M. Stanley in 1935 from United States of America. He crystallized a protein from mosaic affected tobacco plants and showed that the virus was proteinaceous in nature. This discovery brought Nobel Prize to him. Later, Bawden and Pirie (1937) demonstrated that viruses are not only proteins but are nucleoproteins (having nucleic acid and proteins). In 1939, German scientists, Kausche, Pfankuch and Rusca for the first time observed rod-shaped particles of tobacco mosaic virus in newly constructed electron microscope. With the development of modern electron microscope and ultracentrifuges since 1945 onward several viruses could be isolated from their hosts. So far about 700 plant viruses have been reported in agricultural and horticultural plants.

Many important discoveries were made between 1950 to 1975 on virus chemistry and structure, identification of new pathogens like mycoplasma-like oraganisms and viroids. Viruses have several features in common with macromolecule of healthy plants and animal cells. Study on Chemistry and Biophysics of such macromolecules is now termed as *Molecular biology*. Researches in this area have provided us more information and knowledge of virus structure, function and biological behaviour. The science of molecular biology provided us important information on genomics of viruses and made it possible to easily establish whether a virus has ribonucleic acid (RNA) or deoxyribose nucleic acid (DNA).

The causal agents of several plant diseases especially known as *yellows* such as peach yellows, mulberry dwarf etc. could not be established as no pathogen could be isolated from plants showing *yellows* symptoms. These diseases were transmissible similar to that of plant virus diseases. Among such diseases, peach yellows is known since 1888 but no causal organism was established although graft transmission and leafhopper (*Macrosteles fascifrons)* transmission of peach yellows were indicative that this disease is caused by virus-like agent. Symptoms of *yellows* diseases include a general yellowing or reddening of leaves, reduced leaf size, shortening of internodes, and loss of apical dominance. These effects lead to stunting of the plant, proliferation of shoots or roots, witches-brooming, reduced yields, general decline, and sometimes death of the plant. Other symptoms to such diseases are flowers virescence (greening of the flowers), phyllody (conversion of petals and sepals to more leaf-like structures) and sterility of flowers. Brinjal little leaf, sesamum phyllody, peach X and Yellows, apple rubbery wood are some of the examples of such diseases. These diseases will be dealt crop-wise in this book.

A breakthrough relating to *yellows* diseases came in 1967 by the reports of Doi *et al.* from Japan when they visualized some structures in ultrathin sections of tissues of mulberry plants showing dwarfing symptoms. These structures were similar to mycoplasmal infections known to cause diseases in animals and humans. They were wallless, pleomorphic bodies *i.e.* having no definite shape and were tentatively designated as *mycoplasma-like organisms* (MLOs) and were put under the class, *Mollicutes.* MLOs reacted to antibiotics and were also transmissible by insect vectors. In 1994, the organisms associated with *yellows* diseases of plants were given the name as '*phytoplasma*'. It is now known that phytoplasmas cause more than 300 distinct plant diseases. Many of the economically important diseases are those of fruit tree crops such as coconut (root) wilt, peach X-disease, lime witches broom and apple proliferation etc. Phytoplasmas are single-celled organisms that are similar to bacteria but lack a rigid cell wall. Phytoplasmas are obligate parasites – they cannot survive without a host plant. They grow and reproduce in the cytoplasm of host cells, both in insect vectors and in plants. They reproduce asexually by budding. The size of Phytoplasma cells range from 70-1000 nm in diameter, or roughly the size of a plant cell's chloroplasts. Phytoplasmas are restricted in the phloem tissue of infected plants especially phloem sieve tube elements and are transmitted by phloem-feeding insects, most commonly by leafhoppers but some psyllids and planthoppers are also vectors of phytoplasmas. Some phytoplasmas, such as peach X- phytoplasma, may be transmitted by several species of leafhoppers; others phytoplasmas appear to be transmitted by one or only a few species.

There are at least 75 distinct phytoplasmas, currently clustered in 14 groups and 38 subgroups. Although phytoplasmas were first detected by electron microscopy but later, staining, serological, and DNA-based techniques have been developed that make detection of phytoplasmas more rapid. Serological and DNA-based techniques provide the ability to distinguish among different phytoplasmas and permit specific identification and classification. The technology of ultrathin sectioning established the cause of many diseases caused by phytoplasma, fastidious bacteria (citrus greening bacterium) and *Rickettsia-like bacteria* infecting plants. Phytoplasma species are defined by a level of at least 97.5 per cent homology of 16S ribosomal RNA sequences (Lee *et al.*, 1998).

Although viruses and phytoplasmas were well established as causal pathogens of plant diseases but still some diseases behaving as virus like diseases were there where no causal agent could be identified and remained with uncertain aetiology. Among such diseases many were important disorders like potato spindle tuber and citrus exocortis etc. Diener from United States of America reported for the first time that the causal incitant of potato spindle tuber consists of only pathogenic RNA of low molecular weight (110000-130000) and termed it as *viroid* (Diener, 1971). These pathogens were much smaller than viruses (250-400 nucleotides). After the discovery of viroids, several diseases like, citrus exocortis, hop stunt, chrysanthemum stunt, cucumber pale fruit, etc. were found to have viroid etiology. One of the characteristic of viroids is that they induce stunting in infected plants and this characteristic was unknowingly exploited to develop dwarfed citrus trees for several years in countries like Australia. However, now dwarfing rootstocks are known hence this practice has been abandoned.

For the last two decades, an advanced infrastructure has been developed in Indian research institutions and Universities which helped to discover new viruses or virus like pathogens and also to characterize the newly discovered and existing pathogens invading fruit crops in India. Many of such pathogens especially viruses and viroids have been cloned and sequenced. Suitable diagnostics have been developed, both based on proteins and nucleic acid of these pathogens. For example, techniques, like ELISA and PCR have been used routinely for detection of these pathogens (Ahlawat, 2010). These diagnostic technologies provided basic support for development of seed industry, budwood certification programmes and also tissue culture industry in India and elsewhere.

Fruit crops are more susceptible to virus and virus-like infections. These diseases are managed by quarantine measures, eradication of infected plants and weed hosts, crop rotation and use of virus-free certified propagation material and use of pesticides to manage insects which are vector of viruses. Breeding for disease resistance provides a better protection for longer time. For successful breeding for disease resistance, availability of sources of resistance in inter-fertile relatives is an essential component which is lacking in many plant species. Therefore, other options were exploited like the pathogen derived resistance which offers an attractive approach for virus management. In this strategy, the plants are transformed with nucleotide sequences which are derived from the pathogens' genome. The transformation of plants through inducing foreign gene is known as molecular breeding. Molecular breeding has been utilized to develop virus resistant plants in several fruit crops. Virus resistant transgenic crops such as Papaya in Hawaii have been developed for commercialization in several countries like USA, China and Canada. An attempt has been made in the following pages to collect and summarize the information developed on virus and virus-like pathogens on fruit crops worldwide with special emphasis to Indian scenario. This will provide a desired and useful information to fruit industry in general and researchers, growers and students in particular.

REFERENCES

Ahlawat, Y.S. 2012. Virus diseases of citrus and management. Studium Press (India) Pvt. Ltd., New Delhi, 323 pp.

Ahlawat, Y.S. 2010. Diagnosis of plant viruses and allied pathogens. Studium Press (India) Pvt. Ltd., New Delhi, 224 pp.

Bawden, F.C. and Pirie, N.W. 1937. The isolation and some properties of liquid crystalline substances from solanaceous plants infected with three strains of tobacco mosaic virus. *Proc. Royal Soc. London* (Ser. B) 123: 274–320

Beijerinck, M.W. 1898. Over seen contagium vivum fluidum als oorzaak van de vlekziekte der tabaksbladen. Versl Gewone Vergad Wis Natuurk. *Afd. K. Akad. Wet. Amsterdam* 7: 229–235.

Chenulu, V.V. and Ahlawat, Y.S. 1993. Virus and mycoplasma diseases of fruit crops in India. Publications and Information Division, Indian Council of Agricultural research, New Delhi. p. 73.

Diener, T.O. 1971. Potato spindle tuber 'virus'. IV. A replicating, low molecular weight RNA. *Virol.*, 45: 411–428.

Doi, Y., Teranaka, M., Yora, K. and Asuyama, H. 1967. Mycoplasma or PLT group – like micro-organisms found in the phloem elements of plants infected with mulberry dwarf, potato witches broom, aster yellows, or Paulownia witches broom. *Annl. Phytopathol. Soc. Japan*, 33: 259–266.

Hayes, W.B. 1966. Fruits growing in India, (3rd revised edn.) Kitabistan, Allahabad, pp 512.

Kauche, G.A., Pfank, E. and Ruska, H. 1939. Die Sichtbarmachung von pflanzlichem virus im ubermikroskop. *Naturwissenschaften*, 27: 292–299.

Lee, I.M., Gundersen-Rindal, D.E., Davis, R.E. and Bartoszyk, I.M. 1998. Revised classification scheme of phytoplasmas based on RFLP analysis of 16S rRNA and ribosomal protein gene sequences. *Int. J. Syst. Bacteriol.*, 48: 1153–1169.

Mayer, A. 1986. Uber die mosaic krankheit des tabaks. *Landwirtschaftliche versuchsstationen* 32:450–467. (English translation: On the mosaic disease of tobacco. *Phytopathological classics. American Phytopathological Society* 7(1942): 11–24.

Stanley, W.M. 1935. Isolation of crystalline protein possessing the properties of tobacco mosaic virus. *Sci.*, 81: 644–645.

2

VIRUS AND VIRUS-LIKE DISEASES OF TEMPERATE FRUITS

Pome and stone fruits are normally covered under temperate fruits. Pome fruits are members of the plant family *Rosaceae,* sub-family *pomoideae.* They are fruits that have a "core" of several small seeds, surrounded by a tough membrane. Pome fruits are trees, woody shrubs or lianas. Such fruit trees need a period of cold (a chilling requirement) each year before they will flower. In India, majority of temperate fruits are grown in Northern Himalayas or its foothills. Among pome fruits, Apple, pear and quince are prone to viral infections.Virus diseases on both pome and stone fruits will be dealt with separately in this chapter.

A. VIRUS DISEASES OF POME FRUITS

VIRUS DISEASES OF APPLE (*Malus domestica*)

Apple is the main crop among pome fruits and apple trees are extensively cultivated in the hills or in temperate climate in many countries of the world. Apple belongs to family *Rosaceae,* genus *Malus* and species *domestica.* There are more than 7,500 known varieties of apples including crabapples. Apple is believed to have originated in Europe or temperate region of western Asia, although related species known as crab-apples are indigenous to Northern Asia, Europe and America. It is not known when apple was introduced in cooler regions of India. Apples are now commercially grown in Northern Himalaya right from Jammu and Kashmir to Himachal Pradesh to Uttarakhand and also in parts of North Eastern Himalayas. Apple is also cultivated in other hilly tracts of the country. Area under apple cultivation in india is 236 thousand hectares with a production of 1842 thousand tons (*Source:* horticulture Division, department of Agriculture and Cooperation, Govt. of India, 2005-06). Apples are normally propagated by budding or grafting on compatible root stocks.

Virus and virus-like pathogens infecting apple are: apple chlorotic leafspot *Trichovirus*, apple dwarf caused by apple stem pitting *Foveavirus*, apple flat apple *Nepovirus*, apple mosaic and Tulare apple mosaic *Ilarviruses*, apple stem grooving *Capillovirus*, apple stem pitting virus, apple union necrosis and decline *Nepoviruses*. Apples are also affected by apple fruit crinkle viroid, apple scar skin viroid, apple fruit crinkle viroid (Japan) and apple chat fruit, decline, proliferation and rubberywood phytoplasmas are also known to affect apple trees. The diseases caused by these pathogens will be described in the following pages.

APPLE MOSAIC

Apple mosaic virus (ApMV) is known by several names such as: Birch line pattern virus, Birch ring spot virus, Dutch plum line pattern virus, European plum line pattern virus, Hop-A virus, Horse chestnut yellow mosaic virus and Rose mosaic virus. Apple mosaic is one of the oldest and most widely distributed virus disease of apple trees. Yield of severely affected trees may reduce upto 50% and bud setting is also affected in some cultivars. Two types of symptoms of apple mosaic are described–one causing yellow flecking and mottling of leaves and the other is manifested by a diffused ringspot mosaic mottle common on apple trees in Shimla hills (Azad and Sehgal, 1958).

SYMPTOMS

Trees infected with apple mosaic virus show pale yellow to bright yellow spots on leaves (Plate 1). Initially, small chlorotic flecks appear on young leaves which coalesce and form bigger spots as the leaves mature. These spots may become necrotic late in the season. Sometimes, chlorosis is restricted along the leaf veins giving vein - banding effect (Plate 2). Some cultivars of apple may show white banding pattern on leaves. Mosaic symptoms on apple trees are so conspicuous that the affected trees can be distinguished easily and even from distances. Most cultivars of apple are affected but vary in severity of symptoms. Cultivars of delicious group and Jonathan are severely affected but cultivars like McIntosh show mild symptoms (Azad and Sehgal, 1958). On apple mosaic virus infected plum, apricot, peach and almond, leaves show light green, yellowish or bright yellow patterns.

Based on symptoms expression on indicator plants, apple mosaic symptoms have been categorized as mild, intermediate and severe (Bhargava and Bist, 1957; Fulton, 1985). Mild strain shows only a few chlorotic flecks on inoculated indicator plants (apple cv. Lord Lambourne), intermediate strain will show vein - banding effect (Plate-2) and the severe strain causes a very severe mosaic with numerous golden – yellow patches or rings covering the major portion of the leaves (Plate-1).

TRANSMISSION

Apple mosaic disease is transmitted by grafting and budding to most pome and some stone fruits such as plum (Posnette and Ellenberger, 1957). It is also transmitted by mechanical inoculation to several herbaceous hosts. Apple mosaic virus is also transmitted by dodder, *Cuscuta hyalina* Roth from apple to apple seedlings (Ahlawat and Dhingra, 1973). In apple orchards, spread of apple mosaic virus was observed by natural root grafting between roots of healthy and infected trees (Hunter *et al.*, 1958; Dhingra, 1972). Seed transmission of apple mosaic virus has also been reported in seeds of hazelnut (Cameron and Thompson, 1985).

HOST RANGE

Host range of apple mosaic virus is very wide among *Malus domestica* and other *Malus* species. Several species in family Rosaceae, and *Humulus* species are susceptible to apple mosaic virus. About 40 plant species in 14 families are susceptible to apple mosaic virus (Fulton, 1952, 1972). Seven hazelnut cultivars are also susceptible to apple mosaic virus infection (Zentit *et al.*, 2009). Symptoms on some important susceptible hosts are given below:

1. *Apple Varieties, Lord Lambourne, Jonathan, and Golden Delicious*: Bright yellow mosaic symptoms on leaves of inoculated plants.
2. *Malus pumila* (Virginia crab, Spy 227, and Jay Darling): Callusing, swelling, browning, line necrosis and red spotting on leaves at graft union.
3. *Russian Clone:* Chlorotic and ring spots on leaves.

4. *M. platicarpa:* Mosaic and oak leaf pattern on leaves
5. *Amaranthus tricolor*: Small brown local lesions on inoculated leaves.
6. *Catharanthus roseus, Vigna unguiculata*: Systemic chlorotic lines and rings.
7. *Chenopodium quinoa*: Mottling of leaves.
8. *Crotalaria juncea*: Veinal necrosis of leaves.
9. *Cucumis sativus*: Chlorotic local lesions, systemic chlorosis and stunting.
10. *Cyamopsis tetragonoloba*: Necrotic local lesions on inoculated leaves.
11. *Torenia fournieri*: Bright yellow mosaic of leaves.
12. *Humulus* spp.: Mosaic

In India, commercial varieties of apple are indexed against apple mosaic virus by double budding method on indicator plants. Apple cultivars Baldwin, Red Delicious, Red Gold, Rome Beauty, Golden Delicious and stark earliest are diagnostically susceptible to apple mosaic virus (Dhingra and Niazi, 1973). *Cyamopsis tetragonoloba* (L.) and *Cucumis sativus* (W) are used as assay host of the virus.

THE PATHOGEN

Virus particles associated with apple mosaic disease are isometric 25–29 nm in diameter (Fig. 1). Virions contain 16% nucleic acid (RNA) and 84% protein. According to International Committee on Taxonomy of Viruses (ICTV) apple mosaic virus (ApMV) has been classified in genus *Ilarvirus* in family *Bromoviridae* and species *apple mosaic virus*. ApMV is the type member of family *Bromoviridae.*

Genome of apple mosaic virus is a single stranded RNA. Non-genomic nucleic acid is also present in virions which is sub-genomic mRNA and is found in infected cells. There is a virion protein of Mr 25000 (Gonsalves and Fulton, 1977). Nucleotide sequences of apple mosaic virus coat protein gene had no similarity with other *Bromoviridae* coat protein genes. Nucleotide sequence of apple mosaic virus coat protein gene showed no similarity to those of alfalfa mosaic virus, tobacco streak virus, brome mosaic virus and cucumber mosaic virus (Rudaina *et al.*, 1994).

Apple mosaic virus is distantly related serologically to Prunus necrotic ringspot virus but not related to other ilarviruses such as tobacco streak virus, Tulare apple mosaic virus, Prune dwarf virus, Citrus variegation virus, Citrus leaf rugose virus, Asparagus-2 virus and Elm mottle virus (Thomas, 1984). Relationship between ilarviruses and almond mosaic virus has also been reported (Digiaro *et al.*, 1992)

DIAGNOSIS AND MANAGEMENT

Apple mosaic virus can be easily diagnosed in orchard trees by its specific symptoms as described above. It is also detected by graft inoculation on 'GF305" peach seedlings. Leaves of inoculated peach seedlings show light green, yellow rings or spots, bands or oak leaf patterns. Mechanical inoculation to herbaceous hosts can also aid to detect the virus. Apple mosaic virus can be detected on woody *Malus* species/varieties by double grafting and double budding indexing method (Thakur *et al.*, 2011). In addition to ApMV, these hosts can also detect apple stem pitting foveavirus, apple stem grooving capillovirus and apple chlorotic leaf spot trichovirus in mixed infections (Thakur *et al.*, 2011). ApMV is easily detected by serological methods like ELISA (Bhardwaj *et al.*, 1991; Wang *et al.*, 2009; Thakur *et al.*, 2011) and also by nucleic acid based diagnostics like PCR and nucleic acid hybridization techniques (Thokchom *et al.*, 2008). An oligonucleotide microarray for the detection of apple mosaic virus has also been developed. Specific fragment of viruses are PCR amplified by using Cy3-labeled primers and hybridized on to microarray (Lenz *et al.*, 2009). Virus causing mosaic in apple has no vector and hence no field spread has been observed. In the absence of a vector, it is easy to manage apple mosaic virus effectively by propagating from virus-free certified planting material (Posnette and Cropley, 1956; Dhingra and Raychaudhuri, 1970). Virus-free plants are also raised through tissue culture technique (Singh *et al.*, 2008).

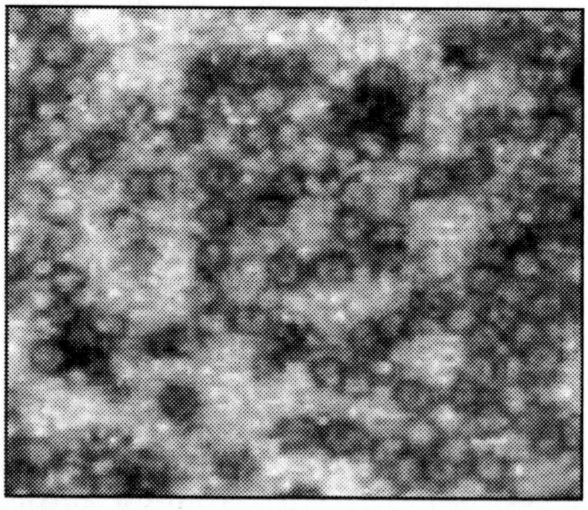

Fig. 1: Particles of Ilarvirus associated with apple mosaic disease

TULARE APPLE MOSAIC

Tulare apple mosaic disease was first reported from Tulare county in California State of United States of America (USA) and after that from Hazel in France. Associated virus with Tulare apple mosaic disease was first isolated from *Malus sylvestris* but later, *Corylus avellana* and *M. sylvestris* were also found to be naturally infected (Fulton, 1971).

SYMPTOMS

Chlorosis of leaves on affected trees is the major symptom of Tulare apple mosaic disease.

TRANSMISSION

Tulare apple mosaic virus is easily transmitted by mechanical inoculation. It is also transmitted by grafting and budding. No seed transmission was observed (Yarwood, 1955).

HOSTRANGE

Tulare apple mosaic virus (TuApMV) is transmitted to *Catharanthus roseus, Corylus avellana, Malus sylvestris, Nicotiana tobaccum and Phseolus vulgaris* (Fulton, 1956). *C. roseus* and *N. tabacum* are used as propagation hosts of the virus. *N. tabacum* and *P. vulgaris* cv Bountiful are used as indicator hosts for TUApMV. Inoculated leaves of *N. tabacum* develop Necrotic lines/rings and oak-leaf pattern and brown local lesions on inoculated leaves are developed on *P. vulgaris* cv Bountiful upon inoculation.

THE PATHOGEN

Tulare apple mosaic virus particles are isometric to quasi-isometric in morphology measuring 33 nm in diameter with icosahedral symmetry (Fulton, 1967). There were three sedimentary components in purified virus preparation. The sedimentation coefficient of the fastest one was $118S_{20w}$, others are 93 and $108S_{20w}$. A260/A280 ratio was 1.36. Thermal inactivation point was 62°C and longevity *in vitro* was only a few hours. International Committee on Taxonomy of Viruses (ICTV) classified this virus to genus *Ilarvirus* in family *Bromoviridae* and species *Tulare apple mosaic virus* (Brunt *et al.,* 1996).

Genome of Tulare apple mosaic virus is tripartite. Three segments have linear positive sense single stranded RNA. RNA-4 is subgenomic.

mRNA is derived from RNA-3 negative strand template. Multiple genome segments are distributed between three different types of particles. Complete genome of Tulare apple mosaic virus has been sequenced (Uyeda and Mink, 1983). The virus has 12 % nucleic acid and 88% of protein. Tulare apple mosaic virus is serologically related to citrus leaf rugose virus (Brunt *et al.*, 1996; Dallwitz, 1980 and Dallwitz *et al.*, 1993).

DIAGNOSIS AND MANAGEMENT

Tulare apple mosaic virus can be detected by the same methods as described for apple mosaic virus. This virus can be effectively managed by propagating plants from virus-free certified planting material (Posnette and Cropley, 1956; Dhingra and Raychaudhuri, 1970).

APPLE CHLOROTIC LEAF SPOT

Apple chlorotic leaf spot has been reported by several names such as Apple latent virus type-1, Plum pseudopox virus, Quince stunt virus and Pear ring pattern mosaic (Cropley, 1969). Apple chlorotic leaf spot virus (ACLSV) is widespread wherever apple is grown. In India, it is widely distributed in apple, apricot and peach plantations of Himachal Pradesh and Punjab states (Rana *et al.*, 2007, 2009). Two different strains have been identified in apple chlorotic leaf spot virus: Apple strain and Peach strain.

SYMPTOMS

Symptoms of apple chlorotic leaf spot disease have been observed on both pome and stone fruits. Some strains cause bark splitting on stem or pseudopox symptoms in apricot, peach and plum. The type of symptoms appeared on different hosts of pome fruits are given below:

- *Malus platycarpa*: Line patterns and chlorotic rings.
- *Malus sylvestris* cv R-12740-7A: Clorotic leaf spots and stem pitting.
- *Malus sylvestris* cv. Spy: Chlorotic spot and stem pitting.
- *Malus sylvestris* Cv. Virginia crab: Stem pitting.

- *Cydonia oblonga* (Quince) C7/1: Chlorotic rings and spots, distortion and yellow spot (Rana *et al.*, 2008).
- *Pyronia veitchii*: Chlorosis and stunting.

Symptoms on Stone fruits under natural conditions

- *Crataegus* species: Chlorotic ringspots
- *Prunus spinosa*: Chlorotic ring spot patterns.
- *Prunus persica*: Dark green mottle, mild chlorotic rings (Rana *et al.*, 2009).
- *Prunus domestica*: Bark split.
- *Prunus armeniaca*: Stock scion incompatibility and pseudopox (Bar-Joseph *et al.*, 1979; Lister and Bar-Joseph, 1981; Rana *et al.*, 2007).
- *Prunus persica* (AM498047 and AM 498050): Leaf mottling (Rana *et al.*, 2009) .
- *Prunus amygdalus*: Infection of apple chlorotic leaf spot virus was established by molecular diagnostics (Barba *et al.*, 1985; Rana *et al.*, 2008).
- *Prunus cerasoides* (wild cherry): (Rana *et al.*, 2008)
- *Pyrus communis*: Ring pattern mosaic.

Apple chlorotic leaf spot virus also infected commercial rootstocks like Malling Merton (MM 106) and Budagovsky (B-9) (Singh *et al.*, 2008).

TRANSMISSION

Apple chlorotic leaf spot virus is transmitted by grafting and by mechanical inoculation. It is also transmitted by natural root grafting between diseased and healthy trees in orchard. It is also reported to be transmitted by nematode vector, *Endorylamoid nematodes* (Fritzshe and Kegler, 1968) but needs confirmation. Apple chlorotic leaf spot virus infects following hosts upon mechanical inoculation.

- *Malus sylvestris* cv R12740-7A: Irregular blotches and malformation of young leaves of inoculated plants.
- *Malus platicarpa* (long Ashton clone): Irregular chlorotic rings, line pattern, small leaves and malformation of inoculated plants.
- *Chenopodium quinoa:* Large etched local lesions which become necrotic, systemic chlorotic spotting, mottling, rings and line patterns.
- *Chenopodium amaranticolor:* Chlorotic local lesions surrounded by necrotic dots on inoculated leaves.

- *Phaseolus vulgaris* cvs. Pinto, Bountiful, Kinghom: Purple brown necrotic spots or rings develop on inoculated leaves, no systemic infection (Cropley, 1969, Sweet, 1980).

Chenopodium quinoa and *Phaseolus vulgaris* are used as assay hosts of Apple chlorotic leaf spot virus (ACLSV).

THE PATHOGEN

Virus particles of apple chlorotic leaf spot virus are filamentous and flexuous. Virions have rod-like features with a clear modal length of 720-740 nm with 12 nm width and have helical symmetry. Pitch of helix is 3.8-3.9 nm. Thermal inactivation point of the virus was 52-55°C, longevity *in vitro* was one day at 20°C and 10 days at 4°C, and dilution end-point was 1:4 (Lister and Bar-Joseph, 1981). ICTV classified apple chlorotic leaf spot virus in order *Tymovirales*, genus *Trichovirus* in family *Betaflexiviridae* and species *apple chlorotic leafspot virus* (Martelli *et al.*, 1994).

Apple chlorotic leaf spot virus (ACLSV) is monopartite, having ssRNA. The complete genome is 7555 nucleotide long (Lister and bar-Joseph, 1981, Al-Rwahnih *et al.*, 2004). 5′-end of the genome has probably a methylated nucleotide cap and 3′-end has a poly (A) tract. The virus encodes for 5 – 7 ORF(s). Genome expression is based on RNA production which can be analysed by dsRNA patterns found in the infected tissue. Combinations of two aminoacids (Ala40 and Phe75 or ser40 and Tyr75) in coat protein of ACLSV are crucial for infectivity (Yaegashi *et al.*, 2007). Viral genome encodes structural and two non-structural proteins (Lister and bar-Joseph, 1981).

DIAGNOSIS AND MANAGEMENT

Chlorotic leaf spot virus of apple is detected by electron microscopy and also by serological and molecular techniques from infected apple tissues. This virus has been detected in apple and peach trees by molecular techniques like ELISA and RT-PCR (Fucus, 1981; Rana *et al.*, 2007, 2009; Thakur *et al.* 2012; Wang *et al.*, 2009). The Kinnaur isolate of ACLSV falls under group A of ACLSV (Rana *et al.*, 2007). Techniques like RT-PCR, IC-RT-PCR, or multiplex IC-RT-PCR have been used to detect ACLSV (Nemchinov *et al.*, 1995). Apple chlorotic leafspot virus and apple mosaic virus can be detected simultaneously in a duplex RT-PCR (Watpade *et al.*, 2012). This virus can be effectively detected by mechanical inoculation to *Chenopodium quinoa, C. amaranticolor, Phaseolus vulgaris* cvs. Kinghom, Pinto and Bountiful and by graft inoculation to Russian apple R-12740-7A. Inoculated plants will develop symptoms as described above. ACLSV can be managed by tissue culture (Singh *et al.*, 2008; Deogratias *et al.*, 1989) and through budwood certification programme.

APPLE STEM GROOVING

Apple stem grooving disease was first reported on *Malus sylvestris* cv. Virginia Crab from United States of America (Lister *et al.*, 1965). This disease is now known to be present in most apple growing countries. The apple stem grooving virus disease is worldwide in its distribution. It has been reported from Australia, Brazil, China, India, Japan, the Netherlands, New Zealand, Portugal, and United Kingdom. In India, the disease has also been reported on both pome and stone fruits (Negi *et al.*, 2008). The synonyms of apple stem grooving virus (ASGV) are: Virginia Crab stem grooving virus, *Chenopodium* dark green epinasty virus and Brown line virus. An isolate of citrus tatter leaf virus was reported to be similar to apple stem grooving virus from Brazil (Lovisolo *et al.*, 1993).

SYMPTOMS

Natural symptoms of apple stem grooving disease appear on apple cultivar Virginia Crab as chlorotic leaf spot on leaves and tree decline. Necrosis and swelling develops in stem at or above the graft union. The virus causes tree decline and graft union necrosis in certain combinations of scion and rootstocks. The virus remains symptomless in several hosts. Two strains have been reported in apple stem grooving virus (ASGV), apple latent virus type-2 isolate C-431 and E-36 isolate (Lister *et al.*, 1965).

TRANSMISSION

Apple stem grooving virus is transmitted by grafting and budding and also by mechanical inoculation. Successful mechanical inoculation can be obtained by using 0.05M phosphate buffer of pH 7-8. The virus is seed borne in seeds of inoculated *Chenopodium quinoa*. No insect vector is known.

HOST RANGE

The susceptible hosts are found in following plant families: *Aizoaceae, Amaranthaceae, Chenopodiaceae, Cucurbitaceae, Labiatae, Leguminosae, Rosaceae, Scrophulariaceae,* and *Solanaceae.* 20 plant species in these families are susceptible to apple stem grooving virus, nine of them are symptomless (Lister *et al.*, 1965, Lister, 1970a and b). Experimentally inoculated plant species show chlorotic or necrotic local lesions, systemic mosaic and necrosis. Following plant species showed visible symptoms upon inoculation.

- *Malus sylvestris* cv. Virginia Crab: Grooves in wood.
- Strain E-36: Graft union becomes brown, fragile, swollen and sunken areas on stem.
- *Chenopodium quinoa*: Etched necrotic local lesions, systemic chlorotic rings and mottle and tip malformation. Some isolates induce epinasty.
- *Nicotiana glutinosa*: Systemic yellow mosaic and line pattern on leaves.
- *Phaseolus vulgaris*: Purple brown spot, chlorotic local lesions, systemic necrosis.

Chenopodium quinoa and *Phaseolus vulgaris* cv. Pinto are used as assay hosts of apple stem grooving virus.

THE PATHOGEN

Apple stem grooving virions are filamentous and flexuous and are not enveloped. The particle size is 600-700 × 12 nm. The axial canal is obscure and basic helix is obvious, pitch of basic helix is 3.4-3.7 nm (Lister and Bar-Joseph, 1981; Lister, 1970a and 1970b). The apple stem grooving virus belongs to genus *Capillovirus* in the family *Flexiviridae*. Thermal inactivation point of the virus was between 60-63°C, longevity *in vitro* was two days, dilution end point was $\log_{10}$ minus 4 and Sedimentation coefficient was $112S_{20w}$ (de Sequeira and Lister, 1969).

The genome size of apple stem grooving virus (ASGV) is 6.5kb. The virus has single stranded RNA. The genome is unipartite and poly A region is present at 3′-end. The genome has been completely sequenced and is 6500 nucleotide long (Yoshikawa *et al.*, 1992; Nickle *et al.*, 2001). The virus has two proteins, one is probably polymerase (M.24000) and other second largest protein (M.36000) which is possibly movement protein (Brunt *et al.*, 1996; Negi *et al.*, 2008). Apple stem grooving virus (ASGV) is serologically related to potato virus T but was not related to apple chlorotic leaf spot virus (Lister, 1986). Indian isolates of apple stem grooving virus have 100% sequence identity with Serbian isolate (Brakta *et al.*, 2012).

DIAGNOSIS AND MANAGEMENT

The virus can be detected by inoculating *Malus sylvestris* cv. Virginia Crab. Inoculated plants will develop grooves in wood. Virginia Crab can also distinguish apple chlorotic leaf spot virus and apple stem grooving virus. Symptoms on inoculated *Chenopodium quinoa*, *Phaseolus vulgaris* and Russian apple R-12740-7A can also be used for disease diagnosis (Negi *et al.*, 2008).The virus can also be detected by serological tests like ELISA (Fucus, 1981; Janeckova and Pluhar, 1987; Wang *et al.*, 2009) and also by molecular technique like Polymerase chain reaction (PCR) and multiplex PCR systems (James,1999). ASGV was detected in 11 different cultivars of apple, nectarine, plum, cherry, quince and apricot from Himachal Pradesh and Jammu & Kashmir region by ELISA and RT-PCR (Kinard *et al.*, 1996, Kummert *et al.*, 1998; Vera *et al.*, 1998; Kundu, 2003; Brakta *et al.*, 2012). Lenz *et al.* (2009) reported the detection of stem grooving virus by microarray technique. Apple stem grooving virus can be effectively managed by heat therapy for 30 days at 36^{0}C (de Sequeira and Posnette, 1969). *In vitro* cultures of apple infected with ASGV grown for 9-12 weeks in media containing quercetin and Ribavirin (10 μg/ml of each) and then free of these chemicals were found free from ASGV when tested in immunocapture RT-PCR. The treated cultures were subcloned, rooted and hardened and planted in field. Such plants were free from the virus even after 10 years of planting. Therefore, *in vitro* chemotherapy with above mentioned chemicals is effective to eliminate ASGV from infected apple plants (Delano, 2009). *In vitro* chemotherapy with quercetin and ribavirin was also effective for the elimination of ASGV from apple (James, 2010). Budwood certification programmes will also be helpful to eradicate this virus.

APPLE FLAT APPLE

Flat apple disease was first reported from California but later it has been reported from several countries where apple is a commercial crop

(Wagon *et. al.*, 1968, Blodgett *et al.*, 1963). Apple flat apple and cherry rasp leaf are the same and important diseases of fruit crops (Parish, 1977).

SYMPTOMS

Flat apple disease is called so because the fruits on affected trees become compressed from both stalk and calyx side and appear flattened. This deformity can reduce fruits quality and taste. Some cultivars show stem pitting, leaf spots, tree decline, and sometimes breakage of the tree upon infection. The fruits on severely affected trees are smaller, flatter and show a tendency for the calyx end to be open. Most severe symptoms appear in apple cv. Red Delicious and related strains. In this cultivar, length of the fruit is significantly reduced and stem cavities become very shallow. At first, only a few fruits show symptoms but eventually, the whole tree may develop such fruits. However, in most cases leaves of infected trees do not show any conspicuous symptom.

In Cherry, this disease is known as cherry rasp leaf disease. The fruit production on affected trees is greatly reduced due to infection by this virus. Tree vigour is also reduced resulting in short life of infected trees.

TRANSMISSION

Apple flat apple disease is transmitted by budding and grafting and also by mechanical inoculation. The virus is also transmitted by vector, the dragger nematode, *Xiphinema americanum*. Since the virus is nematode transmitted the spread is slow and only to neighbouring trees.

HOST RANGE

The symptoms on some apple and herbaceous hosts are given below:

- *Apple cvs Golden Delicious and Newton*: Russet rings
- *Apple cv. Red Delicious* (some strains): leaves roll upward from midrib.
- *Apple cvs. Fuji Empire and Grany Smith*: Show milder symptoms.
- *Chenopodium album* var. album L.: Red colour necrotic lesions on inoculated leaves.

In other susceptible hosts like Paradise apple (*Malus pumila*), sweet cherry (*Prunus avium*), Peach (*Prunus persica*) and *Malus domestica*, symptoms are either rare or no symptoms develop after inoculation. Some weed hosts also act as reservoir of the virus and play great role in virus spread to orchard trees. *Chenopodium album* var. album is used as assay host of the virus which produces local lesions upon mechanical inoculation.

THE PATHOGEN

Apple flat apple is caused by cherry rasp leaf virus. The virus belongs to genus *Nepovirus* in the family *Comoviridae*. The species is *apple flat apple virus*. The virus has a polyhedral morphology.

Apple flat apple virus has positive sense single stranded RNA genome. An apple isolate of the virus was sequenced which had 3274 nucleotides, a single open reading frame (ORF) encoding a putative 108kDa polyprotein. The potential cleavage sites resulted in the production of putative movement protein (41kDa) and three capsid proteins (24,20, 20kDa respectively). The genome organization of cherry rasp leaf (apple isolate) RNA-2 is similar to that of apple latent spherical virus. No other relationship was found except relationship with a range of animal ssRNA positive strand viruses (Picornaviruses) (James and Upton, 2002). RNA-1 consists of 6992 nucleotide (James and Upton, 2005).

DIAGNOSIS AND MANAGEMENT

The flat apple virus can be detected by mechanical inoculations to *Chenopodium album* and also by molecular techniques like PCR (James and Uptron, 2005). The disease can be managed by the use of certified planting material.

APPLE STEM PITTING

Apple stem pitting disease has been reported with various names such as apple spy 227 epinasty and decline, Hawthorn ring pattern mosaic, Pear necrotic spot, Pear stony pit, and Pear vein yellows (Gilmer, 1962; Desvignes and Savio, 1975; Koganezawa and Yanase, 1990; Leone *et al.*, 1998; Yanase *et al.*, 1989). Apple stem pitting has been reported from several countries in Asia, Europe, South Africa, United States of America and former USSR. Recently, the virus has been reported infecting several varieties of apple in Himachal Pradesh, India (Dhir *et al.*, 2008). The yield loss by this disease has been reported upto 60% (Amtjen *et al.*, 2009).

SYMPTOMS

Apple stem pitting virus (ASPV) has been isolated from several orchard and nursery trees of apple cultivars (Kundu, 2003). ASPV is mostly symptomless in apple and pear cultivars (Nemeth, 1986). However, some cultivars show symptoms of dieback, decline, inner bark necrosis, and leaf epinasty in orchard trees. Vein yellowing symptoms are sometimes observed in some cultivars of pear. Symptoms on apple and pear indicators after back transmission from *Nicotiana occidentalis* confirm the identity of apple stem pitting virus with pear vein yellowing virus (Leone *et al.*, 1998). Mild and severe strains have been recognized in ASPV.

TRANSMISSION

The ASPV is transmitted by mechanical inoculation and by grafting. It is not transmitted by insect vector, seed or pollen.

HOST RANGE

The plant species in seven families (*Amaranthaceae, Chenopodiaceae, Cucurbitaceae, Pedaliaceae, Rosaceae, Solanaceae, and Tetragoniaceae*) are susceptible to stem pitting virus infection (Mink *et al.*, 1971; Yanase *et al.*, 1989; Nemeth, 1986; Koganezawa and Yanase, 1990). Experimentally susceptible plant species are *Celosia cristata, Chenopodium murale, C. quinoa, Crataegus, Cucumis sativus, Gomphrena globosa, Malus platycarpa, M. sieboldii, M. sieboldii* var. arborescens , *M. sylvestris, Nicotiana occidantalis (cv 37B), N. occidentalis* sp. Oblique, *Physalis floridana, Pyronia veitchii, Pyrus communis, Sesamum indicum.* Following plant species show symptoms upon inoculation as shown against each.

- *Malus sylvestris* cv. Spy 227: Leaf epinasty and decline
- *M. sylvestris* cv. Virginia Crab: Stem pitting
- *M. sylvestris* cv. Radiant or Sparkler: Leaf epinasty
- *Malus sieboldii:* Die back, inner bark necrosis.
- *Malus sieboldii* var. arborescens: Leaf epinasty, inner bark necrosis
- *Pyrus communis*: Vein yellowing
- *Pyronia veitchii*: Chlorotic spots on leaves and epinasty

- *Crataegus* spp., Sorbus mitchelli: Latent infection
- *Nicotiana occidentalis 37B*: Vein yellowing and necrosis of leaves.
- *Nicotiana occidentalis* species obliqua: Necrotic local lesions

Apple varieties, Royal delicious, Red Fuji, Red Chief, Silver spur, Red Spur, and Well Spur remain symptomless but the virus was detected in these cultivars in India (Dhir *et al.*, 2008). *Nicotiana occidentalis* species oblique, *M. sylvestris* cv. Virginia Crab, Spy 227, Radiant, Sparkler and *M. sieboldii M065* are used as assay hosts for ASPV.

THE PATHOGEN

The purified virus preparation had few virions when observed under electron microscope (Koganezawa and Yanase, 1990). Virions have helical symmetry and are filamentous, flexuous with a modal length of 800 nm and a width of 12-15 nm. The Thermal inactivation point of the virus (TIP) was between 55-60°C. The longevity *in vitro* (LIV) was 0.3-1 day. Apple stem pitting virus belongs to genus *Foveavirus* in the family *Flexiviridae*. The species is *apple stem pitting virus*.

The genome of the virus is unipartite, has linear positive- sense, single stranded RNA which has been fully sequenced (Koganezawa and Yanase, 1990). The sequence of PB66 isolate was 9363 bp in size and consists of 5 open reading frames (Amtjen *et al.*, 2009). Virions have one structural protein M, 48000, coat protein. The complete sequence of the Indian isolate comprised of 9267 nucleotides (nts). Complete sequence showed 82% nucleotide identity to the German isolate (D21829) (Dhir and Hallan, 2012).

DIAGNOSIS AND MANAGEMENT

Apple stem pitting virus is detected by electron microscopy and by inoculation to woody and herbaceous diagnostic host plant species. It can also be detected by RT-PCR, immunocapture PCR and plate-trapped ELISA and by a Real-time PCR assays (Amtjen *et al.*, 2009; Dhir *et al.*, 2008; Kundu, 2003; Kummert *et al.*, 1998; Schwarz and Jeikmann, 1998). Lenz *et al.* (2009) reported the detection of stem pitting virus by microarray technique. Apple stem pitting virus can be managed by sound certification programmes with indexing facilities and also by tissue culture.

APPLE FLAT LIMB

The flat limb disease in apple trees is known to occur in some apple growing countries. In India, it was initially observed in an apple orchard in Shimla hills on apple variety Jonathan and was transmitted by graft inoculation (Ahlawat and Chenulu, 1986). Many rootstocks introduced from Europe were infected with the pathogen that cause this disease (Fridlund and Waterworth, 1989).

SYMPTOMS

Flat apple symptoms are most severe on 'Red Delicious' and related cultivars. The length of the branches is significantly reduced and the stem cavity becomes very shallow. At first, only a few, branches will become symptomatic on lower limbs, but eventually, the entire tree will be involved. Initially hollow and longitudinal cavities appear on branches and stem. Later bark splitting, deformity and twisting occur. Disease appears as flattening of shoots and limbs on 2-3 year's old growth. As the branches become older, the flattening becomes more pronounced and deep furrows develop making the limbs to prone to breaking. In advanced cases, branches of affected trees become twisted, necrotic and brittle

TRANSMISSION

Apple flat limb disease is graft transmitted. No other mode of transmission is known. There is no insect vector. Spread in orchards has not been observed.

HOST RANGE

Symptoms resembling flat limb were observed in apple varieties, James Grieve, Branley's seedling, Ontario, Lord Lambourne, Golden delicious, Jonathan, Wagener, and Golden Pearman. Peach, Quince, Mountain Ash, Cherry and Walnut also showed flat limb symptoms (Ahlawat and Chenulu, 1986; Fridlund and Waterworth, 1989; Wagnon *et al.*, 1968). Anatomical studies in cross sections of both wood and phloem of affected trees showed fibres, tracheids and vessels in wood and fibre-sclereids and sievetubes elements in the phloem failed to differentiate. Wood becomes lignified with thickened parenchyma cells. Deposition of starch and tannin in the pith was enhanced by virus infection (Scurfield and Reinganum, 1964). Flat portion revealed lack of wood development which is probably due to reduction in cambial activity (Ahlawat and Chenulu, 1986). Sections of infected stem gave deep yellow colour as against pale yellow with healthy stem sections (Scurfield and Reinganum, 1964). Apple cv. Gravenstein is used as indicator host of the virus.

THE PATHOGEN

Apple flat limb is a graft transmissible disease. Apple flat limb disease is caused by *Cherry rasp leaf virus* (CRLV). This virus is transmitted by dagger nematodes, *Xiphinema americanum* (Parish, 1976, 1977; James *et al.*, 2001).

DIAGNOSTICS AND MANAGEMENT

Diagnosis of apple flat limb is possible by graft inoculation to sensitive apple cultivar like Gravenstein. The virus can be detected by molecular techniques like RT-PCR (James *et al.*, 2001). The disease can be effectively managed by improved cultural practices and use of virus- free certified planting material.

APPLE LEAF PUCKER

Apple leaf pucker disease has been reported from some European countries (Welsh and Kaene, 1963) and from Himachal Pradesh in India.

SYMPTOMS

On McIntosh apple trees, the fruits commonly bear small depressions, within which the skin is abnormally pigmented, accompanied on some fruits by superficial ring russet patterns. In some orchards, leaves, on trees with mild symptoms accompanied fruits will show networks of ring russetting but in another a tree may display more severe leaf symptoms, with dwarfed, distorted fruits, most of the surface of which bore purple patches, sometimes accompanied by large, expanded cracks (Welsh and Kaene, 1960). These symptoms are prominent on the leaves formed during early part of the season. Sometimes, rings also develop on the leaves. The most characteristic symptoms are severe puckering and distortion of the leaves (Fig. 2). Leaf spots may turn necrotic at later stages of growth. Fruits on affected trees are malformed and show purple blotching and skin russetting in rings similar to that of leaves. The symptoms are highly influenced by weather conditions. If the weather during early part of growing season is warm with sunny days, then the symptoms are mild or no fruit and leaf symptom will appear. However, if the weather is dull and cold during spring, then both leaf and fruit symptoms will be pronounced (Welsh and Kaene, 1960).

TRANSMISSION

The disease is transmitted both by budding and grafting to susceptible apple cultivars. It is not mechanically transmitted and no insect vector is known.

HOST RANGE

Leaf pucker disease of apple infects apple varieties, Golden Delicious, Mc Intosh, James Grieve, warcestor pearman, Northern Sinap, Staymen wine sap, Jonathan and Royal Delicious (Nagaich and Vashisth, 1965a). Apple cv. Golden Delicious is the best indicator host for leaf pucker disease. Inoculated plants will show typical leaf pucker symptoms.

THE PATHOGEN

There are reports that the disease is caused by a virus of potexvirus group but according to Kegler and Proll (1989) the disease could be caused by apple chlorotic leaf spot virus. However, the purification of the virus from apple could not be done as yet. No information on physical and chemical properties of the virus is available.

DIAGNOSIS AND MANAGEMENT

The disease can be diagnosed by grafting on apple variety Golden Delicious which produces both leaf and fruit symptoms. The disease can be managed effectively by the use of healthy planting material.

Fig. 2: Symptoms of apple leaf pucker disease

APPLE ROUGH SKIN

Apple rough skin disease has been reported from some European countries (Van Katwijk, 1956). In India, the symptoms of this disease were first observed on apple variety Robistin Pippin in Shimla hills (Dhingra and Ahlawat, 1979). Apple rough skin disease causes severe losses to growers by making the fruits unsuitable for marketing. No information is available on the virus associated with apple rough skin.

SYMPTOMS

Most characteristic symptoms of the disease develop on fruits which show rough brown corky areas on their skin which may be small or cover large parts of the skin of the fruit. In severe cases, whole skin of the fruit becomes roughened and brown (Fig. 3). Sparse foliage is often seen on trees with rough skin infection (Ahlawat and Chenulu, 1986). Sometimes, these patches are small and have circular shape. In other cases, rough browning rings or elongated strips have been seen. Longitudinal and vertical cracks are seen in inoculated apple var. Robistin Pippin and a large corky layer at the opposite calyx end of the fruit in variety Jamesgrieve (Dhingra and Ahlawat, 1979). Heavily fruited trees show large part of the fruit skin to be rough. Sometimes these rough surfaces show cracks resulting deformity in fruits due to growth retardation. Initially, a few branches will develop such fruits but subsequently all fruits of infected tree will show these symptoms (Van Katwijk, 1956).

TRANSMISSION

The disease is transmitted only by grafting. Disease symptoms normally appear after several years of inoculation (after the infected tree starts fruiting).

HOST RANGE

Following apple cultivars were susceptible to apple rough skin disease (Dhingra and Ahlawat, 1979): Robistin Pippin, Glory Van Holland, Rome Beauty, James Grieve, Belle de Boskoop, Glorie, Sun set, Red gold, Golden Delicious, and Lord Lambourne. Apple variety Belle de Boskoop is used as indicator host for detection of rough skin infection. Red Gold and Rome

Beauty can also be used as indicator hosts of Indian isolates. The first symptom of rough brown corky patches appeared on the fruits of inoculated trees after 2 years of grafting.

DIAGNOSIS AND MANAGEMENT

The symptoms of rough skin disease are quite conspicuous and the disease is not spread by insect vector. Therefore, disease-free mother trees may be identified by indexing on indicator hosts and identified trees can be used for further propagation.

Fig. 3: Symptoms of rough skin on apple fruits. Right is healthy fruit

APPLE STAR CRACK

Star crack, a virus disease, is known to occur on apples in Europe, the United States, Canada, and New Zealand. In India, star crack disease was first observed in Himachal Pradesh on Laxton's Exquisite variety of apple (Ahlawat and Chenulu, 1986). Since the fruits on affected trees develop star-shaped cracks, it makes the fruits unsuitable for marketing resulting heavy loss to growers. There is no information available on the associated virus with apple star crack disease.

SYMPTOMS

The most characteristic symptom of the disease appears on fruits as star-shaped cracks in the skin of fruits, mostly near the calyx end (Fig. 4). The spring growth of affected trees is delayed and such trees develop upright growth of their branches. Die-back of branches is common and profuse branching develops below the dieback affected shoots. Flower and fruits are comparatively less and only few fruits mature which are smaller in size and become lopsided. Symptoms vary widely depending on the apple variety and perhaps the stage of development at which the fruits become affected. Different types of symptoms develop on different apple cultivars even with the same source of inoculum. Some cultivars showed reduced growth, dieback and bud cankers during winter while others showed only mild blister-like symptoms and a few were tolerant to the virus (Campbell and Hughes, 1975). Infected trees of some cultivars develop leaves and flowers after one or two weeks later than healthy trees, whereas, in some apple cultivars leaves and flowers may develop earlier than healthy trees. Fruit symptoms also vary from cultivar to cultivar. In some cultivars fruits showed severe distortion, cracking and reduction in size, whereas, in some cultivars fruits showed only mild pitting. Some cultivars showed only fruit symptoms but no bark or shoot symptoms.

TRANSMISSION

The virus is transmitted by budding or grafting. The symptoms as bark lesions and shoot dieback develop after one year of inoculation. (Campbell and Hughes, 1975) but flower and fruit symptoms commonly develop two years after inoculation.

HOST RANGE

Diagnostic symptoms of the disease develop on Cox's Orange Pippin, Golden delicious and Laxton's Exquisite. Symptoms on these hosts are described below:

Cox's Orange Pippin

Infected plants develop little distortion of the fruits and the corky, star-shaped cracks develop to some extent on the cheeks, but mainly on the calyx end of the fruits.

Golden Delicious and Laxton's Exquisite

Affected fruits are badly distorted, and the cracks may develop on any area.

Two strains have been identified in apple star crack - mild and severe. Trees affected with mild strain of the virus do not usually show symptoms until they bear fruit. But Trees affected with severe strain develop symptoms on both shoots and fruits. Best indicator host of the virus is apple variety Cox's Orange Pippin which develops symptoms on leaves, bark and fruits.

DIAGNOSIS AND MANAGEMENT

The symptoms of apple star crack disease are quite conspicuous and can be easily identified by examining fruits showing star shaped cracks. The disease is not spread by insect vector and hence disease-free mother trees may be identified and can be used as propagation material after indexing on indicator host. The disease must be controlled in the orchard by using virus-free bud and nursery stock. Granny Smith, Lord Lambourne, Sunset and Worcester Pearmain are tolerant to star crack infection and must be used to develop resistant commercial cultivars.

Fig. 4: Star-shaped cracks on apple fruits

APPLE TUMOR

Some trees of apple cultivar Laxton's Exquisite in Shimla hills were observed with tumorous growth on their shoot and branches which was found to be a graft transmissible disorder (Dhingra and Ahlawat, 1979). The incidence of the disease was rare. This disease does not appear to be economically important. There is no information available on the disease inciting agent and studies other than symptoms and transmission.

SYMPTOMS

The main symptom of tumor disease is the appearance of small round swellings at the bifurcation point of two small branches of an affected tree. These outgrowths can extend over a length of ½ to 1 cm and one branch can develop several such tumors.

TRANSMISSION

The disease inciting agent is transmissible by scion grafting to apple seedlings (*Malus baccata* Borkh) where typical symptoms of the disease appear within two years of inoculation (Fig. 5).

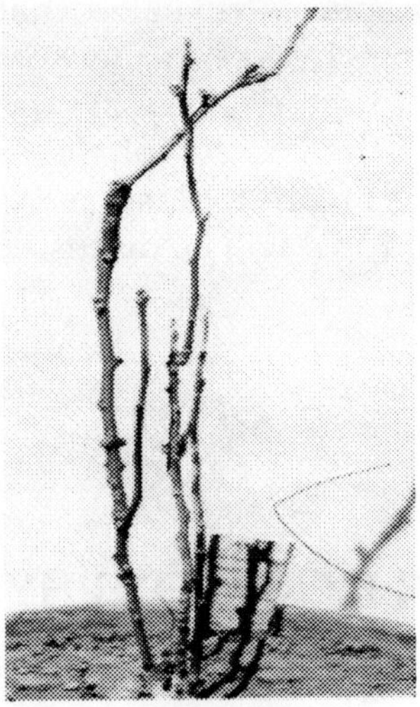

Fig. 5: Tumors on apple seedling

PHYTOPLASMA DISEASES OF APPLE

The disease inciting agent of the so called "yellows" diseases was discovered for the first time in Japan and was designated as Mycoplasma-like organisms (MLOs) (Doi *et al.*, 1967). The molecular studies on these pathogens over the years revealed that MLOs do not resemble to mycoplasmal infections of animals and were redesignated as Phytoplasmas. Apple trees are also susceptible to some phytoplasmal infections. Diseases like apple proliferation, apple decline, apple rubberywood and chat fruit are important disorders of phytoplasma origin in apple.

The phytoplasma bodies are found in sieve tubes of phloem tissue of infected plant. These are systemic in the host but their distribution is uneven. The phytoplasmas are pleomorphic in morphology. Therefore, phytoplasmas are difficult to purify like viruses. Phytoplasmas of temperate fruits can be effectively managed by hot water treatment for a few minutes (6-10 min) in a water tank at 50°C (Kunkel, 1936). Phytoplasmas are sensitive to tetracyclines and hence can be managed by treating infected trees with oxytetracycline-HCl (0.59 to 3 kg/tree) (Raychaudhuri and Maramorosch, 1988; Ahlawat and Chakraborty, 1993). Phytoplasmas belong to class *Mollicutes* in order *Acholeplasmatales.*

Acholeplasmas are thought to be closely related to phytoplasmas. However, like acholeplasmas, phytoplasmas cannot be cultured in synthetic media but the mode of transmission by both is almost same. The genomic studies of *Acholeplasma* species revealed that the determined highly interconnected metabolic networks in the *Acholeplasma* species indicate an unexpected distance to the *Phytoplasma* species which is impressive and weekly depicted by the current phylogeny (Migdoll *et al.,* 2009). However, the taxonomy of phytoplasmas is yet to be finalized. The phytoplasma diseases of apple will be discussed in following pages.

APPLE PROLIFERATION PHYTOPLASMA

Apple proliferation is one of the serious diseases of apple in most of the apple growing countries. The apple proliferation phytoplasma has been reported from apple growing countries in Asia, Africa and European Union. The disease has also been reported from India and South Africa (Ahlawat and Chenulu, 1986; Seemuller, 1990). Apple proliferation disease is now known to be associated with a phytoplasma *'candidatus* phytoplasma mali' (*Ca.* P.mali).

SYMPTOMS

The most reliable symptom on proliferation affected trees is premature development of axillary buds during 2-3 years following infection which give rise to secondary shoots forming witch's broom symptoms near the apex of main shoot. Following infection, the causal phytoplasma spreads within the tree and witch's broom symptoms may appear to most branches of an affected tree. Leaves on affected trees are reduced in size with irregular serration and enlarged stipules. Leaves in some cultivars become chlorotic and show reddening. The affected branches show early defoliation and delayed flowering. A few phylloid flowers develop on Cox's Orange where stamens are converted into petals and petals into leaves. Fruits of apple cv. Jonathan show reduction in size, long and thinner peduncle with cavities and shallow calyx. Enlarged stipules of leaves have also been observed in apple cv. Golden delicious. Affected trees may show lack of vigour, shoots become thin with reddish brown colour. Necrotic areas may develop on bark and some branches may wither (Bovey, 1963).

In India, a similar graft transmissible disease, the little leaf disease of apple was also reported (Nagaich and Vashisth, 1965a) from Himachal Pradesh on apple cultivar Red Delicious. Affected trees showed leaf reduction with short internodes. Affected leaves become mottled and deformed (Fig. 6). Distribution of the disease on an affected tree was irregular. Leaf buds on affected branches open later than the buds on healthy branches. Some of the diseased buds do not open at all and tend to die eventually.

TRANSMISSION

Apple proliferation phytoplasma (*Ca.* P. mali) is transmitted by grafting. Bench grafting was suggested to be the best technique for development of early symptoms (Cornaggia *et al.*, 2012). However, like other phytoplasmal diseases it is also transmitted by parasitic plant-the dodder, (*Cuscuta* species) to periwinkle (*Catharanthus roseus*) under experimental conditions (Heintz, 1986; Ahlawat and Dhingra, 1973). Like many phytoplasmas *Ca.* P. mali is also transmitted by psyllid vectors, *Cacopsylla picta* and *C. melanoneura* which can acquire *Ca.* P. mali at a high rate but the percentage of transmission was very poor (Barbara *et al.*, 2009; baric *et al.*, 2009).

HOST RANGE

Following apple varieties/cultivars are susceptible to *Ca.* P. mali infection: Belle de Boskoop, Gravenstein, Starking, Golden delicious, Winter banana and Rosa de Benejama. Apple varieties Lord Lambourne and Red Delicious are susceptible to little leaf infection in India and are used as indicator host. However, no information is available on aetiology of apple little leaf disease reported from India.

THE PATHOGEN

Apple proliferation disease is caused by wall-less prokaryote *Candidatus* Phytoplasma mali. The genome of *Ca.* P. mali has been completely sequenced and a number of putative candidate genes have been identified. A gene *hflb* is present in multiple copies which reflect its importance for the organism.

Analysis by single strand conformation polymorphism (SSCP) of the 530 bp PCR fragment reveals unexpected variability in this phytoplasma (Schneider *et al.* 2009). *Ca.* P. mali is found in various strains. Using cDNA-AFLP technique, Aldaghi *et al.* 2009 proposed a model to explain the mode of action of the '*Ca.* P. mali' in its host plant, the apple tree.

DIAGNOSIS AND MANAGEMENT

Apple proliferation phytoplasma remains as latent infection in several host plants but with the help of inter-stock grafting latent infections are easily determined. This phytoplasma can also be detected by direct grafting on sensitive indicator *Malus X dawsoniana.* The grafted plants will show reddening of leaves during the following autumn and will show bark splitting and scaling during next spring (Morvan and Castelain, 1975). Monoclonal antibodies have been prepared for several phytoplasmas which helped diagnosis by serological methods like ELISA and immunofluorescence. It is also detected by PCR technology (Seemüller *et al.*, 1998, Lorenz *et al.*, 1995). Rajan and Clark, (1995) successfully detected apple proliferation phytoplasma by IC-PCR in bark tissues of infected apple trees using polyclonal antibodies. All infected samples gave a product which was characteristic of apple proliferation phytoplasma as determined by RFLP analysis.

Although the phytoplasmas are sensitive to tetracyclines but the best method is to use phytoplasma-free planting material. Natural resistance against apple proliferation is found in *Malus sieboldii* which can be exploited to develop resistant rootstocks by crossing and back crossing with M 9 standard stock against this phytoplasma (Claudia *et al.*, 2009; Seemuller *et al.*, 2009).

Fig. 6: Apple proliferation

APPLE RUBBERYWOOD

Apple rubberywood disease (ARW) is found in most of the apple growing countries. In India, rubberywood was first observed in Kumaon hills (Bhargava and Bist, 1960). Later the disease was found to be prevalent in apple plantations of Shimla hills (Ahlawat and Chenulu, 1986). In some orchards, the incidence of the disease was as high as 50%. The distribution of the disease was more common in commercial varieties of apple especially in the delicious group (Van Oosten, 1983).

SYMPTOMS

The characteristic symptom of rubbery wood is abnormal flexibility of stems and branches 1 to 3 years after young trees are infected. The wood becomes soft which can be easily detected while pruning. A vigorous branch just above the ground level develops and outstrips the main shoot. Affected trees develop a 'weeping' habit and their shoots bend and may touch the ground (Fig. 7). Such trees are usually stunted with short internodes (Dhingra and Ahlawat, 1973). There are no acute symptoms on many commercial cultivars but fruit quality and yield was adversely affected. On sensitive cultivars, branches fail to develop properly and remain very pliable and droop under their own weight. The affected limbs are also more prone to cold and frost damage than healthy limbs (Fridlund and Warterworth, 1989). There are no diagnostic symptoms in the fruits or leaves of diseased trees. After 3 to 5 years, trees may again produce normal wood and in some instances the tree may recover and does not show symptoms.

TRANSMISSION

The disease is easily transmitted by grafting. No insect vector is known. It is neither seed transmitted nor by mechanical inoculations.

HOST RANGE

Apple varieties, Allington, Ambri, American Mother, Early Orange, Golden Delicious, Jonathan, John a Red, Lord Lambourne, McIntosh, Red Astrachan, Red Delicious, Rich a Red, Rome Beauty, Royal, Rymer, Stark Jone Grieve, Stark Stema Red, Victory Wine Sap Virginia, Winter

Banana Worcester Pearmain and Zillatoe Graminum were susceptible to Indian isolate of rubberywood (Dhingra and Ahlawat, 1973). Apple cultivars Lord Lambourne and Victory are highly susceptible to ARW infection. These cultivars produce the disease specific rubbery condition after one year of grafting (Dhingra and Ahlawat, 1973). Pear seedlings (*Pyrus communis*) can also be used as indicator host for ARW as it produces leaf symptoms. The symptoms in grafted pear seedlings develop as chlorotic spots on leaves which later become necrotic and occasionally form brown patches. The symptoms can develop within 6 months after inoculation. The pathogen does not produce rubbery condition in pear.

THE PATHOGEN

Initial studies suggested that the pathogen associated with rubberywood was a virus. however, attempts to isolate a virus were not successful. Early electron microscopic and DAPI (4'-6'-diamidino-2-phenylindole) staining studies of phloem from affected trees revealed the presence of phytoplasmas, though later research using both of these techniques failed to confirm the presence of phytoplasmas in rubbery wood diagnosed trees.The rubberywood disease was reported to be associated with mycoplasma-like organisms (Phytoplasma) in ultrathin sections of diseased tissue (Beakbane *et al.*, 1971). But later reports reveal that neither phytoplasmas nor rickettsia-like organisms (RLO) were detected by polymerase chain reaction (PCR) using primers derived from the 16S rRNA gene (Pollini *et al.*, 1995). More recent studies using phytoplasma-specific primers in polymerase chain reaction (PCR) assays indicated the presence of an aster-yellows-type phytoplasma in rubbery wood trees from Italy (Bertaccini, 1997), though their presence was not detected in experimentally inoculated *Lord Lamborne* trees (Poggi Pollini *et al.*, 1995).

DIAGNOSIS AND MANAGEMENT

The disease can be diagnosed by grafting to indicator hosts. It can also be detected when sections of affected branches are stained with 0.02% fluroglucinol in concentrated HCl and observed under the light microscope. The healthy sections will show red colour whereas paler areas in phloem and pericycle will be visible with diseased sections. It would also be observed that the walls of the xylem vessels and trachieds are not lignified but are thickened with cellulose only (Beakbane and Thompson, 1945). Such cells develop in patches which appear as islands of unlignified tissues surrounding by normal cells in sections. French Scientists have described a detection technique using F12/1 Mazzard cherry cuttings to detect the rubbery wood pathogen. Infected grafted chips of tissue from diseased sources induced necrosis and sunken areas near the graft on F12/1. The disease can be realized while pruning the tree hence such trees should be avoided while taking budwood for propagation. The rubberwood phytoplasma can also be detected by molecular techniques (Bertaccini *et al.*, 1997). RW Phytoplasma can also be detected by fluorescent microscopy (Ahlawat, 1996). Heat therapy has been suggested for management of apple rubberywood disease (Campbell, 1968).

Fig. 7: Apple rubberywood symptoms on Golden delicious

APPLE SESSILE LEAF

A new disease of apple (*Malus domestica*) with symptoms of leaf yellowing and shoot proliferation has been reported from Kaisiadorys region of Lithuania. The disease affected golden yellow leaves remain attached to the trunk. This disease has been named as sessile leaf (Jomantiene and Davis, 2005).

SYMPTOMS

The symptoms of sessile leaf differ from those of apple proliferation (AP). Major symptoms are enlarged stipule, witches' broom and bronze-reddish discoloration of leaves (Baric and Dalla-Via, 2004). There appears no information on transmission of the disease.

THE PATHOGEN

DNA extracted from leaves, with and without symptoms, was used as template in PCR with phytoplasma rDNA universal primers P1/P7, as previously described by Valiunas (2003). Phytoplasma infection was confirmed by amplification of 1.8 kbp rDNA fragments from all trees but not from trees without symptoms. RFLP and nucleotide sequence analysis of a 1.2 kbp DNA fragment, amplified by nested PCR using primers 16SF2n/16SR2, revealed that the phytoplasma, termed apple sessile leaf (ApSL) phytoplasma, is a member of the 16SrI-B ('*Candidatus* Phytoplasma asteris') subgroup. This group of phytoplasmas is known to infect oat plants (Jomantiene *et al.*, 2002b). '*Candidatus* Phytoplasma asteris' belongs to class Mollicutes in Phylum *Firmicutes* of Bacteria *Regnum*. It belongs to order *Acholeplasmatales* and family *Achoplasmataceae* with genus *Candidatus* Phytoplasma and species asteris. It has been established that ApSL is different from apple proliferation phytoplasma, *Candidatus* Phytoplasma mali', a member of group 16SrX (Jomantiene and Davis, 2005).

There are several species in this genus like - *Aster yellows phytoplasma - Apricot chlorotic leafroll phytoplasma - Blueberry stunt phytoplasma - Paulownia witches'-broom phytoplasma - Cherry little leaf phytoplasma - Mulberry dwarf phytoplasma - Strawberry phylloid fruit phytoplasma -Ribes nigrum phytoplasma - Peach red leaf phytoplasma - Apple sessile leaf phytoplasma - Chile grape vine yellows phytoplasma - Strawberry phytoplasma FL4 - Arabian papaya phytoplasma - Iranian cherry phytoplasma PCh8 - - Yucatan papaya phytoplasma - Arecanut yellow leaf disease phytoplasma - Banana stunt phytoplasma* and several others.

Other diseases of apple like apple decline and chat fruit earlier considered to be virus diseases are now known to be caused by phytoplasmas (Posnette and Cropley, 1965; Beakbane *et al.*, 1971). Fruits on chat fruit infected trees are dwarfed. Such infected fruits cease growing about mid-season and remain small and green until harvest (Buczacki and Harris, 1998). Affected apple cultivars develop small and deformed fruits which may drop in June. The remaining fruits do not colour properly and may show dark water soaked spots. Many commercial cultivars do not show any symptoms of the disease. The disease is transmitted by budding on apple cultivar Lord Lambourne which may develop disease symptoms three years after inoculation (Cropley, 1989). Apple chat fruit is caused by phytoplasma infection (Beakbane *et al.*, 1971). Detection by budding to indicator hosts is the only method to detect the chat fruit infection.

VIROID DISEASES OF APPLE

Apple trees are also infected by some viroids causing serious diseases and heavy losses wherever apple is grown. In India, viroid diseases are also known to occur in apple plantations (Thakur *et al.*, 1995). As described earlier viroids are infectious RNA of low molecular weight. There are two viroid species infecting apple trees (1) apple scar skin viroid associated with diseases like apple dimple fruit and apple scar skin also known as apple dapple, apple sabi-ka and apple bumpy fruit (2) apple fruit crinkle viroid (AFCVd) associated with apple blister bark and apple fruit crinkle diseases.

APPLE SCAR SKIN

Apple scar skin disease was first reported in China (Manchuria) with the name of "Manshu-sabika-byo". Later, the disease was reported from USA, Italy and United Kingdom but now the disease is known to occur in several apple growing countries. The disease is widespread on apple and Asian pear (Nashi) in some China provinces. Recently, Apple dimple fruit (Apple scar skin viroid) has been reported in apple plantations in Himachal Pradesh, India (Walia *et al.*, 2008). It has become a problem when old orchard are top worked, and the new is susceptible to the viroid that had been symptomless in the original cultivar. Apple scar skin viroid infection is responsible for reducing fruit quality and may cause huge economic losses.

SYMPTOMS

Fruits on viroid affected apple trees show dappling symptoms in the form of small circular spots near the calyx end of the fruits which enlarge as the fruit reaches to maturity. The spots can be easily seen on the background colour of the fruits. The spots may coalesce and may form broader areas of dappling. In severe cases, the spots may become brown and necrotic, and fissures appear on the fruit. The affected fruits are smaller than healthy fruits. The viroid remains symptomless in many cultivars of apple. Pear also appears to be a symptomless carrier of apple scar skin viroid. The viroid is systemic in infected trees (Hadidi *et al.*, 1991).

In India, affeced apple cultivar Golden Delicious trees develop bumby fruits. Affected trees bear sparse foliage, showed lack of vigor and only few fruits were developed on each branch. Initially the fruit surface develops depressions, at times including a "false-sting", which became prominent with bumps upon fruit maturity. Infected trees also had some fruits which, at maturity,.showed hard wart-like swellings with or without wedges or cracking of varying patterns. The fruits grew slowly in the latter part of the season. Apparently, the fruit quality (sweetness and storage life) was also affected (Walia *et al.*, 2008).

TRANSMISSION

The method of field spread is not known. The viroid is graft transmitted and seed borne (Hadidi *et al.*, 1991). It is also transmitted by contaminated operational tools.

HOST RANGE

Host range of apple scar skin viroid is large. Following plant species belonging to families of pomaceous fruits are susceptible: *Chaenomeles japonica, Sorbus aucuparia, Sorbus mougeoti, Sorbus domestica, Sorbus prattii, Pyronia veitchii* and *Cydonia oblonga*. In addition, following *Malus* species are also susceptible and show fruit symptoms: *Malus baccata f. gracilis, M.baccata, M. floribunda, M. hupehensis, M. ioensis, M. micromalus, M. platycarpa, M. pruni, M. pumila, M. perpurea, M. toringo, Malus X robusta,* and *M. yunnanensis* (Desvignes *et al.*, 1999). Apple cv. Indo, Red Delicious and Starkrimson are used as indicator hosts for this viroid. In affected apple cv. Starkrimson, the fruit skin turns red, small rough mottle, sometimes starred around the eye cavity. In addition, two apple (*Malus domestica*) cultivars, Stark's Earliest and Sugar Crab, are also effective indicators of the apple scar skin viroid. In these hosts distinct fruit marking or leaf epinasty symptoms develop within two months after chip budding on one-year-old apple seedlings. Fruits developed on inoculated greenhouse-grown trees showed characteristic symptoms of colour dappling and cracking of the calyx end. Epinasty of young expanding leaves appears on vegetative shoots of both cultivars 6 to 8 weeks after inoculation when grown at 24-hour photoperiods at 18 or 28°C. (Howell and Mink, 1992).

THE PATHOGEN

The diseases popularly known as apple dimple fruit, apple dapple, apple scar skin and bumpy fruit are caused by apple scar skin viroid (ASSVd). Apple scar skin viroid belongs to family *Pospiviroidae*, genus *Apscaviroid* and type species is *Apple scar skin viroid*. This viroid was first sequenced in

Japan in 1987 and recently in India (Walia *et al.*, 2008). Apple scar skin viroid has 330 nucleotides. The clones showed significant sequence variability (94-100%) with each other. Variability was more common in the pathogenic domain of the viroid genome. The Indian isolates showed close affinity with some Chinese and Korean isolates (Walia *et al.*, 2008). Recent evidence, however, suggests that the symptoms observed in Italy were probably induced by another viroid of 306 nucleotides called apple dimple fruit viroid (ADFVd), belonging also to the genus *Apscaviroid,* whose type member is apple scar skin viroid. Following species, serotypes, strains and isolates are included in genus *Apscaviroid:* Apple dimple fruit viroid, Apple scar skin viroid, dapple Apple scar skin viroid – Japanese pear Apple scar skin viroid - pear rusty skin, Dapple apple viroid, Japanese pear fruit dimple viroid, and Pear rusty skin viroid.

DIAGNOSIS AND MANAGEMENT

The apple scar skin viroid is detected by biological indexing on indicator hosts. The viroid was effectively detected in PCR (Hadidi and Yang, 1990). Management is possible using viroid-free certified planting material. The disease is also managed by thermotherapy for 41 and 99 days at 36 to 37°C to be performed at the end of the vegetative period or for 48 days performed on young shoots after a dormant stage of 3 months. This second treatment is superior and produced about 90% of ASSVd-free plants (Desvignes *et al.*, 1999). In infected pears, the viroid is eliminated by *in vitro* thermotherapy (30 to 38°C) or cryotherapy (4°C) followed by apical meristem culture (Postman and Hadidi, 1995).

APPLE FRUIT CRINKLE

Apple fruit crinkle viroid was initially discovered in Japan in 1976 from mature fruits of apple.This viroid is associated with diseases like blister bark and apple fruit crinkle.

SYMPTOMS

The most characteristic symptoms of fruit crinkle viroid disease are: depressions and malformation, crinkling in addition to variations in colour, and dappling of the fruit surface. The crinkling symptoms appear when the fruits are near maturity. These symptoms are less severe in cool summers. The dappling symptoms of the disease resemble to those caused by apple scar skin viroid. The severity of these symptoms varies among cultivars but apple cultivar 'Ohrin' appears to exhibit the most serious deformation with scattered brown necrotic areas in the fruit flesh.

HOST RANGE

Apple fruit crinkle associated viroid (AFCaVd) infects apple trees of cvs. Ohrin, Jonathan, Starking Delicious and NY58-22 crab apple. The symptoms caused by the viroid are severe fruit crinkling and necrosis in the fruit flesh on cv. Ohrin, fruit crinkling and fruit dappling on cv. Jonathan, blister bark on cv. Starking. Delicious group likeNY58-22 showed severe fruit crinkling, necrosis in fruit flesh, fruit dappling and blister bark (Ito and Yoshida, 1998).

TRANSMISSION

Apple fruit crinkle viroid (AFCaVd) is transmitted by chip budding on apple seedlings which had been inoculated with purified AFCaVd by razor slashing. All inoculated plants of susceptible apple cultivar showed characteristic symptoms of the disease.

THE PATHOGEN

Apple fruit crinkle disease is associated with a viroid termed as apple fruit crinkle viroid (AFCaVd). Ito and Yoshida (1998) concluded that AFCaVd is the causal agent of apple fruit crinkle disease and proposed its name as apple fruit crinkle viroid (AFCVd). They also suggested that AFCVd is also associated with a bark disorder of apple cv. Nero 26 earlier reported in Japan. This viroid belongs to family *pospiviroidae,* genus *Apscaviroid* whose type species is *apple scar skin viroid*. Other tentative species in genus *Apscaviroid* is *Apple fruit crinkle viroid* (AFCaVd).

DIAGNOSIS AND MANAGEMENT

Apple fruit crinkle viroid is detected by bidirectional polyacrylamide gel electrophoresis. Apple fruit crinkle viroid is detected by RT-PCR using primers based on the base sequences of AFCVd (Yoshida and Ito, 2002). The diseases caused by this viroid can be managed by sound indexing programmes both biological and molecular followed by certification strategy.

B. VIRUS DISEASES OF PEAR

Pear (*Pyrus communis*) is grown all over the world in temperate and sub-temperate regions. In India, pear is grown both in the hills and plains of North and South India. Pear is very similar to apple in its cultural requirements. Pear fruits are born on spurs, and pruning is required as in apple trees. The Bartlett or Williums is a variety of excellent commercial quality and widely grown in India. Other varieties grown in India are: Winter Nelis, Kieffer and wild spicies like *P. pashia, P. khasiana, P. baccata* etc.

About 15 virus and virus-like diseases are known to affect pear trees or fruits. The important ones are: Flemish beauty corky pit, Pear blister canker, Pear decline (phytoplsma disease), Pear freckle pit, Pear red mottle, Pear ring pattern mosaic, Pear rough bark, Pear stony pit, Pear vein yellows, pear ring mosaic, pear bud drop. Most of the viruses infecting apple also infect pear such as apple chlorotic leaf spot virus, apple stem grooving virus and apple stem pitting virus (Ne *et al.*, 2012). Most of the pear diseases are apparently caused by viruses except the pear decline which has been established as phytoplasma disease. Pear yellows is a filamentous virus similar to apple stem pitting virus (Yanase *et al.*, 1989).

Pear ring mosaic and pear ring pattern mosaic are same disease and caused by apple chlorotic leaf spot virus (Dunez, 1981). Likewise, pear vein yellows and pear red mottle are now known as symptom type of stony pit virus. Based on symptoms, pear stony pit and pear corky pit are found to be closely related diseases. Pear necrotic spot, pear ring pattern mosaic and pear vein yellows produce symptoms on leaves and fruits. Pear stony pit produces symptoms on fruits and pear decline phytoplasma, pear rough bark and blister canker produce symptoms on stem and bark.

PEAR RING DISEASE/PEAR RING PATTERN MOSAIC DISEASE

The pear ring disease is caused by apple chlorotic leaf spot virus (ApCLsV) (see classification in ApCLsV). This disease is also known as pear green spot, pear mosaic or pear ring pattern mosaic.

SYMPTOMS

Trees affected with pear ring disease remain stunted. The hardiness of branches is decreased due to infection. On young leaves of infected trees, symptoms appear as mosaic, green rings, and ringspots. The leaf midribs grow slowly, bend and result in a curled appearance. Pale green ringspots appear on the surface of fruits. In most susceptible varieties, symptoms mainly occur on fruits. Many pear cultivars show diffuse symptoms of indistinct mottle but numerous pale green or greenish yellow spots develop on fruits of infected trees.

TRANSMISSION

Pear ring disease is transmitted by budding and grafting. The causal virus is also transmitted by sap inoculation to herbaceous hosts similar to apple chlorotic leaf spot virus (see host range of apple chlorotic leaf spot virus).

HOST RANGE

Most of the pear varieties do not produce conspicuous symptoms of the disease. However, pear cultivars, Beurre Hardy, Beurre Bosc, Nouveau Poiatean, Marianne and Williams produce diagnostic symptoms of the disease. Pear cv. Beurre Hardy develops variable mosaic symptoms on inoculated plants and used as indicator host. Another indicator host is *Chenopodium quinoa.* Leaves of inoculated *C. quinoa* plants develop local chlorotic spots, rings and lines.

DIAGNOSIS AND MANAGEMENT

Diagnosis of ring disease is done by inoculating indicator hosts which will develop typical symptoms of the disease in due course. The disease can be managed by the use of healthy planting material. Virus-free planting material can be obtained by incubation of bud sticks at 37°C for 2-3 weeks (Posnette, 1957).

PEAR NECROTIC SPOT

Pear necrotic spot disease has been reported from Japan (Kishi *et al.*, 1976). Necrotic spots, reddish purple to black in colour and circular or irregularly shaped, 2–3 mm in diameter appear on the leaves of Japanese pear (*Pyrus serotina* var. Cuita Rehder). These symptoms develop only on mature leaves during early or middle June. Leaves of affected trees turn yellow and defoliate during summer. The most sensitive cultivar is Nijisseiki, one of the best commercial cultivars in Japan.

Seven pear cultivars and five promising hybrids of Japanese pear are susceptible and show symptoms upon graft-inoculation with diseased scion. One cultivar of Chinese pear (*Pyrus bretschneideri* Rehder) and three of pear (*P. communis* Linn.) are also susceptible and showed symptoms. All other cultivars of pear remained symptomless and appear to be tolerant but not immune. This suggests that the causal pathogen is harboured latently in many commercial cultivars of both Japanese pear and other pears. Sixteen Japanese pear hybrid seedlings showed necrotic leaf spots by graft-inoculation. Among them, HN-39 seedling was selected as the most suitable indicator. HN-39 is very sensitive to necrotic spots formation. A filamentous virus is associated with the disease (Yanase *et al.*, 1989) and is related to apple stem pitting virus.

PEAR MOSAIC

A mild mosaic disease of pear has been reported from India in the Kumaon hills, Himachal Pradesh, Punjab and Uttar Pradesh on local cultivars (Nagaich and Vashisth, 1962; Ahlawat and Chenulu, 1986). Pear mosaic disease does not appear to be economically important.

The symptoms on affected trees develop as transient mottle, chlorotic spots or rings, feathery and oak leaf pattern. The disease is transmissible by wedge grafting on pear seedlings and some local varieties. This disease is perhaps similar to the mosaic of pear described from England (Posnette, 1957).

This is the only record of a virus disease on pear in India. No other information is available on this disease.

Another disease of minor importance in pear is pear vein yellows. Faint, yellow vein banding, particularly of the secondary veins on current year's growth or red mottling of various intensities are the most common symptoms of vein yellows. The disease is caused by Apple Stem Pitting Virus which has already been discussed in earlier pages of the book.

PEAR DECLINE

Pear trees affected with decline disease may collapse within a few years of infection (Plate-3). The disease is transmitted by grafting from pear to pear and by dodder (*Cuscuta* species) from pear to periwinkle. In periwinkle, symptoms appear as yellowing or clearing of veins in newly infected leaves and shortening of internodes of the main stem. Infected periwinkle plants remain stunted and flower size is reduced. Pear decline is also transmitted by a psyllid vector, *Cacopsylla pyri* (Plate 3A).

Pear decline is caused by a phytoplasma known as *Candidatus* Phytoplasma pyri. The presence of phytoplasma in periwinkle was detected by using P1/P7 and U3/U5 primer pairs in nested PCR (Caglayan *et al.*, 2009). Pear decline phytoplasma is also detected by PCR using primers to sequences in the 16sRNA gene of pear decline phytoplasma. Using the same technique, the phytoplasma can also be detected even from a single psyllid vector (Davies *et al.*, 1985, 1995). The primers from sequences of 16s RNA gene could not detect other phytoplasmas like aster yellows, clover phyllody, apple proliferation, plum lectonecrosis, apricot chlorotic leaf roll and elm yellows suggesting that the phytoplasma of pear is distinct from these phytoplasmas (OEPP/EPPO (1978).

Pear decline phytoplasma can be detected by PCR amplification of ribosomal and nonribosomal DNA (Lorenz *et al.*, 1995). The pear decline agent survives overwinter in the root of quince strains which are used as rootstocks for pear (Poggi Pollini *et al.*, 1995).

PEAR BLISTER CANKER VIROID

Pear blister canker viroid (PBCVd)-a member of the genus *Apscaviroid*, family *Pospiviroidae (*Hernández *et al.*, 1992a) has been reported in France and Spain, Italy, Greece, Tunisia, Canada and Australia (Desvignes,

1970; Hernández *et al.*, 1992b; Ambrós *et al.*, 1995, Loreti *et al.*, 1997; Kyriakopoulou *et al.*, 2001; Malfitano *et al.*, 2004; Joyce *et al.*, 2004; Hassen *et al.*, 2006; Kaponi *et al.*, 2010). In fact, large scale surveys conducted in the last few years using tissue printing hybridization (TPH) (Lolic *et al.*, 2007), also revealed the presence of PBCVd in Malta (Attard *et al.*, 2007), in Canada (Torchetti *et al.*, 2012) and in Bosnia and Herzegovina (Lolic *et al.*, 2007) suggesting widespread distribution of this viroid. The viroid, in question, infected pear, quince and *Pyrus serotina.* It remained symptomless in several pear cultivars but in susceptible cultivars symptoms developed as pustules or superficial cracks on the epidermis that turned into scattered cankers and scaly bark or deep bark splits. Affected plants died within 5-8 years. PBCVd can be mechanically transmitted experimentally to cucumber (*Cucumis sativus*) in which it induces mild leaf rugosity or no symptoms (Flores *et al.*, 2005). It also infects symptomlessly species of *Pyrus, Chaenomeles, Cydonia* and *Sorbus.* The viroid was transmitted by graft propagation and mechanically. A viroid with a circular single-stranded RNA molecule of 315-316 nt was associated with the disease. PBCVd can be detected by molecular methods including polyacrylamide gel electrophoresis and dot-blot hybridization with radioactive and non-radioactive probes (Ambrós *et al.*, 1995). It was also detected in RT-PCR using primers derived from the reference sequence (Hernández *et al.*, 1992a). Molecular hybridization also detected PBCVd in symptomless plants of the pear cv. Williams (Ambrós *et al.*, 1995). Control is possible only through the use of viroid-free propagating material.

C. VIRUS DISEASES OF QUINCE

Quince (*Cydonia oblonga)* is a fruit grown only to a limited extent in some parts of the world. It is not grown commercially in many countries including India. In India, few plantations of quince are there in Kashmir valley and Himachal Pradesh. It is primarily grown for medicinal value of its seeds. The fruits are not palatable raw, but used to make excellent jelly and preserves. Few people in India know how to use it, so there is no market for quince fruits. Quince is a natural host of several virus and virus-like diseases and is susceptible similar to apple and other hosts in Rosaceae family. Although no original virus has been described in quince yet quince virus diseases are identical to apple and pear. Quince is natural host of apple chlorotic leafspot virus. Quince virus diseases are: Spy epinasty and decline, apple/pear stem grooving, Quince bark necrosis, Quince sooty ring spot, Quince stunt, and Quince yellow blotch.

QUINCE STUNT DISEASE

Quince stunt disease is caused by apple chlorotic leaf spot virus (ApCLSV). However, it appears to be a complex disease of ApCLSV, sooty ringspot and probably apple rubberywood phytoplasma (Kegler, 1977). The disease is transmitted by grafting. Quince stunt disease is known to occur in USA, Canada, New Zealand, South Africa and Europe.

The symptoms of quince stunt disease appear in susceptible quince, C-7/1 and E cultivars. Leaves of affected trees show irregular shaped yellow spot and become curled. Shoots of affected plants grow slowly and eventually die. Quince cvs. C-7/1 and E are used as indicator hosts which develop symptoms as described above.

The stunt disease of quince is diagnosed by inoculation to indicator hosts and can be managed by the use of virus-free planting material. Heat therapy is also useful to provide healthy planting material. This can be done at 37°C for 11 days.

QUINCE RING PATTERN

Ring pattern disease in quince is also caused by apple chlorotic leafspot virus. Most of the quince cultivars remain symptomless. Pale green, yellow rings and line patterns are developed on leaves of inoculated plants of some cvs. of quince, apple, pear and many other rosaceous hosts. The disease is transmitted by grafting.

QUINCE SOOTY RINGSPOT

Sooty ringspot disease is known to be present in USA, Australia, New Zealand, South Africa and Europe. The causal organism of this disease is not known. Apple and pear are the additional hosts of this disease. Quince Cvs E and C-7/1 are symptomatically susceptible.

Symptoms of sooty ringspot appear as young leaf epinasty, necrosis, reddening in leaf cuticle around veins and the yellow spots which give the leaves a sooty mold appearance. Trees may collapse or show decline with a mixed infection of rubberywood phytoplasma and quince sooty ringspot pathogen. The disease is transmitted only by grafting. Quince cv. C-7/1 is used as indicator host of the disease. Symptoms on inoculated indicator host appear as black pigment in leaf cuticle, vein clearing, bark necrosis and decline.

Diagnosis of the disease is done by inoculating indicator host and observing symptoms as described above. The disease can be managed by the use of healthy planting material which can be obtained by heat treatment at 36°C for 3 weeks.

QUINCE FRUIT DEFORMATION DISEASE

It is a graft transmissible disease of quince reported from Europe. It is not of great economic importance. The causal organism is not known. Fruits from infected plants have uneven growth with depressions of varying shapes and sizes on the surface of fruit. Skin depressions are greenish to brown. On leaves, diffuse pale green spots and deformities are often observed. Quince cv. C-7/1 is used as indicator host for diagnosis of this disease.

QUINCE YELLOW MOSAIC

In India, a yellow mosaic disease of quince is known to occur in the states of Himachal Pradesh and Jammu and Kashmir. The main symptoms of the disease are yellow flecking and severe mottling of leaves (Plate-4). Later in the season, yellow patches become brown and necrotic causing the leaves to wither away.

The disease is transmissible by bud and wedge grafting (Nagaich and Vashisth, 1962). The symptoms of quince yellow mosaic resemble those produced by the severe strain of apple mosaic virus on quince. Quince yellow mosaic was considered to be a strain of apple mosaic virus (Ahlawat and Chenulu, 1986).

D. VIRUS DISEASES OF STONE FRUITS

The stone fruits are drupes and belong to the genus *Prunus*. The following species of stone fruits are under cultivation: Apricot (*Prunus armeniaca* or *Armeniaca vulgaris*), Almond (*P. amygdalus*), Cherry, sweet, black, sour, and wild species (*Prunus avium*, *P. serotina*, *P. cerasus*, and others), Chokecherry (*P. virginiana*), Plum (*P. domestica*), of which there are several domestic and wild species, Peach and its variant the nectarine (*P. persica*). Stone fruits are woody trees which makes the diagnosis of viral/ phytoplasma diseases a difficult task. Important pathogens infecting stone fruits are: Ilarviruses (Prunus necrotic ringspot, prune dwarf, line pattern), NEPO viruses (Strawberry latent ringspot, myrobolan latent ringspot, raspberry ringspot), chlorotic leaf spot and sharka diseases of plum and peach mosaic. Phytoplasma Diseases like chlorotic leaf roll (ECA), X disease and little cherry are also important diseases of stone fruits. The virus and virus-like diseases of stone fruits will be discussed in this part of the chapter.

A. VIRUS AND VIRUS-LIKE DISEASES OF APRICOT

Apricot trees are infected by several viruses and phytoplasmas. It includes both characterized and uncharacterized graft transmissible pathogen. Apricot trees are infected by several viruses such as *Nepovirus, Strawberry latent ringspot virus, Tobamovirus, Cucumber green mottle mosaic virus, Ilarvirus, Prunus necrotic ringspot virus, Trichovirus, Apple chlorotic leaf spot virus, Cherry mottle leaf virus, Potyvirus, Tomato ringspot virus* and *Prune dwarf virus*. Some of these viruses have already been described in earler pages. Among uncharacterised graft transmitted diseases are : Asteroid spot (= Peach asteroid spot), Cherry mottle leaf, Chlorotic leaf mottle, Deformation mosaic (associated with an isometric particle), Moorpark mottle, Peach yellow mottle, Pucker leaf, Ring pox (= Spur cherry?) and Stone pitting. Among the phytoplasmal diseases Chlorotic leaf roll (= Apple proliferation) and Witches' broom are the important diseases. Two virus diseases, Plum Pox Potyvirus or sharka (PPV), and Apple Chlorotic Leaf Spot Virus (ACLSV) are commonly occurring in apricot. Sharka virus is known to spread by aphids from tree to tree. Apricot Chlorotic Leaf Roll (ACLR), a disease associated with mycoplasma-like organisms is spread from tree to tree probably by leafhoppers. A viroid has also been detected for the first time in apricot (Liacer, 1995)

In India, ringspot (ringpox), oak-leaf pattern and mosaic viruses have been recorded (Ahlawat and Chenulu, 1986). Apple chlorotic leaf spot virus has been reported on wild and cultivated apricots in Himachal Pradesh, India (Rana *et al.*, 2007). Most of the viruses and phytoplasmas mentioned above are the pathogens of other temperate fruit crops and apricot is an additional host of these pathogens. The important diseases of virus/ phytoplasma nature in apricot will be briefly discussed in this chapter.

APRICOT LATENT

Apricot latent virus (ALV) has been reported from France and later found in several European and Mediterranean countries. The virus remains latent in apricot trees. The virus is graft transmitted to several Prunus species. Typical yellowish-green spots on leaves of peach seedlings appeared after chip-budding of infected apricot buds. Filamentous virus particles were detected by thin sectioning in the cytoplasm of parenchyma cells of infected peach leaves. The virus was mechanically transmitted from peach leaves to *Nicotiana occidentalis* B37 where it caused vein clearing of growing leaves and developed necrosis on inoculated leaves of *N. occidentalis* sp *obliqua* N1 (Rysanek, 2012).

Apricot latent virus particles are filamentous and the virus is a tentative member of the genus *Foveavirus* in *Betaflexiviridae* family of the order

Tymovirales. The virus genome is monopartite positive sense single stranded RNA. ALV can be detected by grafting or budding from infected plants to peach seedling GF 305 (Rysanek, 2012). Antiserum of apricot latent virus and an antiserum to apple stem pitting virus cross reacted in ELISA and ISEM assays with extracts from ALV-infected peach seedlings (Zemtchik *et al.*, 1998). The use of cRNA probe showed cross hybridization with peach asteroid spot and peach sooty ringspot viruses. Sequence analysis of the cDNA clones of these peach viruses showed very high homology with the sequences of apricot latent virus. The viral agents associated with these two peach diseases may be the variants of apricot latent virus (Zemtchik *et al.*, 1998; Gentit *et al.*, 2001). Eradication of the pathogen largely depends on the use of virus-free propagating material and rootstock.

STEM PITTING

A Stem pitting disease in apricot has been reported from south Apulia (Salento peninsula) on cultivar Tirynthos grafted onto myrobalan with cv. Stanley as intermediate stock (Di Terlizzi and Savino, 1995). Affected trees show bushy appearance with small leaves that turned chlorotic during summer. The fruits on affected trees are larger than normal but remain tasteless. Bark splitting of trunk and branches was of common occurrence. The bark was spongy and thicker than normal trees. Pittings are found only in scion but not in rootstock.

The affected trees when tested in ELISA with antibodies of Prunus necrotic ringspot ilarvirus (PNRSV) showed positive result suggesting that the disease is caused by PNRSV. The stem pitting virus is transmitted by mechanical inoculation to similar herbaceous hosts of PNRSV further confirming that the disease is caused by PNRSV. Prunus necrotic ringspot virus will be discussed in following pages.

DEFORMATION AND DISCOLORATION OF APRICOT FRUITS

This disorder is known to exist in South Italy since 1995 in apricot cultivar Tyrinthos grafted on Myrabolan (Alioto *et al.*, 1995). Young fruits are severely misshapen and drop prematurely. The fruits remain green even after ripening. Affected fruits show depression on fruit surface and become lopsided and discoloured. Symptoms on leaves are unevenly distributed as yellowish or reddish spots during late in the summer. Several leaves turn yellow with green spots and lines before they drop. Affected trees are overgrown at the graft union and develop excessive root suckers. A brown area may develop between scion and root stock and bark is thicker than normal. The disease appears to be a synergistic effect of three viruses namely apple chlorotic leaf spot, Prunus necrotic ringspot and prune dwarf. These viruses are often detected from trees showing fruit symptoms by biological and ELISA test (Alioto *et al*, 1995).

APRICOT OAK LEAF PATTERN

Oak leaf pattern symptoms were commonly found in apricot trees in Himachal Pradesh, India (Nagaich and Vashisth, 1962). The disease symptoms were first observed on apricot cultivar 'Blenheim" in Shimla hills. The characteristic symptoms of the disease were the development of oak-leaf pattern and yellowing of the margins of leaf lamina of infected trees (Fig. 8a). Occasionally, veins of the leaf also become yellow.

The disease is transmitted by wedge grafting and also by dodder, *Cuscuta hyalina* (Ahlawat and Dhingra, 1973). Apricot and plum seedlings are susceptible to the oak-leaf pattern virus. The symptoms produced on graft inoculated plum seedlings (*Prunus domestica*) are similar to those of caused by plum line pattern virus (Fig 8b). The virus causing apricot oak leaf pattern symptoms in apricot appears to be a strain of apple mosaic ilarvirus or Prunus necrotic ringspot ilarvirus.

Fig. 8: Oak leaf symptoms: (a) on apricot leaves (b) On plum leaf

APRICOT MILD MOSAIC

Apricot mild mosaic disease was reported from apricot variety "Stark Early Orange" in Himachal Pradesh, India. The leaves of affected trees develop chlorotic spots scattered over the leaf lamina. These spots diffuse and coalesce in due course and form mild spot mottle symptoms (Nagaich and Vashisth, 1962). The disease is easily transmitted by bud grafting to apricot and plum seedlings. Apricot variety Kaisha, New Castle and Zerdalu are susceptible to mild mosaic infection. Symptoms of the virus on plum are similar to that of prunus ringspot virus group (Ahlawat and Chenulu, 1986).

APRICOT MOSAIC

In India, a disease with mosaic symptoms was first observed on apricot variety Ansu in Himachal Pradesh. Bright yellow mosaic symptoms are commonly observed on leaves of affected trees (Fig. 9). These symptoms are entirely different from the mild mosaic symptoms mentioned above on susceptible hosts. Cross protection studies revealed that the mild mosaic can protect the plants against infection of severe mosaic (Dhingra and Ahlawat, 1979). The virus is transmitted by budding and grafting and also by dodder, *C. hyalina* (Ahlawat and Dhingra, 1973). The following hosts are susceptible and produce bright yellow systemic mosaic after one year of inoculation: Apricot seedlings, Apricot varieties, Shakerpara, Safaida, Kaisha, New Castle, peach variety J.H. Hale, Plum varieties Santa Rosa and Green.

Two viruses have been reported from apricot showing bare twigs and unfruitfulness. One is serologically related to strawberry latent ringspot virus with virions measuring 28 nm in diameter. The other one is a filamentous virus identified as yellow (aucuba) strain of cucumber green mottle mosaic virus (Cech *et al.*, 1981). A Japanese apricot isolate of

Fig. 9: Apricot mosaic on apricot cv. Safaida

Cucumber mosaic virus (CMV-Um) was isolated from Japanese apricot (*Prunus mume* Sieb. et Zucc.) which showed leaf margin necrosis and abortion syndrome, together with Prunus necrotic ringspot like ilarvirus in infected apricots (Kuriharajun *et al.*, 2005).

PEACH VEIN CLEARING PHYTOPLASMA IN APRICOT TREES

Some trees of apricot infected by apricot leaf roll phytoplasma showed some atypical symptoms like mild chlorosis, partial rolling of leaves, and vein chlorosis during summer (Cornaggia *et al.*, 1995). The leaf fall is delayed in autumn and fruit number and size is reduced. During winter, swollen bark and bud burst are delayed from 6 to 8 days. The nursery indexing is done on indicator Priana apricot which develops similar symptoms as described above upon inoculation.

In greenhouse inoculated peach seedling, GF 305, the symptoms develop as gradual chlorosis with atrophy of young leaves, vein clearing, then reddening and rolling of older leaves after 100 to 120 days of indexing. The re-growth showed a weeping habit because the wood of affected trees becomes rubbery and badly lignified. No necrosis, gurmmosis was observed in indicator, Gf 305. The examination by DAPI test of GF 305 petioles and young shoots confirmed phytoplasma bodies in phloem tissues, comparable to those of apple chlorotic leaf roll phytoplasma. However, based on symptoms on indicator host it is probably a distinct new disease, tentatively named as peach vein clearing phytoplasma which is natually transmitted in orchard.

APRICOT DECLINE

A severe decline of apricot trees was observed in several cultivars of apricot in South Italy (Pestore *et al.*, 1995). The affected trees showed yellowing and upward rolling of leaves, sparse foliage and witches broom on shoots. Infected trees showed premature break of leaf buds hence the leaves on affected plants develop before flowers in spring. Some flowers can also be seen in winter on affected trees. The leaves may show reddish colour and stiff texture. The veins of leaves are enlarged. Affected leaves drop prematurely showing dieback symptoms. Feeder roots are less and show brush-like appearance. Phytoplasma infection was detected in petioles and midrib of leaves of affected trees by DAPI test (Pastore *et al.*, 1995).

A chlorotic leaf roll disease in apricot occurs in several areas of Europe. Affected trees showed typical leafroll symptoms. A phytoplasma '*Candidatus* Phytoplasma prunorus" was found associated with leaf roll disease (Contaldo *et al.*, 2012).

VIROID-LIKE RNAs ASSOCIATED WITH SOME APRICOT FRUIT DISORDERS

Two apricot viroid diseases were recognized on apricot cultivars "Bulida" and "Tyrinthos' on Myrabolan rootstock. The brown reddish spots and/or concentric rings are developed in leaves on infected "Bulida" trees. It is graft transmissible and resemble to apricot ring pox (Di Serio *et al.*, 1995).

In infected "Tyrinthos" trees, fruits become deformed, showing depressed green spots and some erratic symptoms on leaves. The disease is transmitted by normal grafting and by chip grafting. Viroid like RNAs were observed in infected "Bulida" and "Tyrinthos" trees. The electrophoretic mobility of the isolated RNAs under denaturing condition was similar to hop stunt viroid. This viroid is now known to infect apricots. However, apple chlorotic leaf spot virus was also detected in such trees suggesting that the symptoms of this disease could be caused by both hop stunt viroid and apple chlorotic leaf spot virus (Di Serio *et al.*, 1995).

B. VIRUS DISEASES OF ALMOND

The Almond (*Prunus dulcis*, syn. *Prunus amygdalus* Batsch., *Amygdalus communis* L., *Amygdalus dulcis* Mill.) belongs to the subfamily *Prunoideae* of the family *Rosaceae* and is native to the Middle East. Within *Prunus*, it is classified in the subgenus *Amygdalus*, distinguished from the other subgenera by the corrugated seed shell. Although popularly referred to as a nut, the almond fruit's seed is botanically not a true nut, but the seed of a drupe (a botanic name for a type of fruit).

Almond is a rich source of Vitamin E, containing 24 mg per 100 g. It is also rich in monounsaturated fat, one of the two "good" fats responsible for lowering LDL cholesterol. Health benefits of almonds include improved complexion, improved movement of food through the colon (feces) and the prevention of cancer. In Ayurveda, an ancient system of health care that is native to the Indian subcontinent, almond is considered a nutritive diet for brain and nervous system. Recent studies have shown that the constituents of almond have anti-inflammatory, immunity boosting, and anti-hepatotoxicity effects. The almond is highly revered in some cultures. In India, Almonds are the principal crop of Jammu and Kashmir and Himachal Pradesh although the tree will grow under wide range of climatic conditions. In Jammu and Kashmir, it is designated as the State tree of Kashmir. Most varieties are self sterile hence more varieties should be placed at one place. The so called "paper shell" varieties are most popular.

Among stone fruit species, almond (*Prunus amygdalus* Batsch.) seems to be affected by a relatively low number of virus and virus-like diseases as compared to the 37 recorded from peach and the 39 from sweet cherry (Nemeth, 1986). Almond trees are affected by five virus diseases namely, Calico (genus *Ilarvirus, Prunus necrotic ring spot virus* (PNRSV) (Maliogka *et al.*, 2010), Enation (genus *Nepovirus, Tomato black ring virus* (TBRV), Infectious bud failure (genus *Ilarvirus, Prunus necrotic ring spot virus* (PNRSV), Peach mosaic (Cherry mottle leaf virus), and Peach yellow bud mosaic (genus *Nepovirus, Tomato ringspot virus* (ToRSV). Phytoplasma diseases reported in almond are: Almond brown line and decline, Almond kernel shrivel, Peach yellow leaf roll phytoplasma, European stone fruit yellows, Western X disease. In India, three virus-like disorders *i.e.* Bud failure, Calico and mosaic have been recorded (Ahlawat and Chenulu, 1986) and apple chlorotic leaf spot virus was found to infect almonds (Rana *et al.*, 2008). Infections by hop stunt viroid (HSVd) were recently reported from Spain.

ALMOND MOSAIC

Almond mosaic disease was reported in almond plantations of Kumaon and Shimla hills of India (Bhargava and Bist, 1960; Nagaich and Vashisth, 1965a) and Mediterranean area (Martelli and Savino, 1997). The disease originally reported from Shimla hills was described as "yellow Vein of Almond". The first visible symptom of the disease was yellow flecking on the margin of the leaves formed during early growth. As the leaves become mature, the veins become yellow and the characteristic "yellow net" of veins develops on a portion or entire leaf lamina. The disease in Mediterranean area is characterized by a variety of symptoms ranging from bright chrome-yellow (calico) to chlorotic discolorations, localized necrosis of the leaf blade, leaf curling, bud failure, fasciations, rosetting, stunting and bushy growth.

The disease was transmissible only by grafting. Most of the commercial varieties of almond are susceptible to mosaic virus. In India, the disease has been considered to be identical to peach mosaic (cherry mottle leaf virus) which will be discussed later in this chapter. The host of the virus are several species of *Prunus*, peach and apricot. No other information is available on this disease in India.

Three ilarviruses, apple mosaic (ApMV), prune dwarf (PDV) and prunus necrotic ringspot (PNRSV), are associated with almond mosaic throughout the Mediterranean, and are involved to different extents in its aetiology. Apple chlorotic leaf spot trichovirus (ACLSV) may also be present in mosaic-affected plants, and is sometimes associated with a chlorotic leafroll condition. Two nepoviruses, tomato ringspot (ToRSV) and tomato black ring (TBRV), have been reported from the USA and Europe as the agents of diseases called yellow bud mosaic and enations, respectively (Martelli and Savino, 1997).

Infectivity of the virus was lost by the application of electric current to purified preparations of almond mosaic virus. Similar effect *in vivo* was observed by applying electric current to almond cuttings but only when the final surface temperature of 36–38°C was reached. Buds of treated cuttings grafted on healthy almond seedlings failed to transmit the virus. Electrical current produced an increase of temperature in the cuttings. Increase of temperature may not be the only factor accounting for virus inactivation since almond cuttings heated for 30 min in a water bath at 37–38°C remained diseased (Quacquarelli *et al.*, 1981).

ALMOND YELLOW NET

A yellow net diseases on almonds was first recorded on kagzi almond plants in a stone fruit nursery at Shimla (Nagaich and Vashisth, 1965a). The disease was later observed by Professor Ahlawat from Kulu valley of Himachal Pradesh in India. The disease is characterized by prominent yellowing of all the veins and vein-lets leaving dark green islands (Fig. 10). Affected leaves were deformed. Puckering and reduction of leaf lamina were also seen in some affected trees. The twisting and upward rolling of leaves were also common symptoms in severely affected trees. The fruits on affected trees were smaller and shrivelled with brown skin and deformed. The pericarp of these fruits was also dark and blackish brown with deep longitudinal grooves. The kernels were shrivelled and deformed (Giri *et al.*, 1987).

The disease is graft transmissible from almond to almond and peach only, and not to plum and apricot. In grafted peach seedlings, symptoms developed as yellowing of leaf veins with puckering and distortion.

In ultrathin sections of diseased leaf, an accumulation of inclusions resembling of P2 like protein were observed in the companion cells of phloem elements. These inclusions were absent in sections of healthy leaves. However, the etiology of this disease is still to be investigated.

ALMOND BUD FAILURE DISEASES

The disease normally occurs in association with calico strain of prunus necrotic ringspot virus (PNRSV). Characteristic symptoms of the

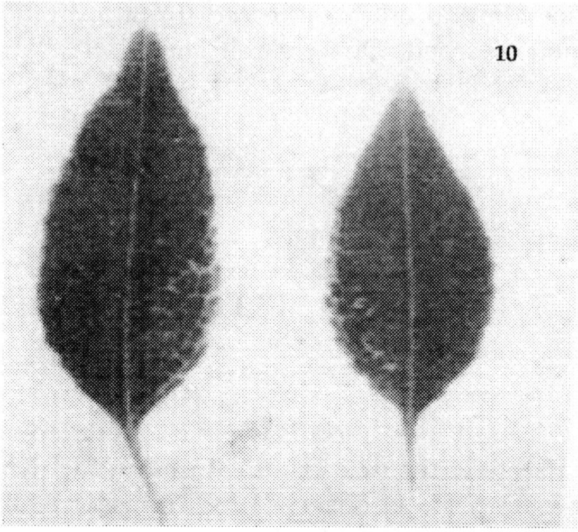

Fig. 10: Yellow net symptoms on almond leaves

disease are : death of flower and leaf buds, and slow growth. Dead lateral buds can remain attached on shoots for long time. Terminal bud usually fail to develop, resulting excessive development of lateral shoots giving bushy appearance.

The disease is transmissible both by bud and scion grafting. The hosts of the virus other than almond are: peach, nectarine, pear, sour cherry, shiro plum and mahaleb cherry seedlings. The disease is caused by Prunus necrotic ringspot virus. The disease can be managed by the use of virus-free healthy planting material.

VIRUSES FROM OTHER HOSTS INFECTING ALMOND

Following viruses have been found to infect almonds from other hosts (Dragoljub *et al.*, 1999): Latent infection of Apple chlorotic leaf spot virus (ACLSV), Tomato ring spot virus (TRSV), Cucumber mosaic virus, Tobacco necrosis virus and Tomato black ring virus. The hosts showing diagnostic symptoms are described below:

- *Tomato ring spot virus (TRSV)*: Causes yellow bud mosaic and stem pitting symptoms in almond. Yellow bud mosaic is identical to peach yellow bud mosaic. The damage by this virus is serious.
- *Tobacco necrosis virus*: Local and necrotic symptoms are developed.

- *Tomato black ring virus*: It is a Nepovirus and cause abundant enations on lower surface of almond leaves. The leaf size is reduced.

Recent studies have shown that Prunus necrotic ringspot virus (PNRSV) and Prune dwarf virus (PDV) are the major contraints in almond production. PNRSV causes the almond calico symptoms sometimes seen during Spring. Most trees carrying the virus do not show any obvious symptoms on the leaves and the level (or titre) of virus in the trees can vary. However, symptoms are common in Price, Ne Plus and GF677 rootstocks. Tested of GF677 rootstocks revealed that the titre in this host was extremely high. In all cases, the symptoms were typical of almond calico, chlorotic (yellow) mottling of the leaves, the Price trees showed a general lack of vigour.

Studies on distribution of PDV in tissues of infected plants using RT-PCR revealed that the virus was distributed in all the meristematic cells, located mostly around the nucleus and form a ring- like structure. PDV was also found in pollen grains. To introduce virus resistance, constructs were made based in the coat protein gene of PDV and prunus necrotic ringspot virus (PNRSV) and include several variations in the fragment used (sense, antisense and a partially deleted gene sequence, plus a sense construct flanqued by nuclear matrix attachment regions). The cloned fragments were first transferred to transformation vectors (pGREEN), then transferred to *Agrobacterium*, and used in *Nicotiana* transformation for testing efficiency (Raquel *et al.*, 2002, 2005). In Spain and Australia PDV is a major problem and causes dwarfing of the almond trees.

PHYTOPLASMA DISEASES OF ALMOND

Four diseases of Phytoplasma origin have been found in almond trees. Almond brown line and decline and almond kernel shrivel are caused by peach yellow leaf roll phytoplasma. European stone fruit yellows phytoplasma and Western-X phytoplasmas also infect almonds. The most studied phytoplasma in almond is witches' broom phytoplasma from Lebanon and Iran.

In Lebanon, Within less than a decade, almond witches' broom phytoplasma (*Candidatus* Phytoplasma phoenicium) killed over a hundred thousand almond trees (Jawdah *et al.*, 2010). Almond trees with symptoms of leaf yellowing, shoot proliferation and dieback were observed in the Bekaa region of Lebanon. Witches'-brooms on affected trees was the most characteristic symptom. The most typical symptom of the disease was the appearance of witches'-broom, small leaves and deformed fruits. The trees died within two years after first symptoms appear. The rapid spread of the disease and its phytoplasma etiology suggested that it may have an insect vector. All symptomatic samples tested by ELISA for the known almond viruses were negative. Tissues from infected almond trees with witches'-broom symptoms gave positive PCR reactions, but not the healthy or symptomless samples. The amplified DNA of the phytoplasma was analyzed by RFLP, and the restriction profiles were different from those published for other phytoplasmas infecting almond trees in Western Europe. Sequence analysis of the amplified DNA showed that it is closely related to the Pigeon pea Witches'-brooms cluster (PPWB) (Jaraush *et al.*, 1998; Schneider *et al.*, 1995).

Experimental transmission by dodder to periwinkle, electron microscopy, molecular studies and epidemiology of the disease revealed the association of the phytoplasma with this disease of almond. This is the first report of a phytoplasma infection in Lebanon and the first report for a PPWB group phytoplasmas in almond trees. Finally, the phytoplasma infecting almonds in Lebanon and Iran has been characterized as, *Candidatus* Phytoplasma phoenicium'(Verdin *et al.*, 2003; Maryam, 2011) The results of field surveys and laboratory assays revealed that this phytoplasma is one of the most lethal diseases of almond. (Choueiri *et al.*, 2001). In Lebanon, Witches' broom phytoplasma has been declared as a regulated pest (Jawdah *et al.*, 2012).

C. VIRUS AND VIRUS-LIKE DISEASES OF CHERRY

Cherries are mostly grown in temperate regions. In India, this crop is mostly grown in Himalayan regions of Jammu and Kashmir, Himachal Pradesh and Uttrakhand states. Sweet cherry is eaten as fresh fruit but sour cherry is used for cooking. Most cherries are self sterile and most of the best varieties are also intersterile. It is therefore important to plant a few trees of poor varieties in order to provide pollination. Several virus, phytoplasma and graft transmissible diseases of cherry are known to affect cherry plantations.

Cherry trees are infected by several viruses: *Tombusvirus, Tomato bushy stunt virus* (TBSV), *Nepovirus, Tomato ringspot virus* (ToRSV), *Cherry rasp leaf virus* (CRLV), *Cherry leafroll virus* (CLRV) and *Tobacco ringspot virus* (TRSV) *Trichoviruses, Apple chlorotic leaf spot virus* (ACLSV) *Ilarvirus, Prunus necrotic ringspot virus* (PNRSV), *Prune dwarf, Closterovirus* and *Sour cherry green ring mottle virus*. However, only two virus diseases have been recorded in India, *cherry variegation virus* and *tatter leaf virus*. Diseases caused by X-

Disease phytoplasma (Cherry albino, Cherry Western X disease), *Cherry lethal yellows phytoplasma* (Cherry lethal yellows) and two diseases by unknown phytoplasma (Cherry decline and Cherry Rosette disease) are important.

There are several other transmissible disorders have been recorded in cherry such as: Amasya cherry disease (Turkey), Black canker, Freckle fruit, Hungarian rasp leaf , Line pattern and leaf curl associated with an isometric particle (= peach leaf pattern and leaf curl), Little cherry associated with a filamentous virus particle, Mottle leaf, Necrotic rusty mottle (= Lambert mottle), Rough bark, Rough fruit, Rusty mottle (American and European), Rusty spot, Short stem, Sour cherry bark splitting, Sour cherry gummosis, Sour cherry line pattern, Sour cherry pink fruit, Sour cherry vein yellow spot, Spur cherry (= apricot ring pox?), Stem pitting, Twisted leaf associated with a filamentous virus, Utah dixie rusty mottle (= peach asteroid spot) and Xylem aberration (TBSV?). Important virus, phytoplasmas and transmissible disorders of cherry will be described on following pages.

GREEN RING MOTTLE

Cherry green ring mottle virus (CGRMV) was first reported on Montmorency sour cherry (*Prunus cerasus*) trees in 1937 in Michigan. The disease is now known to occur in Europe, North America, Australia, New Zealand and South Africa (Nemeth, 1986). The virus, in question, infects several *Prunus* species, including sweet cherry (*P. avium*), sour cherry (*P. cerasus*), oriental flowering cherry (*P. serrulata*), peach (*P. persica*) and apricot (*P. armeniaca*) in fruit-growing regions throughout North America and Europe. The viral infection on several *Prunus* spp. has also been reported in Japan, New Zealand, and some countries in Africa, Europe, and North America (Parker *et al.*, 1976). Presence of CGRMV in ornamental and sweet cherry and peach plants has been reported in China (Zhou and Wang, 2011). The virus also infects plums in North America (Wang *et al.*, 2009). The virus has also been reported from Turkey (Sipahioglu *et al.*, 2007).

The symptoms of the disease appear as leaf yellowing on sour and tart cherry. Sweet cherry plants are symptomless hosts of the virus. Tart cherries cultivars Montmorency or Morello shows corky, brown-discoloured pits or rings in the epidermis which may extend into the flesh of the fruit. Fruits on affected trees are off-flavour. Deep black depressions on fruit associated with cankering of twigs or branches developed with tattered appearance of leaves. Distinct chlorotic or necrotic spots, rings or lines, shot-hole or tottering of leaves are shown by affected trees. Leaves become cupped, distorted, narrowed, yellowed or green mottled followed by defoliation. Interior canopy leaves of Montmorency cherry turn yellow and green mottled and finally drop. Margins of Montmorency or Morello leaves show "constricting chlorosis" caused by restricted expansion of leaves along chlorotic veins.

The virus is transmitted by grafting. CGRMV is a member of the genus *Foveavirus* in the family *Flexiviridae*. It has a single-stranded, positive-sense RNA genome of approximately 8.4 kb (Zhang *et al.*, 1998). The virus was detected in sandwich-ELISA and RT-PCR (Komorowska, and Cieœliñska, 2008; Zhou and Wang, 2011).

PRUNUS STEM PITTING

The leaves of Prunus stem pitting affected cherry trees show upward cupping with yellow or red colour and subsequently dropped. Twigs are associated with thick spongy bark at crown area. Older leaves show irregular green to yellow mottling or interveinal chlorosis and finally drop 3-4 weeks after petal fall. On infected trees, bark and sapwood will be thicker than normal, and wood will have elongated indentations, swellings or pits. Generally, infected trees produce large numbers of small fruit, which tend to ripen early and may drop prematurely.

The virus can be transmitted through grafting and budding of infected tissue. Some weeds, such as dandelion, can spread the virus over long distances through seed. Prunus stem pitting (PSP) is caused by tomato ringspot virus, which can attack all stone fruit species, as well as many other hosts. Tomato ringspot virus is transmitted from weeds to trees by dagger nematodes (*Xiphinema* spp.). The virus was also found in broadleaf weeds. The virus can be detected by dot blot hybridization assays (Hadidi and Powel, 1991). Prior to planting, land preparation should include destruction of weeds, which serve as resevoir hosts of the virus, and, if nematode vectors are detected, fumigation should be considered. Growers should obtain and plant trees derived from a virus-free source. Only Marianna 2624, among commonly used rootstocks for stone fruits, is immune to the *Prunus* stem-pitting virus.

SOUR CHERRY YELLOWS

Symptoms of yellowing of leaves may appear during the first year of an infected tree, but usually symptoms do not develop for several years. The symptoms of sour cherry yellows depend on temperature. Symptom expression is most favourable about 2 weeks after night temperatures of 10 to 15.5°C and day temperatures of 30 to 35°C. Leaves of affected trees show stages of mottling and dropping as the temperature fluctuates. Leaves on affected trees show green and yellow mottle with the green along the midrib and larger veins. These leaves drop, and a second or third stage of symptoms may occur. Older trees showing willowing type of growth from reduction of fruit spurs. Fruits are sparse but large. Yield quality is better but may be reduced by 50%. Affected trees may decline in the presence of *Prunus necrotic ringspot virus*.The disease is associated with *Ilarvirus* (Prune Dwarf Virus) (Fulton, 1959). *Prune dwarf virus* (PDV) is transmitted easily by budding and grafting. It also spreads naturally in the orchard by infected pollen. Thus, there is little spread until trees are old enough to bloom. Spread in sweet cherry is much slower than in sour cherry. Due to pollen transmission, a high percentage of the seeds may carry the virus (Gilmer, 1965; Gilmer *et al.*, 1976). Following recommendations have been given for management of PDV.

- Use nursery stock that has been tested and found to be free of all known viruses.
- Do not introduce infected pollen into healthy orchards during pollination.
- Establish new plantings in blocks, the larger the better, and preferably at some distance from older orchards.
- Rogue infected trees in new virus-indexed orchards, but it is not economical to rogue or replant mature infected orchards.

PRUNUS NECROTIC RINGSPOT

SYMPTOMS

Prunus necrotic ringspot disease is worldwide in distribution. Individual branches or entire tree show delayed foliation, stunted wavy leaves, and shortened blossom pedicels in spring. Leaves develop chlorotic spots, lines, or rings as they emerge. In severe cases, chlorotic areas become necrotic and fall out giving a shot-hole appearance to the leaves (Fig. 12). In India, wild Himalayan cherry (*Prunus cerasoides*) is a natural host of Prunus necrotic ringspot virus (Chandel *et al.*, 2007).

TRANSMISSION

The virus is transmitted by mechanical inoculation; by grafting and through seed upto 80% in *Prunus pennsylvanica*. It is also transmitted by pollen to the seed and to the pollinated plant.

HOST RANGE

Plant species in about 9 families are susceptible to prunus ring spot virus infection: *Antirrhinum majus, Cassia tora, Catharanthus roseus, Chenopodium album, Chenopodium quinoa, Cucumis sativus, Cucurbita maxima, Cucurbita moschata, Cucurbita pepo, Cyamopsis tetragonoloba, Gomphrena globosa, Gossypium hirsutum, Helianthus annuus, Humulus, Lactuca sativa, Lupinus albus, Melilotus albus, Momordica balsamina, Nicotiana benthamiana, Petunia × hybrida, Phaseolus vulgaris, Pisum sativum, Prunus, Prunus cerasus, Prunus persica, Prunus serrulata, Rosa, Sesbania exaltata, Trifolium repens, Vigna unguiculata, Zinnia elegans* (Fulton, 1957). Following hosts are used as assay host of the virus: *Momordica balsamina* (L), *Cucumis sativus* (W), *Cyamopsis tetragonoloba* (L). Following are the diagnostic hosts of the virus (Fulton, 1970):

- *Cucumis sativus* - Large yellow chlorotic local lesions in cotyledons, growing point stunted.
- *Momordica balsamina* - Necrotic local lesions.
- *Cyamopsis tetragonoloba* - Dark necrotic local lesions.
- *Chenopodium quinoa* - Systemic mottle.
- *Prunus serrulata* - Local necrosis and gumming at site of grafts.

- *Lupinus albus* - Systemic mottle.
- *Helianthus annuus* - Systemic chlorotic lines and rings.
- *Cassia tora* - Local and systemic ringspotting.

THE PATHOGEN

Prunus necrotic ring spot virus (PNRSV) is an *Ilarvirus* and belongs to family *Bromoviridae* (Brunt *et al.*, 1996). The Virions are isometric; not enveloped; 23, 25, and 27 nm in diameter; rounded in profile; without a conspicuous capsomere arrangement. The Virions contain 16% nucleic acid; and 84% protein. Three strains have been reported in this virus *i.e.*, cherry rugose mosaic virus, Danish plum line pattern virus and North American plum line pattern ilarvirus. The virus has a genome of single stranded RNA. Total genome size is 8.056 kb. Genome is divided into three parts: 3.662 kb; 2.507 kb and 1.887 kb (Loesch and Fulton, 1975). PNRSV showed a high degree of similarities with those of tobacco streak virus, prune dwarf virus, apple mosaic virus and also alfalfa mosaic virus (Deyin Guo *et al.*, 1995).

DIAGNOSIS AND MANAGEMENT

Prunus necrotic ringspot virus (PNRSV) can be detected in serological tests like ELISA (Mink *et al.*, 1987). It can also be detected in dot blot hybridization tests using a nucleic acid probe derived from the PE-5 isolate, the probe detected both severe and mild isolates of the CH9 and CH30 serotypes equally well using moderate stringency conditions (Crosslin *et al.*, 1992). Biophysical differences have been observed in pathotypes (Crosslin and Mink, 1992). PCR methods can detect both pathotypes and serotypes (Rosemarie *et al.*, 1998). Prunus necrotic ringspot virus in cherry can also be detected by nested PCR (Maliogka *et al.*, 2010). The virus can be managed by certification programmes. It can also be eliminated by combination of thermotherapy, chemotherapy and *in vitro* culture (Deogratias *et al.*, 1989).

PRUNE DWARF

Prune dwarf virus (PDV) was first reported from Israel in several imported sweet cherry cultivars and one peach cultivar (Spiegel *et al.*,

1998). But now Prune dwarf virus (PDV) is probably worldwide in its distribution. Synonyms of prune dwarf virus are cherry chlorotic ringspot virus, peach stunt virus and sour cherry yellows virus.

SYMPTOMS

Symptoms of prune dwarf virus disease vary according to species, virus strain and temperature. The diseases caused by PDV have been described by various names as for example in Prune it has been described by the name of Prune dwarf, Prune mosaic, Willows and shoestring; on peach: Muir peach dwarf and peach stunt; on apricot: gummosis resetting of foliage. Symptoms on cherry are described below:

Sour Cherry

The disease caused by Prune Dwarf Virus in sour cherry is called Sour Cherry Yellows, yellow leaf, chlorotic ringspot and border tree. Leaves on affected trees develop striking pale green to yellow areas 3 to 4 weeks after petal fall. Later, entire leaf turns yellow except for green along larger veins. Leaves soon drop. Growth is willowy, and the number of fruits are reduced. In mixed infection with raspberry ringspot virus, severe rasp leaf symptoms develop, and in association with Prunus necrotic ringspot virus severe stunting and yellows symptoms appear.

Sweet Cherry

Leaves on affected trees may show yellow rings, mottling or shot-holes on leaves, or may not show any symptom. Many strains cause no acute effects, while others induce the formation of blindwood. Symptom development requires warm days and cool nights. Infected trees may not show symptoms for some years. Trees grown in warm climates may never show symptoms.

Symptoms of PDV persist in *Prunus domestica* as leathery strap-like leaves, in *P. cerasus* as leaf yellowing and in *P. persica* as stunting. Cultivars of *Prunus salicina*, plum hybrids, cherry rootstocks 'Mazzard' and Mahaleb', some sweet cherry, apricot and almond cultivars are symptomless.

TRANSMISSION

Prune dwarf virus is spread by budding, grafting, seeds, and pollen (Kenneth *et al.*, 2005). It is transmitted by seed upto 10% in *Prunus cerasus* and also transmitted by pollen to the seed. PDV is also transmitted by mechanical inoculation.

HOST RANGE

In nature, PDV infects almond, apricot, sweet cherry, sour cherry, peach, and plum. Experimentally, Susceptible host species are in 9 families and these are: *Antirrhinum majus, Catharanthus roseus, Crotalaria spectabilis, Cucumis melo, C. sativus, Cucurbita maxima, C. moschata, C. pepo, Cyamopsis tetragonoloba, Datura stramonium, Gossypium hirsutum, Helianthus annuus, Ipomoea nil, Lactuca sativa, Melilotus albus, M. officinalis, Momordica balsamina, Nicotiana benthamiana, N. clevelandii, N. debneyi, N. glutinosa, N. occidentalis, N. rustica, N. sylvestris, N. tabacum, Petunia × hybrida, Phlox drummondii, Prunus cerasus, P. domestica, P. persica , Sesbania exaltata, Thunbergia alata, Tithonia speciosa, Trifolium repens, Zinnia elegans.* Following are the diagnostic herbaceous hosts (Brunt *et al.*, 1996):

- *Cucumis sativus* - Small chlorotic local lesions; systemic mosaic.
- *Cucurbita maxima* - Interveinal chlorosis.
- *Sesbania exaltata* - Small dark local lesions in cotyledons; not systemic.
- *Crotalaria spectabilis* - Small dark local lesions.
- *Momordica balsamina* - Chlorotic local lesions; systemic mottle.
- *Tithonia speciosa* - Prominent chlorotic lines and rings.
- *Phlox drummondii* - Systemic mottle.
- *Thunbergia alata* - Systemic chlorotic rings.
- *Melilotus officinalis* - Systemic mottle.

Sesbania exaltata, Crotalaria spectabilis, Cucumis sativus are used as indicator or assay hosts.

THE PATHOGEN

Virions are isometric and bacilliform; 19-20 nm in diameter; 20 nm in length, or 23 nm in length, or 26 nm in length, or 38 nm in length (some 20 × 73 nm); rounded in profile; without a conspicuous capsomere arrangement. The virus belong to genus *Ilarvirus*, Family *Bromoviridae* and species p*rune dwarf virus*. Genome of PDV consists of positive sense single-stranded RNA. Virions contain 14% nucleic acid and 86% protein.

DIAGNOSIS AND MANAGEMENT

The virus can be detected by transmission to 'GF305' peach seedling, *P. Tomentosa* and *Cucumis sativus*. In addition to biological indexing, double-antibody sandwich (DAS) and triple antibody sandwich (TAS) ELISA tests were successfully used for PDV detection. Reverse-transcription (RT) and immunocapture (IC) polymerase chain reaction (PCR) amplification assays applied to PDV-infected sweet cherry cultivars detected the virus (Parakh *et al.*, 1995). PCR assays were superior to ELISA, season-independent and generally allowed overcoming the uneven distribution of PDV in woody hosts, typical of ilarviruses (Spiegel *et al.*, 1998). Prune dwarf virus in cherry can also be detected by nested PCR (Maliogka *et al.*, 2010). The virus can be managed by sound budwood certification programme. It can also be eliminated by combination of thermotherapy (36-37°C for 15 days or longer), chemotherapy and *in vitro* culture (Deogratias *et al.*, 1989). For management of the disease in sour cherry, establish sour cherry orchards with virus free trees, isolated from older orchards. Rogue out and replant, sour cherry trees in 1 to 5 years old orchards. 10 years old or older orchards do not benefit from rouging out trees.

CHERRY TWISTED LEAF

Twisted leaf disease of cherry was reported from North America. Symptoms of the disease appear as abrupt kink in leaf midrib or petiole. Leaves twist, curl down, remain small, or develop asymmetrically. Tissues on the shortened areas die between spur joints and thus cause spurs to bunch together (Kenneth *et al.*, 2005, Lott and Keane, 1960). Cherry twisted leaf disease spreads through budding and grafting. How disease spreads in fields is unknown. The cherry cultivar, Bing was found to be most severely affected and is used as indicator host. Napoleon, Rainier, and Van may also show severe symptoms. Most other cultivars donot show symptoms when infected. Wild chokecherry (*Prunus virginiana*) was found to be a symptomless host of the apricot ring pox and cherry twisted leaf viruses, and may be a source of infection in orchard trees (Lott and Keane, 1960). These two virus diseases are related to each other and the causal agent of both was serologically related to apple stem pitting virus (Zhang *et al.*, 1992). To prevent spread of the disease, remove infected trees as soon as symptoms appear. Destroy choke-cherry stands near commercial orchards.

NECROTIC RUSTY MOTTLE

Necrotic rusty mottle disease of cherry has been reported from North America, Europe, New Zealand and Japan (Posnette and Cropley, 1964). Affected trees show angular spots or large areas of dead tissue on leaves 3 to 6 weeks after bloom. Spotted leaves develop shot-holes. Severely nectorized leaves drop. Before harvest, remaining leaves turn mottled yellow, with green bands or islands. Buds on terminal shoots fail to open or swell; eventually they die. Bark symptoms consist of shallow areas of decay with gum blisters or deep gum pockets (Kenneth *et al.*, 2005). Severe symptoms develop in cool and late spring season. Leaf symptoms usually are not noticeable during summer and fall when temperatures remain high. The virus associated with necrotic rusty mottle disease spreads through infected bud-wood. However, its spread in the orchard is unknown. Rusty mottle virus infects several *Prunus* species, *P. avium, P. cerasus, P. persica, P. armeniaca, P. mahaleb* and others.

Cherry necrotic rusty mottle virus (CNRMV) is closely related to cherry green ring mottle virus and is an unassigned member of the family *Betaflexiviridae*. The virus particles are flexuous, filamentous with single stranded RNA. The genome of the virus has 8432 nucleotides with seven open reading frames (Rott and Jelkmann. 2001, Li and Mock 2008). In India, the CP gene of JK10 isolate was partially sequenced which consists of 2203 nucleotides. The sequences showed 91% and 84% similarity to the type isolate of CNRMV from Germany at the nucleotide and amino acid level respectively (Awasthi *et al.*, 2012). The virus can be detected by indexing on *P. avium* cv. Sam. For management of the disease, remove affected trees (Cameron and Moore, 1985). Use virus free nursery stock and scion wood. Several pollinating cultivars are symptomless hosts. Hence testing is required before taking budwood for propagation.

CHERRY RUGOSE MOSAIC

Two types of infections develop in rugose mosaic affected cherry trees *i.e.* acute infection and chronic infection. In acute infection, symptoms of necrotic ringspot occur early in the season followed by yellowing, browning, and dead tissues of ring spots. Dead tissues drop out, giving leaves a tattered or lace-like appearance. Fruit symptoms may not appear during first year of infection. Fruits on affected trees may ripe unevenly. In chronic infection, leaves appear with twisted tips. Small outgrowths or enations of dark green tissue may develop along the midrib on lower leaf surface. Severe strains cause uneven maturation of fruit. Green fruits may be cylindrical. Fruits are malformed and ripe later than normal. Mild forms show leaf symptoms, with little or no effect on fruit. Fruit ripening may be delayed 7 to 10 days with no leaf symptoms present. Severe forms exhibit blossom blast or withering, severe leaf deformity, pronounced leaf curl, reduced shoot growth, and reduced tree vigour (Kenneth *et al.*, 2005). Additional symptoms include distorted growth, death of twigs, buds, or young foliage, and stunting of trees. Sometimes only one branch will show tattered leaves. Yield on trees infected with the normal strain is reduced about 5%. Trees infected with the rugose mosaic strain set a very light crop that matures late.

The virus is transmitted by grafting, seed, or pollen from diseased trees. Spread in sweet cherry is much slower than in sour cherry. Due to pollen transmission, a high percentage of the seed may carry the virus (Oliver *et al.*, 2009). The disease is associated with *Prunus necrotic ringspot virus* (PNRSV) - an *Ilarvirus* which has many strains causing various diseases such as necrotic ringspot, tatter leaf, peach ringspot, currant ringspot, cherry rugose mosaic, rose mosaic, and almond calico. Prunus necrotic ringspot virus (PNRSV) occurs worldwide wherever stone fruits are grown and is a serious pathogen of many woody species, including sweet cherry, almond, rose and plum (Hammond and Howell, 2000).

Polymerase chain reaction (PCR) technique has been used to detect rugose mosaic virus (Hammond and Howell, 2000). Diagnostic test like ELISA to detect symptomless or infected trees are used. To manage *Prunus necrotic ringspot virus,* virus-free budding material and rootstock in nurseries and plant certified nursery stock in production orchards are recommended. Remove infected trees before bloom for effective management. Use virus free planting stock and scion wood.

RUSTY MOTTLE

Two types of rusty mottle diseases have been reported; the American rusty mottle and the European rusty mottle. The former is distributed in USA whereas the later is found in several countries of Europe and South Africa (Nemeth, 1986). Older leaves turn a mottled yellow midway between blossom and harvest. Leaves show variable yellow rings and patterns. Affected leaves turn abruptly to yellow and red during fall. Symptoms appear first on leaves in the lower centre of the tree. Two to three weeks before harvest, rusty yellow coloured leaves drop, taking 30 to 70 percent of a tree's leaves. Infected trees decline in vigour; main limbs show die back. Fruit ripening may be delayed on severely affected trees and fruits are flavourless (Kenneth *et al.*, 2005). Cherry is a natural host of the American type virus but experimentally it can be transmitted to apricot, peach, *P. mahaleb, P. serrulata* and *P. virginiana* (Zeller and Milbrath, 1947; Reeves, 1951). Sweet cherry is the host of European type. In infected sweet cherry chlorotic rings, patches and mottle develop on the foliage in May, bright yellow to red rusty spots and lines with islands of green develop on the foliage in July (Ramaswamy, 1975).

Rusty mottle viruses were transmitted by grafting. American Cherry Rusty Mottle Virus is associated with Trichovirus but the etiology of European type is unknown. Both rusty mottle and cherry twisted leaf come from the same *Foveavirus* family. Indexing of American type can be done on Bing cherry (*P. avium* cv. Bing) and of European type on *P. avium* cv. Sam. For management, remove diseased orchard trees and replant virus free trees.

CHERRY LEAF ROLL VIRUS

Cherry leaf roll virus disease has been reported in the Eurasian region and the North American region; Australia, China, New Zealand, Turkey, and the former USSR (Nemeth, 1986). The synonyms and common names of this virus are: elm mosaic, ash mosaic virus, sambucus ringspot, yellow net virus, cherry leaf roll and walnut blackline disease

(Brunt *et al.*, 1996). The virus is not damaging in cherry but it is economically important in walnut (Mircetich and Rowhani, 1984).

SYMPTOMS

Cherry leafroll virus (CLRV) infected cherry trees develop symptoms of delayed flowering, excessive leaf senescence and leaf rolling. Affected young plants may die. In other susceptible plant species chlorotic ringspot, line patterns and/or yellow vein netting may develop. In the early stages of infection, trees set abundant fruit but it is small and ripens late as compared to uninfected trees. Over several seasons, fruit production declines dramatically and the canopy becomes open and sparse. When CLRV occurs in combination with either Prunus necrotic ringspot virus (PNRSV) or Prune dwarf virus (PDV), trees decline rapidly. Initially, only one or two scaffold limbs may show symptoms, but the symptoms will spread to the remainder of the tree (Kenneth *et al.*, 2005).

TRANSMISSION

The leaf roll virus disease of cherry is transmitted by grafting, mechanical inoculation, pollen and seeds. The virus is spread by the use of infected budwood, but not through root grafting in the orchard. The associated virus is seed-borne (0.5 - 35%) in most natural hosts, including American elm, *Betula* spp., *Prunus* spp., rhubarb, *Sambucus* spp. and walnut. Seed-borne up to 100% has been found in many herbaceous hosts (Cooper *et al.*, 1984; Brunt *et al.*, 1996; Quacquarelli and Savino, 1977). There is evidence that at least some strains may be transmitted in pollen to the plant pollinated (Rebenstorf *et al.*, 2006; Bandte and Büttner, 2001; Jones, 1985). Cherry leaf roll virus is also transmitted by nematode vectors; *Xiphinema coxi*, *X. diversicaudatum* and *X. vuittenezi* (Fritzsche and Kegler, 1964; Flegg, 1969).

HOST RANGE

Plant species in more than 9 families are susceptible to cherry leaf roll virus. Cherry, walnut, elm (*Ulmus* spp.), Brich (*Betula* spp.), Ash (*Fraxinus* spp.), elderberry (*Sambucus* spp.), Beech (*Fagus* spp.), rhubarb (*Rheum* spp.), dogwood (*Cornus florida*), and Liac (*Syringa vulgaris*) are natural woody host of cherry leaf roll virus. Experimentally inoculated plants mostly show chlorotic or necrotic local lesions, systemic necrosis or mosaic. Susceptible host species are: *Antirrhinum majus, Apium graveolens, Atriplex hortensis, Berteroa incana, Beta. macrocarpa, B. vulgaris, Betula, Capsicum frutescens, Catharanthus roseus, Celosia argentea, C. cristata, Chenopodium amaranticolor, C. foetidum, C. murale, C. quinoa, Commelina, Cornus florida,*

Cucumis melo, C. sativus, Cucurbita maxima, C. pepo, Delphinium elatum, Glycine max, Gomphrena globosa, Helianthus annuus, Juglans regia, Lathyrus odoratus, Lavatera trimestris, Lobelia erinus, Momordica balsamina, Nicotiana benthamiana, N. clevelandii , N. debneyi, N. megalosiphon, N. occidentalis, N. rustica, N. tabacum, Ocimum basilicum, Olea europaea, Petunia × hybrida, Phaseolus vulgaris, Phlox drummondii, Physalis floridana, Physalis peruviana, Pisum sativum, Prunus avium, P. cerasifera, P. persica, Ptelea trifoliata, Raphanus sativus, Rheum rhaponticum, Rubus idaeus, Rumex obtusifolius, Sambucus racemosa, Senecio vulgaris, Sesamum indicum, Spinacia oleracea, Stellaria media, Tinantia erecta, Torenia fournieri, Tropaeolum majus, Ulmus americana, Zinnia elegans (Schmelzer, 1966; Hansen and Stace-Smith, 1971; Horváth, 1979; Tomlinson and Walkey, 1967; Ahmed and Bailiss, 1975; Walkey and Cooper, 1973, Nemeth, 1986). Symptoms on some of the susceptible hosts are shown below:

- *Juglans regia* - Line pattern and black line.
- *Ulmus americana* - Chlorotic mosaic, ring pattern and dieback.
- *Prunus avium* - Leaf rolling and death. bark necrosis and gumming at point of inoculation are characteristic reactions. Flower pedicels are very short, often half the length of those on healthy trees.
- *Betula* spp., *Sambucus* spp., *Rubus* spp., *Cornus florida* - Chlorotic ringspot, leaf patterns and/or yellow vein netting.
- *Ptelea trifoliata* - Yellow spotting.
- *Olea europaea, Rheum rhaponticum, Berteroa incana, Delphinium elatum, Rumex obtusifolius* - Symptomless.
- *Chenopodium amaranticolor, C. quinoa* - Chlorotic or necrotic local lesions; systemic mottle, necrosis and malformation.
- *Cucumis sativus* - Chlorotic local lesions, occasional systemic mosaic.
- *Nicotiana rustica, N. tabacum* cvs White Burley, Xanthi-*nc* - Necrotic local lesions and rings; systemic necrotic or chlorotic rings and line patterns depending on the virus strain and growing conditions.
- *Olea europaea, Rheum rhaponticum, Berteroa incana, Delphinium elatum, Rumex obtusifolius* - Symptomless (Jones, 1985; Posnette, 1956).

Assay hosts of the virus are: *Nicotiana rustica, N. tabacum* cvs White Burley, Xanthi-*nc,* and *Chenopodium amaranticolor*. The virus is found in several strains:Type (cherry) strain, Elm mosaic strain, Rhubarb strain, Golden elderberry strain, Red elder ringspot strain, Dogwood ringspot strain, Birch strain, Walnut ringspot and walnut yellow vein strains and Blackberry and red raspberry strains (Jones, 1985). The walnut strain of CLRV is the causative agent of blackline disease, a fatal necrosis of the graft union of English walnut trees.

THE PATHOGEN

Virions of leaf roll virus are isometric; not enveloped; 28 nm in diameter; angular in profile; without a conspicuous capsomere arrangement. Virions are found in leaves, roots, meristem and within tubules in anther cells, pollen, ovules and mature seeds, in cytoplasm of infected cherry, birch and almond (Walkey and Webb, 1968; Jones, 1973; Bargen *et al.*, 2012). Cherry leaf roll virus (CLRV) belongs to subfamily *Comovirinae,* family *Secoviridae,* order *Picornavirales,* genus *Nepovirus* and species *cherry leaf roll virus*. Genome consists of single stranded RNA which is linear.The two RNA species are (+)ss RNAs of a rhubarb isolate of *cherry leaf roll virus* (CLRV) have been reported. The larger RNA1 is 7918 nucleotides and the shorter RNA2 6360 nucleotides in size, each genome component comprising a single open reading frame (ORF) (Susanne *et al.*, 2012). Genome of two parts (RNA-1 and RNA-2) are required for infectivity (Jones and Mayo, 1972; Jones and Duncan, 1980; Haber and Hamilton, 1980).

DIAGNOSIS AND MANAGEMENT

The disease diagnosis can be done by observing typical symptoms of the disease. It is also detected by sap inoculation on herbaceous hosts like *Chenopodium quinoa, Cucumis sativa* and *Nicotiana* spp. Woody indicator (GF305) when chip budded with infected *Prunus* spp. show rusetting and slight leaf rolling. Gel double diffusion tests (Cropley, 1960, 1961), Enzyme-linked immunosorbent assay (ELISA) and immunosorbent electron microscopy (ISEM) have been used to detect infection in tissues of natural and experimental hosts (Jones, 1985). Nucleic acid hybridization (Mas *et al.*, 1993) and immuno-capture RT-PCR techniques have also been used to detect the virus (Werner *et al.*, 1987).

Removal of affected trees and herbicide treatment of freshly cut stump of symptomatic trees may help to avoid infection. It is important to note that pruning out only the infected limbs will not control the virus. Tree removal is necessary as the virus will have spread throughout the tree once a diseased limb is observed. Testing of trees for virus before propagation is important. To avoid root grafting proper distance between trees must be maintained.

CHERRY RASP LEAF

Cherry rasp leaf disease was first reported in *Prunus avium* from USA by Bodine and Newton (1942). Flat apple virus is synonym to cherry rasp leaf (CRLV). Rasp leaf disease in cherry is found in the North American and the Pacific regions: Canada, New Zealand, Australia, China and South Africa. It is also found, but with no evidence of spread, in the U.K. (Jones *et al.*, 1985). The symptoms of raspleaf have been observed in many countries (Hansen *et al.*, 1974). In older cherry growing areas, incidence of the disease was as high as 38% (Luepschen *et al.*, 1974).

SYMPTOMS

Symptoms on affected cherry trees appear as prominent lumps or outgrowths or enations on the underside of leaves between lateral veins and along the midrib. As a result, the leaves become narrow, deformed and folded. The face of the leaf has a bumpy texture as a result of the enations on the underside. Leaf narrowing, folding and distortion are common symptoms. The upper leaf surface turns rough and pebbly in texture. Symptoms first appear on lower leaves in newly infected trees. Trees develop an open, bare appearance as spurs and lower branches die. Lateral (tree-to-tree) infection is slow and fruit production is reduced.

TRANSMISSION

Cherry rasp leaf virus is also transmitted by grafting. It is also transmitted by sap inoculation. Seed transmission has been shown to occur in some herbaceous hosts. Virus has been detected in pollen from infected cherry trees, but transmission by pollen has not been confirmed. The dagger nematode, *Xiphenema americanum* is the vector of the disease (Nyland, 1976; Hansen *et al.*, 1974, EPPO/CABI, 1996). The virus does not affect pears, but causes flat apple disease on apple trees (Kenneth *et al.*, 2005). Spread of the virus in the field is generally slow due to the slow movement of the nematode vector (Bodine and Newton, 1942; Hansen *et al.*, 1974; Nyland, 1976; Stace-Smith and Hansen, 1976; Dunez, 1981; Jones and Badenoch, 1982).

HOST RANGE

The principal hosts of rasp leaf virus are cherries and peaches, and also apples. The rootstock species *Prunus mahaleb* is also susceptible. Raspberries are also attacked (Jones *et al.*, 1985, OEPP/EPPO, 1984). Some weeds like species of *Taraxacum, Plantago* and *Balsamorhiza* are naturally infected, but without symptoms. A wide range of herbaceous test plants are also susceptible to cherry rasp leaf virus (CRLV). Symptoms on important hosts are given below:

- *Prunus avium, P. mahaleb* - Affected cherry branches are very frost sensitive (Hansen *et al.*, 1974; Wagon *et al.*, 1968), enations, stunting.
- *Prunus persica* - Enations, stunting, short internodes and decline.
- *Malus sylvestris* - Enations, flat fruit.
- *Balsamorhiza sagittata, Taraxacum officinale, Plantago major, Rubus*– symptomless (Jones, 1987).

Following species are susceptible to CRLV by sap inoculation: *Balsamorhiza sagittata, Chenopodium amaranticolor, Chenopodium quinoa, Cucumis sativus, Cyamopsis tetragonoloba, Malus sylvestris, Physalis floridana, Plantago major, Prunus avium, Prunus mahaleb, Prunus persica, Sesbania exaltata, Taraxacum officinale, Vigna unguiculata (*Hansen *et al.*, 1974). *Cucumis sativus, Cyamopsis tetragonoloba* are the assay hosts of the virus.

THE PATHOGEN

Cherry rasp leaf virus particles are isometric about 30 nm in diameter. Rasp leaf virus belongs to the order *Picornavirales*, family *Secoviridae*, subfamily *Comovirinae*, and genus *Cheravirus*. The disease caused by cherry rasp leaf nepovirus (CRLV) should not be confused with European cherry rasp leaf (Pfeffinger disease), which is caused by a combination of prune dwarf ilarvirus with raspberry ringspot nepovirus or arabis mosaic nepovirus. Cherry rasp leaf virus genome consists of single stranded RNA. Total genome size 13.3 kb. Genome is in two parts; one of 6.5 kb and the 2nd of 4.8 kb (Brunt *et al.*, 1996, OEPP/EPPO, 1984).

DIAGNOSIS AND MANAGEMENT

Diagnosis can be done by transmission to indicator hosts, *Prunus avium cv. Bing, Cucumis sativus, Cyamopsis tetragonoloba, Chenopodium quinoa* and *C. amaranticolor* (ISHS, 1980). Other detection methods are ELISA or confirmatory gel double-diffusion serology. For management, remove infected trees and other trees closeby. Fumigate soil with a nematicide before planting cherry or apple trees. Use virus free nursery stock and scion wood.

CHERRY MOTTLE LEAF

Cherry mottle leaf disease has been reported in Europe, North America and South Africa (Nemeth, 1986).

SYMPTOMS

Symptoms of mottle leaf disease appear as irregular light green to yellow mottling on expanded leaves, which become distorted and puckered. Fruit set is reduced; fruit may be small, ripen late, and lack flavour. Affected trees become rosette due to lack of terminal growth and shortening of internodes. Cherry cultivars Bing and Napoleon (Royal Ann) are severely affected. Cherry cultivars Black Tartarian, Van, Chinook, Lambert, Rainier, and Sam are symptomless.

TRANSMISSION

Cherry mottle leaf virus is transmitted by budding or grafting, mechanical inoculation and by the microscopic scale mite, *Eriophyes inaequalis* (Oldfield, 1970).

HOST RANGE

Following plant species are susceptible to mottle leaf: *Chenopodium amaranticolor, C. quinoa, Prunus armeniaca, P. avium, P. cerasifera, P. mahaleb, P. persica, P. yedoensis* and *P. serrulata* (James, 1992). *Chenopodium quinoa* is an indicator host of the virus (James *et al.*, 1999).

THE PATHOGEN

Cherry mottle leaf virus is a *Trichovirus* (Brunt *et al.*, 1996). Virions are filamentous; not enveloped; usually flexuous; with a clear modal length;

of 760 nm; 10.2 nm wide. Axial canal is obscure. Basic helix is obvious; pitch of basic helix 3.1 nm (James, 1994; James and Mukerji, 1993). The subunit size of cherry mottle leaf virus (CMLV) capsid protein is estimated at 20.5 kDa (James, 1992). Genome of the virus is single-stranded, linear RNA. Total genome size 8.2 kb (James and Mukerjee, 1993). Search of the database revealed some sequence homology of the partially characterized genome of CMLV with that of apple chlorotic leafspot virus (James *et al.*, 1999).

Cherry mottle leaf virus and peach mosaic virus (PMV) were considered to be the same virus (James and Howell, 1993, 1998; James and Mukerjee, 1993). All isolates of PMV tested cross-reacted with CMLV-associated antibodies, both polyclonal and monoclonal (James and Howell, 1998; Oldfield *et al.*,1995). Both the viruses infecting *Prunus* spp. are known to be transmitted by eriophyid mites: PMV by *Eriophyes insidiosus* Kiefer and Wilson (1955), and CMLV by *E. inaequalis* Wilson & Oldfield (Oldfield, 1970). The two viruses are very similar, having filamentous morphology, similar sizes, identical coat protein subunit sizes, and identical double-stranded RNA(dsRNA) profiles (James and Howell, 1998; James and Mukerji, 1996; Oldfield *et al.*, 1995). The two viruses are closely related and may be strains of the same virus.

DIAGNOSIS AND MANAGEMENT

Polyclonal and monoclonal antibodies have been developed against this virus and have been used for serological detection of the virus (James and Mukerji, 1996). The CMLV-associated antibodies cross react with peach mosaic disease–associated virus (PMV) (James and Howell, 1993,1998). All isolates of PMV tested cross-reacted with CMLV-associated antibodies, both polyclonal and monoclonal suggesting that the two viruses are same (James and Howell, 1998; Oldfield *et al.*, 1995). In addition to serological diagnosis (ELISA, Western blotting), CMLV can also be detected by dot blot hybridization and also by RT-PCR (James *et al.*, 1999). *Chenopodium amaranticolor* and *C. quinoa* are used as assay hosts. Symptoms on these hosts upon inoculation develop as local lesions, and systemic epinasty of leaves. Removal of infected trees to prevent new infections has been recommended as a measure for management. Destroy stands of wild cherry near commercial orchards. Use virus free nursery stock and scion wood (Kenneth *et al.*, 2005).

LITTLE CHERRY DISEASE

Little cherry disease was first described in British Columbia in 1933 and is now known to occur worldwide. It has been recorded from several countries of Europe, USA and former USSR (Németh, 1986, Smith *et al.*, 1992).

SYMPTOMS

Little Cherry Disease primarily affects fruit size and quality. Fruit on infected trees are smaller than normal, poorly coloured, and lack in flavour. Fruits are one-half to two-third of normal size, dull red, and may (on certain varieties) be pointed in shape with three flat sides tapering toward the blossom end. Fruit symptoms are not always uniform, and may appear on only part of the tree. Severity of fruit symptoms varies from variety to variety. Infected trees may be slightly less vigorous, but generally there is no noticeable deterioration in the overall health of the tree. A premature leaf reddening occurs in the fall on infected trees of some varieties like Sam, Van, Star and Lapins. Fruit symptoms were most pronounced in the Lambert variety. Cherry trees infected with the little cherry virus 2 produce fruits that are smaller than normal, with poor colour and poor flavour. Symptoms of little cherry virus 1 appear to be less severe than for Little Cherry Virus 2 but this is not yet well documented.

TRANSMISSION

Little cherry disease may be caused by two different viruses, Little Cherry Virus 1 and Little Cherry Virus 2. Little Cherry Virus 2 is vectored by the apple mealybug, *Phenacoccus aceris* (Raine *et al.*, 1986). The vector of Little Cherry Virus 1 is unknown. Both are spread by propagation. The causal pathogen is graft-transmissible and infected plants contain flexuous filamentous virus-like particles (Ragetti *et al.*, 1982). The apple mealybug overwinters as partially grown nymphs in small, white cocoons (approximately 2 mm long) which are hidden in bark crevices. Females emerge during the first warm weather in the spring, and begin feeding on the undersides of twigs. Natural spread to and between sweet cherry trees occurs only in areas where *P. aceris* is present. Root grafting can also transmit the virus to healthy trees adjacent to infected trees. Little cherry virus-1 was transmitted from infected *Prunus avium* F12 rootstocks by dodder, *Cuscuta europea* to *Nicotiana occidentalis* (Jelkmann *et al.*, 2010).

HOST RANGE

Sweet cherry (*Prunus avium*) is highly susceptible host of the disease. Fruit symptoms are also found in sour cherry (*P. cerasus*) and in *P. pensylvanica*. The ornamental cherries, *P. incisa, P. serrulata, P. sieboldii, P. Subhirtella, P. yedoensis, P. emarginata, P. mahaleb* and *P. tomentosa* are often latent hosts of the virus (Welsh and Cheney, 1976). A wide range of ornamental and fruit trees including apple, pear, plum, cherry, currants, maple, birch, willow and white ash are the host of little cherry disease. Following varieties of cherry, sour cherry and rootstock are susceptible to little cherry infection:

- *Lambert*: Leaves show slightly upward curling. Fruits are 1/2 to 2/3 smaller than normal size, angularily pointed and Pink to dark colour with insipid taste. Infected fruits remain hanging out after harvest.
- *Bing*: Leaves slightly curl upward from margins. Mild to moderate reddening in leaves. Infected fruits are similar in shape with abnormal taste.
- *Royal Ann*: Mild marginal leaf upcurl. There is little or no leaf-reddening. Some infected fruits are pointed, flat-sided, and poorly coloured. Size is reduced, taste insipid.
- *Van and Sam*: Moderate to strong reddening and mild marginal upcurl. Reduced size and flavour; little effect on shape and colour.
- *Lapin*: Moderate to strong fall leaf-reddening. Reduced fruit size and flavour.
- *Stella and Sweetheart*: Reduced fruit size and flavour.
- *Mazzard F12/1 Rootstock*: Moderate fall leaf-reddening. Mild marginal upcurl. Fruit symptoms are less clear.

THE PATHOGEN

The little cherry pathogen is probably phloem borne. Virus particles are elongated, flexuous, virus-like rods as well as small spherical or ellipsoid vesicles (diameter about 75 nm); additionally hexagonal tubules (530 x 1450 nm) were detected by electron microscopy in the phloem companion cells (Raine *et al.*, 1975; 1979). The flexuous rods (diameter about 12 nm) were usually arranged in large aggregates and occurred also in the tubules. From the leaves of diseased sweet cherry trees, a double-stranded RNA could be isolated but could not be isolated from latently infected trees (Hamilton *et al.*, 1980; Legrand and Verhoyen, 1986).

DIAGNOSIS AND MANAGEMENT

In field, the disease can be diagnosed at the picking time as most cherries of diseased trees are pointed in shape, imperfectly coloured, small and insipid in taste. A few sweet cherry cultivars also show the interveinal areas of the upper leaf-surface turn red-violet or become bronze-coloured, while the midrib and the main veins retain their green colour for some time. Red-violet discoloration also occurs on the leaves of the cherry rootstock Mazzard F 1211 after inoculation with the little cherry pathogen. Sweet cherry cv. Sam has been recommended as indicator for diagnosis of the disease, but Canindex 1 is better because it reacts more distinctly and earlier in the season (Hansen and Green, 1985). Normally the method of double-budding is used for diagnosis in the glasshouse (Fridlund, 1980). Diagnosis can also be done by observing cross-sections of petioles by UV microscopy. Fluorescent inclusions are observed in the walls of phloem cells of infected trees after staining with acridine orange. It may be of diagnostic value for the detection of latent infections in cultivars of the oriental flowering cherry *Prunus serrulata* (Legrand and Verhoyen, 1985; 1986). Detection is also possible by hybridization assay (Eastwell *et al.*, 1996).

Little Cherry Disease can be controlled by avoiding planting of infected stocks. Removal of all trees known or suspected to be infected with little cherry disease is important to limit spread within the orchard. Do not interplant cherry trees amongst older cherry trees that may be carrying the virus. Do not grow or import ornamental Japanese flowering cherries, as these trees are often symptomless carriers of the little cherry virus. Eliminate cherry seedlings in and near the orchard, as they can be symptomless carriers. Apple mealybug should be kept under control. Since 1977 the eradication programme in the Okanagan region has been improved by insecticide sprays against *P. aceris* (Slykhuis *et al.*, 1980). Testing for latent infections with the little cherry pathogen must be included in the certification schemes for fruit-tree propagation material, to provide healthy young cherry trees (OEPP/EPPO 1991/1992).

VARIEGATION OR MARBLE MOSAIC

The variegation disease was commonly observed in sweet cherry (*Prunus avium*) var. Black heart in Himachal pradesh in india (Azad and Sehgal,

1958; Ahlawat and Chenulu, 1986). The leaves of affected trees show a wide spread variegation similar to that of genetic origin (Fig. 11). A similar transmissible disorder as "cherry white spot has been reported on Rigi variety of cherry in Switzerland and on Black heart variety in Germany (Mallach, 1957). The disease is transmissible by bud grafts from Black Heart cherry to Bird cherry (*P. padus*) and to Bigarrean Jaboulay. The disease is transmissible by eriophyid mite (G.N. Oldfield personal communication). No further information is available on this disorder.

Fig. 11: Cherry leaf showing variegation (left) and marble mosaic on cherry twig and leaf (right)

TATTER LEAF

Tatter leaf disease is found in cherry, peach, and plum. It causes tatter leaf symptoms. Fruit yields may be decreased by 40 to 50 percent, depending on conditions. This is a common disease of cherry in Himachal Pradesh in India. Both cherry and peaches are affected with this disease (Nagaich and Vashisth, 1965). The symptoms of the disease are similar to those described under Prunus necrotic leaf virus. Symptoms generally consist of chlorosis, necrosis, leaf deformity, and stunting. Entire plants may be affected, or only a portion may show symptoms. Areas of symptomatic leaves may become necrotic and fall out; giving the leaves a tattered look. Green fruit may show arcs and rings (Oliver *et al.*, 2009). Bird's cherry (*Prunus padus*) is also susceptible to tatter leaf virus. Initial symptom of the disease on peach and cherry appear as chlorotic ring spots, which later become necrotic. The necrotic tissues fell down leaving the holes in the leaves (Fig. 12). A disease reported as plum ringspot and shot-hole from Himachal Pradesh, Kumaon hills and Kalimpong region (Darjeeling hills) appears to be the same or closely related to tatter leaf of cherry (Bhargava and Bist, 1957).

Fig. 12: Necrotic spots and shot hole on cherry leaf due to PNRSV

The transmission of the disease is through wedge and bud grafting to peach Vars. Alton and Elberta and sweet cherry variety Babcock. Movement of virus in the orchard occurs through transmission by pollen to seed and to pollinated plants. The disease is transmitted mechanically from peach to squash (*Cucurbita maxima*) *var.* Butternut. The affected butternut plants are stunted and the leaves show typical mosaic mottling. However, back inoculations from squash to cherry were unsuccessful. The disease is also transmitted by *Cuscuta hyalina* (Ahlawat and Dhingra, 1973).

The disease is caused by *Prunus necrotic ringspot virus* (PNRSV)- an *Ilarvirus* and belongs to family *Bromoviridae*. There are several strains in this virus that cause variously named diseases, necrotic ringspot, tatter leaf, peach ringspot, cherry recurrent ringspot, cherry rugose mosaic, rose mosaic and almond calico.

Two methods of control have been suggested *i.e.,* Certification programmes that supply disease free material and orchard management practices to prevent tree-to-tree spread. This can be done by prompt removal of diseased plants; avoid the use of commercial beehives that have been working the area and the use of virus-free wood for top working. Obtaining virus free certified planting material from a reputable plant improvement organization is important.

TOBACCO RINGSPOT VIRUS INFECTING CHERRY

Tobacco ringspot virus (TRSV) has been reported in several herbaceous and woody hosts including *Prunus* spp. in different parts of the world. In fruit trees, the virus has been recorded in sweet and ornamental cherries but experimentally it has a wide host range (Uyemoto *et al.*, 1977; Stace-Smith and Hansen, 1974). The symptoms in cherry are delayed in bud opening and flowering with irregular light green leaf blotching.

The virus is transmitted by graft and mechanical inoculation. It is also seed borne in several hosts including *Cucumis sativus*. The vector of the virus is a nematode, *Xiphinema americanum sensu lato* (Stace-Smith, 1985).

Tobacco ringspot virus is a *Nepovirus* with three types of isometric particles, 28 nm in diameter with a bipartite RNA genome (Stace-Smith 1985). However, like many other viruses of Nepovirus group of which it is the type member.

The virus can be detected by grafting to Bing cherry and by mechanical transmission to *Chenopodium quinoa* and also by ELISA. To manage the disease infected plants may be removed and replanted with healthy plants. Control of nematodes is important for keeping the healthy plants to remain healthy.

TOMATO RINGSPOT VIRUS INFECTING STONE FRUITS

Tomato ringspot virus (ToRSV) naturally infects cherries in Northern America and Chile (Stace-Smith, 1984; Auger, 1988). The causal virus causes stem pitting and decline in several *Prunus* spp. including peach, nectarine, cherry, apricot, almond and plum (Schlocker and Tayler, 1976). The leaves on affected plants become chlorotic, curl upward and turn red in autumn and fall prematurely. In stem, the bark become spongy with necrotic areas, longitudinal pitting and grooving on rootstock or scion or on both. ToRSV causes a Prune brown line disease in *Prunus* spp. (Auger, 1988; Fiore *et al.*, 2012).

Tomato ringspot virus has a wide host range in herbaceous and woody plants including ornamentals, apples, small fruits and legumes. The virus is transmitted by grafting and mechanical inoculations. It is also seed borne in several herbaceous hosts. The vectors of the virus are nematodes, *Xiphinema americanum sensu strico, X. Rivesi* and *X. Californicum* (Stace-Smith, 1984). The virus associated with tomato ringspot disease is a *Nepovirus* with isometric particles of 28 nm diameter. The genome of the virus is bipartite (Stace-Smith, 1984).

The virus can be detected by grafting to *Prunus tomentosa* and *P. Persica* 'GF305" and by mechanical inoculation to *Chenopodium quinoa*. The virus can also be detected in ELISA and dot blot hybridization (Hadidi and Hammond, 1988; Hadidi and Powell, 1991). Infected plants must be removed and replaced but, unless measures are taken to destroy nematode vectors, the virus would soon be transmitted to healthy plants in replanted areas.

TWIG NECROSIS OF CHERRY

Twig necrosis disease of cherry has been found in several countries in Europe and Canada. Main symptoms of the disease are: bark necrosis, necrosis of midrib and some lateral veins, twisting of leaves and stunting. Fruits are malformed with necrotic spots. The causal virus infects naturally to sweet cherry, sour cherry, plum, pear and several herbaceous host plants (Novak and Lanzova, 1980). Experimental host range is large. The virus is transmitted by grafting and through soil. Previously tomato bushy stunt virus was reported to be associated with this disease (Novak and Lanzova, 1980) but recent investigations revealed that two tombus viruses, petunia asteroid mosaic virus and carnation Italian ringspot virus are associated with this disease (Koenig and Kunze, 1982; Lesemann *et al.*, 1989; Pfeilstetter *et al.*, 1992). There is uneven distribution of the virus hence no reliable diagnostic technique was suggested.

PHYTOPLASMA DISEASES OF CHERRY

Phytoplasmas are plant pathogenic, wall-less unculturable bacteria classified in the class Mollicutes, that cause diseases in plants. They are transmitted by insects belonging to the families Cicadellidae, Cixidae, Psyllidae, Delphacidae and Derbidae (Davis and Lee, 2000; Weintraub and Beanland, 2006). Following the application of molecular technologies the phylogeny of these prokaryotes was resolved and led to the designation of a new taxon named '*Candidatus* phytoplasma' (IRPCM, 2004; Duduk and Bertaccini, 2011) and to the new trivial name of "phytoplasma", the'Candidatus' term should be dropped only when biological characteristics could be used for their classification *i.e.* after consistent success in attempts to culture them in pure culture in cell free media (Bertaccini *et al.*, 2010).

Twenty eight phytoplasma groups (16Sr groups), and more than 50 subgroups have been identified on the basis of RFLP analysis of 16S ribosomal (r)DNA (Lee *et al.*, 1998; Marcone *et al.*, 2000; Davis and Dally, 2001; Jomantiene *et al.*, 2002a; Wei *et al.*, 2007). Thus far, four 16Sr groups (16SrI, 16SrIII, 16SrV, and 16SrXII) and twelve subgroups have been reported in Lithuania (Jomantiene *et al.*, 2000, 2002a, 2002b; Staniulis *et al.*, 2000; Valiunas *et al.*, 2001a, 2001b; Wei *et al.*, 2007). In Kaunas Region of Lithuania, sweet cherry (*Prunus avium*) and sour cherry (*P. cerasus*) trees showed phytoplasma symptoms (Valinuas *et al.*, 2005, 2009). The RFLP profiles indicated that DNAs were derived from 2 different phytoplasmas. One of them, designated ChD phytoplasma, was found in 11 sweet cherry trees and 2 sour cherry trees and tentatively classified as a member of a new subgroup designated 16SrIII-T (X-disease phytoplasma group). The other phytoplasma, found in 2 sweet cherry trees, was classified in subgroup 16SrI-B (group of *Candidatus* Phytoplasma asteris and related strains) and was designated cherry proliferation phytoplasma. Two strains, PYR and AY-A were identified in *Candidatus* Phytoplasma asteris infecting peach and apricot. These strains were specifically detected by real time LAMP assay (Miura *et al.*, 2012). Thus, in Europe, cherry may be affected by diseases associated with phytoplasmas belonging to groups 16SrI, 16SrIII, 16SrX (apple proliferation group), and 16SrXII (stolbur group).

CHERRY LITTLE LEAF

Phytoplasmal diseases of woody plant species including cherry have been reported in several countries. In Europe, these diseases are responsible for severe damage (Dosba *et al.*, 1991; Lorenz *et al.*, 1994; Poggi Pollini *et al.*, 2001).

Symptoms of shoot proliferation and abnormally small leaves were observed in diseased sour cherry (*Cerasus vulgaris* Mill., syn. *Prunus cerasus* L.) on the Neringa peninsula, Lithuania (Valiunas *et al.*, 2001, 2005, 2009), Hungary (Varga *et al.*, 2001) and Italy (Paltrinieri *et al.*, 2001). The disease was termed cherry little leaf (ChLL). Phytoplasma associated with this disease was identified as a member of group 16SrX (apple proliferation phytoplasma group) (Lee *et al.*, 1998). Phytoplasma strains belonging to subgroups 16SrX-B and 16SrI-B were identified in sour cherries in Hungary (Varga *et al.*, 2001). In Italy, sweet cherries were found to be infected by phytoplasmas belonging to subgroups 16SrI-B, 16SrX-B, 16SrX-C, 16SrXII-A, and to group 16SrIII (Paltrinieri *et al.*, 2001).

A phytoplasma-characteristic 1.2 kb 16S rDNA product was amplified from DNA of diseased leaves, but not from healthy cherry plants using phytoplasma universal primer pairs P1/P7 and R16F2n/R2 in nested PCRs, indicating that diseased plant species was infected by a phytoplasma (Gundersen and Lee, 1996). On the basis of RFLP and nucleotide sequence analysis of amplified 16S rDNA, the ChLL phytoplasma was classified in group 16SrI (aster yellows phytoplasma group), new subgroup 16SrI-Q (I-Q) (Gundersen *et al.*, 1996). Phylogenetic analysis indicated that cherry little leaf phytoplasma is closely related to phytoplasmas classified in subgroups I-B and I-M, but represents a distinct phytoplasma lineage related to '*Candidatus* Phytoplasma asteris. The ChLL phytoplasma differs from phytoplasmas (members of groups 16SrI, 16SrIII, 16SrV, 16SrX, and 16SrXII) previously reported in *Prunus*, including the subgroup I-B phytoplasmas previously reported from cherries in Hungary and Italy. The results revealed geographic distribution of phytoplasmal diseases in cherry, and of known phytoplasma biodiversity (Valiunas *et al.*, 2005).

CHERRY LETHAL YELLOWS

Cherry lethal yellows disease has been reported from China (Zhu and Shu, 1992). It is caused by a phytoplasma closely related to Jujube witches' broom phytoplasma (Lee *et al.*, 1995). The main symptoms of the disease are: yellow discoloration of foliage, premature defoliation, witches' broom and dieback. Fruits on affected trees remain small and do not ripe. The causal phytoplasma is transmitted by grafting on Chinese cherry (*Prunus pseudocerasus*) and possibly to sweet and sour cherry. The phytoplasma can be detected by electron microscopy, fluorescent microscopy and by nested PCR.

X-DISEASE (CHERRY BUCKSKIN DISEASE)

X-disease phytoplasma is known by several names such as Western X phytoplasma, Peach –X phytoplasma, Cherry-X phytoplasma, Cherry buckskin MLO, Eastern peach X disease phytoplasma, Western peach-X phytoplasma, Green Valley X (GVX) phytoplasma, Napa Valley X (NVX) phytoplasma and peach yellow leafroll 1 (WX/PYLR1) phytoplasma. X-disease, also called cherry buckskin, is a major cause of tree decline and serious losses of sweet cherry trees.

SYMPTOMS

Cherry trees affected with X- disease phytoplasma bear small fruits with shorten peduncles, small leaves and also show dieback of the tree. The leaves on affected trees become light green to yellow with a reddish tint and curls upward. Graft union failure was observed on Mahaleb rootstock due to a hypersensitive reaction of the rootstock resulting in necrosis of the phloem. On Mahaleb rootstock, buckskin disease causes pits and grooves in the wood at the graft union. The canopy becomes sparse. Cherry trees decline and the shoots and branches show die back symptoms. The decline is caused by rapid killing of the rootstock cells just below the graft union. Light pink and dark red fruit can occur on the same spurs. Pink fruit tend to be small. Fruits fail to ripe and may have an astringent taste. Initially only a few spurs on one branch are affected. The infection eventually spreads throughout the tree. Terminals are slightly rosetted. Leaves of terminal may be slightly undersized and bronze or rusty green in color. Stipules may be enlarged. Trees on Mahaleb rootstock may collapse and die within one or two seasons (Kenneth *et al.*, 2005; Uyemoto, 1989). On other rootstocks, Colt, Mazzard or Stockton morello, the leaves are smaller, pale green and more upright. The foliage may be sparse. Fruits on affected branches are smaller, lighter in colour, pointed and born on short and thick pedicels.

TRANSMISSION

The disease is transmitted by grafting and budding. The disease can also be transmitted to *Catharanthus roseus* through dodder. The X-disease of cherry is transmitted by leafhopper vectors, *Colladonus montanus* (mountain leafhopper), *Fieberiella florii* (cherry leafhopper) and *Scaphytopius acutus* in persistent way. Mountain leafhopper which survives on sugarbeet is the most efficient vector. In nature, the pathogen survives on wild cherry, Chokecherry (*Prunus virginiana*) and bitter cherry (*P. emarginata*) from where the vector carries infection and infect commercial cherry cultivars. The symptoms of the disease may appear within 2-3 months after inoculation but usually takes 6-9 months.

HOST RANGE

X-disease phytoplasma infects peach, nectarine, sweet and sour cherries, and chokecherry. Symptomless hosts are flowering cherry and plum. Herbaceous hosts include: *Medicago polymorpha, Trifolium species, Taraxacum officinale* and *Melilotus species*. The woody hosts are *Prunus dulcis, P. emarginata, P. virginiana, P. persica, P. saliciana,* and *P. avium*.

THE PATHOGEN

The Western X disease phytoplasma has been classified as *Candidatus* Phytoplasma pruni. This organism belongs to a relatively new class of disease agents which are similar to bacteria but lack cell walls and are limited to growing in the phloem of host plants.

DIAGNOSIS AND MANAGEMENT

The causal Phytoplasma can be detected by budding or grafting the susceptible hosts with infected budsticks. Transmission of the phytoplasma with dodder from woody hosts to periwinkle plants is the best method of diagnosis in glasshouse. X-Phytoplasma can also be detected in extracts of vector species by dot-blot nucleic acid hybridization assays using a ^{32}P-labeled DNA probe specific for X-phytoplasma. Use of disease-free rootstock and scion wood is the most effective management strategy to manage X-disease. Remove wild/alternate host trees such as chokecherry growing near commercial orchards (Kenneth *et al.*, 2005). Leafhopper vector management is also important. Apply a delayed dormant application of oil and an effective insecticide to control overwintering cherry leafhoppers. Commercially produced yellow sticky traps or make traps out of yellow boards or rigid plastic sheets about 5X6 inches, treated on both sides with a sticky material are useful for management. Place 6-10 traps in a cherry orchard hanging on high on cherry trees and check them once a week (Uyemoto *et al.*, 1991, 1998; Douglas and McClure. 1988).

VIRUS AND VIRUS-LIKE DISEASES OF MARASKA SOUR CHERRY (*CERASUS ACIDAS* P. MARASKA)

The most common viruses isolated from Maraska are Ilarviruses – prunus necrotic ringspot virus (PNRV) and prune dwarf virus (PDV). NEPO viruses - Arabis mosaic virus and strawberry latent ringspot virus affect less frequently, while cucumber mosaic virus and sowbane mosaic virus occur sporadically (Šaric and Velagic, 1981).

VIROID DISEASES OF CHERRY

Some strains of Hop stunt viroids (HSVd)-an infectious circular low molecular weight RNA consisting of 297 nucleotides cause dapple fruit of plum and peach. HSVd was also reported to infect sweet (*Prunus avium* L.) and sour (*P. cerasus* L.) cherry trees. The presence of HSVd infections in sweet and sour cherry trees represents a serious threat to the stone fruit industry of Turkey (Gazel *et al.*, 2008). Two small circular RNAs (cherry small circular RNAs, cscRNAs) were isolated from sweet and sour cherry trees in Italy. The two viroid-like RNAs consisted of two groups of variants with similar sequences but differing in size (394–415 and 372–377 nt for cscRNA1 and cscRNA2, respectively) (Barone *et al.*, 2010). The phytoplasma- and viroid-caused infections were detected by molecular techniques and their *in vitro* culture was induced. The fragments of circular RNA of similar sizes were detected in the samples of the infected trees. Apple scar skin viroid (ASSVd) infecting sweet cherry was reported from Greece. The viroid isolated from sweet cherry trees was 327-340 nucleotide long and share 96-99% identity with ASSVd isolates from Indian apples (Kaponi *et al.*, 2010). ASSVd is the type species of the genus *Apscaviroid* in family *pospiviroidae*. HSVd strains can be detected by mechanical inoculation to cucumber 'Suyo' which upon inoculation develop symptoms of streaking, leaf curling and vein clearing. Viroids can also be detected by return gel electrophoresis, dot blot or Northern blot hybridization and RT-PCR (Rustem and Ahlawat, 1999; Roy and Ramachandran, 2003). Induction of *in vitro* culture partly eliminated phytoplasmas but, however, did not eliminate viroids (Stanien *et al.*, 2009).

D. VIRUS DISEASES OF PEACH AND NECTARINE

Peach fruit is rich in proteins, sugar, minerals and vitamins. Peaches are grown in most of the temperate and sub-temperate countries of the world. Introduction of the cultivated peaches into India probably took place in the latter half of the 19th century. Today, it is being grown in the midhill zone of the Himalayas extending from Jammu and Kashmir to Khasi hills 1000—2000 m above mean sea-level. Besides, low chilling peaches are grown in submountainous regions of the country covering hills of south India and in the North-Eastern region of the country. Nectarine is a cultivar group of peach (*Prunus persica*) and is cultivated in Northern Himalayan States of India. The major cultivars of nectarine grown in India are 'Snow Queen' and 'May Fire'.

Important peach varieties recommended for temperate regions are: Alton, World's Earliest, Early White Giant, Redhaven, Stark Red, Gold, Early Candor,Redhaven, Sunhaven Peshwari, Quetta, Saharanpur Prabhat, JulyElberta, Kanto5, Shimizu Hakuto, Sunhaven, Alexander, Crawford Early, Alexander, Parrot Delux, J.H. Hale and Peregrine. In India, following peach cultivars are grown: July Elberta, J.H. Hale, Sharbati, Shan-e-Punjab, Burbank, Partap, Summerset, Worlds Earliest, Early white Giant, Stark red Gold, Snow Queen, Sun Haven, Alton and Red Heaven. Besides these varieties, Stark Earlyglo, Stark Early, White Giant, Starking Delicious and Candor are new promising peaches for midhills of Himachal Pradesh; Floridasun, Dawn Ramber and Dawn Rose for subtropical regions of north India; Safeda Early Cream and SRE 6 for Tarai region of Uttar Pradesh; and Prairie Dawn, Praire Ramber and Praire Rose for cold areas of dry temperate regions.

Peach is commercially propagated through grafting or budding. Peach seedlings are generally used as rootstock, though plum, apricot and almond seedlings can also be used. Peaches are propagated commercially by tongue or cleft grafting and T-budding or ring budding. In hills, tongue grafting during January- February and T-budding during May-June are performed. However, in plains grafting is performed during November-January and budding during April-June and September.

The virus and virus-like diseases infecting peach and nectarine are: Asteroid spot = cherry Utah dixie spot, Bark and wood grooving, Blotch, Calico, Chlorosis, Chlorotic spot, Enation (Peach enation virus), Dark green sunken mottle (genus *Trichovirus,* Apple chlorotic leaf spot virus), Latent mosaic (Peach latent mosaic viroid), Line pattern (genus *Ilarvirus,* Prunus necrotic ringspot virus, genus *Ilarvirus,* Apple mosaic virus (ApMV), Line pattern and leaf curl = cherry line pattern leaf curl (associated with isometric particles), Mosaic (Peach mosaic virus), Mottle, Mule's ear (associated with PNRSV), Necrotic ringspot (genus *Ilarvirus,* Prunus necrotic ringspot virus (PNRSV), Oil blotch, Plum pox = Sharka (genus *Potyvirus,* Plum pox virus (PPV), Prunus stem pitting (genus *Nepovirus,* Tomato ringspot virus (ToRSV), Purple mosaic, Rosette and decline [genus *Ilarvirus,* Prunus necrotic ringspot virus (PNRSV), + genus *Ilarvirus,* Prune dwarf virus (PDV)], Rosette mosaic (Peach rosette mosaic virus), Seedling chlorosis, Shoot stunting = almond enation (genus *Nepovirus,* Tomato black ring virus (TBRV), Star mosaic, and Stubby twig. For many of these diseases peach is a host and such diseases will be covered under the diseases of their respective hosts. Among phytoplasma diseases, important ones are: Apricot chlorotic leaf roll, Little peach (= yellows), Red Suture (= yellows), Rosette, X-Disease, Yellow leaf roll and Yellows. The peach yellows and peach X are important phytoplasma infections and also known to occur in India. Diseases which are economically important to peach will be covered in this chapter.

PEACH MOSAIC

Peach mosaic disease is widely present in Mexico and USA (Hutchins, 1932). The disease is considered important in Mexico, where many symptomatic trees are currently being observed (Oldfield *et al.*, 1995). In India, the disease has been reported from Kumaon and Shimla hills (Ahlawat and Chenulu, 1986).

SYMPTOMS

In USA and Mexico, there is 4-6 days delay in leaf emergence, flowering and maturity on mosaic affected peach trees. Pink broken lines develop on the white petals in warm temperatures.The fruits are irregularly shaped, flattened, colourless with cracked sutures and enlarged pits. Bud necrosis is also common. Some isolates cause mosaic, blotch, calico and necrosis of the leaves, whereas others induce stem pitting and leaf twisting and decline (Plate 5B) (Hutchins, 1932; Hutchins *et al.*,1937, Larsen and Oldfield, 1995; Pine, 1976). In India, mosaic affected trees show stunting, profuse branching and short internodes. Typical yellow mottling can be seen on new growth flushes during spring. By June-July, this mottling disappears slowly and leaves become crinkled (Plate 5 A). The fruits are smaller, malformed and unsuitable for marketing (Ahlawat and Chenulu, 1986).

TRANSMISSION

Peach mosaic is easily graft and bud-transmissible to healthy peach. It is also transmissible by dodder, *Cuscuta hyalina* (Ahlawat and Dhingra, 1973). The vector of the disease is the peach bud mite *Eriophyes insidiosus* (Oldfield *et al.*, 1995) (Plate-6). This mite feeds and reproduces on developing leaf primordia within the bud. Infested buds are swollen and reddened, growth remains retarded, and buds may eventually die. A single infectious mite can transmit the disease to a healthy tree. The virus was mechanically sap-transmitted from symptomatic Elberta leaves to *Chenopodium quinoa*.

HOST RANGE

American Peach mosaic affects only *Prunus* spp.: peach (*Prunus persica*), nectarine (*P. persica* var nectarina), almond (*P. dulcis*), apricot (*P. armeniaca*), *P. besseyi, P. serrulata* and several species of plum. Cherries are not hosts of this disease. Cultivated European and Japanese plums (*P. domestica* and *P. salicina*) are highly susceptible to mosaic infection but express only the leaf symptoms as described above. In apricot, some susceptible cultivars (*e.g.* Blenheim) develop leaf discoloration patterns and distortion and stubby twig symptoms. Almond trees generally show mild foliar and fruit symptoms, with a reduction in yield or nut quality. In addition, *P. mume, P. davidiana, P. insititia, P. americana, P. angustifolia, P. munsonia, P. ceracifera, P. maxima, P. besseyi, P. tangutica* and *P. tomentosa* are also susceptible to peach mosaic (Bodine and Reeves, 1951). Trees are less vigorous, produce less fruit and have more sun burned fruits than healthy trees. The herbaceous host range of the virus includes: *C. amaranticolor, C. quinoa,* and *Nicotiana occidentalis* (James and Howell, 1998).

THE PATHOGEN

A flexuous clostero-like virus was reported to be associated with peach mosaic disease which was isolated from an affected peach tree cv. Elberta and had an estimated size of 752 nm in length and 9.25 nm in width, with striations 3.47 nm in pitch. RNA extracted from purified virus was approximately 8.1 kilobases, and a capsid protein of approximately 27 kDa was found. Polyclonal rabbit antiserum, produced against purified virus, reacted with samples from peach mosaic and cherry mottle leaf-infected plants when used in Western blot analysis (Carman and Thomas, 1998). In serological assays, this virus reacted with monoclonal antibodies prepared for cherry mottle leaf closterovirus (James and Howell, 1993, 1998). However, the latter differs in that it infects cherry and has *E. inaequalis* as its mite vector (Oldfield *et al.*, 1995).

Sequence analysis of the genome of Peach mosaic virus (PcMV) indicates that it is closely related genetically to cherry mottle leaf virus (CMLV) but distinct, with similar genome organization and a member of the genus *Trichovirus*. The genome of PcMV consists of 7,988 nucleotides, excluding a poly(A) tail at the 32 end of the genome. Four putative open reading frames (ORF1 to 4) were identified coding for proteins of 216.3, 47.2, 21.7, and 15.7 kDa, respectively. Also, three noncoding regions were identified, including an intergenic region separating ORF3 and ORF4. Phylogenetic analysis supports classification of PcMV and CMLV as members of the genus *Trichovirus*. They are unique members of this genus with an extra ORF (ORF4). PcMV ORF4 appears to code for a putative nucleic acid-binding (NB) protein which has identity with the NB protein of CMLV and members of the genera *Allexivirus, Carlavirus,* and *Vitivirus*. PcMV and CMLV appear to be the products of recombination between members of the genus *Trichovirus* and a virus group containing the putative NB protein (James *et al.*, 2006).

DIAGNOSIS AND MANAGEMENT

Peach mosaic can be detected in the glasshouse on indicator plants (peach seedlings cvs. Elberta, GF305, Rio Oso Gem) by grafting (Desvignes, 1976) and by sap inoculation to herbaceous hosts, *C. amaranticolor, C. quinoa,* and *Nicotiana occidentalis* (James and Howell, 1998). Indirect enzyme-linked immunosorbent assay (ELISA) using the monoclonal antibody prepared for cherry mottle leaf closterovirus was used as a useful interim detection tool for peach mosaic in America. A polyclonal antibody (PAb) was produced and used in ELISA and Western blot analysis to detect the virus in woody and herbaceous hosts (James and Howell 1998). A reverse transcription-polymerase chain reaction procedure was developed for reliable and specific detection of PcMV (James *et al.*, 2006). Control measures are the use of healthy propagation material and control of the mite vector which will help to reduce the spread of peach mosaic.

Eradication of infected trees was useful to avoid the spread of the disease (Thresh, 1988; Larsen, 1997).

PEACH ASTEROID SPOT AND PEACH SOOTY RINGSPOT DISEASES

Peach asteroid spot (PAS) and peach sooty ringspot (PSRS) are two degenerative diseases of stone fruit trees of unknown aetiology (Grasseau *et al.*, 1999). LA2 isolate of peach asteroid spot disease and Caserta-12 for the peach sooty ringspot disease were transmitted to GF305 peach seedlings and on *Nicotiana occidentalis.* These diseases are graft transmissible from apricot to peach (Gentit *et al.*, 2001).

LA2 a polyadenylated, 1151 bp long cDNA was cloned and sequenced. For isolate Caserta-12 an approximately 4 kbp polyadenylated cDNA was cloned and partially sequenced. These cDNAs encode the last 327 amino acids of the LA2 coat protein and the last 202 amino acids of the Caserta-12 coat protein. The partial coding sequences thus obtained revealed respectively 96% and 97% identity with the sequence encoding the coat protein of *Apricot latent virus* (ApLV), a *Foveavirus* serologically related to ASPV (Gentit *et al.*, 2001). Sequence analysis of the genome revealed a genomic organization similar to that reported for other members of the *Foveavirus* genus, including the triple gene block and a large, 43.6 kDa coat protein. The viral agents associated with peach asteroid spot and peach sooty ringspot diseases might be variants of the recently described ApLV (Gentit *et al.*, 2001).

A filamentous virus was isolated from mechanically transmitted GF 305 seedlings showing asteroid ringspots to *Nicotiana occidentalis*. The electrophoretic dsRNA pattern from symptomatic GF 305 and *N. Occidentalis* showed a major band of about 9.5 kbp. Bundles of elongated virus-like particles were present in the cytoplasm of infected *N. occidentalis*. RT-PCR using primers specific for *Apricot latent virus* (ApLV) RNA amplified a DNA product of about 200 bp from infected GF 305 and *N. occidentalis* plants. Sequence analysis of this fragment showed 90-93% identity in the encoded amino acid sequence with corresponding sequences from different ApLV isolates. An ApLV-specific digoxygenin-labelled riboprobe was produced and used for molecular hybridization tests (Abou Ghanem-Sabanadzovic *et al.*, 2005).

PEACH ROSETTE MOSAIC

Peach rosette mosaic (PRMV) disease was first observed in Georgia in 1890 (McClintock, 1923). There are several diseases of peach similar to that of PRMV that include: "peach rosette" in Europe which is caused by strawberry latent ringspot nepovirus; "peach rosette and decline" in Australia which are due to a combined infection with prune dwarf and prunus necrotic ringspot ilarviruses. In USA, peach rosette phytoplasma also causes the "peach rosette" symptom. The virus is present in USA and Canada but not in India or Italy, as mentioned by Németh (1986). After its discovery, this disease has been reported in several states of United States of America (McClintock *et al.*, 1951). Although peach rosette may cause serious losses in peach trees, but it is considered of minor importance when compared to other diseases of peach such as peach yellows, little peach, phony peach, and peach mosaic. The peach rosette virus has been shown to be distinct from other known peach viruses (Kunkel, 1936).

SYMPTOMS

Peach trees infected with the rosette mosaic virus are characterized by two types of symptoms. The most familiar is the rosette-type of symptom (Plate-7) and the other, sudden wilting and death of infected tree. In rosette affected peaches, shortened internodes, rosetting and mosaic of leaves are the main symptoms. New growth is reduced as internodes become shorter. The leaves are reduced in size and become chlorotic with thickening and clearing of the veins. The first formed leaves develop red spots and fall in early summer, leaving only tufts of younger leaves at the branch tips. Diseased trees seldom bear fruits. When young trees are infected near the growing points, they suddenly wilt and die without producing rosette-type symptoms. In contrast, when young trees are infected near the base, the rosette-type of growth is produced. Rosette symptoms also occur with resumption of growth following infection shortly before dormancy. All rosette-infected trees wilt and die after a short period of time.The virus causes leaf malformation, shortening of cane internodes and crooked cane growth in grapevines. This virus occurs only in Concord and Catawba grapes in Michigan. Symptoms appear 3 to 4 years after infection. The plant canopy is umbrella-like with shortened and crooked inter-nodes. Leaves are misshapen with a flattened base. Clusters are scraggly and may shell berries. Infected vines lack vigor, are prone to winter injury and may die after several years of infection.

TRANSMISSION

The peach rosette mosaic virus (PRMV) is known to be transmitted only by budding and grafting. PRMV was shown to be seed-borne in grapevine cv. Concord up to 9.5%. It was also shown to be seed-borne in *Taraxacum officinale* (Ramsdell and Myers, 1978). PRMV is also transmitted through seed of *Chenopodium quinoa*. Several nematode species have been recorded as vectors. *Xiphinema americanum* (*sensu lato*) has been known for several years to be a vector in USA (Klos *et al.*, 1966) and Canada (Dias, 1975). From the geographic distribution of the virus and species of the *X. americanum* group in North America, it is probable that the species involved is *X. americanum sensu stricto* (EPPO/CABI, 1996c). Other vectors are a species of *Criconemoides* (Klos *et al.*, 1966); this may have been *C.* (= *Macroposthonia*) *xenoplax*, as later referred to by Ramsdell and Myers (1974). *Longidorus diadecturus*, has also been reported as a vecor of PRMV (Allen *et al.*, 1982; Driel *et al.*, 1990). Two other species of *Longidorus*, *L. breviannulatus* and *L. elongatus* occasionally transmitted PRMV (Allen, 1986; Allen and Ebsary, 1988).

HOST RANGE

Susceptible symptomatic hosts of peach rosette virus are: apricot (*Prunus armeniaca*) varieties Moorpark, Royal and Wilson; sand cherry (*P. pumila*); sour cherry (*P. cerasus*); mazzard cherry (*P. avium*); Chickasaw plum (*P. angustifolia*); Japanese plum (*P. salicina*) varieties Kelsey, Ogon, Maynard, and Red June; damson plum (*P. institia*) and almond (*P. amygdalus*). Wilson apricot appears to be a symptomless carrier of the peach rosette virus and, Marianna plum is immune to the peach rosette mosaic virus (McClintock *et al.*, 1951). In addition, several weed species have been shown to be hosts for the virus: *Rumex crispus, Solanum carolinense* and *Taraxacum officinale*.The experimental herbaceous host range is rather narrow. Some plant species of *Chenopodiaceae, Cucurbitaceae, Fabaceae* and *Solanaceae* are infected by mechanical inoculation with sap from infected grapes or peaches. Tobacco (*Nicotiana glutinosa*), periwinkle (*Cathranthus roseus*), tomato (*Lycopersicon esculentum*) and *Chenopodium quinoa* are also susceptible hosts. Chlorotic and necrotic local lesions will develop on sap inoculated leaves of *C. quinoa* within 7-10 days post inoculation.

THE PATHOGEN

Peach rosette mosaic virus is a multicomponent, isometric (28 nm) single-stranded, positive-sense RNA virus of family *Comoviridae*, genus *Nepovirus* and species *rosette mosaic virus*. The genome consists of two RNA species of MW 2.2×10^6 and 2.5×10^6 d. The protein subunit MW is 57×10^4 d. The three components have sedimentation coefficients of 52 s (top), 115 s (middle), and 134 s (bottom). In infected sap of *Chenopodium quinoa,* the thermal inactivation point; (10 min) was 58-68°C, the dilution end point was 10^3-10^5 and the virus retains infectivity for about 15-25 days at room temperature (20°C) (Dias, 1975, and Dias and Cation, 1976). PRMV is serologically unrelated to any other nepovirus.

DIAGNOSIS AND MANAGEMENT

Rosette mosaic virus of peach can be detected by sap inoculation to *C. quinoa* leaves from inoculum made up of PRMV infected young leaves of peaches triturated in 2 ml of 0.05 M potassium phosphate buffer, pH 7.0, containing 1% (v/v) nicotine alkaloid (2 ml/g of tissue). Chlorotic and necrotic local lesions will develop on inoculated leaves in 7-10 days. The virus can also be detected in serological tests like gel-diffusion and ELISA. The only effective control of the peach rosette disease is accomplished by diligent, systematic eradication of infected trees in orchards as well as native wild *Prunus* spp. growing in the proximity of commercial production areas, particularly when symptoms are evident in the spring. Peach rosette is known to develop rapidly and reach serious levels, sometimes affecting entire orchards in 3 to 5 years (Alfieri, Jr., 1967). It is not possible to cure plants once infected with PRMV and, therefore, prevention is the only alternative. The use of virus-free plants emanating from a certification scheme is recommended. The soil in which the material is to be planted should be free from viruliferous nematodes. Soil-applied nematicides can be used to reduce the number of nematodes in the soil.

PRUNUS NECROTIC RINGSPOT

Prunus necrotic ring spot (PNRSV) is worldwide in distribution especially in temperate regions. Prunus necrotic ringspot has also been reported from India (Chandel *et al.,* 2008). This disease has been reported by several names such as European plum line pattern, Hop B virus disease, Hop C virus disease, Peach ringspot, Plum line pattern, Prunus ringspot, Red currant necrotic ringspot, Rose chlorotic mottle, Rose line pattern, Rose vein banding, Rose yellow vein mosaic, Sour cherry necrotic ringspot, Peach necrotic leafspot, Peach tatter leaf, Peach lace leaf and cherry ringspot. Peach and nectarine are the major host of PNRSV. However, in addition to stone fruits, the virus also infects ornamental plants such as Begonias and roses (Sastry, 1966). This disease is responsible for reducing the quality and quantity of the fruits on affected trees (Mink, 1992).

SYMPTOMS

The disease was first reported as peach ringspot virus causing severe ringspot symptoms on peach (Brunt *et al.*, 1996). Prunus necrotic ringspot (PNRSV) is a common disease of stone fruits in family *Rosaceae*. Prunus necrotic ringspot virus (PNRSV) is common in *Prunus* species, and can cause substantial reductions in the growth and yield of peach (Pine, 1964; Schmitt *et al.*, 1977; Topchiiska, 1982; Uyemoto *et al.*, 1992; Wells *et al.*, 1986). In peaches and nectarine, the symptoms appear as chlorotic and necrotic leaf spots on leaves. Necrotic areas can fall down leading to shot hole effect. In some peach cultivars , severe infection induces bark necrosis, cankering and splitting of the trunk. In India, a disease with similar symptoms was reported as necrotic leaf spot from peach and cherry showing ringspot on leaves which later become necrotic (Ahlawat and Chenulu, 1986). After sometime, the necrotic tissues fall down giving shot-hole effect to affected leaves (Fig. 12). This disease appears to be similar to prunus necrotic ringspot reported elsewhere.

TRANSMISSION

The disease is easily transmitted by wedge and bud grafting. PNRSV does not have a vector. Virus is transmitted by mechanical inoculation and also by seeds to over 80% in *Prunus pennsylvanica* but much less in peach. It is also transmitted by pollen (Nyland *et al.*, 1976). PNRSV has been shown to be located on the surface layer of pollen and also inside the pollen (Cole *et al.*, 1982; Kelley and Cameron, 1986). The virus from pollen is transferred to healthy plants by bees or other insects fertilizing flowers (Mink, 1982).

HOST RANGE

Woody hosts of PNRSV are: plum, apricot, cherry, almond, sour cherry, hops, apple and roses. The Indian isolates can infect peach varieties Alton, Elberta; Sweet cherry vars. Babcock and bird cherry (*P. padus*). In India, wild Himalayan cherry proved to be the natural host of PNRSV and may play a significant role in virus epidemiology. The natural and experimental hosts are given below:

NATURAL HOSTS AND SYMPTOMS

Prunus cerasus — Dark necrotic lines and rings, shotholes, and recovery from symptoms.

Peach — Brown lines and rings, become symptomless but symptoms may recover.

Rose — Chlorotic lines and rings, oak leaf patterns, no recovery.

Plum — Only some strains show chlorotic lines and rings and tree decline.

Almond — Bright mosaic symptoms, flower and leaf buds do not grow.

Hops — No symptoms.

Prunus serrulata — Local necrosis and gumming at site of grafts (Zilkah *et al.*, 1995).

The following species were susceptible to experimental virus inoculation: *Antirrhinum majus, Cassia tora, Catharanthus roseus, Chenopodium album, Chenopodium quinoa, Cucumis sativus, Cucurbita maxima, Cucurbita moschata, Cucurbita pepo, Cyamopsis tetragonoloba, Gomphrena globosa, Gossypium hirsutum, Helianthus annuus, Humulus, Lactuca sativa, Lupinus albus, Melilotus albus, Momordica balsamina, Nicotiana benthamiana, Petunia x hybrida, Phaseolus vulgaris, Pisum sativum, Prunus, Prunus cerasus, Prunus persica,* and *Prunus serasus.* Symptoms on some herbaceous host species are given below:

Cucumis sativus — Large yellow chlorotic local lesions in cotyledons, growing point stunted.

Momordica balsamina — Necrotic local lesions.

Cyamopsis tetragonoloba — Dark necrotic local lesions.

Chenopodium quinoa — Systemic mottle.

Lupinus albus — Systemic mottle.

Helianthus annuus — Systemic chlorotic lines and rings.

Cassia tora — Local and systemic ringspotting.

Momordica balsamina, Cucumis sativus and *Cyamopsis tetragonoloba* are used as assay hosts of the virus (Fulton, 1957).

THE PATHOGEN

Prunus necrotic ring spot virus belongs to genus *Ilarvirus* in the family *Bromoviridae*. Virus preparations contain more than one particle components. Virions consist of a capsid with icosahedral symmetry and appear round. The capsids are isometric and has a diameter of 23, 25, and 27 nm. There are 3 sedimenting component(s) found in purified preparations. Mr of the genome constitutes 16% of the virion by weight. The genome is segmented, tripartite and consists of four segments of linear, positive-sense, single-stranded RNA. The complete genome is 8056 nucleotides long. RNA-1 is fully sequenced and complete sequence is 3662 nucleotides long (Loesch and Fulton,1975; Hammond and Crosslin, 1995). The encoded coat protein (CP) is translated from a subgenomic RNA-4. Ilarviruses are interchangeable in initiation of infection (Gonsalves and Fulton, 1977). In India, a full CP (672bp) gene was cloned and sequenced from Kulu and Solan isolates of PNRSV (Chandel *et al.*, 2008). The different isolates of PNRSV fall into three groups (group 1 is known as PV 32, group 2 as PV 96 and group 3 as PE 5). On the basis of multiple alignment results of Indian PNRSV sequences, Indian isolates fall under group PV 96 (Chandel *et al.*, 2007). Proteins constitute about 84% of the particle weight. The viral genome encodes structural proteins and non-structural proteins located in the capsid. The virus shares some serological determinants with that of apple mosaic virus.

DIAGNOSIS AND MANAGEMENT

Grafting to 'GF305' peach seedling, *Prunus tomentosa* and *P. serrulata* cv. Shirofugen or *P. avium* 'F12/1' are used worldwide as a biological indicators for virus diseases of *Prunus* spp. Typical symptomatic necrosis developed when Shirofugen was grafted as a 'scion' onto PNRSV-infected Elberta peach (Zilkah *et al.*, 1995). Mechanical inoculation to cucumber can also detect the virus. Nucleic acid hybridization with a virus-derived cRNA probe was used to detect virus infection (Crosslin *et al.*, 1992). Virus presence can be confirmed by ELISA, immune tissue printing and PCR (Chandel, 2007; Knapp *et al.*, 1995; Scott *et al.*, 1992).

Control of this disease by planting virus-free stock and removal of ELISA-positive trees has been applied to peach (Scott *et al.*, 1986). The Prunus necrotic ringspot virus can be managed by efficient budwood certification programmes. Another possible control strategy is the development of virus resistant peach cultivars through genetic engineering. *In vitro* thermotherapy; shoot tip micrografting has been used to produce virus-free cultures of peach and other stone fruits (Navarro *et al.*, 1982; LaRosa *et al.*, 1995; Juarez *et al.*, 1988). Prunus necrotic ringspot virus (PNRSV)-free Begonia spp. plants were raised from petioles of virus-infected plants using *in vitro* techniques (Verma *et al.,* 2005).

☙•❧

PEACH GUMMOSIS

A disease with excessive gum production in peach and apricot was observed in Shimla and Saproon velley of Himachal Pradesh in India. Affected trees produced enormous quantities of gum on stem and branches. At later stages, the symptoms develop as stem necrosis, bark roughening, dieback of new shoots and lace leaf appearance. The leaves of affected trees become yellow and fell down prematurely causing dieback of twigs. Gum may also ooze from the fruits making them unsuitable for human consumption.

The causal virus is transmissible by budding and grafting to peach varieties July Alberta, J.H.Hale, Alberta and Foster, cherry vars. Lamberta and Bigareauseu, apricot vars. Safaida and Shakerpara, Plum vars. Beauty and Mariposa. New Castle apricot, Green and Santa Rosa plum, Non-Pasail almond are not susceptible. The virus has been tentatively identified as strain of prune dwarf virus (Dhingra and Ahlawat, 1979).

☙•❧

PHYTOPLASMA DISEASES OF PEACH

Peach rosette, peach yellows, little peach, and red suture are known to be phytoplasma diseases of peach trees. Peach yellows disease was first observed in 1791 in Pennsylvania, rosette in 1891 in Georgia, while little peach and red suture were first seen in 1896 and 1911, respectively, in Michigan. Other disease of peach associated with phytoplasma is Peach yellow leaf roll, caused by the western-X phytoplasma, occurs in California. X-disease occurs in the Northern United States and Canada.

Although peach rosette, peach yellows, little peach, and red suture are distinct diseases, they all are caused by phytoplasmas that are very closely related to the western-X phytoplasma. In Europe, the stone fruit yellows phytoplasma (a species distinct from western-X) causes disease(s) in *Prunus*. Synonyms of Peach X-disease are Peach western-X and Peach yellow leafroll. Phytoplasma belongs to Bacteria: Tenericutes: Mollicutes: Phytoplasmas. General principles of control for peach phytoplasmas are: Eradicate known diseased trees. Removal of infected trees, eliminates sources of infection within the orchard, select propagating material from disease free or indexed trees free from phytoplasma. Control of insect vector(s) is important. The eradication of known alternate hosts and Antibiotics treatment by tree injection, have potential for control of phytoplasma diseases.

PEACH 'X' PHYTOPLASMA DISEASE

Peach X phytoplasma disease has been reported from North America (USA and Canada) (Gilmer and Blodgett, 1976). In India, the peach X disease is known to occur in Himachal Pradesh and Kalimpong region of West Bengal (Ahlawat and Chenulu, 1979). The incidence of the disease as high as 70 % was observed in peaches in Himachal Pradesh (Thakur *et al.*, 1998). A high incidence of Peach X disease was also observed in almond plantations of Himachal Pradesh (Ahlawat *et al.*, 1984). Two forms of peach X-disease are known: eastern X-disease and western X-disease. Both these diseases differ slightly on symptoms. These are now considered to be due to strains of the same phytoplasma as monoclonal antibodies to an eastern X strain cross-react with western X strains (Jiang *et al.*, 1989).

SYMPTOMS

The first symptom of the disease in peach appears as yellow spotting and rolling of the leaves (Plate 8A). Subsequently all leaves become chlorotic and fall, leaving a few rosettes at the tips of the shoots. Young trees may die within 1-4 years after first symptom appearance. Affected trees bear little or no fruit. Infected cherry trees on *Prunus mahaleb* rootstocks die rapidly because the rootstock is resistant, and a hypersentivive reaction occurs at the graft union. On other rootstocks, decline is slower (Uyemoto,

1989; Kirkpatrick *et al.*, 1995). Leaves are smaller and red-tinged, sometimes with enlarged stipules; fruits mature late, have short pedicels and bland-flavoured watery flesh. In India, symptoms of X-disease can begin on the leaves of some or all branches, as a slight mottle and reddish-purple spots that eventually die and fell out, creating a shothole appearance. Affected peach and almond trees were characterized by varying degree of anthocyanosis, twig and shoot dieback and general decline. Affected trees look normal during the spring but by June-July the leaves on affected branches roll upward, fall prematurely, leaving a tuft of leaves on shoot tips. The symptoms are either confined to isolated branches or the whole tree may show symptoms. Fruits are small, insipid in flavour, malformed and generally abort without ripening fully. Visible symptoms of the disease started appearing after 7–8 weeks of foliation in the form of chlorosis of leaves with typical longitudinal upward rolling and occasional red spotting. Later, the foliage turned pale to yellow-red in color. Similar symptoms can also be developed in nursery plants if contaminated buds are used (Plate 8B). Similar symptoms with a very high incidence were observed in peach orchards and nurseries in the Rajgarh peach valley of Himachal Pradesh (Thakur *et al.*, 1998).

TRANSMISSION

The disease is readily transmitted by budding or grafting. Maximum transmission was obtained during summer (Rosenberger and Jones, 1977). It is also transmitted by dodder, *Cuscuta reflexa* (Ahlawat and Chenulu, 1986). It is also transmitted by root grafts between adjacent trees. Peach-X Phytoplasma is transmitted by several species of leafhoppers in the genera *Colladonus* and *Scaphytopius* (Granett and Gilmer, 1971) in addition to *Paraphlepsius irroratus* in eastern USA. Efficient vectors of X disease are: *Colladonus montanus* (Plate 9A) and *Scaphytopius acutus* (Plate 9B) but *Colladonus geminatus*, *Fieberiella florii* and *Graphocephala confluens* were also found to be the vector of X disease. In India, the vector of peach X is a leafhopper, *Exitianus fusconervosus* (Ahlawat *et al.*, 1984; Ahlawat and Chakraborty, 1993). The latency period in the insect vector is long: 22-35 days for *C. geminatus* and 45 days for *S. acutus*. Transmission is mainly from wild trees of *Prunus virginiana* but also from infected peach to healthy peach. Herbaceous weeds have also been shown to act as reservoirs in orchards; the vectors overwinter on these weeds and move onto the trees in summer.

HOST RANGE

Peach (*Prunus persica*) is the principal host of peach X-disease phytoplasma, but the pathogen also infects cherries (*Prunus avium* and *P. cerasus*),

P. salicina, almonds (*P. dulcis*), apricots (*P. armeniaca*) and plums (*P. domestica*) (Uyemoto *et al.,* 1991). The phytoplasma can also be sap transmitted to herbaceous hosts like celery (*Apium graveolens*). It has been shown to occur in herbaceous weeds like *Medicago hispida* in orchards. Chokecherry (*P. virginiana*) is a natural reservoir for the pathogen. In India, Peach X phytoplasma infects peach vars. July Alberta, J.H. Hale, Holton and Foster, plum var. Santa Rosa, Cherry, Almond, Celery and Periwinkle (*Catharanthus roseus*). The phytoplasma can be transmitted to periwinkle by dodder. The dodder inoculated plants will show leaf chlorosis (Plate 10) and may show witches broom like symptoms later.

THE PATHOGEN

Peach-X disease is one of the most serious diseases known in peach (*Prunus persica*). Electron microscope studies have shown the presence of typical phytoplasma cells in infected plants (Nasu *et al.,* 1970; Granett and Gilmer, 1971; MacBeath *et al.,* 1972). Based on RFLP analysis of 16S rRNA gene sequences, peach X-disease phytoplasma strains from eastern and western United States and eastern Canada were classified in 16S rDNA RFLP group 16SrIII, subgroup A. Phylogenetic analyses of 16S rRNA gene sequences revealed that the X-disease phytoplasma strains formed a distinct subclade within the phytoplasma clade and supported the hypothesis that they represent a lineage distinct from those of previously described '*Candidatus* Phytoplasma' species. Nucleotide sequence alignments revealed that all studied X-disease phytoplasma strains shared less than 97.5 % similarity of 16S rDNA with previously described '*Candidatus* Phytoplasma' species. Based on unique properties of DNA, it has been proposed that X-disease phytoplasma strain PX11CT1R as representative of a novel taxon, '*Candidatus* Phytoplasma pruni' (Davis *et al.,* 2012).

In California (USA), one form of peach yellow leafroll is caused by X-disease phytoplasma, but similar diseases are caused by unrelated phytoplasmas (Kirkpatrick *et al.,* 1995). Canadian peach X-disease phytoplasma belongs to 16S ribosomal RNA (16Sr) group III of Lee *et al.* (1993), subgroup A (of which it is the type member). Phytoplasmas are also found in *Prunus* in Europe, but tests with DNA probes against X-disease and other phytoplasmas showed that the northern European *Prunus* phytoplasmas are very homogeneous and have affinities with apple proliferation phytoplasma (EPPO/CABI, 1996d), that the southern European Prunus phytoplasmas are related, and that peach X-disease phytoplasma is quite distinct (Ahrens *et al.,* 1993). The name European stone fruit yellows phytoplasma is proposed for the European pathogens (Lorenz *et al.,* 1994; Seemüller and Foster, 1995); this also incorporates apricot chlorotic leaf roll phytoplasma (EPPO/CABI, 1996b).

DIAGNOSIS AND MANAGEMENT

Peach X-disease phytoplasma can be detected by grafting on peach seedlings (cv. Elberta or GF305) and by dodder transmission on periwinkle plants. The phytoplasma can be seen by thin section electron microscopy where pleomorphic bodies can be seen in sieve tubes of phloem tissue (Jones *et al.*, 1974). Serological methods (ELISA, immunosorbent electron microscopy,) have been used for the detection of the pathogen (Sinha and Chiykowski, 1984) in plants and in the leafhopper vectors (Sinha and Chiykowski, 1986; Sinha, 1988). Monoclonal antibodies specific to peach X-disease phytoplasma have been prepared and used for serological detection of the phytoplasma by fluorescent microscopy (Jiang *et al.*, 1989; Douglas, 1986, Thakur *et al.*, 1998). DNA probes have been applied for detection and identification of X-disease in *Prunus.* The phytoplasma has been detected, with DNA probes in *P. irroratus* in Michigan (Rahardja *et al.*, 1992) and in *C. montanus* and *F. florii* in California (Kirkpatrick *et al.*, 1990).

Antibiotic treatment during the month of September by injecting oxytetracycline hydrochloride solution resulted in symptom remission in the subsequent year (Plate 11) (Cooley *et al.*, 1992). In Pennsylvania, wild chokecherry is considered the primary reservoir for X-disease but in India this wild host is not found, hence the origin of the disease in India has yet to be established.

Although phytoplasma disease symptoms may be delayed or alleviated by treatment with certain classes of antibiotics by tree injection (Jones *et al.*, 1974; Rosenberger and Jones, 1977) but this approach is not usually practical. Plants are not really cured, and symptoms reappear when treatments is stopped. For trees, injections into the trunk are most effective. Microinjection capsules of oxytetracycline have been used successfully to treat peach trees (Cooley *et al.*, 1992). Infected plants or dormant propagative tissue can be freed of phytoplasma by heat treatment for 8-10 minutes at 50°C. This technique is used at quarantine facilities when phytoplasma infection is suspected in fruit tree nursery stock being imported into a country. The most promising strategies for avoiding phytoplasma disease are:

- Propagate from seed, or from phytoplasma-free plants.
- Eliminate perennial and biennial weed hosts.
- Rogue out and destroy symptomatic plants.
- Avoid planting susceptible crops next to crops harboring the phytoplasma.
- Control the leafhopper vector in the crop and nearby weeds early in the season (Douglas and McClure, 1988).

- Plant varieties that are more resistant to the disease must be cultivated, if available.

PEACH ROSETTE

Peach rosette disease was first reported in Georgia (USA) in 1881 (Smith, 1891) but now it has spread upto Texas. A severe outbreak occurred in Arkansas in 1977 (Slack and Kim, 1977). A disease resembling peach rosette has also been reported from Italy (Marcone and Ragozzino, 1994).

SYMPTOMS

Symptoms of peach rosette are similar to those caused by the peach rosette mosaic Nepovirus (Dias, 1975; EPPO/CABI, 1996a). Initial symptoms are yellowing and inward folding of leaves. Subsequently there is reduction in shoot elongation with dormant buds producing a compacted growth habit. The internodes become short. Leaves are dense, smaller in size with dark green in colour. Normally, dormant buds break and produce rosette-like symptoms (Plate 12). Fruits usually fall before maturity. The disease can kill a tree in a few months after systemic infection to whole tree. Other fruit trees (almond, plum) show similar symptoms. Infected plum trees are highly stunted (Dunez, 1981).

TRANSMISSION

The disease is transmitted by budding and grafting. The disease was dodder (*Cuscuta pentagona, C. campestris*) transmitted to periwinkle (*Catharanthus roseus*), tomato and *Nicotiana glutinosa* (Scot and Zimmerman, 2001). Although specific vector species of peach rosette are unknown but natural spread has been suspected by a leafhopper vector which survives on an alternate host during winter.

HOST RANGE

Peaches (*Prunus persica*) are the principal host of peach rosette phytoplasma, but the disease is also significant on Japanese plum, *P. salicina* (Kirkpatrick, 1995). Other *Prunus* spp. like almonds (*P. dulcis*), apricots (*P.*

armeniaca), cherries (*P. avium, P. cerasus*) and ornamental and wild *Prunus* spp. can be infected. The wild species *P. angustifolia, P. hortulana* and *P. munsoniana* can act as natural reservoirs of the pathogen (Kenknight, 1976). The phytoplasma can also be artificially transmitted to herbaceous hosts. The causal agent survives mostly on wild plum (*Prunus angustifolia*) from season to season and displays yellowing and reduced growth (Scot and Zimmerman, 2001)

THE PATHOGEN

Phytoplasmas have been detected by electron microscopy in diseased peach and almond trees, and also in inoculated perwinkle plants. The organism measures 80 x 1000 nm and was present in the sieve tubes of phloem tissues of visibly affected leaves, but not in apparently healthy ones. Symptom remission after tetracycline hydrochloride treatment further confirms that the agent is a phytoplasma (Kirkpatrick *et al.*, 1975). According to Kirkpatrick (1995), DNA hybridization studies suggest that peach rosette phytoplasma is related to, but distinct from, peach X-disease phytoplasma (EPPO/CABI, 1996d). It differs from it in host range and geographical distribution. According to Davis *et al.* (2012) the term '*Candidatus* Phytoplasma pruni' be applied to phytoplasma strains whose 16S rRNA gene sequences contain the oligonucleotide sequences of unique regions that are designated in the formally published description of the species. Such strains include X-disease phytoplasma and - within the tolerance of a single base difference in one unique sequence - peach rosette, peach red suture, and little peach phytoplasmas. Peach rosette phytoplasma can be classified as Bacteria; Tenericutes; Mollicutes; Acholeplasmatales; Acholeplasmataceae; *Candidatus* Phytoplasma; 16SrIII (X-disease group).

DIAGNOSIS AND MANAGEMENT

Peach rosette phytoplasma can be tested by inoculating peach seedlings (cv. Elberta or GF305) in field but 4 years are needed for getting observable symptoms. It can also be tested on the same indicators in the glasshouse where symptoms can appear within 3 months after inoculation. The phytoplasma can be detected by nucleic acid hybridization and also by PCR (Scot and Zimmerman, 2001). Details of phytosanitary procedure for fruit tree phytoplasmas are given by EPPO (OEPP/EPPO, 1994). The disease can easily be controlled by destroying affected trees and removing wild *Prunus* spp. near orchards. However, if these measures are not taken, the disease can spread epidemically, as in Arkansas (USA) in 1977 where whole orchards were affected though previously only isolated diseased trees had been seen (Kim and Slack,

1978). The pathogen can be made inactive by treating budsticks in hot water treatment for 8-10 minutes at 50°C.

PEACH YELLOWS

Peach yellows disease was first described near Philadelphia in Pennsylvania as early as 1791. According to CMI (1983), yellows disease of peach has been recorded in Israel, Lebanon, Syria and possibly in Turkey. In Asia, the disease is found in India (Himachal Pradesh, West Bengal). In North America, it is found in Canada (Ontario) and USA. Peach yellows reached in epidemic proportions in the late nineteenth and early twentieth centuries in some places in North America. In late 1970s, in India, a disease on *Prunus padas* characterized by the production of compact tufts or rosettes with smalling of leaves was first observed in Shimla hills and designated as 'Paja rosette' (Dhingra and Ahlawat, 1979). Detailed studies of this disease confirmed it to be similar to peach yellows (Ahlawat and Chenulu, 1981; Ahlawat and Chakraborty, 1993). Peach yellows disease on peach was further studied in Kalimpong region of West Bengal, India.

SYMPTOMS

Peach yellows affected trees show numerous upright growing branches from the main stem or older branches. These branches are so numerous that the tree looks bushy. The leaves on these branches are small but at the base one or two leaves remain abnormally long and straight with inrolled margins and clearing of veins. The growth of affected tree is stunted and internodes of the branches are short. Diseased tree finally appears as a bushy growth with small yellow leafy tufts in contrast to normal trees. The newest expanding leaves are sickle-shaped instead of being in their normal upright position. Premature unfolding of leaf buds is common. The leaves on affected trees continue to grow even after the fall of normal leaves. The larger leaves on affected plants are also mottled. Later, symptoms of wilting and dieback appear resulting in death of affected tree. Fruits on affected trees are larger than normal, insipid and ripe prematurely. The flesh of affected fruits shows reddish streaks and a deep red colour around the pit.

TRANSMISSION

The yellows disease of peach is transmitted by wedge and bud grafting and also by dodder, *Cuscuta reflexa* (Ahlawat and Chenulu, 1986). One known vector of peach yellows is the plum leafhopper, *Macropsis trimaculata* (Kunkel, 1933). The mean incubation period in the vector is 16 days (Pine and Gilmer, 1976).

HOST RANGE

Peach yellows is a serious disease of almond, apricot, cherry, nectarine, peach, plum, and prune (*Prunus* spp.). Peach (*Prunus persica*) is the principal host of peach yellows phytoplasma. Almond (*P. dulcis*), apricot (*P. armeniaca*) and *P. salicina* are also infected. All *Prunus* spp. which have been experimentally graft-inoculated proved to be susceptible. Peach yellows phytoplasma is symptomless in some cultivars of *P. salicina* such as Abundance, Chalco and Chabot (Seemüller and Foster, 1995). The phytoplasma can also be artificially transmitted to herbaceous hosts. Peach seedlings and varieties J.H. Hale and July Alberta were found to be susceptible to the Indian isolate of the peach yellows phytoplasma. Following species of plum are symptomless carriers of Peach Yellows phytoplasma: Native American plum, *Prunus hortulana* and *P. americana*; Common or European plum, *P. domestica*; Japanese plum, *P. salicina*; Myrobalan plum, *P. cerasifera*; Othello plum, *P. cerasifera* var. atropurpurea and Wild goose plum, *P. munsoniana*.

THE PATHOGEN

The associated pathogen with the disease was identified as mycoplasma-like organism now known as Phytoplasma (Jones *et al.*, 1974). According to Kirkpatrick (1995), DNA hybridization studies and sequence analysis suggest that peach yellows phytoplasma is related to, but distinct from, peach X-disease phytoplasma (EPPO/CABI, 1996e). It differs from it in symptoms and vector. From the results of Phylogenetic analyses, it can be concluded that the Apple proliferation (AP), pear decline (PD) and European stone fruit yellows (ESFY) phytoplasmas are coherent but discrete taxa that can be distinguished at the putative species level, for which the names '*Candidatus* Phytoplasma mali', '*Candidatus* Phytoplasma pyri' and '*Candidatus* Phytoplasma prunorum', respectively, are proposed (Seemüller and Schneider, 2004).

DIAGNOSIS AND MANAGEMENT

The phytoplama can be detected by grafting to woody indicators. Serological techniques like fluorescent microscopy and molecular techniques can be used for detection. Infected trees should be destroyed and, if necessary, the vector, *Macropsis trimaculata* should be controlled. Healthy planting material should be used. In this respect, peach yellows was one of the first 'virus-like' diseases to be treated by thermotherapy - dormant trees treated for 10 min at 50°C and buds for 3-4 min at 50°C (Kunkel, 1936).

LITTLE PEACH

The symptoms of little peach disease resemble to some extent to the symptoms caused by yellows and rosette phytoplasmas. In little peach affected trees, only one or two branches may be affected. At the beginning of growing season the foliage is greener and proliferation of leaves on short side branches give a tree bushy appearance. The pits are small and kernels either undeveloped or fail to germinate (Pine and Gilmer, 1976). Apical dominance is lost and all buds on an infected limb develop thin, willowy shoots with small leaves. The shoots are longer than on rosette-affected trees but shorter than on yellows-affected trees. Shoots are upright in growth habits as with yellows-affected trees. No fruit is produced on limbs showing little peach symptoms. Fruit maturity is delayed by 3 weeks and fruit is smaller than normal. Death occurs several years after symptoms appear. The vector(s) of little peach is unknown. Peach rosette, little peach, red suture were induced by a phytoplasma which is closely related to western X-disease phytoplasma (Scott and Zimmerman, 2001). The Taxonomic position is: Bacteria > Tenericutes > Mollicutes > Acholeplasmatales > Acholeplasmataceae > *Candidatus* Phytoplasma > 16SrIII (X-disease group).

RED SUTURE

Symptoms of red suture disease may appear as a dark red suture on prematurely ripen fruits. The flesh in the suture becomes soft and watery. In some cases, a prominent red suture is often observed on a green fruit. The fruits are not edible. The tree, except for fruit, is symptomless at this point. After several years, an infected tree or branch may show symptoms similar to little peach-affected trees. Leaves have a yellowish-green or greenish bronze appearance after a few weeks of petal fall or just before harvest. However, although grafting of red suture transmitted the phytoplasma (as detected by PCR), the disease was not observed in fruiting trees. The disease was transmitted to periwinkle (*Catharanthus roseus*) using dodder (*Cuscuta pentagona*) grown from seed. An infected tree may survive for several years. The vector(s) of red suture is unknown. Peach rosette, little peach, red suture are induced by a phytoplasma which is closely related to western X-disease phytoplasma. The red suture phytoplasma has been sequenced (Scott and Zimmerman, 2001). Taxonomy of the pathogen is: Phylum: *Firmictes,* Class: *Mollicutes,* Order: *Acholeplasmatales,* Family: *Acholeplasmataceae,* Genus: *Candidatus* phytoplasma, Species: *Red suture.*

PEACH YELLOW LEAF ROLL

Symptoms of yellow leaf roll disease are yellowing of the leaf margin and rolling, drying and necrosis of older leaves. Young leaves were tightly rolled and yellow, and proliferation of shoots was evident along branches. Many trees were dead or dying (Jones *et al.,* 2005). A Phytoplasma has been reported to cause peach yellow leaf roll (PYLR) disease (Smart *et al.,* 1996). Two peach diseases in California, western X-disease (WX) and peach yellow leaf roll (PYLR) are caused by two genetically distinct phytoplasmas. Based on symptoms alone, WX and PYLR cannot be reliably distinguished. The results indicated that pear psylla is the primary vector of PYLR in Northern California. The most abundant Phytoplasma bodies were detected in diseased peach trees using strain-specific DNA hybridization or polymerase chain reaction (PCR) (Blomquist and Kirkpatrick, 2002). The PYLR phytoplasma belongs to the 16SrX Apple proliferation group ('*Candidatus* Phytoplasma mali') (Seemüller and Schneider, 2004). However, sequence of phytoplasma rDNA (Accession No. AY 725212) was compared with others in the GenBank database by using BLAST. The greatest similarity (98%) was with papaya dieback phytoplasma (Accession No. Y10095) from Australia, a member of the 16SrXII Stolbur group ('*Candidatus* Phytoplasma australiense') (Firrao *et al.,* 2004).

VIROID DISEASES OF PEACH

PEACH LATENT MOSAIC VIROID

Peach latent mosaic viroid disease has been reported in all peach and nectarine producing areas of the world including Europe, Asia, North America and South America. Natural infection is quite common. The disease remains latent in peach trees for approximately 5-7 years (Hadidi *et al.*, 1997). A relationship of Peach latent mosaic was shown to peach mosaic and peach yellow mosaic virus diseases (Desvignes, 1986).

SYMPTOMS

The symptoms of the disease appears after 5-7 years of infection as necrosis of buds, delayed shoot development, necrotic branches, reduced tree life, flower streaking and ripening deformations. The stones become rounded and enlarged. Circular discoloured areas develop on the fruit skin. Leaf symptoms are rare but in some cases mosaic, blotch, vein banding or calico develops on infected leaves (Hadidi *et al.*, 1997). Extreme chlorosis of peach leaves, stem and fruits are induced by peach latent mosaic viroid. Symptoms produced by severe strains of the viroid appear similar to those described for peach mosaic, peach calico, peach blotch and peach yellow mosaic. Peach and peach hybrids are suscepyible to latent mosaic viroid (Desvignes, 1986).

TRANSMISSION

Peach latent mosaic viroid (PLMVd), the causal agent of peach latent mosaic disease, is transmissible by grafting, mechanically and by contaminated pruning tools (Hadidi *et al.*, 1997; Shamloul *et al.*, 1995). PLMVd was experimentally transmitted at a low percentage by the aphid *Myzus persicae* (Flores *et al.*, 1992), but natural spread by this vector has never been documented. It was also experimentally transmitted with blades dipped in purified viroid preparations (Flores *et al.*, 1990) or surface-contaminated by slashing infected plants (Hadidi *et al.*, 1997). PLMVd is pollen-transmitted but is not transmitted through seeds or roots of infected peach plants (Barba *et al.*, 2007). This viroid is not transmitted by natural root grafting.

THE PATHOGEN

Analysis by polyacrylamide gel electrophoresis of nucleic acid extracts from different peach samples, healthy or infected with peach latent mosaic disease, demonstrated the association of this disease with an RNA exhibiting the electrophoretic properties typical of circular viroid molecules (Flores *et al.*, 1990). This viroid-like RNA migrated in denaturing gels between the positions of the circular forms of the 371 base citrus exocortis viroid (CEV) and 297 base hop stunt viroid (HSV) (Flores and Llacer, 1988). This viroid family is characterized as having chloroplastic viroids with hammerhead ribozymes (Flores *et al.*, 2006). Peach latent mosaic viroid is a 336-338nt circular RNA which has a branched formation (Shamloul *et al.*, 1995). This branched formation is stabilised by a pseudoknot between two kissing loops (Flores *et al.*, 2006; Hernández and Flores, 1992). PLMVd infections were found on peach trees in Montenegro with sequences 337 and 338 nucleotides long which shared 90.3 to 100% identity. (Mavriè Pleško *et al.*, 2012). Peach latent mosaic viroid is a type species from the family *Avsunviroidae,* genus *Pelamoviroid.*

DIAGNOSIS AND MANAGEMENT

The disease can be detected on peach seedling GF 305 by grafting (Liacer, 1998). The viroid can be detected successfully by PAGE, nucleic acid hybridization, RT-PCR and multiplex PCR from contaiminated trees (Shamloul *et al.*, 1995; Ragozzino *et al.*, 2003; Loreti *et al.*, 1999). This viroid can also be detected by reverse transcription loop mediated isothermal amplification (RT-LAMP) method (Boubourakas *et al.*, 2010). The viroid may be eliminated from some cultivars by micrografting (shoot tip grafting) (Barba *et al.*, 1995).

E. VIRUS DISEASES OF PLUM

Plums are delicious and valued fruit in many parts of South East Asia. Some plum varieties are self-fertile, but many require a compatible plum tree nearby for pollination to occur. Plum trees have a short and very distinct pollination period. There are different types of plum such as Japanese plums and European plums. Japanese Plums are: Au Amber,

Early Golden, Methley, Shiro, Au Rosa, Santa Rosa, Black Amber, Wickson, Satsuma, Vanier, Queen Rosa, Fortune, Elephant Heart, Ruby Queen and Friar. The European Plums are: Castleton, Stanley, Long John, Victory Plum, Bluefre, Vision Plum, and President. Plum fruits are of great importance in northern India. The varieties of plum are mainly grown better in the submountain tracts. Plum trees are infected by several viruses. Many of them are the original viruses of other temperate fruit crops and plum is a host for them. Originally, a few viruses have been recorded on plum. The important ones are bark split, line pattern, sharka or plum pox, white spot and mosaic. A phytoplasmal disease known as leptonecrosis is also known to occur in plum. These diseases are briefly described below:

SHARKA DISEASE OR PLUM POX

Plum pox virus disease is the most important disease not only in plum but also in other stone fruit species of the genus *Prunus* including peaches, nectarines, plums, apricots, almonds, cherries and ornamental plants. This disease, also known as Sharka, was first reported occurring in plums in Bulgaria in 1915, reaching the former Yugoslavia in 1935 and Hungary in about 1941. After about 1950 the disease spread more rapidly reaching Germany in 1956, Poland and Russia in 1961, and France in 1970. The disease appeared in England in 1965. Plum pox invaded in Spain in 1984. Throughout Europe plum pox is considered the most devastating disease of stone fruits and it has been estimated that over 100 million European trees are infected. In 1992, plum pox symptoms were first detected in the western hemisphere in Chile (Roy and Smith, 1994). In October 1999, plum pox virus (PPV) (D-strain) was identified for the first time in North America in peaches; and in 2000, PPV-D was detected in peaches growing in the Niagara area of Ontario, Canada. In the summer of 2007, one PPV infected tree was identified in Michigan and at two sites in western New York near Lake Ontario (Moorman and Gildow, 2007). The disease is now wide spread in Bulgaria, Croatia, Germany, Greece, Hungary, Poland, Romania, Slovakia and also introduced to Azores, Bosnia-Herzegovina, Egypt, Former USSR including Central Asia, India, Lithuania. The disease is restricted in distribution in Albania, Austria, Canada, Cyprus, Czech Republic, France, Italy, Luxembourg, Moldova,

Norway, Portugal, Southern Russia, Slovenia, Spain, Syria, Turkey, Ukraine, United Kingdom, United States (Levy *et al.*, 2000). In India, Plum pox was reported in 1994 (Thakur *et al.*, 1994). PPV does not kill trees, but can drastically reduce yields.

SYMPTOMS

The symptoms of the plum pox disease develop both on leaves and fruits of most stone fruit trees and also on flowers of certain varieties of peaches. Symptoms in leaves are yellowing of veins and/or yellow to light green rings on the surface of the leaf. Brown, dead areas develop on leaves and fall out, leaving "shot holes" appearance. The skin of the fruit may develop lightly pigmented rings or line patterns. Necrotic or brown areas may appear and the fruit may become deformed and irregular. There is an early fruit drops. Plums show more conspicuous symptoms than other species, and are considered good indicator hosts. In apricots, leaves show yellow or light green ring spots, blotches, netting, or vein yellowing and have rings on the surface of the seed, but these rings are not visible on the external skin of the fruit. The fruits on infected trees are often misshapen and blemished with reduced sugar contents. In addition of leaf and fruit symptoms, infected peach trees also show colour breaking of flower petals where dark pink stripes are developed. Leaves develop yellow ringspots or netting, veinal chlorosis, speckling, or blotches. Leaves may also become deformed, with brown patches. Yellow or red rings or blotches develop on the skin of the fruit. Symptoms on leaves of sour and sweet Cherry (*Prunus cerasus* and *P. avium*) develop as yellow rings or blotches. Fruits develop yellow or brown rings with notched marks. Some *Prunus* species are characterized by yellow line patterns and blotches, or necrotic ring symptoms on expanded leaves (Brunt *et al.*, 1999). Symptoms vary considerably with the host plant species, cultivar, age and nutrient status of the plant, and environmental conditions. In addition, different strains of PPV may cause different degrees of disease severity. Stones of apricots and some plum cultivars show pale rings or spots.

TRANSMISSION

Plum pox virus is transmitted by aphid vectors and also by mechanical inoculation and by grafting but not by contact between hosts. It is also transmitted in seeds of apricots, peach and plum (Nemeth and Kölber, 1983), and also by pollen (Nemeth, 1986). Several aphid species can transmit plum pox in non-persistent manner. Following aphid species have been recorded the vector of PPV: *Aphis craccivora, A spiraecola, Brachycaudus helichrysi, B. cardui, Myzus persicae, M. varians, Myzus (Phorodon) humuli*. The principal natural vector(s) are *Brachycaudus helichrysi, Myzus persicae, Myzus (Phorodon) humuli*. One of the most efficient vectors, the green peach aphid (*Myzus persicae*) colonizes peaches and other stone fruits; and the second most efficient vector, the spirea aphid (*Aphis spiraecola*), is a common migrant into stone fruit orchards (Wallis *et al.*, 2005). Aphids can also acquire PPV after feeding and probing on fruits harvested from infected trees and transmitted the virus to healthy *Prunus* seedlings (Gildow *et al.*, 2004). Long distance spread usually occurs as a result of the movement of infected nursery stock or propagative materials. Once a plant is infected the virus becomes systemic.

HOST RANGE

Plum pox virus (PPV) infects plum, peach, nectarines, apricot, almond, cherry, and wild *Prunus* species, privet (*Ligustrum vulgare*), walnut (*Juglans regia*) (Baumgartnerova, 1997) and euonymus (*Euonymus europea*). Under laboratory conditions, PPV infects large number of wild native and weed species, including herbaceous plants. The following plant species have been found to be susceptible to PPV strains under natural and experimental conditions (Damsteegt *et al.*, 2007):

Prunus armeniaca — Pale rings are developed on fruits and stones and fruits become deformed.

P. cerasifera — Mostly the infection is latent but occasional mild pale green mottling of leaves develop.

P. domestica — Leaves show pale green mottling, light green or yellow rings, spots or lines, some brown necrotic spots; fruits show blue, shallow depressions or ringlike discolouration, deeply engraved rings and spots of the skin, flesh turns brown and gum like, fruit drop; shoots split and die back.

P. glandulosa, P. persica — Vein clearing, chlorotic blotching and malformation of leaves; fruits develop pale rings and diffuse bands on the skin, some malformation.

P. insititia, P. spinosa, P. salicina — Symptomless.

Prunus armeniaca — Pale rings are developed on fruits and stones and fruits become deformed.

P. cerasifera — Mostly the infection is latent but occasional mild pale green mottling of leaves develop.

The following plant species were susceptible under experimental conditions: *Capsella bursa-pastoris, Celosia argentea, Chenopodium ambrosioides,*

Chenopodium capitatum, Chenopodium foetidum, Chenopodium foliosum, Chenopodium murale, Chenopodium quinoa, Cyamopsis tetragonoloba, Emilia sagittata, Gomphrena globosa, Humulus lupulus, Hyoscyamus niger, Lupinus albus, Lycopersicon esculentum, Melilotus albus, Melilotus officinalis, Nicandra physalodes, Nicotiana acuminata, Nicotiana benthamiana, Nicotiana bigelovii, Nicotiana clevelandii, Nicotiana debneyi, Nicotiana glutinosa, Nicotiana megalosiphon, Nicotiana occidentalis, Nicotiana rustica, Nicotiana sylvestris, Nicotiana tabacum, Petunia x hybrida, Physalis floridana, Physalis peruviana, Pisum sativum, Prunus armeniaca, Prunus cerasifera, Prunus domestica, Prunus glandulosa, Prunus insititia, Prunus japonica, Prunus mahaleb, Prunus maritima, Prunus persica, Prunus salicina, Prunus sibirica, Prunus spinosa, Prunus tomentosa, Ranunculus arvensis, Ranunculus sardous, Senecio vulgaris, Sesbania exaltata, Sorbus domestica, Stellaria media, Trifolium pratense, Trifolium repens, Vicia sativa, Vicia villosa, Zinnia elegans (Kassanis and Sutic', 1965; Kegler, 1977; Németh, 1986; Van Oosten, 1970; Sutic',1961).

Following plant species are used as diagnostic host species. The symptoms develop on them upon inoculation are described below:

Chenopodium foetidum — Yellow spots, some with necrotic centres or necrotic spots; not systemic.

Nicandra physalodes — Black-brown necrotic local lesions.

Prunus domestica cvs Italian Prune and Poz'egac'a — Chlorotic rings, bands and spots.

P. japonica — Diffused chlorotic spots.

P. maritima — Chlorotic spots, vein necrosis.

P. sibirica — Green spots and faint line pattern.

P. tomentosa — Epinasty and malformation of young leaves, chlorotic spots, necrotic spots of older leaves (Damsteegt *et al.*, 1997).

Sorbus domestica — Yellow spots, leaf chlorosis.

THE PATHOGEN

Plum pox virus belongs to the genus *Potyvirus* in the family *Potyviridae*. Virus particles are filamentous, flexuous with a clear modal length of 660-770 nm and a width of 12.5-20 nm (Bovey, 1971). The Mr of the genome constitutes 7% of the virion by weight. The genome is monopartite, linear, positive-sense, single-stranded RNA and has been sequenced. The viral genome encodes structural proteins and non-structural proteins. Virions consist of 3 structural protein(s) (Kerlan and Dunez, 1979). Various strains/

pathotypes have been identified in PPV (Sutic' *et al.*, 1971; Marenaud and Mazy, 1976). In Europe, plum pox virus is known to occur in different forms or strains (PPV-D, PPV-M, PPV-C, and PPV-EA). Most of the PPV isolates occurring in North and South America are related to the PPV-D strain. The PVM strain was also splitted into two subclusters (Glasa *et al.*, 2012). Cytoplasmic inclusions are amorphous X-bodies and pinwheels. Inclusions are needle shaped (Ples'e *et al.*, 1969) and are protein paracrystals (Bovey, 1971). Some well known relatives of PPV are potato virus Y, bean common mosaic virus, maize dwarf mosaic virus, and watermelon mosaic virus. All these viruses are transmitted by aphids. The virus is serologically related to potato Y virus, but distantly (Grüntzig and Fuchs, 1979).

DIAGNOSIS AND MANAGEMENT

Plum pox disease can be identified by observing diagnostic symptoms on leaves and fruits under field conditions. Under experimental conditions, it can be detected by inoculation to indicator hosts as described above. *Prunus tomentosa* has also been reported as a diagnostic host for the detection of plum pox virus (Damsteegt *et al.*, 1997). The virus can also be detected efficiently by ELISA and PCR technology (Wetzel *et al.*, 1991,1992, Poggi Pollini *et al.*, 1997, James *et al.*, 2003). It can also be detected by heminested-PCR and PCR-ELISA (Olmos *et al.*, 1997) and also by immune tissue printing (Knapp *et al.*, 1995).

There is no cure or treatment for the disease once a tree becomes infected. Therefore, it is necessary to destroy PPV-infected trees once identified because an infected tree will serve as a virus source for all surrounding susceptible species. Once the disease becomes established, control and prevention measures for plum pox include field surveys, use of certified nursery materials, control of aphids and elimination of infected trees in nurseries and orchards. Sources of resistance exist in *Prunus*, but are not common. Since few sources of resistance to PPV have been identified, the transformation of plum with viral genes, such as coat protein gene offers a complementary approach for developing PPV-resistant stone fruit cultivars. C5, a transgenic clone of *Prunus domestica*, containing the PPV coat protein (CP) gene, has been described as highly resistant to PPV in greenhouse tests, displaying characteristics typical of post-transcriptional gene silencing (PTGS) (Scorza *et al.*, 1998). C5 trees exposed to natural aphid vectors in the field remained uninfected after 4 years while susceptible transgenic and untransformed trees developed severe symptoms within the first year (Ravelonandro *et al.*, 1998; Jarosova *et al.*, 2010). These results indicate both stability and efficiency of PTGS-based PPV resistance in plum under field conditions (Ralph Scorza *et al.*, 1994; Jean-Michel Hily *et al.*, 2004). Recently, plum pox virus-resistant plum,

Honey-sweet has been approved by the US regulatory agencies for commercial production in US (Pappu, 2009). The efficacy of the RNAi strategy for producing PPV resistant stone fruits has also been explored (Dolgov *et al.*, 2010; Monticelli *et al.*, 2012)). *In vitro* micrografting or thermotherapy followed by meristem culture are useful for eliminating plum pox virus (Gabova, 1995). Horticultural mineral oil treatment (Sunspray ultrafine 1%) can reduce the incidence of plum pox as the aphid vectors may not like to feed on sprayed plants (Vidal *et al.*, 2012).

PLUM MOSAIC

A mosaic disease of plum is quite prevalent in plum plantations in Himachal Pradesh, India (Azad and Sehgal, 1958; Nagaich and Vashishth, 1965b; Ahlawat and Chenulu, 1986). The disease is characterized with typical mosaic symptoms on leaves of affected trees and has been considered of complex nature (Fig. 13). The only means of mosaic transmission is by grafts. The disease was transmitted to peach seedlings, plum vars. Green Gage, Becky Smith, Beauty, Green and Satsuma plums. The virus can be detected by a colour test designed for the detection of polyphenols accumulating in plant tissues (Lindner *et al.*, 1950). Similarly, chemical test designed for the detection of polyphenols accumulating in plant tissues (Reeves, 1951) was also positive with plum mosaic (Azad and Sehgal, 1958).

Fig. 13: Plum leaf showing mosaic symptoms

The virus which Sutic and Juretic (1970) isolated from plum tree was identified to be an isolate of sowbane mosaic virus (SMV). The serological tests carried out by the agar double diffusion tests and the absorption tests revealed that the investigated isolate is indistinguishable from an American strain of SMV, a Yugoslav strain of SMV and also from SMV isolated from grapevine in Yugoslavia.

PLUM CREAMY WHITE SPOT

Initial symptoms of this disease appear as small pale yellow to white spots on leaves of Santa Rosa plum tree (Fig. 14). These spots may coalesce and form larger white areas. The symptoms are so conspicuous that affected trees can be marked from a distance. The disease is transmissible by grafting and host range is restricted to plum only (Nagaich and Vashisth, 1965a). No further information is available on this disease.

Fig. 14: Plum leaves showing creamy white spots

PLUM RINGSPOT AND SHOT-HOLE

Posnette in 1952 reported in annual report of the East Malling Research Station about a ringspot and shot-hole disease of plum. A similar disease has been reported from India (Dhingra, 1972). The leaves of affected trees develop yellow rings with green centre. These spots turned to brown and necrotic late in the season. The necrotic areas fell down giving a shot-hole effect to leaves (Fig. 12).

The disease is transmitted by chip grafting from plum to plum and to other stone fruits but it is not mechanically transmissible to squash cv. Butter nut which makes it different from cherry tatter leaf reported by Nagaich and Vashisth (1965b).

Susceptible Symptomatic Hosts

- *Prunus domestica* cvs. Green, seeded plum, *P. salicina* cv. Santa Rosa: Brown spots, rings, necrosis and shot hole.
- *P. persica* cvs. J.H. Hale, July Alberta, Nectarine, seeded peach: Bigger brown spots, ring and necrosis giving rise to shot hole condition.
- *P. arminiaca* cv. seeded apricot: Brown spots, necrotic lesions producing numerous shot hole.
- *P. arminiaca* cv. New Castle and Shakerpara: Necrotic lesions producing large sized shot hole.
- *P. puddum* (Paja): Acute necrosis on leaves with a little shot hole condition.

PLUM LINE-PATTERN

Plum line pattern virus is known by several names such as peach line pattern virosis virus, Prunus virus 10, plum American line pattern virus. Line pattern disease on plums has been recorded from all places where plum is grown. In India, the disease has been recorded from Himachal Pradesh, Kumaon hills and Kalimpong areas of eastern Himalayas (Azad and Sehgal, 1958; Bhargava and Bist, 1957; Nagaich and Vashisth, 1965b; Ahlawat and Chenulu, 1986; Varma *et al.*, 1991). Two different viruses, the American Plum line pattern and European plum line pattern have been described which are serologically distinct viruses. In Europe, line pattern in plum is caused by strains of prunus necrotic ringspot ilarvirus or apple mosaic ilarvirus, which are not serologically related to American plum line pattern virus (APLPV) (Seneviratne and Posnette, 1970; Scot and Zimmerman, 2001). However, in the western USA, plum line pattern is caused by a strain of Prunus necrotic ringspot ilarvirus. In fact, line pattern symptoms have been reported from many other areas of the world but aetiology in most cases has not been documented. Line pattern symptoms in apricot or peach may be caused by any of these viruses. The virus is a species of subgroup 5 of the *ilarvirus* genus (Rybicki, 1995), although it has no serological relationship with Prune dwarf ilarvirus, Prunus necrotic ringspot ilarvirus or apple mosaic ilarvirus (Fulton, 1984).

SYMPTOMS

The most characteristic symptoms of the disease are yellow vein banding in part or whole of the leaf lamina, with occasional development of oak-leaf pattern (Figs. 15 and 16). In early summer, the yellow colour of these patterns fades to a creamy-white, and new leaves emerging after June are symptomless. On plum cvs German Prune and Grand Duke, fine, irregular, yellowish lines predominate. On most cultivars of peach, in spring and early summer, there are fine irregular, pale-green, wavy bands on each side of the main veins of the leaves. Some leaves develop a network of fine lines, or a golden net pattern, fine confluent rings, veinbanding, or an oak-leaf pattern (Fig. 16). Symptoms usually disappear in summer.

TRANSMISSION

The virus is easily transmissible by bark patch grafting, budding and mechanical means. *Cuscuta campestris* transmitted the virus from *Nicotiana megalosiphon* to *Petunia*. The Indian isolate of plum line pattern disease is transmitted by grafting and budding on plum seedlings and vars. Green Gage, Santa Rosa, Green, peach seedlings, almond and apricot. It is also transmitted by dodder, *Cuscuta hyalina* (Ahlawat and Dhingra, 1973). Plum line pattern is not seed borne. Several ilarviruses are transmitted by pollen but plum line pattern may not be pollen transmitted (Digiaro *et al.*, 1991).

HOST RANGE

Under natural conditions the virus infects *Prunus domestica, P. serrulata, P. cerasus, P. cerasifera, P. americana x P. salicina,* and *P. persica*. The symptoms on these hosts are line patterns and chlorotic rings.

Experimentally infected hosts mainly show symptoms of chlorotic or necrotic local lesions; systemic veinbanding, mottles and ringspots. The following species were susceptible to experimental virus infection: *Catharanthus roseus, Celosia argentea, Chenopodium murale, Chenopodium quinoa, Crotalaria juncea, C. spectabilis, Cucumis sativus, Datura metel, Datura stramonium, Gomphrena globosa, Lactuca sativa, Nicandra physalodes, Nicotiana benthamiana, Nicotiana clevelandii, Nicotiana debneyi, Nicotiana megalosiphon, Nicotiana occidentalis, Nicotiana sylvestris, Nicotiana tabacum, Petunia x hybrida, Phaseolus vulgaris, Physalis floridana, Prunus americana* x *P. salicina, Prunus cerasifera, Prunus cerasus, Prunus domestica, Prunus persica, Prunus serrulata, Sesbania exaltata, Vigna unguiculata, Vigna unguiculata* ssp. *cylindrica, Zinnia elegans* (Paulsen and Fulton, 1968; Seneviratne and Posnette, 1970; Kirkpatrick and Fulton, 1976; Németh, 1986; Smith, 1972). Symptoms on some of the diagnostic host species upon inoculation are given below (Paulsen and Fulton, 1968; Kirkpatrick *et al.*, 1964). Symptoms on some herbaceous hosts are also given below:

Vigna unguiculata ssp. *cylindrica* — Tiny red local lesions; systemic red vein necrosis.

Cucumis sativus, Chenopodium quinoa — Small local lesions, not systemic.

Nicotiana megalosiphon, N. occidentalis — Chlorotic or diffuse necrotic local lesions, systemic ringspots or mottles.

Physalis floridana — Inconspicuous chlorotic local lesions, systemic chlorotic veinbanding ringspots, or oak-leaf

Petunia x hybrida — Chlorotic or necrotic local lesions, systemic necrosis.

Some isolates also infect *Gomphrena globosa* and *Lactuca sativa*. Assay hosts of the virus are: *Vigna unguiculata* ssp. *cylindrica* and *Nicotiana megalosiphon*.

THE PATHOGEN

Association of a virus with plum line pattern disease was reported by Kirkpatrick *et al.* (1964). Particles of the virus are quasi-isometric in four different sizes (26, 28, 31 and 33 nm) corresponding to different sedimentation coefficients (Fulton, 1984). Virus is an ICTV approved species of the genus *Ilarvirus* in family *Bromoviridae*. The genome is segmented, tripartite (segements are distribute among 3 particle types of different size), and consists of probably three segments of linear, positive-sense, single-stranded RNA. The Mr of the genome constitutes 14% of the virion by weight. The encapsidated nucleic acid is mainly of genomic origin, but virions may also contain subgenomic RNA. RNA-4 is a subgenomic mRNA derived from RNA-3 negative strand template. The multipartite genome segments are distributed between 3 different types of particles. The largest particles contain each one molecule of RNA-1 (sedimenting component B). The medium sized particles contain each one molecule of RNA-2 (sedimenting component M). The smallest particles contain one molecule each of RNA-3 and RNA-4 (sedimenting component T). Proteins constitute about 86% of the particle weight. The viral genome encodes structural proteins and non-structural proteins located in the capsid. The virus is not serologically related to Prunus necrotic ringspot, prune dwarf, tobacco streak, Asparagus 2, citrus leaf rugose, citrus variegation, elm mottle, Tulare apple mosaic, apple mosaic, lilac ring mottle and spinach latent viruses.

DIAGNOSIS AND MANAGEMENT

The virus can be detected by mechanical inoculation to the herbaceous indicators *Nicotiana megalosiphon* and *Vigna cylindrica*, both of which develop chlorotic or necrotic local lesions and ringspots followed by a systemic chlorotic mottle with necrosis on new growth. Such symptoms must be confirmed by a serological test. The virus can be effectively detected by serological test, ELISA or by an agar gel double diffusion test (Németh, 1986; Seneviratne and Posnette, 1970; Smith, 1972; Kirkpatrick and Fulton, 1976; Fulton, 1982). The disease can be managed by using healthy planting material through budwood certification programmes.

Fig. 15: Plum leaves showing line pattern symptoms

Fig. 16: Plum leaves showing oak leaf pattern

PLUM BARK NECROSIS/STEM PITTING ASSOCIATED VIRUS

Plum bark necrosis and stem pitting disease was first observed on a 'Black Beauty' plum (*Prunus salicina*) in the United States of America in 1986 and later reported in several other countries in Europe and recently in China (Cui *et al.*, 2011). An *Ampelovirus* in family *Closteroviridae* was identified as the putative causal agent of the disease. The complete RNA genome of plum bark necrosis stem pitting-associated virus (PBNSPaV) was cloned and sequenced and was determined to be 14,214 nts long (Al Rwahnih *et al.*, 2007). The virus infects many stone fruit species and causes decline, gummosis, flattening of scaffold branches, and stem necrotic pits in some diseased trees like peach (*P. persica*), nectarine (*P. persica* var. nucipersica), plum (*P. domestica*), ornamental plum (*P. cerasifera*), sweet cherry (*P. avium*), and flowering cherry (*P. serrulata*) (Al Rwahnih *et al.*, 2007; Marini *et al.*, 2002). Peach GF305 seedlings inoculated by side grafting with budwoods from two PBNSPaV-positive ornamental peach plants developed grooving symptom on the stems of the seedlings and the virus was detected by RT-PCR (Cui *et al.*, 2011). Recently it has been shown that the virus, in question, has a broader diversity by way of developing virus-specific detection technique like RT-PCR (Marais *et al.*, 2012).

PLUM LEPTONECROSIS PHYTOPLASMA

European stone fruit yellows disease was first described in Italy by the name of plum leptonecrosis (Goidanich, 1933). Leptonecrosis is devastating in Japanese plum and apricots in Northern Italy. Within seven years, 100% of the Japanese plum trees became symptomatically infected. Plum cv Ozark Premier is an indicator for plum leptonecrosis. Based on the results of PCR analysis, DAPI staining and on the reaction of the top-grafted Ozark Premier indicators, 50% of the European plum trees, despite their healthy appearance, were shown to be infected with plum leptonecrosis. The detectable presence and graft transmissibility of the plum leptonecrosis phytoplasma in the asymptomatic European plum trees means that the European plum trees are not resistant to the infection but that they are tolerant (Carraro *et al.*, 1998). The disease is transmissible by grafting and dodder (Loi *et al.*, 1995). Though no vector is known to this phytoplasma but natural and experimental transmission of the disease has been observed (Carraro *et al.*, 1992).

16S rRNA gene of leptonecrosis phytoplasma has been PCR amplified, cloned and sequenced (1202bp). The sequence analysis confirmed the relationship of leptonecrosis phytoplasma with apple proliferation phytoplasma (Malisano *et al.*, 1996). The two phytoplasmas can be distinguished when oligonucleotides were labelled with digoxigenin and hybridized in the presence of tetramethylammonium chloride, to 16S rDNA fragments amplified from apple and plum leaf samples (Malisano *et al.*, 1996). Various oligonucleotide primers, derived from the 16S rRNA were used to detect phytoplasmas associated with apple proliferation (AP) and plum leptonecrosis (PLN). These primers in PCR assays gave a higher yield of products from PLN infected tissue than from other phytoplasma infected tissue (Carraro *et al.*, 1998). These phytoplasmas were detected from their respective hosts using universal primers in polymerase chain reaction (PCR) assays. A precise diagnosis of the causal agent of plum leptonecrosis disease is useful and important for epidemiological studies of the disease (Carraro *et al.*, 1998).

OTHER VIRUS-LIKE DISEASES OF PLUMS

In India, some diseases were observed on plum trees in several orchards. There is no information on these diseases except their graft transmission. Among them important ones are: plum marginal chlorosis (Fig. 17a), plum veinal chlorosis (Fig. 17b) and plum leaf roll (Fig. 18) (Mishra, 1966).

Fig. 17: (a) Plum leaves showing marginal chlorosis, (b) Plum leaves showing veinal chlorosis

Fig. 18: Plum leaf roll

A fruit disease of 'Angeleno' plum in the Emilia-Romagna region of central Italy characterized by the presence of a high number of discolored areas with an isodiametric shape (1–2 mm in diameter), irregular border, and scattered in an irregular manner on the fruit but more numerous in the petiole area. Discolored areas may merge to form bigger spots 6–8 mm in diameter. The disease is affecting up to 50% in an orchard with as many as 60–70% of fruit showing disease symptoms. The name "plum spotted fruit" was proposed to describe this disease (Giunchedi *et al.*, 1998). RT-PCR assays and nucleotide sequences showed similarity in peach latent mosaic viroid and plum spotted fruit diseases. The peach latent mosaic viroid isolate was cloned and sequenced after transmission to *Prunus tomentosa* by grafting (Giunchedi *et al.*, 1998).

REFERENCES

Abou Ghanem-Sabanadzovic, N., Abbadi, H., Al Rwahnih, M., Castellano, M.A. and Myrta, A. 2005. Identification and partial characterization of an isolate of apricot latent virus from Palestine. *J. Plant Pathol.*, 87: 37–41.

Ahlawat, Y.S. and Chenulu, V.V. 1981. Occurrence of peach yellows in India. *Curr. Sci.*, 50: 234–235

Ahlawat, Y.S. 1996. Detection of phloem and xylem limited fastidious prokaryotes. *In*: Detection of Plant Pathogens Ahlawat, Y.S., Ramachandran, P., Malathi,

V.G. and Verma, J.P. (*Eds.*) Centre for Advanced studies in Plant Pathology, IARI, New Delhi pp. 144–152.

Ahlawat, Y.S. and Dhingra, K.L. 1973. Dodder transmission of some temperate fruit viruses in Shimla hills. *Indian Phytopath.* 26: 748–749.

Ahlawat, Y.S. and Chenulu, V.V. 1979. New record of peach 'X' in India. *Indian J. Hort.*, 36: 321–322.

Ahlawat, Y.S. and Chenulu, V.V. 1986. Virus and mycoplasma diseases of temperate fruits in India and their management. *Rev. Tropical Plant Pathology,* 2: 197–220.

Ahlawat, Y.S. and Chakraborty, N.K. 1993. Mycoplasma-like diseases of fruit crops in India. *In*: Management of Plant Diseases caused by fastidious prokaryotes. Raychaudhuri, S.P. and Teakle, D.S. (*Eds.*) Associated Publishing Co., New Delhi

Ahlawat, Y.S., Chenulu, V.V. and Varma, A. 1984. X-disease-a potential threat to stone fruit cultivation in India. *In*: Proc. Nat. Sem. Mycoplasma infections in animals, man and plants. CSAV, Mathura, India pp. 143–146.

Ahmed, A.H. and Bailiss, K.W. 1975. Virus infection of Delphinium in Britain. *J. Hort. Sci.,* 50: 47–54.

Alioto, D., Stavolone, L and Ragozzino, A. 1995. Deformation and discoloration of apricot fruits. *Acta Hort.*, (ISHS) 386: 118–121

Al-Rwahnih, M. Turturo, c., Minafra, A, Saldarelli, P., Myrta, A., Palla, V. and Savino, V. 2004. Molecular variability of apple chlorotic leaf spot in different hosts and geographic regions. *J. Plant Path.*, 86: 117–122.

Al Rwahnih, M., Uyemoto, J.K., Falk, B.W. and Rowhani, A. 2007. Molecular characterization and detection of plum bark necrosis stem pitting-associated virus. *Arch. Virol.*, 152: 2197–2006.

Aldaghi, Majid, Massart, Sebastein, Bertaccini, assunta, Jijakli, M. Haissam and Lipoivre, Philippe. 2009. Identification of host genes potentially implicated in the *Malus pumila* and *Candidatus* Phytoplasma mali interactions. *In*: 21st interl. Conf. on virus and other graft transmissible fruit crops. Berichte aus dem JKI 148, 2009 p. 43.

Alfieri, Jr., S.A. 1967. Peach rosette. Plant pathology Circular No. 58 Florida Department of Agriculture, Division of Plant Industry.

Allen, W.R. 1986. Effectiveness of Ontario populations of *Longidorus diadecturus* and *L. breviannulatus* as vectors of peach rosette mosaic and tomato black ring viruses. *Canadian J, Plant Pathol.*, 8: 49–53.

Allen, W.R. and Ebsary, B.A. 1988. Transmission of raspberry ringspot, tomato black ring and peach rosette mosaic viruses by an Ontario population of *Longidorus elongatus. Canadian J. Plant Pathol.* 10: 1–5.

Allen, W.R.; Van Schagen, J.G. and Eveleigh, E.S. 1982. Transmission of peach rosette mosaic virus to peach, grape, and cucumber by *Longidorus diadecturus* obtained from diseased orchards in Ontario. *Canadian J. Plant Pathol.* 4: 16–18.

Ambrós, S., Desvignes, J.C., Llácer, G. and Flores, R. 1995. Pear blister canker viroid: Sequence variability and causal role in pear blister canker disease. *J. Gen. Virol.*, 76: 2625–2629.

Amtjen, Anja, Jelkmann and Wilhelm. 2009. Investigation on virus occurrence in different tissues throughout the year and sequence variability of apple stem pitting virus. *In*: 21st Internl. Conf. on virus and other graft transmissible fruit crops.Berichte aus dem JKI 148, p. 36.

Amtjen, Anja, Jelkmann and Wilhelm. 2009. Towards generation of full length cDNA clone of apple stem pitting virus. *In*: 21st Internl. Conf. on virus and other graft transmissible fruit crops. Berichte aus dem JKI 148. p. 65.

Ahrens, U.; Lorenz, K.H. and Seemüller, E. 1993. Genetic diversity among Mycoplasma-like organisms associated with stone fruit diseases. *Mol .Plant Microbe Interactions,* 6: 686–691.

Attard D., Afechtal M., Agius M., Matic S., Gatt M., Myrta A. and Di Serio F. 2007. First report of Pear blister canker viroid in Malta. *J. Plant Pathol.,* 89: 871.

Auger, J. 1988. Tomato ringspot virus (Tom-RSV) associated with brown line disease of prune trees in Chile. *Acta Hort.* 235: 197–204.

Awasthi, P., Noorani, M.S., Hallan, V., Ram, R., and Zaidi, A.A. 2012. Genome sequence of a sweet cherry isolate(JK 10) of cherry necrotic rusty mottle virus (CNRMV) from India. *In*: XXI Nat. Conf. Immunobiology and Management of viral diseases in 21st Century held at IVRI, Mukteshwar, p. 77.

Azad, R.N. and Sehgal, O.P. 1958. Some virus diseases of temperate fruits. *Indian Phytopath.,* 11: 159–164.

Bandte, M. and Büttner, C. 2001. Occurence, transmission and diagnosis of cherry leafroll nepovirus - a literature review. *Pflanzenschutzberichte,* 59: 1-19.

Barba, M., de sanctis, F. and Cupidi, A. 1985. Distribution of almond viruses in central Italy. *Phytopathologia Mediterranea* 24: 267-269.

Barba,M., Raggozzino, E. and Faggioli, F. 2007. Pollen transmission of peach latent mosaic viroid. *J. Plant Pathol.* 89: 287-289

Barba, M., Cupidi, A., Loreti, S., Faggioli, F. and Martino, L. 1995. *In vitro* micrografting: A technique to eliminate peach latent mosaic viroid from peach. *Acta Hort.* 386: 531–535.

Barbara, J., Fuchs, Annette, Krezal, Gabi, Jarausch, Wolfgang. 2009. Analysis of the acquisition and multiplication efficiency of different strains of *Ca*. Phytoplasma mali by the vector, *Cacopsylla picta*. *In*: 21st Internl. Conf. on virus and other graft transmissible fruit crops. Berichte aus dem JKI 148, p. 45.

Bargen, S.V., Dierker, L., Rott, M., Langer, J. and Buttner, C. 2012. Genome organization of cherry leaf roll virus and analysis of function of virus- encoded proteins. *In*: Proc.22nd Interl. Conf. Virus and other graft transmissible diseases of fruit crops. Held in Rome, p. 57.

Bar-Joseph, M., Garnsey, SM and Gonsalves, D. 1979. The Closteroviruses: A distinct group of elongated plant viruses. *Adv. Virus Res.,* 25: 93–168.

Baric, S. and Dalla-Via, J. 2004. A new approach to apple proliferation detection: A highly sensitive real-time PCR assay. *J. Microbiol. Methods,* 57: 135–145.

Baric, sanja, Oettl, sabine and Dalla Via, Josef. 2009. Infection rate of natural psyllid Populations with *Candidatus* 'Phytoplasma mali' in south Tyrol (Northern Italy). In: 21st Internl. Conf. on virus and other graft transmissible fruit crops.Berichte aus dem JKI 148, p. 47.

Barone, M.; Covelli, L.; Di Serio, F.; Carrieri, R.; Garcia Becedas, M.T.; Ragozzino A. and Alioto, D. 2010. Close similarities between Cherry chlorotic rusty spot disease from Italy and Cherry leaf scorch from Spain. *In*: Proc. 21st International Conference on Virus and other graft transmissible diseases of fruit crops. *Julius-Kuhn-Archiv.,* 427: 122–126.

Baumgartnerova. H. 1997. Walnut - a new host of sharka virus?. *In*: Proc. Middle European meeting 1996 plumpox. Budapest, pp. 104–106

Beakbane, A.B. and Thompson, E.C. 1945. Abnormal lignification in the wood of some apple trees. *Nature,* 156: 145–6.

Beakbane, A.B., Mishra, M.D., posnette, A.F. and slater, C.H.W. 1971. Mycoplasma-like organisms associated with chat fruit and rubbery wood diseases of apple, *Malus domestica* Borkh., compared with those in strawberry with green petal disease. *J. Gen. Microbiol.* 66: 52–62.

Bertaccini, A., Vibio M., Janeckova M., and Franova-Honetslegrova, J. 1997. Molecular detection of phytoplasmas in apple with rubbery wood symptoms in the Czech Republic. XVIIth International Symposium on Virus and Virus-like Diseases of Temperate Fruit Crops", Program Abstracts, 17:102.

Bertaccini, A., Contaldo, N., Calari, A., Paltinieri, S., Windsor, H.M. and Windsor, D. 2010. Preliminary results of exonic growth of phytoplasma from micropropagated infected periwinkle shoots. pp. 147-153. In: Abs. 18th Cong. IOM Chianciano Terme, Italy.

Bhardwaj, S.V., Khosla, K., Thakur, P.D., Modgil, M. and Sharma, D.R. 1991. Detection of apple mosaic virus using indirect ELISA from H.P., India. *In*: Virology in the tropics. (*Eds.*) Rishi, N., Ahuja, K.L., Singh B.P., Mahlotra Publishing House, New Delhi. pp. 615–619

Bhargava, K.S. and Bist, N.S. 1957. Three virus diseases of hill fruits in Kumaon. *Curr. Sci.*, 26: 324–325.

Bhargava, K.S. and Bist, N.S. 1957. Line pattern-a virus disease of plums in Kumaon. *In*: Proc. Indian Sci. Cong. Asson. Calcutta, 44th session, Part III (Abs) p. 47

Bhargava, K.S. and Bist, N.S. 1960. Fruit tree virus diseases investigations in Uttar Pradesh. *In*: Proc. 4th symp. On virus diseases of fruit trees in Europe. Lynghy, tiddskr, Planteavl. 65 Sacrnummer.

Blodgett, E.C., Aichele, M.D., Coyier, D.L. and Milbrath, J.A. 1963. Flat apple disease. *Plant Dis. Reptr.*, 47: 769–771.

Blomquist, C.L. and Kirkpatrick, B.C. 2002. Identification of phytoplasma taxa and insect vectors of peach yellow leaf roll disease in California. *Plant Disease*, 86: 759–763.

Bodine, E.W. and Newton, J.H. 1942. The rasp leaf of cherry. *Phytopathology*, 32: 333–335.

Bodine, E.W. and Reeves, E.L. 1951. *In*: virus diseases of stone fruits. USDA handbook p.10.

Bojan, D. and Bertaccini, A. 2011. Phytoplasma classification: Taxonomy based on 16S ribosomal gene, is it enough? *Phytopathogenic Mollicutes*, 1: 3–13.

Boubourakas, I.N., Fucuta, S., Luigi, M., Figgioli, F., Barba, M. and Kyriakopoulou, P.E. 2010. Improvement of the reverse transcription loop mediated isothermal amplification (RT-LAMP) method for the detection of peach latent mosaic viroid (PLMVd). *In*: Proc. 21st Interl. Conf. Virus and other graft transmissible diseases of fruit crops. *Julius-Kuhn-Archiv.*, 427: 65–69.

Bovey, R.1963. Apple proliferation. *In*: Virus diseases of apple and pears. *In*: Technical communication, *Commonwealth Bureau of horticulture and Plantation Crops*, 30: 63–67

Bovey, R. 1971. Etude au microscope electronique de quelques alterations de structure produities par le virus de la sharka (Plum pox virus) dans les cellulas infectees. *Annls. de Phytopath.*, 3: 225–232.

Brakta, A., Thakur, P.D. and Handa, A. 2012. Detection of viruses infecting apple in north-western Himalayan region. *In*: 22nd Interl. Conf. virus and other graft transmissible diseases of fruit crops. p. 29.

Buczacki, S. and Harris, K. 1998. Pests, Diseases & Disorders of Garden Plants. *BioInfo* (UK)

Brunt, A.A., Crabtree, K., Dallwitz, M.J., Gibbs, A.J., Watson, L.. and Zurchur, E.J. (*eds.*) 1996. Plant viruses on Line' description and list from the VIDE database version 20th August 1996 'URL '*http//biology.anu.edu.au.group/mes/vide/*

Brunt, A.A., Crabtree, K., Dallwitz, M.J., Gibbs, A.J., Watson, L. and Zurcher, E.J. (*eds.*) 1996. Plant viruses online: descriptions and lists from the VIDE database.

Cherry Rasp leaf virus. Version: 20th August 1996.' URL. http:// biology.anu.edu.au/Groups/MES/vide/.

Brunt, A.A., Crabtree, K., Dallwitz,M.J., Gibbs, A.J., Watson, L. and Zurcher, E.J. (*eds.*) 1999. Plum pox potyvirus. *In*: Plant Viruses Online: Descriptions and Lists from the VIDE Database. Version: 16'th January, 1997. URLhttp:// biology.anu.edu.au/ Groups/ MES/vide/

Caglayan, Kadriye, Gazel, Mona, Ulubas Serce, Cigdem and Can, Feza. 2009. Experimental transmission trials by *Cacopsylla pyri* collected from pear decline infected orchards in Turkey. *In*: 21st Internl. Conf. Virus and other graft transmissible diseases of fruit crops. p. 44

Cameron, H.R. and Thompson, M. 1985. Seed transmission of apple mosaic virus in hazelnut. *Acta Hort.* (ISHS) 193: 131–132

Cameron, H.R. and Moore, D.L. 1985. Reduction in spread of necrotic rusty mottle with removal of affected trees. *Phytopathology,* 75: 1311.

Campbell., I. 1968. Heat sensitivity of some apple viruses. *Tagungsberichte Deutsche Akademie derL* and *wirtschaftswissenscluifien zu Berlin,* 97: 311–315.

Campbell, A.I. and Hughes, L.F.1975. Symptoms of star crack virus on the fruit and shoot rowth of apple cultivars. *ISHS Acta Hort.,* 44: 245–250 (IX International Symposium on fruit tree viruses.)

Carmen, G. and Thomas, M.P. 1998. Purification and characterization of peach mosaic virus. *Plant Disease,* 82: 905–908

Carraro, L., Osler, R., Refatti, E., Favali, M.A. 1992. Natural diffusion and experimental transmission of plum leptonecrosis. *Acta Hort.* 309: 285-290.

Carraro, L., Loi, N., Ermacora, P. and Osler, R. 1998. High tolerance of European plum varieties to plum leptonecrosis. *European J. Plant Path.,* 140: 141–145.

Carraro, L., Nemchinov, L. and Hadidi, A. 1998. PCR detection of pome and stone fruit phytoplasmas from active or dormant tissue. *Acta Hort.,* 472: 731-735

Cech, M., Filigarova, M., Pozdena, J. and Branisova, H. 1981. Strawberry latent ringspot and cucumber green mottle mosaic viruses in apricots with the bare twig and unfruitfulness disease syndrome. *Acta Hort.* (ISHS), 94: 391–396.

Chandel, V., Rana, T., Hallan, V. and Zaidi, A.A. 2007. Wild Himalayan cherry (*Prunus cerasoides*) as a natural host of Prunus necrotic ringspot virus in India. *Plant Disease,* 91: 1686.

Chandel, V., Rana,T., Hallan, V. and Zaidi, A.A. 2008. Occurrence of Prunus necrotic ringspot virus on nectarine (*Prunus persica*) in India. *OEPP/EPPO Bulletin,* 38: 223–225.

Choueiri E, Jreijiri S, Verdin E, Bové J. M. and Garnier, M. 2001. First report of a phytoplasma disease of almond (*Prunus amygdalus*) in Lebanon. *Plant disease,* 85: 802.

Claudia, B., Ciccoti, anna Maria, salvadori, Antonella, Jarausch, Wolfgang, grando, Stella. 2009. *In vitro* screening of interspecific hybrids (*Malus* spp.) for Resistance to apple proliferation. p. 44. *In*: 21st interl. Conf. on virus and other graft transmissible fruit crops. *Berichte aus dem JKI* 148.

Cole, A., Mink, G.I. and Regev, S. 1982. Location of Prunus necrotic ringspot virus on pollen grains from infected almond and cherry trees. *Phytopathology,* 72: 1542–1545.

CMI 1983. Distribution map of Plant diseases.No. 60, (Edition 3) CAB International, Wallingford, U.K.

Contaldo, N., Mitrovic, J., Paltrinieri, S., Duduk, B. and Bertaccini, A. 2012. Phytoplasma associated with apricot leafroll disease. *In*: Proc. 22nd Interl. Conf. Virus and other graft transmissible diseases of fruit crops. p. 106.

Cooley, D.R., Tattar, T.A. and Schieffer, J.T. 1992. Treatment of X-disease of peaches using oxytetracycline microinjection capsules. *Hort. Science,* 27: 235–237.

Cooper, J.I., Massalski, P.R. and Mary-Lou Edwards.1984. Cherry leaf roll virus in the female gametophyte and seed of birch and its relevance to vertical virus transmission. *Ann. Appl. Biol.* 105: 55–64.

Cornaggia, D., Gentit, P., Boyé, R. and Desvignes, J.C. 1995. A new phytoplasma disease of apricot trees: The peach vein clearing. *Acta Hort.* (ISHS) 386: 448-453

Cornaggia, D., Grasseau, N. and Gentit, P. 2012. Efficacy of apple proliferation phytoplasma transmission by vegetative propagation technique and impact on its epidemiology. *In*: Abs. 22nd Interl. Conf. Virus and other graft transmissible diseases of fruit crops. p. 118

Cropley, R. 1960. Serological detection of a virus in Cherry trees with a leaf roll disease. *Nature,* 188: 875–876.

Cropley, R. 1961. Cherry leaf-roll virus. *Ann. Appl. Biol.,* 49: 524–529.

Cropley, R. 1969. *Comm. Bur. Hort. Pl. Crops. Tech. Commun. Suppl.,* 2/3/4: 10.

Cropley, R. 1989. Apple chat fruit. *In*: Virus and Virus-like Diseases of Pome Fruits and Simulation Noninfectious Disorders (Fridlund, P.R. *Ed.*), SP0003, Washington State Cooperative Extension and U.S. Department of Agriculture, Pullman, Washington pp. 43–50.

Crosslin, J. M. and Mink, G. I. 1992. Biophysical differences among Prunus necrotic ring spot ilarviruses. *Phytopathology,* 82: 200–206.

Crosslin, J. M., Hammond, R. W. and Hammerschlag, F.A. 1992. Detection of Prunus necrotic ringspot virus serotypes in herbaceous and *Prunus* hosts with a complementary RNA probe. *Plant Disease,* 76: 1132–1136.

Cui, H.G., Hong, N., Xu, W. X., Zhou, J. F. and Wang, G. P. 2011. First Report of Plum bark necrosis stem pitting-associated virus in Stone Fruit Trees in China. *Plant Disease,* 11: 1483.2-1483.2

Damsteegt, V. D., Scorza, R., Stone, A. L., Schneider, W. L., Webb, K., Demuth, M. and Gildow, F.E. 2007. Prunus host range of Plum pox virus (PPV) in the United States by aphid and graft inoculation. *Plant Disease,* 91: 18–23.

Damsteegt, V.D., Waterworth, H.E., Mink, G.I., Howell, W.E., and Levy, L. 1997. *Prunus tomentosa* as a diagnostic host for the detection of plum pox virus and other Prunus viruses. *Plant Dis.,* 81: 329–332.

Davies, D.L., Adams, A.N. and Glenister, M. 1985. Mycoplasma-like organisms (MLOs) in pears and apples." Report East Malling Research Station for 1984 Programme, 6: 151.

Davies, D.L., Barbara, D.J. and Clark, M.F.1995. The detection of MLOs associated with pear decline in pear trees and pear psyllids by polymerase chain reaction. XVI International Symposium on Fruit Tree Virus diseases. *Acta Hort.* (ISHS) 386: 484–488.

Davis R.E. and Lee I.M. 2000. Phytoplasma. *In*: Lederberg, J. (*Ed.*). Encyclopedia of Microbiology, 2nd Ed, Academic Press, San Diego, CA, USA. pp. 640–646

Davis R.E. and Dally E.L. 2001. Revised subgroup classification of group 16SrV phytoplasmas and placement of flavescence dorée-associated phytoplasmas in two distinct subgroups. *Plant Disease,* 85: 790–797.

Davis, R.E., Zhao, Y., Dally,E.L., Lee, I.M., Jomantiene, R and Douglas S.M. 2012. *Candidatus* Phytoplasma pruni', a novel taxon associated with X-disease of stone fruits, *Prunus* spp.: Multilocus characterization based on 16S rRNA, *secY*, and ribosomal protein genes. *International J. Of systemic and Evolutionary Microbiol.* Published online July 13, 2012, doi: 10.1099/ijs.0.041202-0 IJSEM July 2012 ijs.0.041202-0

Delano, J. 2009. Confirmation of the elimination of apple stem grooving virus from apple trees by *in vitro* chemotherapy. *In*; 21st Internl. Conf. on virus and other graft transmissible fruit crops. Berichte aus dem JKI 148, P. 25

Deogratias, J.M., Dosba, F. and Lutz, A. 1989. Eradication of Prune dwarf virus, Prunus necrotic ringspot virus and apple chlorotic leafspot virus in tissue cultured sweet cherry. *Can. J. Plant Pathol,* 11: 332-336.

Deogratias, J.M., Dosba, F. and Lutz, A. 1989. Eradication of Prune dwarf virus, Prunus necrotic ringspot virus, and apple chlorotic leaf spot virus in sweet cherries by combination of chemotherapy, thermotherapy and *in vitro* culture. *Canadian J. Plant Path.,* 11: 337–342.

de Sequeira, O.A. and Lister, R.M. 1969. Purification and relationships of some filamentous viruses from apple. *Phytopathology,* 59: 1740–1749.

de Sequeira, O.A. and Posnette, A.F. 1969. *Commonw. Bur. Hort. Pl. Crops. Tech. Commun.,* 30: 2/3/4, 76*a*.

Desvignes, J.C., 1970. Les maladies à virus du poirier et leur détection. *CTIFL Documents,* 26: 1-12.

Desvignes, J.C. 1976. The virus diseases detected in greenhouse and in field by the peach seedling GF 305 indicator. *Acta Hort.,* 67: 315–323.

Desvignes, J. C. 1986. Peach latent mosaic and its relation to peach mosaic and peach yellow mosaic virus diseases. *Acta Hort.,* 193: 51–57.

Desvignes, J.C. and Savio, A. 1975. Cydonia c 7-1 and Pyronia veitchii two complementary indicators. *Acta Hort.*44: 139.

Desvignes, J. C., Grasseau, N., Boyé, R., Cornaggia, D., Aparicio, F., Di Serio, F. and Flores, R. 1999. Biological properties of apple scar skin viroid: Isolates, host range, different sensitivity of apple cultivars, elimination, and natural transmission. *Plant Dis.* 83: 768–772.

Deyin Guo, Edgar Maiss, Gunter Adam and Rudolf Casper. 1995. Prunus necrotic ring Spot Ilarvirus: Nucleotide sequence of RNA 3 and the relationship to other Ilarviruses based on coat protein comparison. *J. Gen. Virol.* 76:1073-1079.

Dhir, S., Tomar, M., Thakur, P.D., Ram, R., Hallan, V. and Zaidi, A.A. 2008. Apple stem pitting virus on apple in India. *In*: (Abs.) XVIII National Conference of Indian Virological Society, December 11-13, 2008. PGIMER, Chandigarh.

Dhir, S. and Hallan, V. 2012. Complete genome sequence analysis of apple stem pitting Foveavirus from India and development of its diagnostics. p. 40 *In*: Abs. XXI National Conf. on Immunobiology and Management of viral diseases in 21st century held at IVRI, Mukteshwar Campus 8-10 November, 2012.

Dhingra, K.L. 1972. Transmission of apple mosaic by natural root grafting. *Indian J. Hort.* 29: 348–350.

Dhingra, K.L. 1972. Plum ringspot and shot hole virus-a new record for India. *Indian Phytopath.* 25: 588–590

Dhingra, K.L. and Raychaudhuri, S.P. 1970. Sound cultural measures can control virus diseases of apple. *Indian Hort.* 14:3-6

Dhingra, K.L. and Niazi, F.R. 1973. Indexing of apple varieties for the presence of Apple mosaic virus. *Indian Phytopath.,* 26: 315–318.

Dhingra, K. L. and Ahlawat, Y.S. 1973 Apple rubberywood virus in Shimla hills. *Indian Phytopath.,* 26: 343–346

Dhingra, K.L. and Ahlawat. Y.S. 1973. New virus diseases of temperate fruits crops in Himachal Pradesh. *In*: Summer Instt. Pl. Virol. IARI, New Delhi. pp. 60–61.

Dhingra, K.L. and Ahlawat, Y.S. 1979. Additional virus diseases of temperate fruit crops in India. *Indian J. Hort.,* 36: 198–202.

Dias, H.F. 1975. Peach rosette mosaic virus. CMI/AAB Descriptions of Plant Viruses No. 150. Association of Applied Biologists, Wellesbourne, UK.

Dias, H.F. and Cation, D. 1976. The characterization of a virus responsible for peach rosette mosaic and grape decline in Michigan. *Canadian J. Bot.* 54:1128-1139

Digiaro, M., Savino, V., di Terlizzi, B. and Matrtelli, G.P. 1992. The relationship of ilarviruses to almond mosaic. *Adv. Hort. Sci.*, 6: 161–166.

Digiaro, M., Di Terlizzi, B. and Savino, V. 1991. Ilarviruses in apricot and plum pollen. *Acta Hort.*, 309: 93–98.

Di Serio, F., Ragozzino, A. and Alioto, D. 1995. Viroid-like RNAs associated with some apricot fruit disorders. *Acta Hort.* (ISHS), 386: 551–553

Di Terlizzi, B. and Savino, V. 1995. A stem pitting of apricot. *Acta Hort.* (ISHS), 386: 115–117.

Doi, Y., Teranaka, M., Yora, K. and Asuyama, H. 1967. "Mycoplasma or PLT-group-like organisms found in the phloem elements of plants infected with mulberry dwarf, potato witches' broom, aster yellows or paulownia witches' broom". *Ann. Phytopathol. Soc. Japan*, 33: 259–266.

Dolgov, S., Mikhaylov, R., Serova, T., Shulga, O. and Firsov, A. 2010 Pathogen-derived methods for improving resistance of transgenic plums (*Prunus domestica* L.) for plum pox virus infection. *In*: Proc. 21st International Conf. On virus and other graft transmissible diseases of fruit crops. *Julius-Kuhn-Archiv.*, 427: 330–333.

Dosba F., Lansac M., Mazy K., Garnier M. and Eyquard J.P. 1991. Incidence of different diseases associated with mycoplasma-like organisms in different species of *Prunus*. *Acta Hort.*, 283: 311–320.

Douglas, S.M. 1986. Detection of mycoplasma-like organisms in peach and chokecherry with X-disease by fluorescence microscopy. *Phytopathology*, 76: 784–787.

Douglas, S.M. and McClure, M.S. 1988. New integrated approach for controlling X-disease of stone fruits. *Bull. Connecticut Agric. Exp. Station* No. 854.

Dragoljub, D., Sutic, Richard Earl Ford, Malisa and Toric, T. (*ed.*) 1999. Handbook of plant virus diseases. CRC Press, p. 553.

Driel, L. van Potter, J.W. and Ebsary, B.A. 1990. Distribution of virus vector nematodes associated with peach and other fruit crops in Essex County, Ontario. *Canadian Plant Disease Survey*, 70: 23–26.

Duduk, B. and Bertaccini, A. 2011. Phytoplasma classification: Taxonomy based on 16S ribosomal gene, is it enough? *Phytopathogenic Mollicutes* 1:3-13.

Dunez, J. 1981. Exotic virus and virus-like diseases of fruit trees. *Bull. OEPP/EPPO* 11: 251–258.

Eastwell, K.C.; Bernardy, M.G. and Li, T.S.C. 1996. Comparison between woody indexing and a rapid hybridisation assay for the diagnosis of little cherry disease in cherry trees. *Ann. Appl. Biol.*, 128: 269–277.

EPPO/CABI.1996. *Xiphinema americanum sensu lato*. *In*: Quarantine pests for Europe. 2nd edition (*Eds.* by Smith, I.M., McNamara, D.G., Scott, P.R. and Holderness, M.). CAB International, Wallingford, UK.

EPPO/CABI. 1996a. Apple proliferation phytoplasma. *In*: *Quarantine pests for Europe*. 2nd edition (*Ed.* by Smith, I.M., McNamara, D.G., Scott, P.R. and Holderness, M.). CAB International, Wallingford, UK.

EPPO/CABI. 1996b. Apricot chlorotic leafroll phytoplasma. *In*: *Quarantine pests for Europe*. 2nd edition (*Ed.* by Smith, I.M.; McNamara, D.G.; Scott, P.R. Holderness, M.). CAB International, Wallingford, UK.

EPPO/CABI. 1996c. Peach rosette mosaic nepovirus. *In*: *Quarantine pests for Europe*. 2nd edition (*Ed*. by Smith, I.M., McNamara, D.G., Scott, P.R., Holderness, M.). CAB International, Wallingford, UK.

EPPO/CABI. 1996d. Peach X-disease phytoplasma. *In*: *Quarantine pests for Europe*. 2nd edition (*Ed*. by Smith, I.M., McNamara, D.G., Scott, P.R., Holderness, M.). CAB International, Wallingford, UK.

EPPO/CABI. 1996e. Peach yellows phytoplasma. *In*: *Quarantine pests for Europe*. 2nd edition (Ed. by Smith, I.M.; McNamara, D.G.; Scott, P.R.; Holderness, M.). CAB International, Wallingford, UK.

Flegg, J.M. 1969. Tests and potential nematode vectors of cherry leaf roll virus. p. 155–157. In: *Annl. Rep. E. Malling Res. Stn.*, 1968.

Fiore, N., Rivera, L., and Zamorano, A. 2012. Molecular characterization of Tomato Ringspot virus (ToRSV) strain responsible for "Prune Brown line" disease. p.23. *In*: Proc. 22nd Interl. Conf. Virus and other transmissible Diseases of fruit crops. Held in Rome June 3-8, 2012.

Firrao, G., Andersen, M., Bertaccini, A., Boudon, E., Bove, J.M., Daire, X., Davis, R.E., Fletcher, J., Garnier, M., Gibbs, K.S., Gundersen-Rindal, D.E., Harrison, N., Hiruki, C., Kirkpatrick, B.C., Jones, P., Kuske, C.R., Lee, I.M., Liefting, L., Marcone, C., Namba, S., Schneider, B., Sears, B.B., Seemuller, E., Smart, C.D., Streten, C.and Wang, K. 2004. '*Candidatus* Phytoplasma', a taxon for the wall-less, non-helical prokaryotes that colonize plant phloem and insects. *Interl. J. System. and Evol. Microbiol.* 54: 1243–1255.

Flores, R. and Llácer, G. 1988. Isolation of a viroid-like RNA associated with peach latent mosaic disease. *Acta Hort.* (ISHS) 235: 325–332

Flores, R., Hernández, C., Desvignes, J.C. and Llácer, G.1990. Some properties of the viroid inducing peach latent mosaic disease. *Res. Virol.*, 141: 109–118

Flores, R., Hernàndez, C., Avinent, L., Hermoso, A., Llàcer, G., Joarez, J., Arregui, J.M., Navarro, L. and Desvignes J.C., 1992. Studies on the detection, transmission and distribution of peach latent mosaic viroid in peach trees. *Acta Hort.*, 309: 325–330.

Flores, R., Hernández, C., Martínez de Alba, A.E., Dars, J.A. and Di Serio, F. 2005. Viroids and viroid-host interactions. *Ann. Rev. Phytopath.* 43: 117–139.

Flores, R., Delgado Sonia, Rodio María-Elena, Ambrós Silvia, Hernández Carmen, di Serio Francesco. 2006. Peach latent mosaic viroid: not so latent. *Molecular Plant Pathology*, 7: 209–221.

Fridlund, P.R. 1980. Glasshouse indexing for fruit tree viruses. *Acta Phytopathologica Academiae Scientiarum Hungaricae*, 15: 153–158.

Fridlund, P.R. and Warterworth, H.E., 1989. "Apple rubbery wood." *In*: Virus and Virus-like Diseases of Pome Fruits and Simulation Noninfectious Disorders (Fridlund, P.R. *Ed*.), SP0003, Washington State Cooperative Extension and U.S. Department of Agriculture, Pullman, WA. Publication SP0003. pp. 118–126.

Fritzsche, R. and Kegler, N. 1964. Die obertrangung des blattrollvirus der kirsche durch nematoden. *Naturwissenschaften*, 51: 299

Fritzsche, R. and Kegler, H. 1968. Nematoden als vektoren von viruskrankheiten der obstgewachse. *Tag Ber. dt. Akad. Landw. Berlin*, 97: 289–295.

Fucus, E. 1981. Serological detection of apple chlorotic leaf spot virus and apple stem grooving virus in apple trees. *Acta Hort.*, 94: 69–73.

Fulton, R.W. 1952. Mechanical transmission and properties of rose mosaic virus. *Phytopathology*, 42: 413–416.

Fulton, R.W. 1956. Nonidentity of apple mosaic and tobacco streak viruses. *Phytopathology*, 46: 694.

Fulton, R.W. 1957. Comparative host ranges of certain mechanically transmitted viruses of Prunus. *Phytopathology,* 47: 215–220.

Fulton, R.W. 1959. Purification of sour cherry necrotic ringspot and Prune dwarf viruses. *Virology,* 9: 522–535.

Fulton, R.W. 1967. Purification and some properties of tobacco streak and Tulare apple mosaic virus. *Virology,* 32: 153–163.

Fulton, R.W. 1970. Prunus necrotic ringspot virus. CMI/AAB Descriptions of Plant Viruses. No. 5.

Fulton, R.W. 1971. Tulare apple mosaic virus. CMI/AAB description of Plant viruses.

Fulton, R.W.1972. Apple mosaic virus. CMI/AAB Description of plant viruses No. 83. CABI, Kew, UK.

Fulton, R.W. 1982. Ilar-like characteristic of American plum line pattern virus and its serological detection in Prunus. *Phytopathology,* 72: 1345–1348

Fulton, R.W. 1984. American plum line pattern virus. *CMI/AAB Descriptions of Plant Viruses* No. 280. Association of Applied Biologists, Wellesbourne, UK.

Fulton, R.W.1985. tobacco streak virus. AAB description of plant viruses 307. Association of Appl. Biologists, Wellesbourne, Warwick, England.

Gabova, R.N. 1995. Chaemotherapy treatment of *Prunus* spp. genotype infected by plum pox virus. *Plant Science* (Sofia), 32: 16–18

Gazel, M., Ulubaþ, C. and Çaðlayan, K. 2008. Detection of *hop stunt viroid* in sweet and Sour cherry trees in turkey by RT-PCR. *Acta Hort.* (ISHS), 795: 955–958.

Gentit, P., Foissac, X., Svanella-Dumas, L., Peypelut, M. and Candresse, T. 2001. Characterization of two different apricot latent virus variants associated with peach asteroid spot and peach sooty ringspot diseases. *Archives of Virology,* 146: 1453–1464.

Gentit, P., Foissac, X., Svanella-Dumas, L. and Candresse, T. 2001. Variants of *apricot latent foveavirus* (alv) isolated from south european orchards associated with peach asteroid spot and peach sooty ringspot diseases. *Acta Hort.* (ISHS), 550: 213–220.

Gildow, F.E., Damsteegt, V.D., Stone, A.L., Schneider, W., Luster, D.G. and Levy, L. 2004. PlumPox in North America: Identification of aphid vectors and a potential role for fruit in virus spread. *Phytopathology,* 94: 868–874.

Gilmer, R.M. 1962. Pathological anatomy of rapid decline of Spy 227 apple. *Phytopathology* 52: 1027–1030.

Gilmer, R.M. 1965. Additional evidence of tree to tree transmission of sour cherry yellows virus. *Phytopathology,* 55: 482–483.

Gilmer, R.M, Nyland, G. and Moore, J.D. 1976. *In*: Virus diseases and noninfectious disorders of stone fruits in North America. USDA Handbook 437. pp. 179–190.

Gilmer, R.M. and Blodgett, E.C. 1976. X-disease. *In*: Virus diseases and non-infectious disorders of stone fruits in North America. (Gilmer, R.M., Moore, J.D., Nyland, G., Welsh, M.F. and Pine, T.S. *Eds.*). Agriculture Handbook No. 437, USDA, Washington, D.C., USA. pp. 145-155.

Giri, B.K., Mishra, M.D. and Ahlawat, Y.S. 1987. Yellow net -a potential threat to almond cultivation in India. *Indian .J. Plant Pathol.,* 5: 126–128.

Giunchedi, L., Poggi-Pollini, C., Gentit, P., Nemchinov, L. and Hadidi, A. 1998. Plum spotted fruit: A disease associated with peach latent mosaic viroid. *Acta Hort.* (ISHS), 472: 571–580.

Glasa, M., Candresse, T. and the SharCo consortium. 2012. A large scale study of plum pox virus genetic diversity and of its geographical distribution. *In*: Proc.22nd Interl. Conf. Virus and other graft transmissible diseases of fruit crops. p. 38.

Goidanich, G. 1933. Un deperimento dei susini Bolletino Regia Stazione. *Patologia Vegetale Roma,* 13: 160–173.

Gonsalves, D. and Fulton, R.W. 1977. Activation of Prunus necrotic ringspot virus and rose mosaic virus by RNA-4 components of some ilarviruses. *Virology,* 81:398.

Granett, A.L.; and Gilmer, R.M.1971. Mycoplasmas associated with X-disease in various *Prunus* species. *Phytopathology,* 61: 1036–1037.

Grasseau, N., Macquaire, G., Boye, R., Cornaggia, D. and Desvignes, J.C. 1999. Peach red marbling and peach sooty ringspot, two new virus-like degenerative diseases of Prunus. *Plant Pathology,* 48: 395–401.

Grüntzig, M. and Fuchs, E. 1979. Untersuchungen zur serologischen verwandschaft von beet mosaic virus, potato virus Y, bean yellow mosaic virus und plum pox virus. *Arch. Fur Phytopathol. Pflanzenschutz, Berlin,* 15: 135–139.

Gundersen D.E. and Lee I.M. 1996. Ultrasensitive detection of phytoplasmas by nested- PCR assays using two universal primer pairs. *Phytopathologia Mediterranea,* 35: 144–151.

Gundersen D.E., Lee I. M., Schaff, D.A., Harrison, N.A., Chang, C.J., Davis, R.E. and Kingsbury, D.T. 1996. Genomic diversity and differentiation among phytoplasma strains in 16S rRNA groups I (aster yellows and related phytoplasmas) and III (X disease and related phytoplasmas). *International J. System. Bacteriology,* 46: 64–75.

Haber, S. and Hamilton, R.I. 1980. Distribution of determinants for symptom production, nucleoprotein component distribution and antigenicity of coat protein between the Two RNA components of cherry leaf roll virus. *J. gen. Virol.,* 50: 377–389.

Hadidi, A. and Hammond, R.W. 1988. Construction of molecular clones for identification and detection of tomato ringspot and arabis mosaic viruses. *Acta Hort.,* 235: 223–230.

Hadidi, A. and Yang, X. 1990. Detection of pome fruit viroids by enzymatic cDNA Amplification. *J. Virol. Methods,* 30: 261–269.

Hadidi, A. and Powell, C.A. 1991. Complementary DNA cloning and analysis of RNAs of a Prunus stem pitting isolate of tomato ringspot virus. *Mol. Cell Probes,* 5: 337–344

Hadidi, A., Hansen, A. J., Parish, C. L., and Yang, X. 1991. Scar skin and dapple apple viroids are seed-borne and persistent in infected apple trees. *Res. Virol.,* 142:289-296.

Hadidi, A., Giunchedi, L., Shamloul, A.M., Poggi-Pollini, C. and Amer, M.A. 1997. Occurrence of peach latent mosaic viroid in stone fruits and its transmission with contaminated blades. *Plant Dis.,* 81: 154–158.

Hamilton, R.I., Dodds, J.A. and Raine, J. 1980. Some properties of a nucleic acid associated with little cherry disease. *Acta Phytopathologica Academiae Scientiarum Hungaricae,* 15: 75–77.

Hammond, R and Howell, W. 2000. Control of cherry rugose mosaic disease enhanced by new assay technique. Published in Good Fruit Grower Trade Journal Hammond, R.W. and Crosslin, J.M. 1995. The complete nucleotide sequence of RNA 3 of a peach isolate of Prunus necrotic ringspot virus. *Virology,* 208: 349–353.

Hansen, A.J. and Stace-Smith, R. 1971. Properties of a virus isolated from golden elderberry Sambucus mgra aurea. *Phytopathology,* 61: 1222-1229.

Hansen, A.J., Nyland, G., McElroy, F.D. and Stace-Smith, R. 1974. Origin, cause, host range and spread of cherry rasp leaf disease in North America. *Phytopathology,* 64: 721–727.

Hansen, A.J. and Green, L. 1985. Canindex 1, a superior indicator cultivar for little cherry disease. *Plant Disease,* 69: 11–12.

Hassen, I.F., Roussel, S., Kummert, J., Fakhfakh, H., Marrakchi, M. and Jijakli M.H., 2006. Development of a rapid RT-PCR test for the detection of peach latent mosaic viroid, pear blister canker viroid, hop stunt viroid and apple scar skin viroid in fruit trees from Tunisia. *J. Phytopathol.,* 154: 217–223.

Heintz, W. 1986. *Cuscuta odorata* –an effective vector for Mycoplsma-like organisms (MLO). *Nachrichenblatt des deutchen Pfalanzenschutzdienstes* 38: 138- 141.

Hernández, C. and Flores, R. 1992. Plus and minus RNAs of peach latent mosaic viroid self cleave *in vitro* via hammerhead structures. *Pro. National Acad. Sci. USA* 89: 3711–3715.

Hernández, C., Elena, S.F., Moya, A. and Flores R., 1992a. Pear blister canker viroid is a member of the Apple scar skin vi- roid subgroup (apscaviroids) and also has sequence ho- mologies with viroids from other subgroups. *J. Gen.Virol.* 73: 2503-2507.

Hernández, C., Llácer, G., Desvignes, J.C. and Flores, R. 1992b. Evidence supporting a viroid etiology for Pear blister canker viroid. *Acta Hort.,* 309: 319–324.

Horváth, J. 1979. New artificial hosts and nonhosts of plant viruses and their role in the identification and separation of viruses XIII. Nepovirus group (CLRV sub-group). *Acta phytopath. Acad. Sci. Hung.,* 14: 319.

Howell, W.E. and Mink, G.I.1992. Rapid biological detection of apple scar skin viroid. *Acta Hort. (ISHS),* 309: 291–296.

Hunter, J.A., Chamberlain, E.E. and Atkinson, J.D. 1958. Note on transmission of Apple mosaic by natural root grafting. *N.Z. J. Agric. Res.,* 1: 80–82.

Hutchins, L.M. 1932. Peach mosaic - a new virus disease. *Science* 76, 123.

Hutchins, L.M., Bodine, E.W. and Thornberry, H.H. 1937. Peach mosaic, its identification and control. *USDA Circular* No. 427.

IRPCM 2004. e, Candidatus Phytoplasma f, a taxon for the wallless, non-helical prokaryotes that colonise plant phloem and insects. *Internl. J. System. Evolut. Microbiol.* 54: 1243–1255.

ISHS 1980. Detection of viruses and other graft-transmissible virus-like diseases of fruit trees. *Acta Phytopathol. Acad. Sci. Hungaricae* 15: 407-413.

Ito, D. and Yoshida, K. 1998. Reproduction of apple fruit crinkle disease symptoms by apple fruit crinkle viroid. *Acta Hort.* (ISHS), 472: 587–594.

James, D. 1992. Partial characterization of a closterovirus-like particles transmitted from Cherry (*Prunus avium*) infected with several viral diseases. *Acta Hort.,* 309: 39–44

James, D. 1994. Cherry mottle leaf: The fulfilment of Koch's postulates. *In*: Proc. Interl. Symp., Rose rosette and other Eriophyid mite transmitted plant disease agents of uncertain etiology. Epstein, A.H. and Hill, J.H. (*Eds.*) Iowa State University, Ames. pp. 17–20.

James, D. 1999. A simple and reliable protocol for the detection of apple stem grooving virus by RT-PCR and a multiplex PCR assay. *J. Virol. Methods,* 83: 1–9.

James, D. 2010. Confirmation of the elimination of apple stem grooving virus from apple trees by *in vitro* chemotherapy In Proc. 21st Interl. Conf. Virus and other graft transmissible diseases of fruit crops. *Julius-Kuhn-Archiv.,* 427: 47–50.

James, D. and Howell, W.E. 1993. Comparison of cherry mottle leaf virus and a virus associated with peach mosaic disease. *In*: *6th Internatl. Cong. Plant Pathol.* Montréal (CA). p. 304.

James, D. and Howell, W.E. 1998. Isolation and partial characterization of a Filamentous virus associated with peach mosaic disease. *Plant Dis.,* 82: 909–913.

James, D. and Howell, W.E. 1998. Identification of a flexuous virus associated with Peach mosaic disease. *Acta Hort.* (ISHS), 472: 285–290

James, D. and Mukerji, S. 1993. Mechanical transmission, identification, and Characterization of a virus associated with mottle leaf in cherry. *Plant Dis.*, 77: 271–275.

James, D., and Mukerji, S. 1996. Comparison of ELISA and immunoblotting techniques for the detection of cherry mottle leaf virus. *Ann. Appl. Biol.*, 129: 13–23.

James, D. and Upton, C. 2002. Nucleotide sequence analysis of RNA-2 of a flat apple isolate of cherry rasp leaf virus with regions showing greater identity to animal picornaviruses than to related plant viruses. *Archives of Virology*, 147: 1631–1641.

James, D. and Upton, C. 2005. Genome segment of RNA-1 of a flat apple isolate of cherry rasp leaf virus nucleotide sequence and RT-PCR detection. *Archives of Virology*, 150: 1469–1476

James, D., Jelkmann, W. and Upton, C. 1999. Specific detection of cherry mottle leaf Virus using digoxigenin-labeled cDNA probes and RT-PCR. *Plant Dis.*, 83: 235–239.

James, D., Howell, W.E. and Mink, G.I. 2001. Molecular evidence of the relationship between a virus associated with flat apple disease and cherry rasp leaf virus as determined by RT-PCR. *Plant Dis.*, 85: 47–52.

James, D., Varga, A., Thompson, D. and Hayes, S. 2003. Detection of a New and unusual Isolate of Plum pox virus in Plum (*Prunus domestica*). *Plant Dis.*, 87: 1119–1124.

James, D., Varga, A., Croft, H., Rast, H., Thompson, D. and Hayes, S. 2006. Molecular characterization, phylogenetic relationships, and specific detection of peach mosaic virus. *Virology*, 96: 137–144

Janeckova, M. and Pluhar., Z. 1987. Results of ELISA procedure used for the detection of apple stem grooving virus (ASGV) in 24 isolates of ASGV, separated by heat therapy. *Hort. Sci.* (Prague) 14: 86–91.

Jarausch, W., Lansac, M., Saillard, C., Broquaire, J.M. and Dosba, F. 1998. PCR assay for specific detection of European stone fruit yellows phytoplasmas and its use for epidemiological studies in France. *Eur. J. Plant Pathol.* 104: 17–27.

Jarosova, J., Gadiou, S., Polak, J., Ravelonandro, M., Scorza, R. and Kumar, J.K. 2010. Evaluation of transgenic *Prunus domestica* L., clone C5 resistance to plum pox virus. *In*: Proc. 21st International Conf. On virus and other graft transmissible diseases of fruit crops. *Julius-Kuhn-Archiv*, 427: 330–333.

Jawdah, Y.A., Abou-Fakhr, E., Sobn, H., Molino Lova, M., Vercesi, A. and Bianco, P.A. 2010. Almond witches' broom phytoplasma (*Candidatus* Phytoplasma phoenicium): A real threat to almond, peach and nectarine. *In:* Proc.21st Interl. Conf. Virus and other graft transmissible diseases of fruit crops. *Julius-Kuhn-Archiv*, 427: 418–420

Jawdah, Y. A., Molino Lova, Bianco, P.A., Choueiri, E., Fakhr,R., Hajj-Hassan, S., Hydar, L. and Aiachi, R. 2012. Almond witches' broom phytoplasma officially declared as a regulated pest in Lebanon. *In*: Proc. 22nd Interl. Conf. Virus and other graft transmissible diseases of fruit crops. p. 107.

Jean-Michel Hily, Ralph Scorza, Tadeusz Malinowski, Barbara Zawadzka and Michel Ravelonandro. 2004. Stability of gene silencing-based resistance to Plum pox virus in transgenic plum (*Prunus domestica* L.) under field conditions. *Transgenic Research* 13: 427–436.

Jelkmann, W., Hergenhahn, F. and Berwarth, C. 2010. Transmission of little cherry virus-1 (LChV-1) by *Cuscuta europea* to herbaceous host. *In*: Proc. 21st Interl. Conf. Virus and other graft transmissible diseases of fruit crops.*Julius-Kuhn-Archiv.*, 427: 272–274.

Jiang, Y.P., Chen, T.A., Chiykowski, L.N. and Sinha, R.C. 1989. Production of monoclonal antibodies to peach eastern X-disease agent and their use in disease detection. *Canadian J. Plant Pathol.*, 11: 325–331.

Jomantiene, R., Davis R.E., Antoniuk, L. and Staniulis, J. 2000. First report of phytoplasmas in soybean, alfalfa, and *Lupinus* sp. in Lithuania. *Plant Disease,* 84: 198.

Jomantiene, R., Davis, R.E., Valiunas, D. and Alminaite A. 2002a. New group 16SrIII phytoplasma lineages in Lithuania exhibit rRNA interoperon sequence heterogeneity. *Europ. J. Plant Pathol.*, 108: 507–517.

Jomantiene, R., Davis, R.E., Alminaite, A., Valiunas, D. and Jasinskaite R., 2002b. First report of oat (*Avena sativa* L.) as host of a phytoplasma belonging to group 16SrI, subgroupA. *Plant Dis.*, 86: 443.

Jomantiene, R. and Davis, R.E.2005. Apple sessile leaf: A new disease associated with a '*Candidatus* Phytoplasma asteris' subgroup 16SrI-B phytoplasma in Lithuania. *Plant Pathology* 54: 237.

Jones, A.L., Hopper, G.R. and Rosenberger, D.A. 1974. Association of mycoplasma-like bodies with little peach and X-disease. *Phytopathology,* 64: 755–756.

Jones, A.L., Hopper, G.R. and Rosenberger D.A. and Chevalier, J. 1974. Mycoplasma-like bodies associated with peach and periwinkle exihibiting symptoms of peach yellows. *Phytopathology* 64: 1154-1156.

Jones, A.T. 1973. A comparison of some properties of four strains of cherry leaf roll virus. *Ann. Appl. Biol.*, 74: 211–217.

Jones, A.T. 1985. Cherry leaf roll virus. AAB Descriptions of Plant Viruses No. 306. Association of applied biologists, Wellesbourne, Warwick, UK.

Jones, A.T. 1987. Cherry rasp leaf virus in *Rubus*. *In*: Virus diseases of small fruits (*Ed*. Converse, R.H.). *USDA Agriculture Handbook* No. 631.

Jones, A.T. and Badenoch, S. 1982. Recent studies on virus and virus-like diseases of *Rubus* in Scotland. *Acta Hort.*, 129: 49–58.

Jones, A.T. and Mayo, M.A. 1972. The two nucleoprotein particles of cherry leaf roll virus. *J. gen. Virol.*16: 349-358.

Jones, A.T. and Duncan, G.H. 1980. The distribution of some genetic determinants in the two nucleoprotein particles of cherry leaf roll virus. *J. gen. Virol.*, 50: 269–277.

Jones, A.T.; Mayo, M.A.; and Henderson, S.J. 1985. Biological and biochemical properties of an isolate of cherry rasp leaf virus from red raspberry. *Ann. Appl. Biol.* 106:101-110.

Jones, P., Arocha,Y., Antesana, O., Montilliano, E. and Franco, P. 2005. First report of an isolate of '*Candidatus* Phytoplasma australiense' associated with a yellow leaf roll disease of peach (*Prunus persica*) in Bolivia. *Plant Dis. Reptr.*, 11: 13.

Joyce, P., Rodoni, B., Crosslin, J.M., Howell, W.E. and Eastwell K.C., 2004. Detection of Pear blister canker viroid in pome fruit orchards across Australia. *Acta Hort.*, 657: 373–378.

Juarez, J., Arregui, J.M., Camarasa, E., Cambra, G., Llacer, G., Ortega, C., Ortega, V. and Navarro, L. 1988. Recovery of virus free peach trees from selected clones by shoot tip grafting *in vitro. Acta Hort.* 235:177-182

Kaponi, M.S., Luigi, M., Barba, M. and Kyriakopoulou, P.E., 2010. Molecular characterization of Hellenic variants of Apple scar skin viroid and Pear blister canker viroid in pome fruit trees. *In*: Jelkmann W. and Krczal, G. (*Eds.*). 21st International Conference on Virus and other Graft Transmissible Diseases of Fruit Crops, Neustadt 2009. *Julius-Kühn-Archiv* 427: 366-372.

Kaponi, M.S., Luigi, M., Barba, M., Sano, T. and Kyriakopoulou, P.E. 2010. First report and molecular analysis of apple scar skin viroid in sweet cherry. *In*: Proc.21st International Conference on Virus and other Graft Transmissible Diseases of Fruit Crops. *Julius-Kuhn-Archiv* 427: 361–365.

Kassanis, B. and Sutic, D. 1965. Some results of recent investigations of Sharka (Plum Pox) virus disease. *Zast. Bilja,* 16: 387–405

Kegler, H. 1977. ICTVdB virus description. Plum pox virus. p. 315. *In*: *Pflanzliche Virologie* Klinkowski, M. *et al.*, Akademie-Verlag, Berlin.

Kegler, H. and Proll, E. 1989. *Arch. Phytopathol Pflanzenschutz* 25: 435.

Keifer, H. H., and Wilson, N. S. 1955. A new species of eriophyid mite responsible for. The vection of peach mosaic virus. *Bull. Calif. Dep. Agric.*, 44: 145–146.

Kelley, R.D. and Cameron, H.R. 1986. Localization of Prune dwarf and Prunus necrotic ringspot viruses associated with sweet cherry pollen and seed. *Phytopathology,* 76: 317–322.

Kenknight, G. 1976. Peach rosette. *In*: Diseases and non-infectious disorders of stone fruits in North America. (R.M. Gilmer, J.D. Moore, G. Nyland, M.F. Welsh and Pine, T.S. *Eds*.). USDA Agriculture Handbook No. 437. pp. 73–76

Kenneth, C., Eastwell, Gary, A., Grove, Dennis, A., Johnson, Gaylord, I., Mink., Ralph, S., Byther, Ronald, P., Covey Jr., and Robert Parker. 2005. Field guide to sweet cherry diseases in Washington

Kerlan, C. and Dunez. J. 1979. Biological and serological differentiation of strains of sharka virus. *Annales de Phytopathologie,* 11: 241–250.

Kim, K.S. and Slack, D.A. 1978. Peach rosette in Arkansas. *Plant Dis. Reptr.,* 62: 310–311.

Kinard, G.R., Scott, S.W. and Barnett, O.W. 1996. Detection of apple chlorotic leaf spot and apple stem grooving viruses using RT-PCR. *Plant Dis.,* 80: 616–621.

Kirkpatrick, B.C. 1995. Peach rosette. *In*: Compendium of Stone Fruit Diseases. APS Press, St Paul, USA.

Kirkpatrick, B. C. 1995. Peach yellows. *In*: Compendium of Stone Fruit Diseases. J.M. Ogawa, E.I. Zehr, G.W. Bird, D.F. Ritchie, K. Uriu and J.K. Uyemoto, *eds.* APS, St.Paul, MN. p. 57.

Kirkpatrick, H.C., Cheney, P.W. and Linder, R.C. 1964. Mechanical transmission of plum line pattern virus. *Plant Dis. Reptr.* 48: 616.

Kirkpatrick, H.C. and Fulton, R.W. 1976. Plum line pattern. *In*: Virus diseases and noninfectious disorders of stone fruits in North America. *Eds.* Gilmer, R.M., Moore, J.D., Nyland, G., Welsh, M.F. and T.S. Pine. Agriculture Handbook No.437, USDA Washington, D.C., USA

Kirkpatrick, H.C. and Fulton, R.W. 1976. *In*: Virus Diseases and Non Infectious Disorders of Stone Fruits in North America. USDA Hdbk No. 437, p. 166.

Kirkpatrick, H.C.; Lowe, S.K. and Nyland, G. 1975. Peach rosette: The morphology of an associated mycoplasma-like organism and the chemotherapy of the disease. *Phytopathology* 65, 864-870.

Kirkpatrick, B.C., Fisher, G.A., Fraser, J.D. and Purcell, A.H. 1990. Epidemiological and phylogenetic studies on western X-disease mycoplasma-like organism. *Zbl. Bakt., Abt.* 2(20): 287–297.

Kirkpatrick, B.C., Uyemoto, J.K. and Purcell, A.H. 1995. X-disease. *In*: Compendium of Stone Fruit Diseases. Ogawa, J.M., Zehr, E.I., Bird, G.W., Ritchie, D.F., Uriu, K. and Uyemoto, J.K. *Eds.* APS Press, St. Paul, MN. pp. 57–58.

Kishi, K., Takanashi, K. and Abiko, K. 1976. Pear necrotic spot, a new virus disease in Japan. *Acta Hort.* (ISHS), 67: 269–274

Klos, E.J., Fronek, F., Knierem, J.A. and Cation, D. 1966. Peach rosette mosaic transmission and control. *Michigan Agricultural Experimental Station Quarterly Bulletin,* 49: 287–293.

Knapp, E., Da Camara Machado, A., Puhringer, H., Wang, Q., Hanzer, V., Weiss, H., Weiss, B., Katinger, H. and Laimer Da Camara Machado, M. 1995. Localization of fruit tree viruses by immune tissue printing in infected shoots of *Malus* and *Prunus* sp. *J. Virol. Methods,* 55:157–173.

Koenig, R. and Kunze, L. 1982. Identification of tombusvirus isolates from cherry in southern Germany as petunia asteroid mosaic virus. *Phytopathol. Z.* 103:361-368.

Koganezawa, H. and Yanase, H. 1990. A new type of elongated virus isolated from apple trees containing the stem pitting agent. *Plant Dis.,* 74: 610–614.

Komorowska, B. and Cieœliñska, M. 2008. Detection and characterization of some isolates of cherry green ring mottle virus. *Acta Hort.* (ISHS) 781: 65-70

Kummert, J., Marinho, V.L.A., Rufflard, G., Colinet, D. and Lepoivre, P. 1998. Sensitive detection of apple stem grooving and apple stem pitting viruses from infected apple trees by RT-PCR. *Acta Hort.* 472: 97-104.

Kundu, Jiban K. 2003. The occurrence of apple stem pitting virus and apple stem grooving virus within field – grown apple cultivars evaluated by RT-PCR. *Plant Protect. Sci.,* 39: 88–92.

Kunkel, L.O. 1936. Immunological studies on the peach diseases, yellows, rosette, and little peach. *Phytopathology,* 26: 201–219.

Kunkel, L.O. 1933. Insect transmission of peach yellows. *Contrib. Boyce Thompson Instit.* 5: 19-28.

Kurihara Jun, Tsuda Shin'ya, Kirita Masanobu and Tomaru Keiichi. 2005. Cucumber Mosaic virus Isolated from *Prunus mume* (Japanese Apricot) Belongs to Subgroup I but is distinct in Serology. *J. Agric. Science, Tokyo Nogyo Daigaku,* 50: 64–70.

Kyriakopoulou P.E., Giunchedi L. and Hadidi A., 2001. Peach latent mosaic and pome fruit viroids in naturally infected cul-818 Pear blister canker viroid in Turkey. *J. Plant Pathol.,* 92: 813-819.

LaRosa, K.H., Hammond, R., Crosslin, J.M., Hazel, C. and Hammerschlag. 1995. Monitoring *Prunus* necrotic ringspot virus infection by hybridization with a cRNA Probe after *in vitro* Micrografting. *J. Amer. Soc. Hort. Sci.,* 120: 928–931.

Larsen H.J. 1997. Impact and control of peach mosaic in North America. *In*: 17th International Symposium on Virus and Virus-like diseases of Temperate Fruit Crops, ISHS, Bethesda (US). p. 725

Larsen H.J. and Oldfield G.N. 1995. Peach mosaic. *In*: Compendium of Stone Fruit Diseases. APS, St Paul (US).

Lee, I.M., Hammond, I.W., Davis, R.E. and Gunderson, D.E. 1993. Universal Amplification and analysis of pathogen 16S rDNA for classification and identification of Mycoplasma-like organisms. *Phytopathology,* 83: 834–842.

Lee, I.M., Zhu, S., Gundersen, D.E., Zhang, C. and Hadidi, A. 1995. Detection and identification of a new phytoplasma associated with cherry lethal yellows in China. *Phytopathology,* 85: 1179.

Lee I.M., Gundersen-Rindal D.E., Davis R.E., Bartoszyk I. 1998. Revised classification scheme of phytoplasmas based on RFLP analyses of 16S rRNA and ribosomal protein gene sequences. *Internl. J. Syst. Bacteriology,* 48: 1153–1169.

Legrand, G. and Verhoyen, M. 1985. Détection du little cherry disease chez le cerisier par des observations au microscope à lumiére U.V. *Mededelingen van de Faculteit Landbouwwetenschappen Rijksuniversiteit Gent,* 50: 1283–1288.

Legrand, G. and Verhoyen, M. 1986. Use of different methods to detect little cherry disease in ornamental cherry trees. *Acta Hort.* 193: 283–289.

Lenz, O., Petrzik, K. and Spak, J. 2009. The Microarray detecting six fruit tree viruses. *In*: 21st Internl. Conf. on virus and other graft transmissible fruit crops. *Berichte aus dem JKI* 148 p. 27.

Leone, G., Lindner, J.L., Van der Meer, F.A. and Schoen, C.D. 1998. Symptoms on apple and pear indicators after back transmission from *Nicotiana occidentalis* confirm the identity of apple stem pitting virus with pear vein yellowing virus. *Acta Hort.,* 472: 61–65.

Lesemann, D.E., Kunze, L. Krischke, G. and Koenig, R. 1989. Natural occurrence of carnation Italian ringspot virus in cherry tree. *J. Phytopathol.,* 124: 171–174.

Levy, L., Damsteegt, V. and Welliver, R. 2000. First report of Plum pox virus (Sharka disease) in *Prunus persica* in United States. *Plant Dis.,* 84: 202.

Li., R. and Mock, R. 2008. Characterization of flowering cherry strain of cherry Necrotic rusty mottle virus. *Archives of Virology* 155: 973–975.

Liacer, G. 1995. Virus, Viroid and Mycoplasma diseases of apricot. *Acta Hort.* (ISHS), 384: 511–520.

Llácer, G. 1998. General aspects of peach latent mosaic disease. *Acta Hort.* (ISHS), 472: 561–564.

Lindner, R.C., Weeks, T.E. and Kirkpatrick, H.C. 1949. A qualitative chemical test for some stone fruit virus diseases. *Phytopathology,* 39: 1059–1060.

Lindner, R.C., Kirkpatrick, H.C and Weeks, T.E. 1950. A simple staining technique for detecting virus diseases in some woody plants. *Science* 112: 119–120.

Lister, R.M. 1970a. Apple stem grooving virus. C.M.I./A.A.B. Descriptions of Plant Viruses. No. 31. Commonw. Mycol. Inst./Assoc. Appl. Biol., Kew, Eng.

Lister, R.M. 1970b. Apple stem grooving virus. *In*: Brunt, A.A., Crabtree, K., Dallwitz, M.J., Gibbs, A.J., Watson, L. (*Eds.*) Viruses of Plants, Wellingford, U.K., CAB International, pp. 105–198.

Lister, R.M. 1986. Plant viruses on line. Description and list from the VIDE database. Apple stem grooving virus.

Lister, RM and Bar-Joseph, M. 1981. Closteroviruses. *In*: Handbook of Plant Virus Infections and Comparative Diagnosis, *Ed*. Kurstak. E. Elsevier/North- Holland, Biomedical Press, Amsterdam. p. 809

Lister, R.M., Bancroft, J.B. and Nadakavukaren, M.J. 1965. Some sap-transmissible viruses from apple. *Phytopathology,* 55: 859–70.

Loesch, L.S. and Fulton, R.W. 1975. Prunus necrotic ringspot virus as a multicomponent system. *Virology,* 68: 71–78.

Loi, N., Carraro, L., Musetti, R., Pertot, I. and Osler, R. 1995. Dodder transmission of two different MLOs from plum trees affected by leptonecrosis. *Acta Hort.,* 386: 465–470.

Lolic, B., Afechtal, M., Matic, S., Myrta, A., Di Serio, F. 2007. Detection by tissue-printing of pome fruit viroids and characterization of Pear blister canker viroid in Bosnia and Herzegovina. *J. Plant Pathol.,* 89: 369–375.

Lorenz, K.H., Dosba, F., Poggi-Pollini, C., Llácer, G. and Seemüller, E. 1994. Phytoplasma diseases of *Prunus* species in Europe are caused by genetically similar organisms. *Zeitschrift für Pflanzenkrankheiten und Pflanzenschutz* 101: 567–575.

Lorenz, K.H., Schneider, B., Ahrens, U. and Seemuller, E. 1995.Detection of the apple proliferation and pear decline phytoplasmas by PCR amplification of ribosomal and nonribosomal DNA. *Phytopathology,* 85: 771–776.

Loreti, S., Faggioli, F. and Barba, M. 1997. Identification and characterization of an Italian isolate of pear blister canker viroid. *J. Phytopath.* 145: 541–544.

Loreti, S., Faggioli, F., Cardoni, M. Mordenti, G., Babini, A.R., Poggi Pollini, C.M. and Barba, M. 1999. Comparison of different diagnostic methods to detect peach latent mosaic viroid. *OEPP/EPPO Bull,* 29: 433–438

Lovisolo, O., Accotto, G.P., Masenga, V. and Colariccio, A. 1993. An isolate of apple stem grooving virus associated with Cleopatra mandarin's fruit intumescence. *Fitopatol. Bras.* Vol. 28, N0.1.

Lott, T.B. and Keane, F.W.L. 1960. Twisted leaf and ring pox viruses found in chokecherries near diseased orchards. Twisted leaf virus indigenous in chokecherry. *Plant Dis. Reptr.,* 44: 326–7: 328–30.

Luepschen, N.S., Harder, H.H., Rohrbach, K.G. and Sisson, K.G. 1974. Sweet cherry rasp leaf incidence in Colorado. *Plant Dis. Reptr.,* 58: 26–27.

MacBeath, J.H., Nyland, G. and Spurr, A.R. 1972. Morphology of mycoplasma-like bodies associated with peach X-disease in *Prunus persica. Phytopathology* 62: 935–937.

Malfitano M., Barone M., Ragozzino E., Alioto D. and Flores R., 2004. Identification and molecular characterization of Pear blister canker viroid isolates in Campania (Southern Italy). *Acta Hort.,* 657: 367–371.

Maliogka, V.I., Charou, A., Efthimiou, K., Katsiani, A.T., Chatzivassiliou, E.K. and Katis, N.I. 2010. Identification of ilarviruses in almond and cherry fruit trees. Using nested PCR assays. *In*: Proc. 21st Interl. Conf. Virus and other graft transmissible diseases of fruit crops. *Julius-Kuhn-Archiv.,* 427: 281–283

Malisano, G., Firrao, G., and Locci, R. 1996. 16S rDNA derived oligonucleotide probes for differential diagnosis of plum leptonecrosis and apple proliferation phytoplasmas. *EPPO Bull.,* 26: 421–428.

Mallach, N. 1957. Ubertragbarkeit (Im Sinone einer virose) der Weissflecken krankheit Der kirsche. *Pflanzenschutz,* 9: 116–117.

Marais, A., Faure, C., Bergey, B., Gentit, P. and Candresse, T. 2012. Characterization of divergent plum bark necrosis stem pitting-associated virus isolates and development of a polyvalent PBNSPa-specific detection test. *In*: Proc. 22nd Interl. Conf. Virus and other graft transmissible diseases of fruit crops. Held in Rome June 3-8, 2012. Abs. p. 45.

Marcone, C. and Ragozzino, A. 1994. [Peach rosette: A new disease associated with mycoplasma-like organisms (MLOs)]. *Informatore Agrario,* 50: 71–74.

Marcone, C., Lee, I.M., Davis, R.E., Ragozzino, A. and Seemüller, E. 2000. Classification of aster yellows-group phytoplasmas based on combined analyses of rRNA and tuf gene sequences. *Internl. J. System. Evolu. Microbiol.* 50: 1703-1713.

Marenaud, C. and Mazy, K. 1976. Observations sur la diffusion et la detection du Vurusde la sharka. *Phytoma,* 280: 20-24.

Marini, D.B., Zhang, Y.P., Rowhani, A. and Uyemoto, J.K. 2002. Etiology and host range of a closterovirus associated with plum bark necrosis-stem pitting disease. *Plant Dis.,* 86: 415–417.

Martelli, G.P. and Savino, V. 1997. Infectious diseases of almond with special reference to the Mediterranean area. *EPPO Bull.,* 27: 525–534.

Martelli, G.P., Candresse, T. and Namba, S. 1994. Trichovirus, a new genus of plant viruses. *Arch. Virol.,* 134: 451–455

Maryam, G. Z. 2011. Phytoplasmas associated with *Almond* witches' broom disease: An overview. *African J. Microbiol. Res.* 5: 6013-6017

Más, P., Sánchez-Navarro, J.A., Sánchez-Pina, M.A., Pallás, V. 1993. Chemiluminescent and colorigenic detection of cherry leaf roll virus with digoxigenin-labeled RNA probes. *J. Virol. Methods*, 45: 93–102.

MavrièPleško, I.,ViršèekMarn, M., Miladinoviè, Z. and Zindoviè, J. 2012. First Report of *Peach latent mosaic viroid* in Peach Trees in Montenegro. *Plant Dis.*, 96: 150.

McClintock, J.A. 1923. Peach rosette, an infectious mosaic. *J. Agr. Res.*, 24: 307–316.

McClintock, J.A., Kunkel, LO. and Thornberry, H.H. 1951. Peach rosette mosaic. USDA Handbook No.10. p. 276.

Migdoll, Alexander M., seemuler, Erich, Reinhardt, Richard, kube, Michael. 2009. First insight into the genome of the rich equipped acholeplasma species highlight the genome condensation to the related phytoplasmas. *In*: 21st Internl. Conf. on virus and other graft transmissible fruit crops. *Berichte aus dem JKI* 148, p. 43.

Mink, G.I.. 1982. The possible role of honeybees in long distance spread of Prunus necrotic ringspot virus from California into Washington sweet cherry orchards. pp. 85–91. *In*: Plant Virus Epidemiology (ed. By R.T. Plumb & J.M.Thresh), Blackwell Scientific Publications, Oxford.

Mink, G.I. 1992. Prunus necrotic ringspot virus. *In*: E.J.Kumar, H.S. Chaube, U.S. Singh, A.N. Mukhopadhyay (Eds.). Plant diseases of international importance. Vol. 3. Diseases of fruit crops. Prentice Hall, Englewood Cliffs, N.J. p. 335-356.

Mink, G.I., Shay, J.R., Gilmer, RM and Stouffer, R.F. 1971. Latent viruses in apple II symptoms in woody indicators and strain variation. *Search Agric.*, 1: 9–15.

Mink, G. I., Howell, W. E., Cole, A. and Regev, S. 1987. Three serotypes of Prunus necrotic ringspot virus isolated from rugose mosaic-diseased sweet cherry trees in Washington. *Plant Dis.* 71: 91-93.

Mircetich, S.M. and Rowhani, A. 1984. The relationship of cherry leafroll virus and blackline disease of English walnut trees. *Phytopathology*, 74: 423–428

Mishra, M.D. 1966. Plum leaf roll – a new virus disease in India. In: Proc. 53rd Session Indian Sci. Cong. Chandigarh (Abs), p. 429.

Miura, C., Keim, T., Watanab, T., Nij, T., Maejim, K., Oshimad, K. and Namba, S. 2012. Detection of two strains of "*Candidatus* Phytoplasma asteris" from peach and apricot using loop-mediated isothermal amplification. *In*: Abs. 22nd Interl. Conf. Virus and other graft transmissible diseases of fruit crops. Held in Rome June 3-8, 2012. p. 167.

Monticelli, S., DiNicola-Negri, E., Gentile, A.Damiano, C. and Ilardi, V. 2012. Production and *in vitro* assessment of transgenic plums for the resistance to plum pox virus. *In*: Proc. 22nd Interl. Conf. Virus and other graft transmissible diseases of fruit crops. Held in Rome June 3- 8, 2012 p. 73.

Moorman, G. W. and Gildow, F.E. 2007. Plum pox virus. *In*: Ornamentals. Penn. State Agri. Sci. Extension Plant Disease Fact Sheets.

Morvan, G. and Castelain, C. 1975. New evidences of the high sensitivity of *Malus dawsoniana* to apple proliferation disease and of its value for testing apple trees. *Acta Hort.*, 44: 175–182.

Nagaich, B.B. and vashisth, K.S. 1962. On the occurrence of virus diseases of temperate fruits in India. *Indian Phytopath.*, 15: 222–224.

Nagaich, B.B. and Vashishth, K.S. 1965a. Additional viral diseases of temperate fruits in Shimla hills. *Indian Phytopath.* 18: 288–290.

Nagaich, B.B. and Vashishth, K.S. 1965b. Observations on the viral diseases of plum in India. *Indian J. Agric. Sci.* 35: 157–161

Nasu, S., Jensen, D.D. and Richardson, J. 1970. Electron microscopy of mycoplasma-like organisms associated with insect and plant hosts of peach western X-disease. *Virology* 41: 583–595.

Navarro, L., Llacer, G., Cambra, M., Arregui, J.M. and Juarez, J. 1982. Shoot-tip grafting *in vitro* for elimination of viruses in peach plants (*Prunus persica* Batsch). *Acta Hort.*, 130: 185–191.

Ne,H., Yansu, S., Xiaofang, M., Guojun, H., Liping, W. and Guoping, W. 2012. Incidence of viral diseases of pear plants and the molecular Characterization of three pear viruses in China. *In*: 22nd Internl. Conf.on virus and other transmissible diseases of fruit crops. Held in Rome, p. 24.

Negi, A., Kumar, Y., Hallan, V and Zaidi, A.A. 2008. Molecular characterization of apple stem grooving virus from pome and stone fruits in India. *In*: XVIII Nat. Conf. Indian virol. Soc. December 11-13, 2008, PGIMER. Chandigarh.

Nemchinov, L., hadidi, A., Foster, J.A., candresse, T. and Verderevskaya, T. 1995. Sensisitive detection of apple chlorotic leaf spot virus from infected apple or peach tissue using RT-PCR, IC-RT-PCR, or multiplex IC-RT-PCR. *Acta Hort.* (ISHS), 386: 51–62

Nemeth, M. 1986. Virus, mycoplasma and rickettsia diseases of fruit trees. Martinus Nijhoff publishers, Dordrecht, The Netherlands. p. 840

Nemeth, M. and Kolber, M. 1983. Additional evidence on seed transmission of plum pox virus in apricot, peach, and plum, proved by ELISA. *Acta Hort.*, 130: 293–300.

Nickle, O., Fajardo, T.V.M., Jelkmann, W. and Kuhn, G.B. 2001. Sequence analysis of the capsid protein gene of an isolate of apple stem grooving virus, and its survey in Southern Brazil. *Fitopatologia Brasileira,* 26: 655–659.

Novak, J.B. and Lanzova, J.1980. Some diseases of fruit trees in which the tomato bushy stunt virus occurs and new natural hosts of this virus. *Acta Phytopathol. Acad. Sci. Hungary*, 15: 323–327

Nyland, G., Gilmer, R.M., and Moore, J.D. 1976. "Prunus" ringspot group. *In*: Virus diseases and non-infectious disorder of stone fruits in North America (Gilmer, R.M., Moore, J.D. and Pine, T.S. *Eds.*) Agriculture handbook No. 437 USDA Washington D.C. pp. 104–132.

Nyland, G. 1976. Cherry rasp leaf. *In*: Virus diseases and non-infectious disorders of stone fruits in North America. Agriculture Handbook No. 437, US Department of Agriculture, Washington D.C., USA. p.219-221.

OEPP/EPPO. 1978. EPPO data sheets on quarantine organisms No. 95, Pear decline (mycoplasma). *OEPP/EPPO Bulletin,* 8: 2.

OEPP/EPPO. 1984. Data sheets on quarantine organisms No. 127, Cherry rasp leaf virus. *OEPP/EPPO Bulletin* 14: 5–7.

OEPP/EPPO. 1991/1992. Certification schemes. Virus-free or virus-tested fruit trees and rootstocks. *OEPP/EPPO Bulletin,* 21: 267–278; 22: 253–284.

OEPP/EPPO 1994. Quarantine procedures No. 57. MLOs in fruit trees and grapevine. *OEPP/EPPO Bulletin,* 24: 339–342.

Oldfield, G.N. 1970. Mite transmission of plant viruses. *Annu. Rev. Entomol.,* 15: 343–380.

Oldfield, G.N., Creamer, R., Gispert, C., Osorio, F., Rodriguez, R. and Perring, T.M. 1995. Incidence and distribution of peach mosaic and its vector, *Eriophyes Insidiosus* (Acari: Eriophyidae) in Mexico. *Plant Dis.,* 79: 186–189.

Oliver, J.E., Freer, J., Andersen, R.L., Cox, K.D., Robinson, T.L. and Fuchs, M. 2009. Genetic diversity of *Prunus necrotic ringspot virus* isolates within a cherry orchard in New York. *Plant Dis.,* 93: 599–606.

Olmos, A., Cambra, M., Dasi, M. A., Candresse,T., Esteban, O., Gorris, M.T. and Asensio, M. 1997. Simultaneous detection and typing of plum pox potyvirus (PPV) Isolates by heminested-PCR and PCR-ELISA. *J. Virol. Methods,* 68: 127-137.

Paltrinieri S., Martini M., Stefani E., Pondrelli M., Bertaccini A. and Fideghelli C. 2001. Phytoplasma infection in peach and cherry in Italy. *Acta Hort.,* 550: 365–370.

Pappu, H.R. 2009. Virus resistant transgenic plants: from research and development to commercialization. *In*: 5th Interl. Conf. Plant Pathol. In the globalized era. Nov.10-13, 2009, New delhi, India. p. 39.

Parakh, D.R., Shamloul, A.M., Hadidi, A., Scott, S.W., Waterworth, H.E. Howell, W.E. and Mink, G.I. 1995. Detection of prune dwarf ilarvirus from infected stone fruits using reverse transcription-polymerase chain reaction. *Acta Hort.* 386: 421–430.

Parish, C.L. 1976. A comparison between the causal agents of cherry rasp leaf and flat apple diseases. *Acta Hort.* (ISHS), 67: 199–202.

Parish, C.L.1977. A relationship between flat apple disease and cherry rasp leaf disease. *Phytopathology,* 67: 982–984.

Parker, K. G., Fridland, P. R. and Gilmer, G. M. 1976. Green ring mottle. In-Virus Diseases and Noninfectious Diseases of Stone Fruit in North America. USDA, Agriculture Handbook no. 437, Washington, DC, USA, pp. 193–199.

Pastore, M., Pennone, F., Marcone, C., Cristinzio, G. and Ragozzino, A. 1995. Detection of Mycoplasma-like Organisms (MLOs) in apricot by fluorescence microscopy (D.A.P.I.) in the South of Italy. *Acta Hort.* (ISHS) 386: 506–510.

Paulsen, A.Q. and Fulton, R.W. 1968. Hosts and properties of a plum line pattern virus. *Phytopathology* 58: 766–772.

Pfeilstetter, E., Zinkernagel, V., and Kunze, L. 1992. Occurrence of petunia asteroid mosaic (PAMV) and carnation Italian ringspot (CIRV) viruses in cherry orchards in northern Bavaria. *Acta Hort.,* 309: 345–353.

Pine, T.S. 1964. Influence of necrotic ringspot virus on growth and yield of peach trees. *Phytopathology* 54: 604–605.

Pine, T.S. 1976. Peach mosaic. *In*: Virus Diseases and Noninfectious Disorders of Fruits in North America. USDA Agriculture Handbook No. 437. pp. 61–70.

Pine, T.S. and Gilmer,R.M. 1976. Peach Yellows. *In*: Diseases and non-infectious disorders of stone fruits in North America. (R.M. Gilmer, J.D. Moore, G. Nyland, pp. 91–95.

Plese, N., Rilovic, M. and Weischer,M. 1969. Novi domadari intracelularne inkluzije virus sarke. *Zastita Bilja,* 104: 143–150.

Poggi Pollini, C., Giunchedi, L. and Bissani, R. 1997. Specific detection of D- and M-Isolates of plum pox virus by immunoenzymatic determination of PCR products. *J. Virol. Methods,* 67: 127–133.

Poggi Pollini, C., Bissani, R. and Giunchedi, L. 2001. Occurrence of European stone fruit yellows phytoplasma (ESFYP) infection in peach orchards in northern-central Italy. *J. Phytopathol.,* 149: 725–730.

Poggi Pollini, C., Giunchedi, L., Seemuller, E., Filippini, G. and Vindimian, G. 1995, Etiological studies of apple rubbery wood disease," XVI Interl. Symp. on Fruit Tree Virus Diseases. *Acta Hort.,* 386: 503-505.

Poggi Pollini, C., Bissani, R. and Giunchedi, L. 1995. Overwintering of pear decline agent in some quince rootstocks. XVI International Symposium on Fruit Tree Virus disease *Acta Hort* (ISHS), 386: 496–499.

Posnette, A.F. 1956. The leaf roll virus of sweet cherry. *Euro. J. Plant Pathol.* 62: 49–50.

Posnette, A. F. 1957. Virus diseases of pears in England. *J. Hort. Sci.* 32: 53–61.

Posnette, A. F. and Cropley, R. 1956. Apple mosaic viruses. Host reaction and strain interference. *J. Hort. Sci.* 31: 119–133.

Posnette, A. F. and Cropley, R. 1964. Necrotic rusty mottle virus disease of sweet cherries in Britain. *Plant Pathology,* 13: 20–22.

Posnette, A.F. and Cropley, R. 1965. Field experiments with chat fruit virus disease of Apple. *Ann. Appl. Biol.,* 55: 439–445.

Posnette, A.F. and Ellenberger, C.E.1957. The line-pattern virus disease of plums. *Ann. appl. Biol.* 45: 74–80.

Postman, J. D., and Hadidi, A. 1995. Elimination of apple scar skin viroid from pears By *in vitro* thermotherapy and apical meristem culture. *Acta Hort.* 386: 536–543.

Quacquarelli, A. and Savino, V. 1977. Cherry leaf roll virus in walnut (*Juglans regia*) II. Distribution in Apulia and transmission through seed. *Phytopatol. Medit.,* 16: 154–156.

Quacquarelli, A., Gallitelli, D., Savino, V. and Piazzolla, P. 1981. The use of electric current (r.a.c.e.) for obtaining mosaic-free almonds. *Acta Hort.* (ISHS) 94: 251–256.

Ragetti, H.W.J., Elder, M. and Schroeder, B.K. 1982. Isolation and properties of filamentous viruslike particles associated with little cherry disease in *Prunus avium. Canadian J. Botany,* 60: 1235–1248.

Ragozzino, E., Faggioli, F., Alioto, D. and Barba, M. 2003. Detection and differentiation of peach latent mosaic viroid and hop stunt viroid in stone fruit trees in Italy using multiplex RT-PCR. *Phytopatol. Medit.,* 42: 79–84.

Rahardja, U., Whalon, M.E., Garcia-Salazar, C. and Yan, Y.T. 1992. Field detection of X-disease mycoplasma-like organism in *Paraphlepsius irroratus* using a DNA probe. *Environ. Entomology,* 21: 81–88.

Raine, J., Weintraub, M. and Schroeder, B. 1975. Flexuous rods and vesicles in leaf and petiole phloem of little-cherry diseased *Prunus* spp. *Phytopathology,* 65: 1181–1186.

Raine, J., Weintraub, M. and Schroeder, B. 1979. Hexagonal tubules in phloem cells of little cherry infected trees. *J. Ultrastruct. Res.,* 67: 109–116.

Raine, J., McMullen, R.D. and Forbes, A.R. 1986. Transmission of the agent causing little cherry disease by the apple mealybug *Phenacoccus aceris* and the dodder *Cuscuta lupuliformis. Canadian J. Plant Pathol.* 8: 6-11.

Rajan, J. and Clark, M.F.1995. Detection of apple proliferation and other MLOs by Immuno-capture PCR (IC-PCR). *Acta Hort* (ISHS), 386: 511–514.

Ralph Scorza, Michel Ravelonandro, Ann M. Callahan, John M. Cordts, Marc Fuchs, Jean Dunez and Dennis Gonsalves. 1994. Transgenic plums (*Prunus domestica* L.) express the plum pox virus coat protein gene. *Plant Cell Reports* 14: 18–22.

Ramaswamy, S. 1975. Dual infection of rounded cherry with ring mottle and european rusty mottle viruses. *Acta Hort.* (ISHS) 44: 65–68.

Ramsdell, D.C. and Myers, R.L. 1974. Peach rosette mosaic virus symptomatology and nematodes associated with grapevine "degeneration" in Michigan. *Phytopathology,* 64: 1174–1178.

Ramsdell, D.C. and Myers, R.L. 1978. Epidemiology of peach rosette mosaic virus in a Concord grape vineyard. *Phytopathology,* 68: 447–450.

Rana, T., Chandel, V., Hallan, V. and Zaidi, A.A. 2007. Molecular evidence for apple chlorotic leaf spot virus on wild and cultivated apricot in Himachal Pradesh, India. *J. Plant Pathol.,* 89(3): S69–S76.

Rana, T.,Chandel, V.,T., Hallan, V. Thakur, P.D., Handa, A and Zaidi, A.A. 2007. Molecular evidence of apple chlorotic leaf spot virus in Himachal Pradesh. *Indian J. Virol* 18: 70-74.

Rana,T., Chandel, V., Hallan, V. and Zaidi, A.A. 2008. Wild Himalayan cherry (*Prunus cerasoides*) as new host of apple chlorotic leaf spot virus. *Path.*, 38: 73–77.

Rana,T., Chandel, V., Hallan, V. and Zaidi, A.A. 2008. *Cydonia oblonga* as reservoir of apple chlorotic leaf spot virus in India. *Plant Pathology*, 57: 393.

Rana,T., Chandel, V., Hallan, V. and Zaidi, A.A. 2008. Characterisation of apple chlorotic leaf spot virus infecting almonds in India. *Australian Plant Disease Notes*, 3: 65–67.

Rana,T., Chandel, V., Hallan, V. and Zaidi, A.A. 2009. Molecular evidence for the presence of apple chlorotic leaf spot virus in infected peach trees in India. *Scientia Horticulturae* 120: 296–299.

Raquel, H., Silva, C., Batista, R., Margarida Santos, A., Lourenço, T., Mascarenhas, M., Nolasco, G. and Oliveira, M.M. 2002. Virus diseases in Portuguese almond trees: Diagnosis, *in situ* detection and genetic engineering for virus resistance. *Acta Hort. (ISHS)*, 591: 581–584

Raquel, H., Lourenço, T., Moita, C., Costa, M., Silva, S. and Oliveira, M.M. 2005. Strategies to introduce resistance to prune dwarf virus in almond. *In*: Oliveira M.M. and Cordeiro, V. (*Eds.*) . XIII GREMPA Meeting on almonds and pistachios = XIIIème Réunion du GREMPA *sur 'amandier et le pistachier . Zaragoza : CIHEAM-IAMZ*, 3: 385-389

Ravelonandro, M., dunez, J., Labonne, G. and Scorza, R. 1998. Challenging transgenic plums expressing potyvirus coat protein genes with viruliferous aphids. *Acta Hort.* 472: 413-420.

Raychaudhuri, S.P. and Maramorosch, K. 1988. Mycoplasma diseases of crops: Basic and applied aspects / Karl Maramorosch, S.P. Raychaudhuri, editors Springer-Verlag, New York

Rebenstorf, K., Candresse, T., Dulucq, M.J., Büttner, C. and Obermeier, C. 2006. Host species- dependent population structure of a pollen-borne plant virus, Cherry leaf roll virus. *J. Virol.*, 80: 2453–2462.

Reeves, E.L. 1951. Rusty mottle. *In*: Virus diseases and other disorders with virus like symptoms of stone fruits in North America. Agriculture handbook No. 10, USDA Washington D.C. pp. 112–115.

Reeves, R.M. 1951. Histochemical tests for polyphenols in plant tissues. *Stain Technol.* 26: 91–96

Rosenberger, D.A. and Jones, A.L. 1977. Symptom remission in X-diseased peach trees as affected by date, method, and rate of application of oxytetracycline-HCl. *Phytopathology* 67: 277–282.

Rosenberger, D.A. and Jones, A.L.1977. Seasonal variation in infectivity of inoculum from X-diseased peach and chokecherry plants. *Plant Dis. Reptr.* 61: 1022–1024.

Rosemarie W. Hammond and James M. Crosslin. 1998. Virulence and molecular polymorphism of Prunus necrotic ringspot virus isolates. *J. Gen. Virol.* 79: 1815–1823.

Rott, M.E. and Jelkmann W. 2001. Complete nucleotide sequence of cherry necrotic rusty mottle virus. *Archives of Virology* 146:395-401

Roy, A. and Ramachandran, P. 2003. Occurrence of a hop stunt viroid (HSVd) variant in yellow corky vein disease of citrus in India. *Curr. Sci.*, 85: 1608–1612.

Roy, A.S. and Smith, I.M. 1994. Plum pox situation in Europe. *EPPO Bull.* 24: 515–523.

Rudaina, H, Alrefai, Shiel Patrick, J., Leslie, L., Cleora, J., D'Arcy, Berger Philip, H. and Korban, S.S. 1994. The nucleotide sequence of apple mosaic virus coat protein gene has no similarity with other Bromoviridae coat protein genes. *J. Gen. Virol.* 75: 2847–2850.

Rustem, F.A. and Ahlawat, Y.S. 1999. Yellow corky vein – a viroid disease of citrus in India. *Indian J, Agric. Sci.* 69: 789–794

Rybicki, E.P. 1995. *Bromoviridae* family. *Archives of Virology, Supplement,* 10: 450–457.

Rysanek, G.P. 2012. Apricot latent virus – *Review. Hort. Sci.*(Prague) 39:144-148.

Šaric, A. and Velagic, Z. 1981. Viruses and virus-like diseases of maraska sour cherry (*Cerasus acida* ssp. maraska). *Acta Hort.* (ISHS) 94: 367–370

Sastry, K.S. 1966. Yellow vein mosaic of *Rosa bourbiana. Indian Phytopath.* 19: 316.

Schlocker, A. and Tayler, J.A. 1976. Yellow bud mosaic. *In*: Virus Diseases and Noninfectious Disorders of Stone Fruits in North America (R.M. Gilmer, J.D. Moore, G. Nyland, M.F. Welsh and T.S. Pine eds.) Agriculture Handbook No. 437, USDA, Washington, DC. USA pp. 156–165.

Schmelzer, K. 1966. Ultersuchungen an viren der zier-und wildgeholze 5. Mitteilung: Virosen an populous und sambucus. *Phytopath. Z.* 55: 317–351.

Schmitt, R.A., Williams, H. and Nyland, G. 1977. Virus diseases can decrease peach yields. *Cling Peach Quarterly* 13:17-19.

Schneider, B., Seemüller, E., Smart, C.D. and Kirkpatrick, B.C. 1995. Phylogenetic classification of plant pathogenic mycoplasma-like organisms or phytoplasmas. *In*: Molecular and diagnostic procedures in mycoplasmology. Vol-1, Razin, S. and Tully, J.G. (*Eds.*). Academic Press, San Diego, CA. pp. 369–380

Schneider, B., seemuller, E. and Jarauswch, W. 2009. Strain differentiation of *Candidatus* Phytoplasma mali by SSCP and sequence analysis of the HFLB gene. *In*: 21st Internl. Conf. on virus and other graft transmissible fruit crops. *Berichte aus dem JKI* 148: 45

Schwarz, K. and Jeikmann, W. 1998. Detection and characterization of European apple stem pitting virus sources from apple and pear by PCR and partial sequence analysis. *Acta Hort.* 472: 75–85.

Scorza, R., Callahan, A.M., Levy, L., Damsteegt, V., and Ravelonandro, M. 1998. Transferring potyvirus coat protein genes through hybridization of transgenic plants to produce plumpox resistance in plums (*Prunus domestica* L.) *Acta Hort.* 472: 421–428.

Scott, S.W. and Zimmerman, M.T. 2001. Peach rosette, little peach, and red suture are diseases induced by a phytoplasma closely related to western X-disease. *Acta Hort.* (ISHS), 550: 351–354.

Scott, S.W. and Zimmerman, M.T. 2001. American plum line pattern is a distinct ilarvirus. *Acta Hort.*, 550: 221–228.

Scott, S.W., Barnett, O.W. and Baker, R.B. 1986. Some facets of the ecology of Prunus necrotic ringspot virus in peach trees in South Carolina. Proc. Workshop Epidem; Plant Virus Dis., Orlando, Fla. p. 43.

Scott, S.W., Bowman-Vance, V. and Buchman, E.J. 1992. The use of nucleic acid probes for the detection of Prunus necrotic ringspot virus and Prune dwarf virus. *Acta Hort.* 309: 79-83

Scurfield, G. and Reinganum, C.1964. Some anatomical and chemical effects of "flat limb" virus on apple var. Gravenstein. *Australian J. agric. Res.*, 15: 548–559.

Seemuller, E.1990. Apple proliferation. PP. 66-67. In: Compendium of apple and pear diseases APS, St. Paul. MN. USA.

Seemüller, E.and Schneider, B. 2004. 'Candidatus Phytoplasma mali', '*Candidatus* Phytoplasma pyri' and '*Candidatus* Phytoplasma prunorum', the causal agents of apple proliferation, pear decline and European stone fruit yellows, respectively. *Int J Syst Evol Microbiol.* 54: 1217–1226.

Seemüller, E. and Foster, J.A. 1995. European stone fruit yellows. *In*: Compendium of stone fruit diseases. American Phytopathological Society, St. Paul, USA. pp. 59–60.

Seemuller, E., bisognin, C., grando, S., Schnieder, B., Velasco, R. and Jarausch, W. 2009. Breeding of rootstocks resistant to apple Proliferation diseases. *In*: 21st interl. Conf. on virus and other graft transmissible fruit crops. *Berichte aus dem JKI* 148. (Abs) p. 46.

Seemüller, E., Kison, H., Lorenz, K.-H., Schneider, B., Marcone, C., Smart, C.D. and Kirkpatrick, B.C. 1998. Detection and identification of fruit tree phytoplasmas by PCR amplification of ribosomal and nonribosomal DNA. COST 823: New Technologies to improve phytodiagnosis. Advances in the detection of plant pathogens by polymerase chain reaction. pp. 56-66.

Seneviratne, S.N. and Posnette, A.F. 1970. Identification of viruses isolated from plum trees affected by decline, line pattern and ringspot diseases. *Annl. Appl. Biol.* 65: 115–125.

Shamloul, A.M.,Minafra, A., Hadidi, A., Guinchedi, L, Waterworth, H.E. and Allam, E.K. 1995. Peach latent mosaic viroid: nucleotide sequence of an Italian isolate, sensitive detection using RT-PCR and geographic distribution. *Acta Hort.,* 386: 522–530.

Singh, R.M., Ram, R., Hallan, V., and Zaidi, A.A. 2008. Elimination of apple mosaic virus and apple chlorotic leafspot virus from apple rootstocks by tissue culture and chemotherapy. *In*: XVIII National Conference of Indian Virological Society, December 11-13, 2008. PGIMER, Chandigarh.

Sinha, R.C. 1988. Serological detection of mycoplasma-like organisms from plants affected with yellows diseases. *In*: Tree mycoplasmas and mycoplasma-like diseases (*Ed*. by Hiruki, C.), University of Alberta Press, Edmonton, Canada. pp. 143–156.

Sinha, R.C. and Chiykowski, L.N. 1984. Purification and serological detection of mycoplasma-like organisms from plants affected by peach eastern X-disease. *Canadian J. Plant Pathol.,* 6: 200–205.

Sinha, R.C. and Chiykowski, L.N.1986. Detection of mycoplasma-like organisms in leafhopper vectors of aster yellows and peach X-disease by immunosorbent electron microscopy. *Canadian J. Plant Pathol.,* 8: 387–393.

Sipahioglu, H.M., Usta, M. and Ocak, M. 2007. Occurrence of Cherry green ring mottle virus in Turkey. *New Dis. Reports* 15:4

Slack, D.A. and Kim, K.S. 1977. Peach rosette disease. *Arkansas Farm Research,* 26: 15.

Slykhuis, J.T., Yorston, J., Raine, J., McMullen, R.D. and Lis, T.S.C. 1980. Current status of little cherry disease in British Columbia. *Canadian Plant Dis. Survey,* 60: 37–42.

Smart, C.D., Schneider, B., Blomquist, C.L., Guerra, L.J., Harrison, N.A., Ahrens, U., Lorenz, K.H., Seemüller, E. and Kirkpatrick, B. 1996. Phytoplasma-specific PCR primers based on sequence of the 16-23S rRNA spacer region. *Appl. Environ. Microbiol.* 62: 2988-2993.

Smith, E. F. 1891. The peach rosette. *J. Mycol.* 6: 143-148.

Smith, K.M. 1972. A textbook of plant virus diseases (3rd edition), pp. 390-392. Longman, London, UK.

Smith, I.M., McNamara, D.G., Scott, P.R. and Harris, K.M. (Eds). 1992. Quarantine pests for Europe. Data sheet on quarantine pests for rhe European Communities and for the European and Mediterranean Plant Protection Organization. CAB International Wallingford, UK, and European and Mediterranean plant protection organization, Paris, France.

Spiegel, S., Rosner, A., Tam, Y., Zilkah, S., Faingersh, E., Rotbaum, A. and Krizbai, L. 1998. Detection of prune dwarf virus in sweet cherry in israel. *Acta Hort.* (ISHS) 472: 249–256.

Stace-Smith, R. 1984. Tomato ringspot virus. CMI?AAB Description of plant viruses No. 290. Commonwealth Agricultural Bureaux, Kew, UK.

Stace-Smith, R. 1985. Tobacco ringspot virus. AAB Description of plant viruses No. 309. Association of applied Biologists, Wellesbourne, Warwick, UK.

Stace-Smith, R. and Hansen, A.J. 1974. Occurrence of tobacco ringspot virus in sweet cherry. *Can. J. Bot.*, 52: 1647–1651.

Stace-Smith, R. and Hansen, A.J. 1976. Cherry rasp leaf virus. *CMI/AAB Descriptions of Plant Viruses* No. 159. Association of Applied Biologists, Wellesbourne, UK.

Staniene, G., Stanys, V., Vinskiene, J., Abraitis, R., Jomantiene, R., Valiunas, D. and Abraitiene, A. 2009. *In vitro* culture of phytoplasma-and viroid-infected sweet cherry (*Prunus avium* L.). *Zemdirbyste* (Agriculture) 36: 129–140.

Staniulis J., Davis R.E., Jomantiene R., Kalvelyte A. and Dally E.L. 2000. Single and mixed phytoplasma infections in phyllody- and dwarf-diseased clover plants in Lithuania. *Plant Dis.* 84: 1061-1066.

Susanne von Bargen, Juliane Langer, Jenny Robel and Artemis Rumbou. 2012. Complete nucleotide sequence of *Cherry leaf roll virus* (CLRV), a subgroup C nepovirus. *Virus Research*, 163: 678–683.

Sutic', D. 1961. Assay of transmission of sharka virus disease by sap inoculation to herbaceous plants. *Tidssdrrifit for Planteavl.*, 65: 138–146.

Sutic, D and Juretic, N. 1970. Occuring of sowbane mosaic virus in plum tree. ISHS *Acta Hort.* 67: x International Symposium on fruit tree virus diseases (Abs).

Sutic', D., Jordovic', M., Rankovic', M. and Festic', H. 1971. Comparative studies of some sharka (Plum pox) isolates. *Annls. De Phytopath.*, 1: 184–192.

Sweet, J.B. 1980. Hedgerow hawthorn (*Crataegus* spp.) and blackthorn (*Prunus spinosa*) as hosts of fruit tree viruses in Britain. *Ann. appl. Biol.* 94: 83-90.

Thakur, P.D. and Handa, A. 1999. Phytoplasma diseases of temperate fruits. *In*: Diseases of Horticultural crops-Fruits (Eds) L.R. Verma and R.C. Sharma. Indus Publishing Co. New Delhi pp. 296–300.

Thakur, P.D., Bhardwaj, S.V., Garg, I.D., Sharma, D.R. and Khosla, K. 1994. Plum pox virus on stone fruits from India,-a new record. *Plant Dis. Res.* 9: 100–102.

Thakur, P.D., Ito, T. and Sharma, J.N. 1995. Natural occurrence of a viroid disease of apple in India. *Indian J. Virol* 11: 73-75.

Thakur, P.D., Handa, A., Chowfla, S.C. and Krczal, G. 1998. Outbreak of a phytoplasma disease of peach in the northwestern himalayas of india. *Acta Hort.* (ISHS) 472: 737–742

Thakur, P.D., Handa, A. and Brakta, A. 2011. *In*: Abstract of annual meeting of Indian Virology Society held at HAU, Hissar, January 29, 2011. p.52.

Thakur, P.D., Handa, A., Pramanick, K.K., Y.P. Sharma and Tomar, M. 2012. Molecular detection of latent apple chlorotic leaf spot virus in elite mother plants of apple. *Indian J. Virol.*, 23: 359–363

Thokchom, T., Rana, T., Hallan, V., Ram, R. and Zaidi, A.A. 2008. Molecular characterization of Indian strain of apple mosaic virus isolated from apple (*Malus domestica*). *In*: 18th National Conf. Indian Virol. Soc. Chandigarh, December 11-13, 2008 (Abs). P. 52

Thomas, B.J. 1984. Epidemiology of three viruses infecting the rose in the United Kingdom. *Ann. Appl. Biol.* 105: 213–222.

Thresh, J.M. 1988. Eradication as a virus disease control measure. *In*: Control of Plant Diseases: Costs and Benefits (*Ed.* Clifford, B.C. and Lester, E.), Blackwell Scientific Publications, Oxford (GB). pp. 155–194.

Tomlinson, J.A. and Walkey, D.G.A. 1967. The isolation and identification of rhubarb viruses occurring in Britain. *Ann. appl. Biol.,* 59: 415.

Topchiiska, M.L. 1982. Effect of Prunus necrotic ringspot virus (PNRV) and prune dwarf virus (PDV) on some biological properties of peach. *Acta Hort.* 130: 307–312.

Torchetti, E.M., Birch, C., Cooper, M., Masters, C., Arocha, Y., Di Serio, F. and Michelutti, R. 2012. Detection and identification of pear blister canker viroid occurring in pear trees in Canada. In: Proc.22nd Interl. Conf. Virus and other graft transmissible diseases of fruit crops. Held in Rome June 3-8, 2012. p. 25.

Uyeda, I. and Mink, G.I. 1983. Relationship among some ilarviruses: proposed revision of subgroup A. *Phytopathology,* 73: 47–50.

Uyemoto, J.K. 1989. Union aberration of sweet cherry on *Prunus mahaleb* rootstock associated with X disease. *Plant Dis.,* 73: 899–902.

Uyemoto, J.K., Welsh, M.F. and Williums, E. 1977. Pathogenicity of tobacco ringspot virus in cherry. *Phytopathology* 67:439-441.

Uyemoto, J.K., Kirkpatrick, B.C. and Cummins, J.N. 1991. Susceptibility of selected cherry clones and related species to Western X-disease. *Hort Science,* 26: 1510–1511.

Uyemoto, J.K., Asai, W.K. and Luhn, C.F. 1992. Ilarviruses: Evidence for rapid spread and effects on vegetative growth and fruit yields of peach trees. *Plant Dis.,* 76: 71–74.

Uyemoto, J.K., Bethell, R.E., Kirkpatrick, B.C., Munkvold, G.P., Marois, J.J. and Brown, K.W. 1998. Eradication as a control measure for x-disease in california Cherry orchards. *Acta Hort.* (ISHS), 472: 715–722

Valiunas, D. 2003. Identification of phytoplasmas in Lithuania and estimation of their biodiversity and molecular evolutionary relationships. Vilnius, Lithuania: Institute of Botany, Ph.D. Thesis.

Valiunas, D., Alminaite, A., Staniulis, J., Jomantiene, R. and Davis, R.E. 2001a. First report of aster yellows-related subgroup IA phytoplasma strains in carrot, phlox, sea- lavender, aconitum, and hyacinth in Lithuania. *Plant Dis.* 85: 804.

Valiunas, D., Alminaite, A., Staniulis, J., Jomantiene, R. and Davis, R.E. 2001b. First report of alder yellows phytoplasma in the Eastern Baltic Region. *Plant Dis.* 85: 1120.

Valiunas, D., Jomantiene, R., Davis, R.E. 2005. A *candidatus* phytoplasma asteris'-related phytoplasma associated with cherry little leaf disease represents a new subgroup,16SrI-Q. *Phytopathology* 95: S106.

Valiunas, D., Jomantiene, R. and Davis, R.E. 2009. Establishment of a new phytoplasma subgroup, 16sri-q,to accommodate a previously undescribed phytoplasma found in diseased cherry in Lithuania. *J. Plant Pathol.* 91: 71–75

Varga, K., Kölber, M., Nemeth, M., Ember, I., Erdos, Z., Biro, E., Paltrinieri, S., Martini, M. and Bertaccini, A. 2001. Identification of phytoplasmas infecting sour cherry in Hungary. *Acta Hort.* 550: 383–388.

Varma, A., Raychaudhuri, S.P., Pandey, P.K., Ahlawat, Y.S. and Chakraborty, N.K. 1991. Viral and Mycoplasmal diseases of trees in India. *Intl. J. Tropical Plant Dis.* 9:15-33.

Van Katwijk, W. 1956. Rough Skin of apples. *European J. Plant Path.* 62: 46-49.

Van Oosten, H.J. 1970. Herbaceous host plants for the sharka (Plum pox) virus. *Neth. J. Pl. Path.* 76: 253–260.

Van Oosten, J. J. 1983. Growth, yield and fruit quality of virus-infected and virus-free Golden Delicious apple trees, 1968-1982. *Acta Hort.* 130: 213-220.

Vera, L., Marinho, A., Kummert, J., Rufflard, G., Colinet, D. and Lepoivre, P. 1998. Detection of apple stem grooving virus in dormant apple trees with crude extracts as templates for one-Step RT-PCR. *Plant Dis.* 82: 785–790

Verdin, E., Salar, P., Danet, J.L., Choueiri, E., Jreijiri, F., El Zammar, S., Gelie, B., Bové, J. and Garnier, M. 2003. '*Candidatus* Phytoplasma phoenicium' sp. nov., a novel phytoplasma associated with an emerging lethal disease of almond trees in Lebanon and Iran. *Internl. J. Syst. Evolu. Microbiol.* 53: 833–838.

Verma, N., Raja Ram and Zaidi , A. A. 2005. *In vitro* production of Prunus necrotic ringspot virus-free begonias through chemo- and thermotherapy. *Sci. Hort.*103:239-247.

Vidal, E., Zagrai, L., Milusheva, S., Bozhkova, V., Tasheva-Terzieva, E., Kamenova, I., Zagrai, I. and Cambra, M. 2012. *In*: Proc. 22nd Interl. Conf. Virus and other graft transmissible diseases of fruit crops. p. 74.

Wagnon, H.K., Traylor, J.A., Williums, H.E. and Wiener, A.C. 1968. Investigation of cherry rasp leaf virus disease in California. *Plant dis. Reptr.* 52: 618–622.

Walia, Y., Kumar, Y., Rana, T., Hallan, V., Ram, R., Thakur, P.D., Sharma, U., Handa, and Zaidi, A.A. 2008. Molecular characterization and variability analysis of apple scar skin viroid. *In*: XVIII National Conference of Indian Virological Society, December 11-13, 2008. PGIMER, Chandigarh, India

Walkey, D.G.A. and Webb, M.J.W. 1968. Virus in plant apical meristems. *J. Gen. Virol.* 3: 311-313.

Walkey, D.G.A. and Cooper, J.I. 1973. *Rep. nat. Veg. Res. Stn.* p. 100.

Wallis, C., Fleischer, S. Luster, D. and Gildow, F.E. 2005. Aphid (Homoptera: Aphididae) species composition and potential aphid vectors of Plum pox virus in Pennsylvania peach orchards. *J. Econ. Entomol.* 98: 1441–1450.

Wang, L., Michelutti, R., Zunnoon-khan, S., Wang, G., Hong, N., Matic, S and Myrta, A. 2009. Detection of four viruses by ELISA and RT-PCR and cluster analysis of apple chlorotic leaf spot (ACLSV) in apple and pear at the Canadian clonal Genebank. *In*: 21st Internl. Conf. on virus and other graft transmissible fruit crops.*Berichte aus dem JKI* 148 (Abs). pp. 25–26

Wang, L., Hong, N., Wang, G.P., Michelutti, R and Zhang, B.L. 2009. First report of *cherry green ring mottle virus* in Plum (*Prunus domestica*) in North America. *Plant Dis.* 93:1073

Watpade, S., Baswaraj, K., Parmanick, K.K., Thakur, P.D. and Handa, A. 2012. Simultaneous detection of apple chlorotic leaf spot virus and apple mosaic virus in apple by duplex RT-PCR. *In*: XXI Nat. Conf. Immunobiology and Management of viral diseases in 21st century. Held at IVRI, Mukteshwar 8-10 Nov., 2012. P.76.

Wei, W., Davis, R.E., Lee, I.M. and Zhao Y. 2007. Computer-simulated RFLP analysis of 16S rRNA genes: Identification of ten new phytoplasma groups. *Internatl. J. Syst. Evolu. Microbiol.* 57: 1855–1867.

Weintraub, P.G. and Beanland, L.A. 2006. Insect vectors of phytoplasmas. *Ann. Rev. Entomol.* 51: 91–111.

Wells, J.M., Kirkpatrick, H.C and Parish, C.L. 1986. Symptomatology and incidence of Prunus necrotic ringspot virus in peach orchards in Georgia. *Plant Dis.* 70: 444–447.

Welsh, M.F. and Pine, T.S. (*eds.*). A griculture Handbook No. 437, USDA, Washington, D.C., USA.

Welsh, F.A. and Kaene, F.W.L. 1960. The virus disease 'leaf pucker' of Apple and associated fruit symptoms. *Proc. Washington State Hort. Association*, 55: 114–116.

Welsh, F.A. and Kaene, F.W.L.1963. Apple leaf pucker and associated fruit disorders. *In*: Tech. Commic. *Bul. Hort. E. Malling*, 30: 13–15

Welsh, M.F. and Cheney, P.W. 1976. Little cherry. *In*: Virus diseases and noninfectious disorders of stone fruits in North America. *Eds.* Gilmer, R.M., Moore, J.D., Nyland, G., Welsh, M.F. and Pine. T.S. Agriculture Handbook No. 437, USDA, Washington DC, USA. pp. 231–237

Werner, R., Muehlbach, H.P. and Buettner, C. 1987. Detection of cherry leaf roll nepovirus (CLRV) in birch, beech and petunia by immuno-capture RT- PCR using a conserved primer pair. *Euro. J. Forest Path.* 27: 309–318.

Wetzel, T., Candresse, T., Ravelonandro, M.,and Dunez, J. 1991. A polymerase chain reaction assay adapted to plum pox potyvirus detection. *J. Virol. Methods* 33: 355–365.

Wetzel, T., Candresse, T., Macquaire, M., Ravelonandro, M. and Dunez, J. 1992. A highly sensitive immunocapture polymerase chain Reaction method for plum pox potyvirus detection. *J. Virol. Methods* 39: 27–37.

Yaegashi, H. Isogai, M., Tajima, H., Sano, T., Yoshikawa, N. 2007. Combinations of two aminoacids (Ala40 and Phe75 or ser40 and Tyr75) in the coat protein of apple chlorotic leaf spot virus are crucial for infectivity. *J. Gen. Virology* 88: 2611–2618.

Yanase, H., Koganezawa, H and Fridlund, P. 1989. correlation of pear necrotic spot with pear vein yellows and apple stem pitting, and a flexuous filamentous virus associated with them. Fruit tree virus diseases XIV, *Acta Hort.* 235: 157.

Yarwood, C.E. 1955. Mecanical transmission of an apple mosaic virus. *Hilgardia* 23: 613–628.

Yoshida, K. and Ito, D. 2002. Diagnosis of apple fruit crinkle disease using RT-PCR. *Tohoku Nogyo Kenkyu Seika Joho,* 16: 173–174.

Yoshikawa, N., Sasaki, E., Kato, M. and Takahashi, T. 1992. The nucleotide sequence of apple stem grooving capillovirus genome. *Virology* 191: 98–105.

Zeller, S.M. and Milbrath, J.A. 1947. Mild rusty mottle of sweet cherry (*Prunus avium*). *Phytopathology* 37: 177–184.

Zemtchik, E.Z., Verderevskaya, T.D and Kalashyan Yu.A. 1998. Apricot latent virus: Transmission, purification and serology. *Acta Hort.* 472: 153–158.

Zentit, Pascal, Brans, Yoann, ramat and Charlotte. 2009. Susceptibility of a range of hazelnut cultivars to apple mosaic ilarvirus. *In*: 21st Internl. Conf. on virus and other graft transmissible fruit crops. *Berichte aus dem JKI* 148, (Abs) p. 25.

Zhang, Y. P., Kirkpatrick, B.C., Smart, C.D. and Uyemoto, J.K. 1998. cDNA cloning and molecular characterization of cherry green ring mottle virus. *J. Gen. Virol.* 79: 2275–2281.

Zhang, Y.P., Mink, G.I., Tiffany, M.G. and Howell, W.E. 1992. Isolation of viruses associated with cherry twisted leaf, apricot ring pox and apricot pit pox diseases and their relationship to apple stem pitting virus. Abs. Annu. Meeting APS, Portland, OR, USA.

Zhou, J.F., Wang, G.P., Kuang, R.F., Wang, L.P and Hong, N. 2011. First Report of *Cherry green ring mottle virus* on Cherry and Peach Grown in China. *Plant Dis.* 95: 1319.

Zhu, S. and Shu, X. 1992. A mycoplasma-like organism on cherry. *Acta Phytopath. Sinica* 22: 25–28.

Zilkah, S., Faingersh, E., Rotbaum, A., Spiegel, S. and Stein, A. 1995. Symptoms of prunus necrotic ringspot virus on micrografted *Prunus serrulata* cv. shirofugen shoot cultures. *Acta Hort.* (ISHS), 386: 183–186

3

VIRUS DISEASES OF SHELL FRUITS

Among the shell fruits, walnut, Hazelnut, pecan and chestnut are important plantations. The common walnut (*Juglans regia)* is considered indigenous to the region from Persia to Kashmir. Most of the walnuts grown in India are either seedlings or grafted trees. The best-known member of the genus is the common walnut (*J. regia,* literally "royal walnut"), native from the Balkansin southeast Europe, southwest and central Asia to the Himalaya and southwest China. Walnut is the dominant species which is widely cultivated for its delicious nuts. In India, it is used widely as a *prasad* (offering) to Mother Goddess Vaisnav Devi and generally, as a dry food in the season of festivals such as Diwali. Walnuts are also used in Traditional Chinese Medicine. They are said to tonify kidneys, strengthen the back and knees, warm and hold Qi in lungs and help kidneys to grasp the Qi, moisten the intestines and move stool. It is believed to manage asthma and is prescribed to be taken between bouts of asthma, but not for acute asthma. Walnut is infected by some viral and phytoplasma diseases. Among them blackline, line pattern, mosaic are important virus disease and witches broom or bunch phytoplasma disease is the only known phytoplasmas infecting walnut. These diseases are briefly described below:

A. VIRUS DISEASES OF WALNUT

BLACKLINE DISEASE OF WALNUT

Blackline (BL) disease of English walnut trees (*Juglans regia*) is widely distributed and causes serious losses in trees, propagated on *J. hindsii* or Paradox rootstock (*J. hindsii, J. regia*) in several walnut-growing regions.

Blackline causes symptoms in the tree canopy that are similar to those caused by soilborne pathogens such as *Phytophthora* and *Armillaria,* nutrient deficiencies, and scion-rootstock incompatibilities. The first symptoms of the disease are poor terminal growth, yellowing and drooping leaves, and premature defoliation, particularly in the top branches. Most characteristic symptom of blackline disease appears as dark lines between the stock and grafted scion. The black line starts at the graft union but eventually completely girdle the tree, killing the scion in 2 to 6 years. The affected trees have poor growth. The leaves turn yellow and drop early during leaf fall. Severely affected trees show dieback and eventually die.

The causal virus is transmitted through grafting, by pollen and seed (Mircetich *et al.*, 1980). It is seed transmitted in rhubarb and other herbaceous hosts. The vector(s) are nematodes. Black line disease is caused by a strain of the cherry leaf roll nepovirus (CLRV–W) and affects only rootstocks (Cooper, 1980, 1981). The virus particles are isometric, about 26 nm in diameter and in thin sections show characteristics of a NEPO virus (Mircetich *et al.*, 1981). The causal virus infects several herbaceous hosts such as Rhubarb (*Rheum rhabarbarum*). Positive diagnosis of blackline requires careful examination of the union between scion and rootstock. No practical method is available to detect blackline before symptoms appear. Trees infected with blackline cannot be cured. Avoid interplanting with cultivars that develop pistillate bloom during the pollen-shedding period of cultivars already present. Carefully select grafting material for top-working established trees from healthy, blackline-free English walnut trees. Be sure to use grafting material from a blackline-free source.

WALNUT LINE PATTERN

Line pattern disease of walnut was reported from Italy and U. K. (Quacquarelli and Savino, 1977). The characteristic symptoms of the disease appear as pale green or yellow spots with broad lines and stripes on the leaves of affected trees. These symptoms usually develop around and along the leaf veins. The line pattern disease is graft transmissible and use of healthy planting material is the only way of the management of this disease. Line pattern disease is caused by walnut yellow vein strain of cherry leaf roll virus (Savino *et al.*, 1977). A similar syndrome has been reported from Bulgaria (Christoff, 1958) but detailed studies have not been done.

WALNUT MOSAIC

Mosaic disease in walnut has been reported from Italy (Quacquarelli and Savino, 1977) and U.K. (Cooper, 1980, 1981). Walnut is an original host of mosaic disease. The apricots and peaches are the experimental hosts of mosaic disease. The symptoms of the disease are conspicuous on leaves as chlorotic discolouration, yellowish spots or rings and line patterns. Occasionally, necrotic fecks are also seen on leaves and fruits (Savino *et al.*, 1976). A disease similar to walnut mosaic has been reported from India (Ram and Rathore, 1987). The disease is transmitted by graft and also by sap inoculation, pollens and through immature seed (Savino *et al.*, 1976). Seed infection can be reduced by storing seeds at 4-5°C prior to germination (Cooper, 1980).

WALNUT RINGSPOT

Ringspot disease in walnut, *Juglans regia* has been observed by visual examination of leaves and fruits of walnut trees in Slovakia. The virus was transmitted mechanically to *Chenopodium quinoa, Chenopodium amaranticolor, Nicotiana clevelandii, Phaseolus vulgaris.* cv. Bountiful. The thermal inactivation point of the virus lies between 48 and 50°C and the dilution end-point between 10^{-1} and 10^{-2}. Antiserum was produced which had a titer 1:256. The disease causing yellow mosaic and ringspot symptoms in walnut trees is caused by cherry leaf roll Nepovirus (Mircetich *et al.*, 1980). Ringspot isolates gave a positive immunological reaction with the Mircetich's antiserum against the cherry leaf roll virus obtained from walnut tree (Baumgartnerová, 1992).

WALNUT BUNCH OR WITCHES' BROOM

Walnut bunch disease is widely distributed in Southern USA. Walnut bunch is a yellows-type of witches'-broom disease occurring on Japanese, Manchurian, English, native black walnut, and butternut. The common symptoms of the disease are numerous bunches or witches broom growth on twigs of infected trees. The shoots become wiry or thin. The leaf size is reduced and leaves become chlorotic. The yield of fruits is greatly reduced and premature death of affected trees is also common. The witch's brooms normally begin growing earlier in the spring than the normal foliage, and continues growing later in the fall. Infected walnut trees will not produce a normal crop. The disease is transmitted by bark patch grafting

(Carling and Millikan, 1974). The hosts of the disease are: *Juglans nigra, J. sieboldiana* and *J. cinarea* in addition to *J. regia*. An electron-microscope study of the phloem tissues of walnut revealed the presence of a phytoplasma in diseased but not in healthy leaf and stem phloem cells. The phytoplasma bodies are pleomorphic, vary in length from 150 to 1000 nm, and are enclosed in a plastic trilaminar unit membrane. Spherical or oval shapes predominate, but tubular forms were also evident near sieve plates. Some sections clearly showed the presence of ribosome and DNA-like material within Phytoplasma cells. The apparent absence of virus particles, together with the consistent association of phytoplasmas in diseased but not in healthy walnuts, suggests phytoplasma as a likely causal agent. The possible relationship of walnut bunch and pecan bunch disease is suggested (Seliskar, 1976). For management of the disease, prune out the witch's brooms; if more appear, tree must be removed.

B. VIRUS DISEASES OF PECAN (*Carya illinoinensis* or *illinoensis*)

Pecan is a native of United States of America and is widely grown in USA. "Pecan" is from an Algonquian word, meaning a nut requiring a stone to crack. In India, pecan are grown to a very limited scale in Himalayan and submountain regions The nuts of Pecan are edible, with a rich, buttery flavor. They can be eaten fresh or used in cooking, particularly in sweet desserts and also in some savory dishes. One of the most common desserts with the pecan as a central ingredient is the pecan pie, a traditional southern U.S. recipe. Pecans are also a major ingredient inpraline candy. In addition to the pecan nut, the wood is also used in making furniture and in hardwood flooring. Diseases affecting pecans are caused by bacteria, fungi, nematodes, phytoplasma, and physiological problems. Although viruses have not been shown to occur on pecans, but a bunch disease has been found to infect pecan trees.

PECAN BUNCH DISEASE

Bunch disease of pecans is a disease that affects weak pecan trees. It is most often associated with pecans grown along alluvial river bottoms.

Seedling pecans appear to be a source of inoculum. Trees affected with bunch disease show the bunching symptom caused by excessive growth of slender succulent twigs from lateral buds that normally remain dormant on the main limbs. In moderately affected trees, one or several branches will show the "bunch" growth symptom. Bunching in severely affected trees may involve all main limbs which produce thick masses of sucker-like growth. It is not systemic within the tree but localized on individual limbs. The shoots on the affected limbs are brittle and short-lived. The leaves on affected shoots are larger than normal and are flexible. Diseased trees have lower yields and inferior nut quality.

Symptoms can occur on a healthy tree that is grafted with a diseased scion. Zinc deficiency can also cause bunching symptoms and can be confused with bunch disease. The two conditions can be distinguished on the basis of foliage symptoms. Zinc deficient leaves are smaller and very rigid with chlorosis develops between the veins. Bunch disease of pecans is an infection caused by a phytoplasma in subgroup III (Seliskar *et al.*, 1974). The organism has been consistently associated with diseased tissue and has not been identified in healthy trees. The disease may be similar to bunch disease of walnut (Seliskar, 1976).

There is no effective control for bunch disease. Early detection of the first symptom of bunch and pruning out of the affected branch may prevent spread of the disease throughout the tree. Bunch disease can be controlled by pruning, removing infected trees and planting resistant cultivars. Pruning can be effective only when a few scattered bunches are present. Whle pruning, cuts should be made several inches below the bunch symptoms to ensure that all the infected area is removed. When the tree is severely affected, it should be destroyed to protect nearby healthy trees from infection.

C. VIRUS DISEASES OF CHESTNUT (*Castanea* Spp)

Chestnut (*Castanea*) is a genus of eight or nine species of deciduous trees and shrubs in the beech family *Fagaceae,* native to temperate regions of the Northern Hemisphere. The name also refers to the edible nuts they produce. The Chestnut belongs to the same *Fagaceae* family as the Oak and Beech. There are four main species, commonly known as European, Chinese, Japanese and American Chestnuts.

- *European species*: Sweet Chestnut (*Castanea sativa*) (also called "Spanish Chestnut" in the US) is the only European species of chestnut.
- *Asiatic species*: *Castanea crenata* (Japanese Chestnut), *Castanea mollissima* (Chinese Chestnut), *Castanea davidii* (China), *Castanea henryl* (Chinese chinkapin, also called Henry's Chestnut – China) and *Castanea seguinii* (also called Seguin's Chestnut - China).
- *American species*: These include *Castanea dentata* (American Chestnut - Eastern states), *Castanea pumila* (American- or Allegheny Chinkapin, also known as "Dwarf Chestnut" - Eastern states), *Castanea alnifolia* (Southern states), *Castanea ashei* (Southern states), *Castanea floridana* (Southern states) and *Castanea paupispina* (Southern states).

Chestnuts should not be confused with either Horse Chestnuts (genus *Aesculus*), or Water Chestnut (family *Cyperaceae*); these are unrelated to *Castanea* and are named for producing nuts of similar appearance but of no notable edibility in the case of the former, and tubers of similar taste from an aquatic herbaceous plant in the case of the latter. Other trees commonly mistaken for the Chestnut tree are the Chestnut Oak (*Fagaceae, Quercus prinus*) and the American Beech (*Fagus grandifolia*). An antiviral composition comprising an effective amount of α-Aescin obtained from the Indian Horse Chestnut *(Aesculus indica)* against Cucumber mosaic virus which attacks a great variety of vegetables, ornamentals, weeds and other plants (Singh *et al.*, 2007).

CHESTNUT MOSAIC

The mosaic disease of chestnut was introduced in Europe from Japan with the importation of *Castanea crenata* (Desvignes and Lecocq, 1995). The incidence of the disease was observed upto 90% in some orchards.

The infected chestnut trees show necrotic lesions in the bark and wood which turn into cankers, or incompatibilities when two cultivars of different sensitivity are assembled through grafting. The Symptoms of the disease appear as chlorotic mottling, vein yellowing, and mosaic on leaves of susceptible cultivars. Epinasty and partial limb atrophy are also common. There is wilting of apex and necrosis at the graft union.The hybrids such as Marigoule Ca 15 and Maravel Ca 74 are susceptible to the mosaic virus. The later variety was propagated *in vitro* at large number for trading as it is a Phytophthora resistant rootstock. This variety has been used to detect the mosaic infection in glasshouse.

The mosaic disease is graft transmitted and is also transmitted by a vector,Oak aphid, *Myzocallis castanicola* (Desvignes and Cornnagia, 1996; Desvignes, 1998). Experimentally, the mosaic virus has been successfully transmitted to *Quercus* species and *Corylus avellana*. In these species, the symptoms appear as chlorotic mottle, vein yellowing, and mosaic on leaves. The inoculated plants remain dwarfed. *C. sativa* and *C. crenata* and their hybrids are tolerant. Cultivars of the *Castanea crenata* were sensitive. Some cultivars of *C. sativa* such as Belle Epine, Campanese, Olargues were less sensitive, but others were tolerant. Among the *C. crenata* x *C. sativa* hybrids like Maraval, Marigoule were sensitive but hybrids like Bouche de Bétizac, Bournette are less sensitive and *Q. pubescens* and *Q. ilex* appeared as tolerant (Desvignes, 1998). The mosaic disease can be eliminated by heat treatment.

HORSE CHESTNUT YELLOW MOSAIC

A disease with yellow mosaic symptoms on leaves of horse chestnut trees (*Aesculus carnea* and *A. hippocastanum*) has been described by Sweet and Barbara, 1979. The causal virus proved to be an ilarvirus by way of reactions in woody indicator plants grafted with diseased horse chestnut and by mechanical inoculation to herbaceous test plants. The associated virus with yellow mosaic reacted positively with antisera to apple mosaic virus but not with antisera to its serotype Prunus necrotic ringspot virus, or to Prune dwarf virus. Yellow mosaic was induced in horse chestnut seedlings grafted with tissues from herbaceous hosts infected with horse chestnut isolates or with the European plum line pattern isolate of apple mosaic virus. Virus was detected by enzyme-linked immunosorbent assay (ELISA) in embryo and endosperm of immature seed from infected trees but not in mature seed, or progeny seedlings (Sweet and Barbara, 1979, 2008).

D. VIRUS DISEASES OF HAZELNUT

Hazelnuts are species of the genus *Corylus*. The major cultivated species are *Corylus avellana*, *C. maxima*, *C. americana* and *C. colurna*. Hazelnut is the nut of the hazel and is also known as a cobnut. It is roughly spherical to oval, about 15-25 mm long and 10-15 mm in diameter, with an outer fibrous husk surrounding a smooth shell. The nut falls out of the husk when ripe, about 7-8 months after pollination. The kernel of the seed is edible and used raw or roasted, or ground into a paste. The seeds have a thin, dark brown skin which has a bitter flavour and is sometimes removed before cooking. Hazelnuts are produced in commercial quantities in Europe, Turkey, and in the American states of Oregon and Washington.

Hazelnuts are extensively used in confectionery to make praline and also used in combination with chocolate for chocolate truffles and products such as Nutella. Hazelnut oil is strongly flavoured and used as a cooking oil. Hazelnuts are rich in protein and unsaturated fat. Moreover, they contain significant amounts of thiamine and vitamin B6, as well as smaller amounts of other B vitamins. Additionally, 1 cup (237 ml) of hazelnut flour has 20 g of carbohydrates, 12 g of which are fibres.

There are many cultivars of the Hazel, including 'Barcelona', 'Butler', 'Casina', 'Clark' 'Cosford', 'Daviana', 'Delle Langhe', 'England', 'Ennis', Fillbert, 'Halls Giant', 'Jemtegaard', 'Kent Cob', 'Lewis', 'Tokolyi', 'Tonda Gentile', 'Tonda di Giffoni', 'Tonda Romana', 'Wanliss Pride', and 'Willamette'. Some of these are grown for specific qualities of the nut including large nut size, and early and late fruiting cultivars, whereas other are grown as pollinators. The majority of commercial Hazelnuts are propagated from root sprouts. Some cultivars are of hybrid origin between Common Hazel and Filbert.

Hazelnut is infected both by viruses and phytoplasma. Among viral infections, a mosaic disease caused by ilarviruses (apple mosaic virus, prunus necrotic ringspot virus and Tulare apple mosaic virus) is the major infection. Among phytoplasma diseases, Filbert stunt (unknown etiology but suspected to be caused by a phytoplasma) and hazelnut yellows caused by phytoplasma are important.

HAZELNUT MOSAIC

Apple mosaic virus (ApMV) infecting hazelnuts is one of the most important pathogens and has been reported in many countries, such as in Italy, France, Spain, the UK, the USA, Poland and Turkey. Hazelnut mosaic virus is responsible for causing an average loss of 28.2 % in Turkey and severe loss in Spain (Akbas and Degirmenci, 2009; Aramburu and Rovira 2000).

SYMPTOMS

In Turkey, the mosaic disease is widely distributed in hazelnut plantations (Akbas *et al.*, 2004). The leaves on infected trees showed general yellowing, yellow rings and lines, yellow flecking, oak leaf pattern and conspicuous vein banding. Hazelnut trees may also be symptomless (Akbas *et al.*, 2004). Mostly mosaic symptoms are caused by apple mosaic virus. Prunus necrotic ringspot virus produce few symptoms on its own but it has synergistic effect when present with apple mosaic virus.

TRANSMISSION

The mosaic disease is graft transmitted. Mosaic virus is also transmitted by mechanical inoculation and root grafting. It is also transmitted by seed and pollen of hazelnut (Cameron and Thompson, 1985).

HOST RANGE

Apple mosaic virus (ApMV) infects over 65 species in 19 different families both experimentally or naturally. Natural hosts of ApMV include apple, rose, hazelnut, horse chestnut, raspberry, birch, hops, stone fruits, blackberry and strawberry. Some isolates are transmitted by mechanical inoculation to *Chenopodium amaranticolour, C. quinoa* and *cucumis sativus.* Inoculated plants of *C. quinoa* developed chlorotic local lesions and *C. sativus* showed systemic chlorosis and stunting.

THE PATHOGEN

Natural infection of hazelnut showing mosaic symptoms is caused by three ilarviruses (apple mosaic virus, prunus necrotic ringspot virus and Tulare apple mosaic virus). The infection takes place either by individual virus or by virus complexes, which determines the disease severity. These viruses have already been described in chapter-2 of this book. Apple mosaic virus (ApMV) belongs to the genus; *Ilarvirus;* in family *Bromoviridae.* Virus particles are isometric 25 and 29 nm in diameter (and probably a third size). Virus particles contain 16 % nucleic acid and 84 % protein. ApMV has three genomic RNA (RNA 1, RNA 2 and RNA 3) and one sub-genomic RNA (RNA 4 species). Apple mosaic ilarvirus is most important pathogen in mosaic complex and may have an effect on yield.

DIAGNOSIS AND MANAGEMENT

All the three ilarviruses can be detected by Enzyme-linked immunosorbent assays (ELISA) using their specific antibodies. However, the apple mosaic ilarvirus was dominant in infected hazelnut trees in Turkey (Akbas *et al.*, 2004) and also in Spain (Aramburu and Rovira, 2000) when trees tested by ELISA. ELISA was an important tool to test the germplasm (Postman and Mehlenbacher, 1994). The infection of these ilarviruses can be detected by RT-PCR in addition of ELISA detection. Seedlings testing positive for ApMV were nearly all symptomless. All these viruses can be detected simultaneously by one tube RT-PCR (Akbas and Degirmenci, 2010). Heat therapy of infected plants for 21 or more days at temperatures alternating every 4 hours between 30 and 38°C, followed by shoot-tip grafting onto healthy *Corylus* rootstocks resulted in the elimination of ApMV from 26 infected European hazelnut cultivars (Postman and Mehlenbacher, 1994).

HAZELNUT LINE PATTERN

The line pattern disease is normally found with mosaic symptoms and was first reported from Bulgaria and then from Italy. The disease occurs naturally in hazelnut. The symptoms of the disease are: mosaic chlorotic mottle in lines and stripes along the veins of leaves of infected trees. The disease is graft transmitted. Apple mosaic ilarvirus was detected in ELISA in hazelnut plants showing ring and line pattern symptoms (Ragozzino, 1980). In addition, apple mosaic virus was associated with hazelnuts showing line pattern symptoms and Prunus necrotic ring spot virus (PNRSV) with trees showing ring spot symptoms (Barba and Quacquarelli, 1984). The disease can be managed by the use of healthy planting material.

HAZELNUT PHYTOPLASMA DISEASES

A few diseases of hazelnut are known to be caused by phytoplasmas such as Maclatura lineare, hazelnut stunt and decline.

HAZELNUT MACULATURA LINEARE

Hazelnut maculatura lineare (HML) has been reported from Italy. The symptoms appear as elongated spots on leaves of infected trees. The disease is of low economic importance. No vector of the disease is known. Studies revealed that the HML disease is caused by a phytoplasma. Propagation should not be done from infected sources (OEPP/EPPO Bull., 2004).

YELLOWING AND DECLINE OF EUROPEAN HAZEL

In Italy, commercial orchards of hazelnut are severely affected by yellowing and decline. Phytoplasma infection was detected from these trees by nested PCR. The amplified product obtained by nested primers was specific for the fruit tree phytoplasmas of apple proliferation group. Restriction site analysis showed that the trees are infected with apple proliferation, pear decline and European stone fruit tree phytoplasmas and aster yellows agent. The phytoplasmas were also identified by hybridization with suitable oligonucleotide probes (Marcone *et al.*, 1996).

HAZELNUT STUNT SYNDROME

Stunt syndrome was first reported from several commercial hazelnut orchards in North-west Oregon in USA in 1950. The disease is of minor importance and its spread had been very slow (Postman *et al.*, 2001).

SYMPTOMS

Severely affected trees show slight chlorosis with reduced leaf size. The internodes are reduced in size. The yield of nuts is greatly reduced and affected trees showed dieback of branches.

TRANSMISSION

The disease is transmitted by grafting. There are enough evidence that the disease spreads by natural root grafting in hazelnut orchards in Oregon (Postman *et al.*, 2001)

THE PATHOGEN

Leaf samples from affected trees were tested for phytoplasma infection by use of nested polymerase chain reactions primed by phytoplasma universal primer pairs P1/P7 and F2n/R2 for amplification of phytoplasma 16S ribosomal DNA sequences. Phytoplasma characteristic 1.2 kbp DNA sequences were amplified from all affected field-tree samples. Restriction fragment length polymorphism (RFLP) patterns indicated the presence of clover yellow edge phytoplasma (16 Sr111-B) and another phytoplasma, putatively a member of a new subgroup in the X-disease group (16Sr111). The presence of Clover yellow edge (CYE) phytoplasma in the Oregon samples was the first report of this phytoplasma infecting *Corylus*

(Jomantiene *et al.*, 2000). The phytoplasma isolated in Oregan was distinct from those associated with decline and yellows (Postman *et al.*, 2001).

DIAGNOSIS AND MANAGEMENT

The phytoplasmas associated with stunt syndrome were detected both symptomatic (stunted) and symptomless trees by PCR. Removal of diseased trees and also adjacent trees can eliminate the disease from an orchard. Failure to remove trees adjacent to infected trees resulted in the eventual spread throughout the orchard, apparently by root grafting.

VIRUS DISEASES OF KIVI

Pelargonium zonate spot virus (PZSV) -a member of *Anulavirus* genus in family *Bromoviridae* infects several herbaceous hosts like tomato, pepper, artichoke and some weeds in Italy, France, USA and Israel. PZSV was found to infect kivi fruits (*Actinidia deliciosa*) in Italy. Symptoms appeared on affected plants were: chlorotic and necrotic rings on leaves and deformation of berries. The virus was mechanically transmitted to *Chenopodium quinoa, Nicotiana benthamiana, N. glutinosa* and *N. tabacum*. Virus particles of 30 nm in diameter were observed from infected *C. quinoa*. Molecular analysis revealed that the virus was an isolate of PZSV (Biccheri *et al.*, 2012).

Other viruses infecting kivi fruit plants are Apple stem grooving virus, two vitiviruses and a crinivirus closely related to citrus leaf blotch virus belonging to family *Betaflexiviridae*. Cucumber mosaic virus and Alfalfa mosaic virus were also detected in kivi plants in New Zealand and Italy (Blouin *et al.*, 2012). Two tobamoviruses, Ribgrass mosaic virus and turnip vein clearing virus are probably present worldwide in kivi fruit (Pearson *et al.*, 2010). Cherry leaf roll virus also infects kivifruits causing symptoms of leaf spot, fruit malformation , bark cracking and cane wilting (Blouin *et al.*, 2012). A phytoplasma belonging to stolbur phytoplasma was reported infecting kivi plants in Italy (Paltrinieri *et al.*, 2012). Phytoplasmas of aster yellows and apple proliferation groups were also observed in mixed infection with stolbur phytoplasma.

VIRUS DISEASES OF PERSIMMON

A disease showing veinlet necrosis of leaves was observed on Japanese persimmon (*Diospyros kaki*) in Italy. The virus was identified as cryptic virus with a bipartite genome. The virus name was proposed as persimmon cryptic virus (Morelli *et al.*, 2012). Two dsRNA with an estimated length of 1.5 Kbp were identified and characterized from leaves of affected plants. RNA-1 had 1510 bp and RNA-2 1491 bp. Phylogenetic analysis of RNA dependent RNA polymerase (RdRP) sequence showed highest amino acid identity with black raspberry cryptic virus (63%). Whereas, CP putatively encoded by dsRNA-2 shared highest identity with mulberry cryptic virus-1 (49%). The virus was detected in RT-PCR with virus-specific primers. Immunosorbent electron microscopy revealed the virus particles to be isometric about 30 nm in diameter (Morelli *et al.*, 2012).

REFERENCES

Akbas, B. and Degirmenci, K. 2010. Simultaneous detection of apple mosaic virus in cultivated hazelnut by one tube RT-PCR. *Aferican J. Biotech.*, 9: 1753-1757.

Akbas, B. and Degirmenci, K. 2009. Incidence and natural spread of apple mosaic virus on hazelnut in the west black sea coast of Turkey and its effect on yield. *J. Plant Path.* 91: 767–771.

Akbas, B, Ilhan, D. and Atlamaz, A. 2004. A prelinary survey for hazelnut (*Corylus avellana*) viruses in Turkey. *In*: 6th International congress on hazelnut, Tarragona p. 94.

Aramburu, J. and Rovira, M. 2000. Incidence and natural spread of apple mosaic ilarvirus in hazel in North-east Spain. *Plant Pathology*, 49: 423–427.

Barba, M. and Quacquarelli, A. 1984. Two Ilarviruses associated with hazelnut mosaic. *Annali dell'Instituto sperimentale di patologia vegetale di Roma* 9: 97–101.

Baumgartnerová, H. 1992. The occurrence of walnut ringspot on *Juglans regia* L. in Slovakia. *Biologia Plantarum*, 34: 471–474.

Biccheri, R., Babini, A.R., Blouin, A., Lanzoni, C., Pisi, A., Poggi-Pollini, C., Credi, R., Laghi, L., Rocculi, P., Autonell, C.R., Pearson, M.N. and Ratti, C. 2012. Pelargonium zonate spot virus infecting kivi fruit plants in Itali. *In*: 22nd Interl. Conf. virus and other graft transmissible diseases of fruit crops. Rome June 3-8, 2012. p. 27.

Blouin, A.G., Chavan, R.R., Pearson, M.N., MacDiarmid, R.M., Ratti, C., Biccheri, R., Lebas, B., Veerakone, S. and Cohen, D. 2012. A review of viruses infecting Actiridia. *In*: 22nd Interl. Conf. virus and other graft transmissible diseases of fruit crops. p. 26.

Cameron, H.R. and Thompson, M. 1985. Seed transmission of apple mosaic virus in hazelnut. *Acta Hort.*, 193: 131.

Carling, D.E. and Millikan, D.F. 1974. Graft Transmission of an Infectious Entity Associated with Bunch Disease in Juglans. *Pro. American Phytopath. Soc. 1.*

Christoff, A., 1958. Die Obstvirosen in Bulgarien. *Phytopath. Z.*, 31: 381–436.

Cooper, J.I. 1980. The prevalence of cherry leaf roll virus in *Juglans regia* in the United Kingdom. *Acta Phytopathol. Acad. Sci. Hung.*, 15: 139–145.

Cooper, J.I. 1981. The prevalence of cherry leaf roll virus in *Juglans regea* in the United Kingdom. *Acta Hort.*, 94: 139–146.

Desvignes, J.C. and Lecocq, G. 1995. New knowledges on the chestnut mosaic virus disease. *Acta Hort.* (ISHS), 386: 578–584.

Desvignes, J.C. and Cornaggia, D. 1996. Transmission par le puceron Myzocallis castanicola = Chestnut mosaic virus: Transmission by the aphid *Myzocallis castanicola* on chestnut tree. *Phytoma-La Defense des vegetaux*, 481: 39–41.

Desvignes J.C. 1998. Sweet chestnut incompatibility and mosaics caused by the chestnut mosaic virus (ChMV). Proceeding of the Second International Symposium on chestnut, Bordeaux (FR): *Acta Hort.* 494: 451–458.

Jomantiene, R., Postman, J., Montano, H., Maas, J., Davis, R. and Johanson, K. 2000. First report of clover yellow edge phytoplasma in corylus (hazelnut). *Plant Dis.*, 84: 102.

Marcone, C., Ragozzino, A. and Seemuller, E.1996. Association of phytoplasma with the decline of European hazel in Southern Itali. *Plant Pathology*, 45:857–863.

Mircetich, S.M., Sanborn, R.R. and Ramos, D.E. 1980. Naural spread, graft transmission and possible etiology of walnut black line disease. *Phytopath.*,70: 962–968.

Mircetich, S.M., Dezoeten, G.A. and Lauritis, J.A. 1981. Etiology and natural spread of blackline disease of english walnut trees. *Acta Hort.* (ISHS), 94: 147–152.

Morelli, M., De Stradis, A., La Notte, P., Merkuri, J., Boscia, D. and Minafra, A. 2012. Detection and molecular characterization of a novel cryptovirus from persimmon (*Diospyros kaki*). *In*: 22nd Interl. Conf. virus and other graft transmissible diseases of fruit crops. p. 5.

OEPP/EPPO Bull. 2004. Scheme for the production of healthy plants for planting. Organisation Europeenne et Mediterraneenne pour la protection des plantes. European and Mediterranean Plant Protection Organization 34: 149–153.

Paltrinieri, S., Piergiacomi, M., Ardizzi, S., Contaldo, N., Biondi, E., Lucchese, C. and Bertaccini, A. 2012. Phytoplasma detection and identification in kivi plants and possible correlation with Pseudomonas syringae pv actinidiae presence. *In*: 22nd Interl. Conf. virus and other graft transmissible diseases of fruit crops. Rome June 3-8, 2012. (Abs) p. 30.

Pearson, M.N., Cohen, D., Chavan, R., Blouin, A.G. and Cowell, S.J. 2010. Molecular characterisation of viruses from Kiwifruit. In; Proc. 21st interl. Conf. Virus and other graft transmissible diseases of fruit crops. July 5-10, 2009, Neustadt, Germany. Julius-Kuhn-Archiv 427: 87.

Postman, J.D. and Mehlenbacher S.A. 1994. Apple mosaic virus in hazelnut germplasm. *Acta Hort.* (ISHS), 351: 601–610.

Postman, J.D., Johson, K.B., Maas, J.L. and Davis, R.E. 2001. The 'oregon hazelnut stunt syndrome' and phytoplasma associations. *Acta Hort.* (ISHS) 556:407–410.

Quacquarelli, A. and Savino, V. 1977. Cherry leaf roll in walnut. II. Distribution in Apulia and transmission through seed. *Phytopath. Mediterranea*, 16: 154.

Ragozzino, A. 1980. Hazel mosaic. *Acta Hort.*, 94: 375–378.

Ram, R. D. and Rathore, D. S. 1987. Occurrence of walnut mosaic virus in India. *J. Maharashtra Agric. Univ.*, 12: 113–114.

Savino, V., Quacquarelli, A. and Gallitelli, D. 1976. Occurrence of two sap transmissible viruses in walnut. *Mitt. Biol. Bund. Aust. Ld.u. Forstw.* 170: 23–27.

Savino, V., Quacquarelli, A., Gallitelli, D., Piazzola, P., and Martelli, G.P. 1977. II.Virus deHaccartocciamen to fogliare del ciliegio nel Noce. I. Identificazione e caratterizzazione. *Phytopathologia Mediterranea*, 16: 96–102.

Seliskar, C.E. 1976. Mycoplasma-like organism found in the phloem of bunch-diseased walnuts. *Forest Sci.*, 22: 144–148.

Seliskar, C.E., Kenknight, G.E., and Bourne, C.E. 1974. Mycoplsma-like organism associated with pecan bunch disease. *Phytopathology* 64: 1269–1272.

Singh, B., Katoch, M., Ram, R., Zaidi, A. 2007. A new antiviral agent from indian horse chestnut (*Aesculus indica*). *European Patent EP*1489910.

Sweet, J.B. and Barbara, D.J. 1979. A yellow mosaic disease of horse chestnut (*Aesculus* spp.) caused by apple mosaic virus. *Annl. Appl. Biol.*, 92: 335–341.

COLOURED PLATES 1 TO 65

Plate 1: Variable symptoms of severe strain of apple mosaic

Plate 2: Symptoms of vein banding strain of apple mosaic

Plate 3: Pear trees showing decline

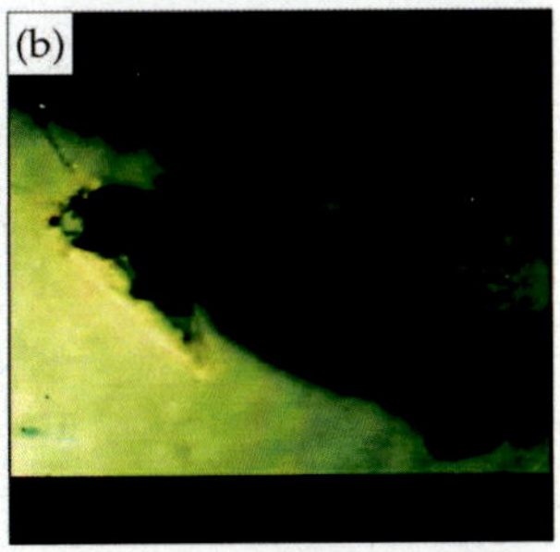

Plate 3a: Adult pear psylla

Plate 4: Quince mosaic disease

Plate 5a: Peach mosaic on peach

Plate 5b: Healthy (right) and decline peach due to mosaic

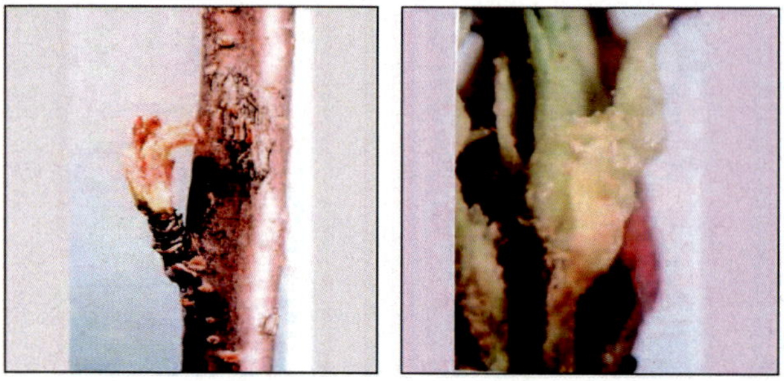

Plate 6: Eriophyid mites in peach buds (Courtesy G.N. Oldfield)

Plate 7: Rosette mosaic on peach

Plate 8a: Peach X disease

Plate 8b: Peach X infection in nursery

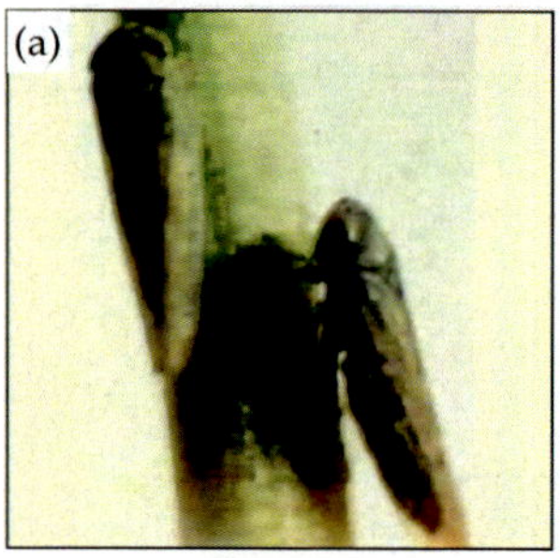

Plate 9a: *Colladonus montanus*

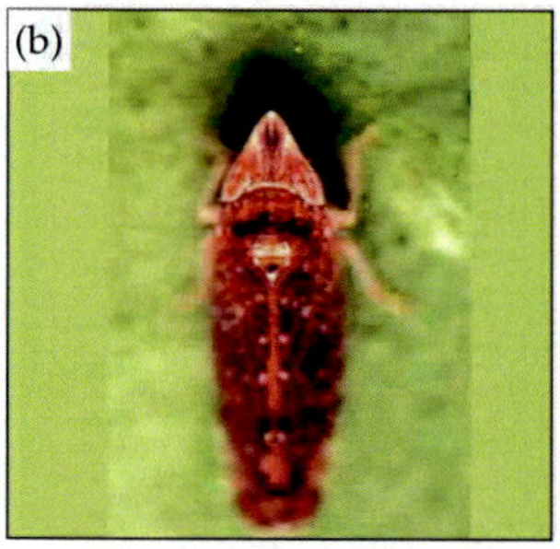

Plate 9b: *Scaphytopius acutus* (Courtesy G.N. Oldfield)

Plate 10: Peach X on periwinkle, healthy: left

Plate 11: Treatment of peach X disease by pressure injection

Plate 12: Peach rosette disease

Plate 13a: Banana bunchy top disease

Plate 13b: Bunchy top in tissue culture plant

Plate 14: Bunchy top vector-*Pentalonia nigronervosa*

Plate 15: Streaks on banana leaf due to BBTV

Plate 16: Bunchy top on hill banana

Plate 17: BBTV on banana cultivar Nendran (Plantain)

Plate 18: Symptoms of bract mosaic on banana bract

Plate 19: Symptoms of bract mosaic on pseudostem

Plate 20: Banana streak virus symptoms on banana leaf (see black dot on right leaf)

Plate 21: Streaks on banana leaf due to BSV

Plate 22: Banana mosaic symptoms

Plate 23: Bark scaling symptoms of psorosis on infected trees

Plate 24: Decline symptoms in Kagzi lime due to Tristeza disease

Plate 25: Vein flecking symptoms of tristeza virus on Kagzi lime

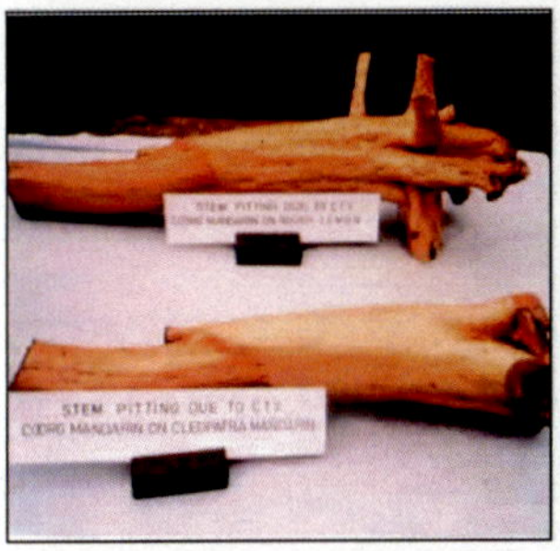

Plate 26: Stem pitting due to CTV

Plate 27: *Toxoptera citicida* – vector of CTV

Plate 28: Vein corking symptoms due to CTV

Plate 29: Infectious variegation

Plate 30: Symptoms of citrus Leprosis

Plate 31: Leprosis symptoms on twigs

Plate 32: Citrus mosaic disease

Plate 33: Woody gall

Plate 34: ICRSV symptoms on Kinnow mandarin

Plate 35: Vein flecking (left) and ring spot on Mosambi leaves

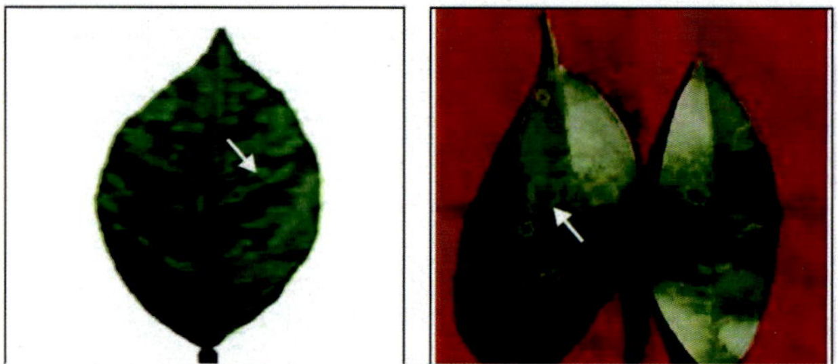

Vein clearing on Mosambi

Ring spot symptom on Kinnow

Plate 36: Symptoms of ICRSV on Citrus cultivars (*Contd...*)

Ring spot symptom on Mosambi

Necrotic symptom on mandarin

Yellow patches on rough lemon

Plate 36: Symptoms of ICRSV on Citrus cultivars

Plate 37: Symptoms of ICRSV on inoculated *C. amaranticolor, C. quino* and French bean leaves

Symptoms of citrus mosaic disease on leaves and fruits

Plate 38: Yellow mosaic sympt. on leaves & fruit

Plate 39: Sympt. of yellow mosaic on various citrus hosts

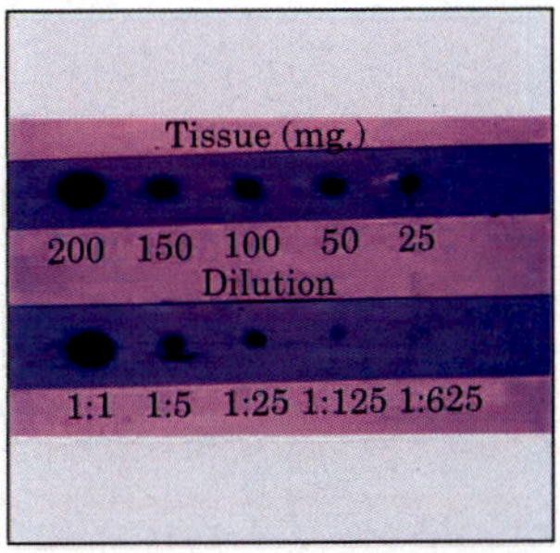

Plate 40: DNA hybridization on tissue and extract dilution

Plate 41: Pummelo yellow vein mosaic

Plate 42: Symptoms of yellow vein clearing on Etrog citron leaves

Plate 43: Water soaking veins due to PYVMV

Plate 44: PYVMV of bean cv. Gheusami

Plate 45: PYVMV of bean cv. Singtami

Citrus blight in Florida

Plate 46a: A blighted tree in Florida

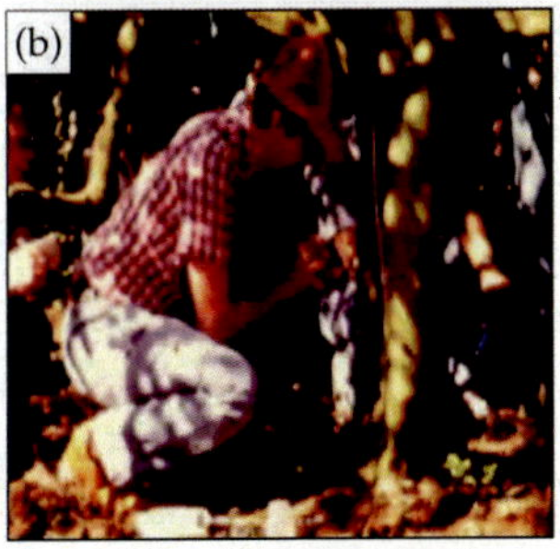

Plate 46b: Demonstration of pressure injection for water uptake

Plate 47: Cristacortis symptoms

Plate 48: Concavities due to blind pocket disease

Plate 49: Impietratura infected fruits and their section

Plate 50: Dieback affected trees in orchard

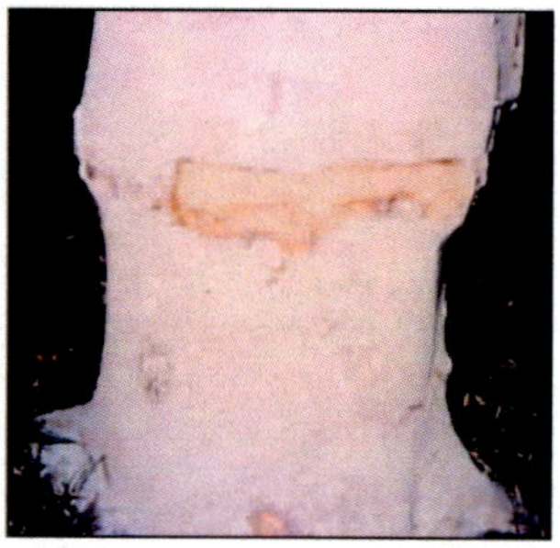

Plate 51: Symptoms of bud union crease

Plate 52: HLB infected tree

Plate 53a: Mottling of leaves due to HLB

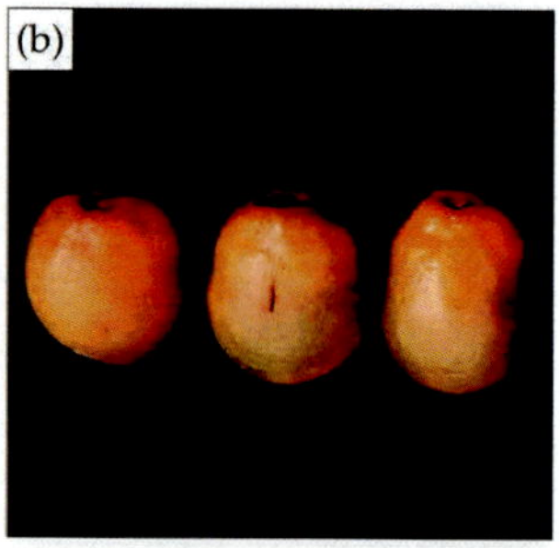

Plate 53b: Discoloration of fruits due to HLB

Plate 53c: Aborted seeds in fruits of HLB affected trees

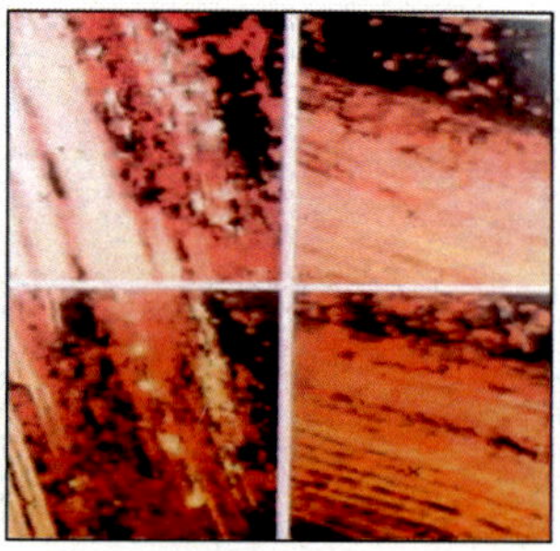

Plate 54: Detection of HLB in immunofluores-cence test

Plate 55a: Citrus stubborn affected tree

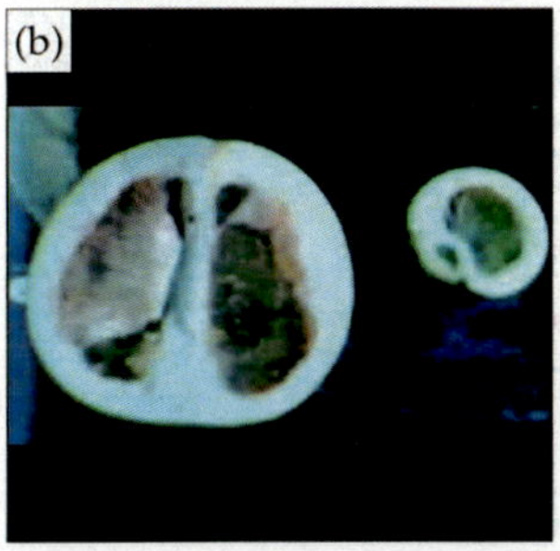

Plate 55b: Curved columella due to stubborn

Plate 56a: Witches broom infected lime tree in India

Plate 56b: Witches' broom inoculated lime seedlings

Plate 57: Rubberywood infected trees at Akola in Maharashtra

Plate 58a: Exocortis symptoms on rootstock

Plate 58b: Exocortis symptoms on Etrog citron (left)

Plate 59: Yellow corky vein symptoms on Kagzi lime

A. On *Citrus aurantifolia*

B. On Etrog citron

C. On Kinnow mandarin
D. On Rangpur Lime

Plate 60: Symptoms of yellow corky vein on different citrus spp. (*Contd...*)

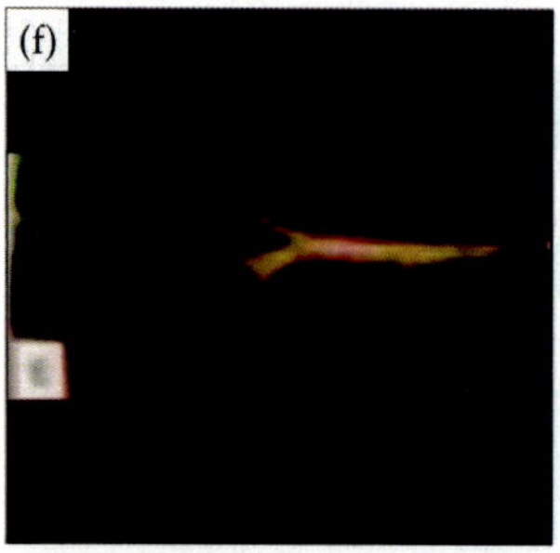

E. On twig of Kagzi lime

F. On *C. reticulata* cv. Nagpur mandarin

Plate 60: (*Contd...*)

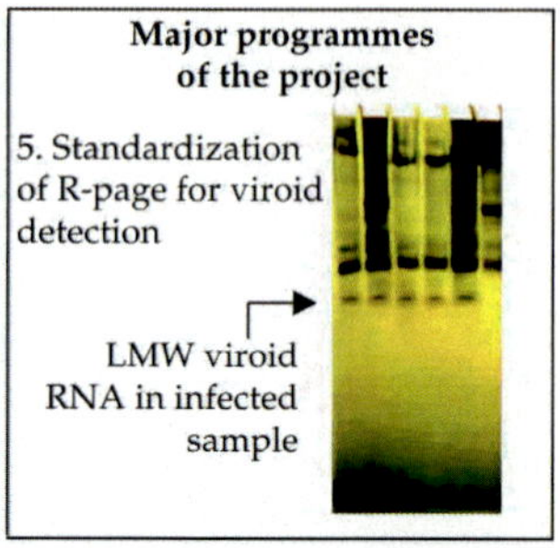

Plate 61: R-PAGE detection of yellow corky vein viroid

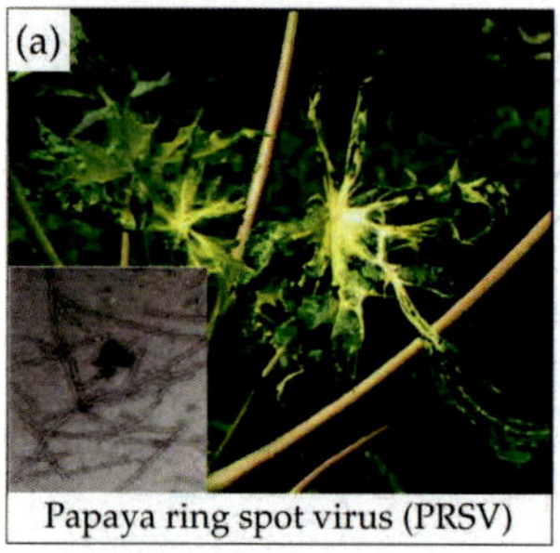

Plate 62a: Papaya ringspot (PRSV)symptoms

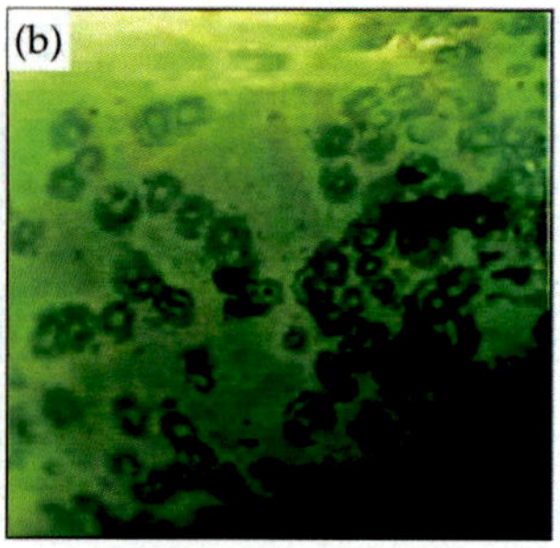

Plate 62b: PRSV on unmatured fruit

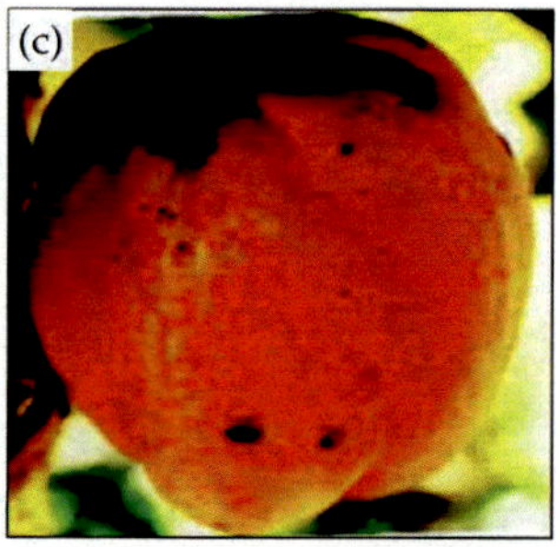

Plate 62c: PRSV on matured fruit

Plate 63: Papaya apical necrosis

Plate 64: Witches broom of Ber (jujube)

Plate 65: Leaf Roll disease of grapevine

4

VIRUS DISEASES OF TROPICAL AND SUBTROPICAL FRUIT CROPS

In tropical and subtropical regions of the world large number of fruit plant species are grown such as mango, citrus, banana, guava, papaya, litchi, grapes, fig, custard apple, cashew, jackfruit, sapota or sapodilla (chiku), jujube, loquat, pomegranate, falsa, avocado, Wood apple, Cape gooseberry, strawberry, date, palm, mulberry, olive, passion fruit, persimmon, tree tomato and a huge number of other arid and small fruits. Fortunately, no virus or virus-like diseases have been reported on large number of fruit trees such as mango, litchi and guava etc.

Tropical fruits play a unique role in India's economy and nutrition security. However, they are subject to disease problems more so because of the wide range of environmental factors. India is the second largest producer of fruits after China, with a production of 44.04 million tonnes of fruits from an area of 3.72 million hectares. A large variety of fruits are grown in India, of which mango, banana, citrus, guava, grape, pineapple are the major ones in tropics. Apart from these, fruits like papaya, sapota, annona, phalsa, jackfruit, ber, pomegranate are also grown in tropical and subtropical regions. Although fruits are grown throughout India, the major fruit growing states are Maharashtra, Tamil Nadu, Karnataka, Andhra Pradesh, Bihar, Uttar Pradesh and Gujarat. This chapter will deal with virus and virus-like diseases on fruit trees where such information is available.

A. VIRUS DISEASES OF AVOCADO

Avocado (*Persea species*) belongs to family *Lauraceae*. The avocado probably originated in southern Mexico but was cultivated from the Rio Grande to central Peru before the arrival of Europeans. The common species and their region of cultivation are: Guatemala (*Persea nubigena* var. *guatamalensis* L. Wms.), Mexican (*P. americana* var. *drymifolia* Blake), West Indian (*P. americana* Mill. var. *americana*). Hybrid forms exist between all three types. The related species are Coyo (*Persea schiedeana* Nees) and Anay (*Beilschmiedia anay* Kosterm). Avocados do well in the mild-winter areas but some hardier varieties can be grown in the cooler parts. West Indian varieties thrive in humid, tropical climates. Guatemala types are native to cool, high-altitude tropics and are hardy. Mexican types are native to dry subtropical plateaus and thrive in a Mediterranean climate. Avocados need some protection from high winds which may break the branches. There are dwarf forms of avocados suitable for growing in containers. There is only one report of a virus like disease avocado from California (Zentmyer, 1955, Wallace, 1958). The disease is known as 'sun blotch' and studied in detail initially by Wallace.

AVOCADO (*PERSEA AMERICANA*) SUN BLOTCH

The sunblotch disease was initially considered as a virus disease (Horne and Parker, 1931) even before the nature of the disease inciting agent was demonstrated. The disease has been recorded in many avocado-growing countries including Australia, Israel, Peru, South Africa, USA and Venezuela. Sunblotch (avocado sunblotch viroid) and trunk pitting (graft transmissible agent) are two virus-like diseases that affect avocado trees.

SYMPTOMS

Symptoms of the disease resembled sunburn on avocado tree (*Persea americana* Mill.) hence the name sunblotch was coined by Coit, 1928. The characteristic symptom of the disease is streaking and spotting of twigs and shoots. The streaks on leaves are yellow in colour but in young trees the streaks may be white or even colourless. On inoculated young seedlings the streaks are pink or red. The first symptom on young budded trees appears as a few small yellow spots without any definite arrangements. Sometimes, the streaked areas are depressed and some vigorously growing shoots with depressed streaks parallel to each other may develop on field trees which may give the tree a fluted appearance. Occasionally, infected trees may develop cluster of leaves with white or pink mottle or variegation. The fruits may also develop streak symptoms usually extending from stem end. Infected trees set fewer fruits which are small and misshaped, thus reducing the yield. Some cultivars/species of avocado are symptomless carrier of the sun blotch infection.

TRANSMISSION

The disease is graft transmitted either by contaminated scion or rootstock. No natural spread was noted either by an aerial vector or natural root grafting. The disease is seed transmitted resulting contamination of rootstock seedlings (Wallace and Drake, 1953). Under experimental conditions, low frequencies (1.8-3.1%) of pollen transmission *via* bees (*Apis mellifera*) were obtained (Desjardins *et al.*, 1979). It is transmitted by mechanical inoculation but with difficulty to avocado and cinnamon (Dale *et al.*, 1982). Plants can also be infected by slashing the stem with a razor blade wetted with a partially purified preparation (Desjardins *et al.*, 1980).

HOST RANGE

The disease is transmitted by graft inoculation only to members of the family *Lauraceae* including *Persea americana, Cinnamomum camphora, Cinnamomum zeylanicum* and *Ocotea bullata* (da Graça and van Vuuren, 1980). Avocado cultivar Hass Seedlings when graft inoculated, yellow, orange or white streaks or spots on the stems and petioles develop after 2 months to 3 years after inoculation and may also show variegation and distortion of the leaves. Symptom development was accelerated by incubating at high temperatures (30-32°C). Guatamalan cultivars (*e.g.* Hass) was found to be most suitable for propagation and purification of pathogen (Palukaitis *et al.*, 1981).

THE PATHOGEN

Recent investigations have revealed that avocado sun blotch disease is not caused by a virus but by a viroid. The scientific classification of the associated viroid was put in family *Avsunviroidae*, genus *Avsunviroid* and Species *Avocado sunblotch viroid.*

Nature of infective agent as viroid was established by Palukaitis *et al.*, 1979. Later the viroid was isolated from infected samples and its nature was established (Palukaitis and Symons, 1980; Palukaitis *et al.*, 1981; Dale and Allen, 1979; Allen *et al.*, 1981). An infective single-stranded circular RNA molecule of 247 nucleotides was associated with the disease. Avocado sun blotch viroid (ASBVd) is the smallest known viroid having 247 nucleotides. The titre of viroid varies from tree to tree. Symptomless trees have high titre of viroid than symptomatic trees. ASBVd RNAs are circular with positive and negative polarity which suggests that the replication proceeds through a symmetric pathway with two rolling circles where these two circular RNAs are the templates. Sunblotch viroid (247 nucleotide residues with molecular weight of 0.8 x 10^5; (Symons, 1981) is considerably smaller than potato spindle tuber viroid (359 residues) (Gross *et al.*, 1978) and chrysanthemum stunt viroid (356 residues) (Haseloff and Symons, 1981) but it has 18% sequence homology with them (Symons, 1981). It is also smaller than citrus exocortis viroid (371 residues) (Visvader *et al.*, 1982; Gross *et al.*, 1982) but is similar in size to the smallest RNA component of cadang-cadang viroid (Haseloff *et al.*, 1982).

DIAGNOSIS AND MANAGEMENT

Indexing on susceptible hosts can be used to detect viroid infection. When infection results from rootstock seedlings, the first symptom usually appears as a very small yellow spot close to but above the bud union. Because of the length of time to symptom expression, avocado sunblotch viroid (ASBVd) can be more conveniently diagnosed by nucleic acid hybridization with complementary DNA (Palukaitis *et al.*, 1981; Allen and Dale, 1981). RNA was extracted from plants infected with avocado sunblotch viroid (ASBVd) and was analyzed by electrophoresis in polyacrylamide gel. The ASBVd related fraction was eluted from the gel, labelled with [^{32}P] using polynucleotide kinase and used as a probe for hybridization with a purified ASBVd-RNA preparation dot spotted on nitrocellulose paper. Positive self-hybridization indicated a high degree of internal complementarity. Dot spots of whole cell RNA and of leaf sap from ASBVd infected plants were shown to hybridize with the labelled probe. This hybridization procedure proved to be 16–64 times more sensitive in diagnosing ASBVd when compared with polyacrylamide gel analysis (Rosner *et al.*, 1983). The sun blotch viroid can also be effectively detected by PCR. The best method of management is to use viroid-free planting material. To achieve this both rootstock and scion should be indexed.

A TOBAMO VIRUS INFECTING AVOCADO

Avocado sunblotch is the only viroid known to infect avocado. A tobamovirus related to tobacco mosaic virus was isolated from symptomless infected avocado seedlings in Israel, but its distribution and importance is unknown. It can be readily distinguished from avocado sunblotch viroid because it is mechanically transmissible to *Datura stramonium, Chenopodium amaranticolor* and *Nicotiana glutinosa* (Alper *et al.*, 1978).

B. VIRUS DISEASES OF BANANA

Banana evolved in the humid tropical regions of South East Asia with India as one of its centres of origin. Modern edible varieties have evolved from the two species – *Musa acuminata* and *Musa balbisiana* and their natural hybrids, originally found in the rain forests of S.E. Asia. During the seventh century AD its cultivation spread to Egypt and Africa. At present, banana is being cultivated throughout the warm tropical regions of the world between 30°N and 30°S of the equator.

Banana is a very important economically profitable plantation crop of India having a high degree of export potential. Banana (*Musa* sp.) is the second most important fruit crop in India next to mango. Its year round availability, affordability, varietal range, taste, nutritive and medicinal value makes it the favourite fruit among all classes of people. In India, it is called by several names such as 'kela' and 'Amrit' in Hindi, in Bengali, it has several names such as 'Kaula', 'Champa', 'Chinese Champa' or 'Dhakkai' and it is called as 'Valai' in Tamil and 'Vala' in Malayalam. While most banana species grow best in the warmer climates, there are varieties which can be better grown in cold climates. Banana leaves are huge - depending on the variety, they can be up to 2' wide and 9' long and are used as food plates in South India.

Banana plants are monoecious. They are cultivated primarily for their fruit, and to a lesser extent for the production of fibre and as ornamental plants. The banana stem is actually a pseudostem and not a true stem. Each pseudostem can produce a bunch of green bananas which when ripened often turn yellow or sometimes red. A variety discovered recently in a rainforest in Asia bears purple banana. After bearing, pseudostem is removed and it is naturally replaced by another pseudostem. The fruit actually comes from the female flowers, which, strangely enough, develop without pollination. The fruit grows in a cluster, called a hand and each hand had several fingers (bananas). Not all varieties/ species are edible for human consumption. The commonly eating banana seen in stores is the 'Cavendish' variety. Banana and plantains are grown in about 120 countries. Major banana producing countries are: Brazil, Ecuador, China, Colombia, Costarica, India, Indonesia, Mexico, Phillipines, and Thailand.

Internationally, banana ranks first in the terms of volume and second after citrus fruit in terms of value. In India, banana ranks first in production and third in area among fruit crops. It accounts for 13% of the total area and 33% of the production of fruits. Global production of banana was about 76436 thousand tons of which India contributed 24.5%.

Area under banana production in the world is 4306 thousand ha. In India, area under banana is 565.1 thousand ha with a production of 18701.9 thousand tons (Indian Horticulture database 2006). In 2009-10, the area under banana cultivation was increased to 748 thousand ha with a production of 26996 thousand tons. Worlds total export of banana accounted to 14.7 million tones in 2002 (FAO 2003). Worlds' banana exports are valued over USD 4 billion per year. In India, banana was exported 8655 tons in 2002-3 which was increased to 11475 tons during 2006-07 (APEDA database 2006-07).

In India, major varieties of banana under cultivation are Dwarf Cavendish, Robusta, Poovan (Mysore AAB), Rasthali (Silk AAB), Ney Poovan (AB), (AAB), Nendran (Plantain AAB), Karpuravalli (AAB), Udayam (AAB) and Monthan (AAB). Recently, new varieties like Co. 1 (Coimbatore), Hy.1, Hy. 2, Gandevi Selection have been developed and brought under cultivation in India. Viruses infecting banana are the major threats in achieving the target production and productivity.

Banana is a rich source of carbohydrate and is rich in vitamins particularly vitamin B6, Vitamin C and potassium. It is also a good source of potassium, phosphorus, calcium and magnesium. The fruit is easy to digest, free from fat and cholesterol. Banana powder is used as the first baby food. It helps in reducing risk of heart diseases when used regularly and is recommended for patients suffering from high blood pressure, arthritis, ulcer, gastroenteritis and kidney disorders. Processed products, such as chips, banana puree, jam, jelly, juice, wine and halwa can be made from the fruit. The tender stem, which bears the inflorescence is extracted by removing the leaf sheaths of the harvested pseudostem and used as vegetable. Plantains or cooking bananas are rich in starch and have a chemical composition similar to that of potato. Banana fibre is used to make items like bags, pots and wall hangers. Rope and good quality paper can be prepared from banana waste. Banana leaves are used as healthy and hygienic eating plates. Banana is known to be infected by four major viruses, bunchy top, bract mosaic, cucumber mosaic and banana streak. Abaca mosaic potyvirus also infects banana. All these viruses will be dealt with in detail in this chapter.

BANANA BUNCHY TOP

Among virus diseases of horticultural plants, banana bunchy top is the most serious and devastating disease. It is also called as 'Strangles", curly top or cabbage top.The origin of bunchy top of banana is not known. In 1879, it was first observed in Fiji on Cavendish banana and local plantations (Magee, 1927) and by 1894 many Fiji plantations were abandoned because of severity of infection (Wadlaw, 1972). The disease was reported from Egypt in 1901, Australia and Sri Lanka in 1913. Banana bunchy top disease is now widely distributed in several countries in Africa, Oceania, Asia and in USA (Dale, 1987, Frison and Putter, 1989) (CMI map 19). In India, around 1940, bunchy top disease was observed in Kerala, Maharashtra, Tamil Nadu, Orissa, and Assam states (Kamat and Patel, 1957; Capoor, 1959). This disease was listed under diseases of national importance in 1943. The disease is now spread to all the banana growing states of India. A loss of Rs. 40 crores annually has been estimated due to this disease in Kerala state alone. In India, the disease is distributed in all banana growing regions and some of the elite cultivars like hill banana (Plate 16) and Virupakshi has been wiped out in Sirumalai and Palani hills of Tamil Nadu.

SYMPTOMS

All growth stages of the banana plants are susceptible to bunchy top infection. The leaves of affected plants become smaller and narrow with wavy margin. The smaller leaves bunch together at the top of pseudostem giving rosette or bunchy look to the infected plants (Plate 13A). Dark green streaks of variable length are present on the midrib, leaf veins and petiole of the leaves, extending down in the pseudostem (Plate 15) (Thomas *et al.*, 1994). However, the streaks on petiole of Cavendish subgroup may be rare or absent (Magee, 1953). The initial leaf symptoms may be observed as soon as the first leaf appears above ground. The margin of such leaves become wavy and they roll upward. The leaves of affected plants also show marginal chlorosis and brittle texture. The young leaves are upright and erect in habit. Sometimes, transverse wrinkling occurs along the length of lamina. Leaf stalk is reduced in length. The affected plants are highly stunted and remains between 9 inches to two feet in height.

The most characteristic symptoms of the disease develop on young leaves as dark green streaks often broken in dots when viewed against light. These streaks are about 0.75 mm wide and vary in length upto 2.5 cm. A powdery bloom covers the midrib and pertole, if this is rubbed off, the dark green streaks can be clearly seen. Pale white streaks may be seen along the secondary veins. The presence of interrupted dark green streaks along the secondary veins of the lamina or along the midrib or petiole, are considered to be the most definite and reliable symptoms of bunchy top (A.S. Summanwar personal communication). In the primary infection which comes through infected suckers, the plants remain small and stumpy and hardly any bunches are obtained from such plants

(Raychaudhuri and Nariani, 1989). However, in the secondary infection which comes in the standing crop, a few bunches of small fruits may develop but the opening bunches are choked by the pseudostem and may split them instead of emerging in normal fashion. The bunches are of reduced size and fruits are unmarketable. The lower hand of the bunch often dies off. The root system of infected plants show decay but perhaps bunchy top is not responsible for root decay (Raychaudhuri and Nariani, 1989). Severely infected plants may produce no fruits or the bunch may not emerge from the pseudostem. The diseased plants do not die but remains in the field and act as a source of inoculum for virus spread by aerial vectors.

TRANSMISSION

The disease is of systemic nature and primary spread is through use of infected planting material (suckers/rhizomes). All the rhizomes/daughter suckers produced by a diseased plant carry the virus. The virus is not sap transmissible. The disease is transmitted through banana aphid, *Pentalonia nigronervosa* (Plate 14) (Magee, 1927, 1940). The virus is probably retained when the vector moults. The vector multiplies on upper leaf sheath and petiole of banana plants. The transmission is of semi-persistent type. The vector requires long acquisition feeding period of several hours (4-18h) to become viruliferous and once acquired the virus the vector start transmitting the virus within few hours to a few days and will go on transmitting the virus through its adult life for a period of 15-20 days. After inoculation, the symptoms appear within 7-8 weeks. Rao (1980) reported successful transmission of bunchy top disease through other aphid species like *Aphis gossypii, A. spiraecola* and *M. persicae* under glasshouse conditions to *Cucumis sativus, C. melo, Gynura aurantiaca* and *Catharanthus roseus*.

HOST RANGE

Susceptible host species are found in the Family *Musaceae*. The following host species were susceptible to experimental transmission by vector: *Musa sapientum* (cv. Cavendish), *M. fehi, M. textilis* and *Heliconia* sp., plantain, abaca, wild banana, bird of paradise and ginger. Inoculated plants show chlorosis and stunting. The banana cultivars *viz.*, Virupakshi, Nendran (Plate 17), Robusta, Neypoovan and Monthan are susceptible to bunchy top infection in India and cause considerable losses. All the commercial banana varieties are susceptible to the disease with varying degree of intensity. *Colocasia esculenta* was found to be a symptomless reservoir host for bunchy top in the first generation (Ram and Summanwar, 1984). Typical mosaic symptoms were observed in subsequent generations. *C.*

esculenta is grown as intercrop among banana plantations in some banana growing region of Maharashtra, India. Hence it serves as reservoir of the virus and also the host of vector, *P. nigronervosa* (Marathe *et al.*, 1982). In addition, *Canna indica* and *Hedychium coronarium* have also been suggested to act as alternate hosts of the virus (Su *et al.*, 1993). Diagnostic host species is *Musa sapientum* cv. Cavendish which develops chlorosis and stunting upon inoculation.

THE PATHOGEN

Wu and Su (1990) purified bunchy top virus and claimed it to be ssRNA virus with the virion size of 20-22 nm in diameter. Dale *et al.* (1986) obtained ds DNA from bunchy top infected plants. However, Harding *et al.* (1991) purified the Virus-like particles from banana plants infected with banana bunchy top disease. The virus particles were found to be isometric with a diameterof 18 to 20 nm with a density of 1·28 to 1·30 g/ml in cesium sulphate. Associated with these particles were a ssDNA of about 1 kb and one major protein of M_r 20100 (Harding *et al.*, 1991). The virus particles were found in phloem tissues of infected plants. The virus particles from Indian isolates were isometric and averaged 18-20 nm in diameter (Selvarajan *et al.*, 2002). ICTV has now classified banana bunchy top virus as: genus: *Babuvirus*, family: *Nanoviridae*,and Type species: *banana bunchy top virus* (Brunt *et al.*, 1996).

The genome of banana bunchy top virus (BBTV) is multicomponent single-stranded DNA. The genome consists of six segments of circular ssDNA components. Further, BBTV DNA-1 has previously been shown to be present in all BBTV isolates tested from ten countries. The genome of banana bunchy top virus (BBTV) has been fully sequenced (Dale *et al.*,1986; Harding *et al.*, 1993; Brunt *et al.*, 1996). Complete sequence is 1300 nucleotides long. The sequences of BBTV DNA-1 from the different isolates divided into two groups, the South Pacific group and the Asian group. It has been established that BBTV DNA-2, 3, 4, 5 and 6 are present in all isolates tested from eight countries suggesting that BBTV DNA-1 to 6 are integral components of the BBTV genome (Karan *et al.*, 1997a). The sequences of BBTV DNA-6 from different isolates also divided into two groups, the South Pacific group and the Asian group (Karan *et al.*, 1997b). Interestingly, there was far greater sequence variability between the sequences within the Asian group than those within the South Pacific group of isolates suggesting that either BBTV has been present in bananas in this region for a longer period or there has been more than one introduction of BBTV into bananas from another host in this region. Mild and severe strains have been reported in BBTV (Selvarajan *et al.*, 2004). Viruses similar to BBTV are: Coconut foliar decay virus (CFDV) and Subterranean clover stunt Virus (SCSV). Association of a nanovirus with foorkey disease of large cardamom in India had 83% similarity of replication associated gene of BBTV (Mandal, 2010).

DIAGNOSIS AND MANAGEMENT

BBTV is found to occur in latent form in some cultivars hence visual infection will not help to diagnose the disease. In India, sero-diagnostic kits were developed for BBTV detection and are being used to detect the virus in DAC-ELISA and DIBA (Selvarajan *et al.*, 2002). The double antibody sandwhich (DAS) – ELISA detected *Banana bunchy top virus* in both single and group of IInd instar nymphs and adults of the banana aphid *Pentalonia nigronervosa* from infective colony but not in Ist instar nymphs. The virus was not detected in the nymphs and adults from non-infective colony. The virus was persistent in the aphid vector throughout their survival period (Thiribhuvanamala *et al.*, 2005). Monoclonal and polyclonal antibodied are commercially available for virus diagnosis in ELISA (Frison and Putter, 1989). The virus can also be detected by polymerase chain reaction (PCR) in plants and also in aphid vector. A product of 600bp was amplified in RT-PCR from Indian isolates (Selvarajan *et al.*, 2002, 2008). Youngest unopened leaf rolled in the pseudostem, the centre core region and meristematic regions are better for PCR detection. Technology was also developed to detect BBTV by immunocapture PCR, by DNA-DNA hybridization and simultaneous detection of BBTV with other DNA or RNA viruses of banana (multiplex PCR) (Selvarajan *et al.*, 2008, 2010, 2011). Non-radioactive nucleic acid probes were developed for all DNA components using digoxigenen PCR based kits for tissue culture raised banana. A colour technique was developed for detection of banana bunchy top and infectious chlorosis in banana suckers with the application of 2,3,5 triphenyl tetrazolium chloride (Summanwar and Marathe, 1982). It is interesting that banana plants are being raised in tissue culture now and are being used for planting virus-free banana plantations commercially. However, some viruses may escape elimination through tissue culture if proper care is not taken for identification of mother plants and size of the tissue to be used for propagation (Plate 13B). To avoid dissemination of viruses through tissue culture raised banana plants, it is highly desirable to use the diagnostic kits and tecnology to identify virus-free mother plants and also testing of the plants after hardening as per scientific protocols so that only healthy planting material is reached to the farmers. However, care should be taken:

- To use disease-free planting material which may be obtained by meristem culture combined with heat therapy (Ramjos and Zamora, 1990; Wu and Su, 1991).

- Control of aphid vector by using either soapy water or insecticidal soap. Sprays must be done thoroughly on petiole, furled leaves, whorls and on young suckers.
- Recommended insecticides and doses must be sprayed at 2-3 months interval.
- Removing and rouging of infected banana plants is important to avoid the spread of the virus. Affected plants must be chopped off, dry and then burn. The rhizomes of affected plants should be killed with suitable weedicides (Magee, 1940, 1953).
- In India, better yielding commercial banana clones devoid of viruses identified in: Grand Naine, Robusta, Poovan, Ney Poovan, Kanthali and hill banana at National Research Centre for Banana, Tiruchirapally, Tamil Nadu, India.

BANANA BRACT MOSAIC

The banana bract mosaic disease was first reported from the Philippines where it was spreaded very fast covering the banana plantations within very short period (Magnaye and Espino, 1990). In India, the disease is commonly found on French plantain in Kerala state and is called as 'Kokkan' ((Samraj *et al.*, 1996, Dhanya *et al.*, 2007). It was also observed in Coimbatore and Trichi region of Tamil Nadu and Pune region of Maharashtra (Rodoni *et al.*, 1997). The disease has also been recorded in Bangladesh, Sri Lanka (Thomas *et al.*, 1997), and suspected to be present in Pakistan, Costa Rica and several African countries including Uganda, Ghana, Zanzibar and South Africa (Anonymous, 1992; Rivera *et al.*, 1993; Tushmereirwe *et al.*, 1996). Bract mosaic disease was observed only in a few plants of banana in Coimbatore and Trichi region of Tamil Nadu and Pune region of Maharashtra in a joint survey of senior author with some Australian Scientists in 1995-96. However, during the last decade the disease has spreaded very fast in banana growing regions of Kerala, Tamil Nadu, Karnataka and Andhra Pradesh (Singh *et al.*, 2000; Cherian *et al.*, 2002; Kiranmai *et al.*, 2005; Mokashi *et al.*, 2005). The associated virus can cause yield losses up to 40% (Bateson and Dale, 1995; Espino *et al.*, 1990). The yield loss caused by this virus in Kerala (India) was 70% and 40% in cvs. Robusta and Nendran respectively (Cherian *et al.*, 2002). The disease poses considerable quarantine risk to other banana growing region.

SYMPTOMS

The characteristic symptoms of the bract mosaic disease appear as dark broad discontinuous streaks on the bract of the banana inflorescence (Plate 18) and green or brown spindle shaped streaks that are irregularly scattered along the petioles (Magnaye and Espino, 1990). Shortening of bunch internodes is also a characteristic symptom. On removal of dead leaf sheaths, large dark coloured strips of varying length with mottled discolouration on the pseudostem can be seen (Plate 19). Greenish to brownish broad, irregularly scattered spindle streaks develop along the petiole sometimes with raised veins. Fruits fails to fill on infected plants in India. Susceptible host species are found in the Family *Musaceae*. In India, Natural infection of bract mosaic was observed on banana cultivars *Musa balbisiana*, Klue Terapod, Cavendish, Pisang Awak and Bluggoe (Rodoni *et al.*, 1997).

TRANSMISSION

The virus is transmitted through contaminated banana suckers. The virus is also transmitted by aphids, *Pentalonia nigronervosa, Rhopalosiphum maidis,* and *Aphis gossypii*. It was demonstrated that Banana bract mosaic virus (BBrMV) is a non-circulative virus. The aphid failed to transmit BBrMV to abaca. BBrMV was not mechanically transmitted to banana and to 14 other plant species belonging to 7 families. The virus was also not transmitted through soil (Munez, 1992).

THE PATHOGEN

The virus particles of banana bract mosaic virus are elongated with helical symmetry. The capsid is filamentous, flexuous with a clear modal length of 750 nm. The virus has a +tive sense single stranded RNA as its genome and belongs to family *Potyviridae*, genus-*potyvirus* and species – *banana bract mosaic virus* (Anonymous, 1992). The virus particles recovered from purified preparations were flexous rods measuring 750 nm in length and 11 nm in width (Munez, 1992). The virions have a 38 kDa coat protein that react with general potyvirus antisera (Batson and Dale, 1995).

The genome of BBrMV is monopartite. Only one particle size of linear, positive-sense, single-stranded RNA was recovered. The viral genome encodes structural proteins and non-structural proteins. The virus was cloned and sequenced from BBrMV isolates from banana cultivars *Musa balbisiana*, Klue Terapod and Pisang Awak from India. Sequence comparison of the C-terminus of the coat protein-coding and 3′ untranslated regions revealed that the Indian isolates of BBrMV had greater than 96.6 and 97.2% homology with a Philippine isolate at the nucleotide and amino acid levels (Rodoni *et al.*, 1997). The coat protein gene of the virus is 71% identical to that of *Maize dwarf mosaic virus*, and its 3′UTR is most similar to that of ornithogalum mosaic virus (40%) (Bateson and Dale, 1995). The virus reacts to the antiserum prepared to the conserved core regions of a mixture of potyviruses. The virus does not show serological relationships to *Sugarcane mosaic virus*.

DIAGNOSIS AND MANAGEMENT

The diagnosis of the disease can be done on the basis of specific symptoms it produces in banana. The virus can also be detected in electron microscopy (Rodoni *et al.*, 1997), by serological tests like ELISA and also by RT-PCR (Espino *et al.*, 1990; Bateson and Dale, 1995; Rodoni *et al.*, 1997; Sankaralingam *et al.*, 2006). A multiplex, immunocapture PCR (M-IC-PCR) was developed for the simultaneous detection of three viruses from crude sap extracts of banana and plantain (*Musa* spp.). A reverse transcription step was required for *Banana bract mosaic virus* and *Cucumber mosaic virus*, which have ssRNA genomes. The detection of *Banana bunchy top virus* (ssDNA genome) was not adversely affected by inclusion in this step. All the three viruses could be detected simultaneously from a mixed infection. Identification and detection of individual viruses was achieved through the visualisation of discretely sized PCR amplicons by gel electrophoresis. Alternatively, a colourimetric microplate detection system utilising digoxigenin-labelled virus-specific probes was used. The latter assay was up to five times more sensitive than detection by gel electrophoresis and between 25 and 625 times more sensitive than ELISA for the detection of various viruses. Careful selection of PCR primers was necessary to ensure the detection of a wide range of virus isolates and to avoid detrimental interactions between heterologous templates and primers (Murray *et al.*, 2001).

The bract mosaic potyvirus can be eliminated by tissue culture techniques which have become popular during the last decade. The methods suggested for vector management for banana bunchy top can also be applied for bract mosaic virus. Bract mosaic disease of banana can be mitigated in cultivar Ney Poovan with application of higher doses of fertilizer. However, the fertilizer doses have no effect in ratoon crop (selvarajan *et al.*, 2009).

BANANA MILD MOSAIC

Banana and Plantains (*Musa* AAB) are important sources of food and income for millions of people in many developing countries. Banana mild mosaic virus (BanMMV) is a previously undescribed filamentous virus infecting *Musa* spp. Banana mild mosaic has a very wide geographical distribution and its presence was reported from Guadeloupe and Martinique, French Guiana and Reunion. It is an unassigned species in family *Flexiviridae*. The symptoms of the disease in plantain showed mild chlorotic streak. Electron microscopy of symptomatic tissue extracts revealed the presence of filamentous virus-like particles approximately 800 nm long (Reichel *et al.*, 2003). The complete genome has been sequenced from PCR clones and consists of 7352 nucleotides, encoding five open reading frames (ORFs) of 205, 25.5, 12.4, 8.0 and 26.8 kDa, respectively. Banana mild mosaic is most closely related to *fovea* and *carlaviruses*, based on phylogenetic analysis using a portion of the viral replicase gene. Analysis of other parts of the genome revealed similarities with fovea-, carla and potexviruses, but the virus was not clearly aligned to any previously recognised genus (Gambley and Thomas, 2001). BanMMV was detected by a method known as polyvalent degenerate oligonucleotide RT-PCR (PDO-RT-PCR) assay. PDO inosine-containing primers were found to be well suited to the detection of BanMMV, despite its high molecular diversity, but not to that of the highly conserved Banana virus X (BVX), for which species-specific primers were designed. A polyclonal anti-BanMMV antiserum was raised and successfully used for the immunocapture (IC) of BanMMV viral particles from leaf extracts, leading to the development of a PDO-IC-RT-nested PCR assay (Teycheney *et al.*, 2007; Reichel *et al.*, 2003). This virus is most probably spread mainly inadvertently in infected planting material because it is not widely recognized as a potential pathogen of banana. It was suggested that virus indexing of mother plants for *in vitro* plant production could be used to avoid movement of this virus.

BANANA STREAK

Banana streak virus disease is known to occur in most of the banana growing countries and has been considered to be the most important disease of banana. It has been reported from Africa, Europe, Latin America, Carribean and Asia Pacific region (Diekmann and Putter, 1996; Lockhart and Jones, 1999). The disease was also reported from Uganda (Tushmereirwe *et al.*, 1996), Australia (Daniells *et al.*, 1995), Cuba (Javer *et al.*, 2009) and Nigeria (Dahal *et al.*, 1999). In South East Asia, The disease has been reported from India on cvs. BRS-1 (AAB) and Mysore Poovan (AAB) and confirmed by electron microscopy and PCR (Cherian *et al.*, 2004). The disease has also been reported on banana cv. Mysore (AAB) in Taiwan which was introduced from Australia (Su *et al.*, 1997; Braithwaite, 1995). Banana streak virus (BSV) is a badnavirus that causes a leaf streak disease of banana and plantain (*Musa* spp.). Symptom expression is sporadic but fruit yields are often significantly reduced (Lassoudièrre, 1974; Dahal *et al.*, 2000; Daniells *et al.*, 2001). It has a deleterious effect on the productivity of infected plants as well as being a major constraint to *Musa* breeding programmes and germplasm dissemination.

SYMPTOMS

The diagnosis of the disease is often unreliable as the symptoms are variable and can be confused with those of banana mosaic caused by cucumber mosaic virus. The most characteristic foliar symptoms of banana streak disease are chlorotic streaks or spindle shaped patterns which are first chlorotic and become necrotic with time. In older leaves black streaking can be seen. The leaf lamina may also be narrower, thicker and become torn. Other symptoms include stunting of the plant, constriction of the bunch on emergence (choking), altered phyllotaxis and detachment and splitting of the outer leaf sheaths of the pseudostem. In severe instances, cigar leaf necrosis, internal necrosis and collapse of the pseusodstem is observed (Lockhart, 1995). Most common symptoms of the disease appear as chlorotic streaks and black streaks and dots on the leaves of infected banana (Plate 20). Infected plants also show mild mottle symptoms including diamond-shaped chlorotic spots and fine striate streaks on their leaves (Plate 21). Different cultivars of banana showed different symptoms. The Peichiao of Taiwan Cavendish (AAA) produced severe typical streak symptoms, while only mild streak symptoms developed on Mysore (AAB) banana plants. No typical streak symptoms developed on Taiwanese and exotic cultivars of Latundan banana (AAB). The infected Latundan plants were symptomless or produced mild chorotic mottling. Typical symptoms include yellow streaks on leaves, splitting of the pseudostem and abnormal emergence of the bunch from the middle of the pseudostem, bearing fewer fruits of smaller size (Jones and Lockhart, 1993).

The artificial inoculation onto Cavendish tissue culture plantlets induced severe streak symptoms, including uniform vein yellowing, and chlorotic banding and stripes (Braithwaite, 1995). Temperature can affect symptom development in Cavendish plants (Dahal *et al.*, 1998). Symptoms of varying severity developed on leaves emerging in different seasons on the same plant. Leaves which emerged during the cold winter did not produce symptoms or produced very mild symptoms, while leaves emerging during spring produced distinct symptoms. Milder symptoms develop or masked on leaves growing during the hot summer. Dark colored stripes on the pseudostem, as caused by bract mosaic virus (BBrMV) (Magnaye and Espino, 1990), were observed in Mysore plants showing streak symptoms. The maturity of banana plants also affected symptom expression. Milder symptoms developed in larger or older plants. The symptoms on young and spindle leaves were distinct and severe and became milder or indistinct towards maturity of the leaf. *Musa accuminata* clones show severe streak symptoms upon infection. Infection can cause delay in fruit bunch emergence, harvest and can reduce bunch weight (Lockhart and Jones, 1999; Daniells *et al.*, 2001).

TRANSMISSION

The virus is transmitted by vegetative propagation. It is not transmitted by mechanical inoculation (Lockhart and Jones, 1999). It has been reported to be seed transmitted (Daniells *et al.*, 1995). However, report of seed transmission of BSV must now be considered inconclusive, as the possibility that infection in the seedlings arose from expression of integrated viral DNA cannot be ruled out. *Banana streak virus* (BSV) is naturally transmitted by several mealybug species probably in a semi-persistent manner (Lockhart and Jones, 1999; Kubiriba *et al.*, 2001) and by activation of endogenous sequences integrated in the genome of *Musa balbisiana*, one of the progenitors of synthetic and natural hybrids (Dallot *et al.*, 2001). In glasshouse experiments, *Planococcus citri* and *Saccharicoccus sacchari*, both known to colonise banana, have been shown to be capable of transmitting BSV (Lockhart, 1995). Two biotypes short and long tailed *P. citri* are known to occur and multiply on citrus plants. Both the biotypes transmit BSV with equal efficacy. The young instars play a more important role in BSV transmission than adults and old instars. The alate adults with degenerated mouth parts cannot transmit BSV. Acquisition feeding period for the citrus mealybug with BSV is around 24h. The BSV isolate is transmitted by the citrus mealybug (*Planococcus citri*) to Cavendish (AAA) tissue-culture plantlets which produced severe streak symptoms within three to four weeks after inoculation feeding.

HOST RANGE

Banana streak virus (BSV) has been experimentally transmitted only to plants within the family *Musaceae* using mealybug vectors. Many different banana genotypes are susceptible to BSV infection, including *Musa acuminata* (*Musa* AA and AAA groups) and *M. balbisiana* (*Musa* BB group) types, as well as hybrids of these two species (*Musa* AAB, ABB, ABBB groups) (Harper *et al.*, 1999a; Lockhart and Jones, 1999; Geering *et al.*, 2000). *Enseteventricosum* is both an experimental and natural host of this virus (Lockhart, 1995). The following plant spp. are not susceptible to BSV: *Avena sativa, Hordeum vulgare, Nicotiana benthamiana, Nicotiana glutinosa, Nicotiana tabacum, Nicotiana x edwardsonii, Zea mays (*Lockhart, 1986).

THE PATHOGEN

Banana streak virus (BSV) was first described by Lockhart (1986). The virus has been purified and purified preparations had nonenveloped bacilliform-shaped particles approximately 130-150 nm in length with a diameter of 30 nm (Lockhart, 1986; Lockhart and Jones, 1999). The Banana streak virus (BSV) belongs to badnaviruses (Lockhart, 1986). The Banana streak virus is a VI RNA Reverse Transcribing virus in Family: *Caulimoviridae,* Genus: *Badnaviruses,* Species: *banana streak virus.*

The virus is highly variable and show genetic diversity (Cherian *et al.,* 2004; Harper *et al.,* 2004; Geering *et al.,* 2000). The concentration of BSV replicating in the three cultivars was detected by PCR followed by electrophoresis analysis. The highest content of BSV was detected in the Cavendish plants, with a lower content in Mysore plants and the lowest content in plants of Latundan. Badnaviruses in general are restricted to the cytoplasm and are not associated with any inclusion bodies or membrane-bound structures (Hull *et al.,* 2000).

Banana streak virus has a circular double stranded DNA (dsDNA) genome of approximately 7.4kb (Lockhart and Jones, 1999; Harper and Hull, 1998). The genome is not segmented and contains a single molecule of circular DNA (Lockhart, 1990). The genome is fully sequenced, complete sequence is 7389 nucleotides long (Harper and Hull, 1998). Incomplete sequences, comprising parts of ORF III and the intergenic region, are available for BSV-Cav, BSV-Mys and BSV-GF (Geering *et al.,* 2000). Encapsidated badnavirus DNA typically has two discontinuities, one in each strand. With all badnaviruses, the discontinuity in the (-) sense

DNA is adjacent to the 3'-end of the cytosolic initiator methionine ($tRNA^{met}_i$) binding site (Medberry *et al.*, 1990). The partial polyprotein gene of ORF III of Indian isolates of BSV has also been cloned and sequenced (Accn. Nos. AY101188 and AY101189) (Cherian *et al.*, 2004). The genome has terminally redundant sequences. The terminally redundant sequences have direct terminal repeats; are reiterated internally in inverted form. The viral genome encodes structural proteins and non-structural proteins. The virus codes for an RNA-dependent DNA polymerase. Replication involves a reverse transcription step. The isolates of BSV are highly variable (Cherian *et al.*, 2004; Jaufeerally-Fakim *et al.*, 2006).

An unusual attribute of BSV-OL is that its DNA is integrated in the B-genome of cultivated *Musa* (Ndowora *et al.*, 1999; Harper *et al.*, 1999b; Geering *et al.*, 2001). There is evidence to suggest that with some *Musa* hybrids, infection develops *de novo* following activation of this integrated DNA. Integration is not an essential step in the replication cycle of the virus, and DNA has possibly become integrated over time by a process of illegitimate recombination (Ndowora *et al.*, 1999). Sequencing of plant nuclear genomes reveals the widespread presence of integrated viral sequences known as EPRVs (endogenous pararetroviruses). Banana is one of the three plant species known to harbor infectious EPRVs. *Musa balbisiana* carries integrated copies of *banana streak virus* (BSV), which are infectious by releasing virions in interspecific hybrids. *Banana streak virus* (BSV) is causing increasing concern in almost every producing area of banana and plantain (*Musa* spp.) worldwide. This situation appeared partially linked to some breeding lines and micropropagated hybrids. A complete BSV sequence integrated into the genome of a triploid plantain has been recently characterised. It has been hypothesised that it could give rise to infectious virus *via* recombination (Dallot *et al.*, 2001). Philippe Gayral (2008) described for the first time the full genomic and genetic organization of the two EPRVs of BSGfV present in cv. PKW, in reply to the challenge facing both scientists and breeders to identify and generate genetic resources free from BSV.

The virus shows serological relationships with most of the badnaviruses except cacao swollen shoot. The antisera of BSV Morocco isolate, BSV Rawanda isolate, BSV Trinidad Mysore isolate reacted positively with the antigen of citrus yellow mosaic badnavirus from India (Ahlawat *et al.*, 1996) suggesting the serological relationship of the two mealybug transmitted viruses. Sugacane bacilliform virus (SCBV) also infects a range of monocot hosts, including banana in which it elicits typical banana streak disease symptoms (Bouhida *et al.*, 1993; Lockhart, 1995). Some badnavirus isolates from diseased sugarcane and banana plants are closely related serologically (Lockhart, 1998; Lockhart and

Autrey, 1988) and it is possible that there is movement of the virus between these two crops. *Sugarcane bacilliform virus* has been experimentally transmitted to bananas by agro-inoculation and by mealybugs and, in this host, elicits symptoms typical of BSV infection (Bouhida *et al.*, 1993; Lockhart, 1998).

DIAGNOSIS AND MANAGEMENT

Banana streak virus (BSV) can be detected by electron microscopy (Lockhart, 1986; Cherian *et al.*, 2004). BSV is moderately immunogenic and individual antiserum often failed to detect some BSV isolates. The Bacilliform particles were detected by immunosorbent electron microscopy (ISEM) (Font *et al.*, 1998) and in ELISA (Thottappilli *et al.*, 1998). ISEM is generally a more sensitive assay than ELISA (Thottappilly *et al.*, 1998; Lockhart, 1995). It is important that the mother plants to be selected for micropropagation should be indexed by PCR and IC-PCR tests and IC multiplex PCR as these tests are more reliable for BSV detection (Dallot *et al.*, 2001; Sharman *et al.*, 2001; Le Provost *et al.*, 2006, Selvarajan *et al.*, 2008). Immunocapture PCR should be used to distinguish between integrated and episomal forms of the viral DNA, especially in *Musa* AxB hybrids (Harper *et al.*, 1999a). BSV detection can be successfully achieved by polymerase chain reaction (PCR) amplification with primer pairs followed by electrophoresis analysis of PCR products. Primers from highly conserved region of BSV could be used for detection of BSV more reliably. BSV is easily amplified from *M. balbisiana* and Poovan.). Diploid clones of banana appear to be resistant to BSV.

Roguing is probably a good management practice for BSV as the rate of spread by vector, mealy bug is slow and dissemination of the virus is mostly by inadvertent use of infected planting material. In many instances, infection is thought to develop *de novo* from viral DNA integrated in the host's nuclear genome (Ndowora *et al.*, 1999; Harper *et al.*, 1999b). All *Musa* AxB hybrids are considered to have integrated BSV-OL DNA, hence the use of tissue culture to propagate bananas is the best option as it will also help to identify the expression of these integrated viral sequences (Dallot *et al.*, 2001). Plants initiated into tissue culture should be indexed for virus, and a sample of cloned plantlets also tested for virus on at least two occasions over a six month period (Diekmann and Putter, 1996). BSV is a limiting factor in the development of new *Musa* hybrids by breeding programs across the world. Infection in these new hybrids is often thought to originate from viral sequences that are integrated in the host's nuclear genome (Ndowora *et al.*, 1999; Harper *et al.*, 1999b).

BANANA MOSAIC OR INFECTIOUS CHLOROSIS

Mosaic or infectious variegation disease was first reported in Australia (Magee, 1940). Later the disease was recorded in South and Central America, the Caribbean, India and the Philippines (Bouhida and Lockhart, 1990; Joshi and Joshi, 1974; Mali and Deshpande, 1976; Mali and Rajegore, 1980; Palukaitis *et al.*, 1992). The disease is now cosmopolitan in its distribution. Banana Chlorosis which have been variously described as infections chlorosis, heart rot, mosaic disease and virus sheath rot are now regarded as single disease caused by strains of the cucumber mosaic virus (Magee, 1940; Singh *et al.*, 1995; Kiranmai *et al.*, 1998).

In India, the disease was first reported from Maharashtra (Capoor and Varma, 1968). This disease appears to be in epidemic form in certain parts of Maharashtra and Tamil Nadu. Cucumber Mosaic Virus (CMV) infecting banana was found in alarming proportions in all the districts of Tamil Nadu on the 'Robusta' (Dwarf Cavendish) and 'Poovan' (Mysore) cultivars (Mohan and Lakshmanan, 1988). The disease is known to occur in most of the banana growing states of the country (Mali and Deshpande, 1976; Joshi and Joshi, 1974; Estelitta *et al.*, 1996; Kiramani *et al.*, 1998). High incidence of mosaic disease caused by CMV was reported on Poovan (20-80%) and on Robusta (10-68.5%) (Mohan and Lakshmanan, 1988). In Marathwada region of Maharashtra, the incidence of cucumber mosaic virus (CMV) in banana varied from 1.5% to 27.5%. The incidence data revealed that the percent incidence of CMV was more in ratoon crop (27.5%) followed by sucker grown plantation (22.8%), whereas, in tissue culture raised plants the incidence was less (16.8%). Among the different cultivars, highest incidence was found in Ardhapuri (27.5%) followed by Basrai (16.5%), Shrimanti (16.0%) and was not observed in the Grand naine (Aglave *et al.*, 2007). In Kerala, intercropping banana with cucurbitaceous vegetables is a common practice, leading to an increase in disease incidence (Cherian *et al.*, 2004).

SYMPTOMS

Leaves on severely affected plants show broad chlorotic or yellow patches which cover most of the leaf portion (Plate 22). Other symptoms are chlorosis, mosaic and heart rot in banana. The symptoms of the disease are sometimes confused with zinc deficiency and/or with the symptoms of Banana streak virus. The first symptom appears as lemon yellow streak or patches on young folded leaves or newly expended leaves. These chlorotic patches later turn brown in colour. The leaves show inward rolling and malformation in some cases. CMV produced characteristic symptoms of Chlorosis, chlrorotic streaks or flecking followed by necrotic streaks. Leaves on severely affected plants show broad chlorotic yellow patches which cover most of the leaf portion. Initially chlorotic spots develop on leaves without any definite arrangement. Chlorosis and mosaic are the common symptoms produced by banana mosaic virus on different banana cultivars. Affected plants are comparatively stunted and seldom produce a fruit bunch. Banana mosaic virus has been found infecting banana plants grown in plastic houses and showing chlorotic striping on leaves and stunting (Bouhida and Lockhart, 1990).

TRANSMISSION

The disease is mainly transmitted through suckers from the diseased plants. The virus is transmitted by sap inoculation to the members of Chenopodiaceae, Cucurbitaceae, Fabaceae, and Solanaceae. The disease is also mechanically transmissible from banana to cucumber but not from cucumber to banana back. The virus is also transmitted by aphid vectors, *Aphis craccivora* and *A. gossypii.* Both species efficiently transmitted the virus in a non-persistent manner within 2 to 5 min. of acquisition and 5 to 30 min. inoculation feeding. *A. gossypii* is more efficient vector in transmitting the virus than *A. Craccivora*. Transmission of CMV causes severe mosaic and leaf deformations in inoculated cucumber plants. Single aphid could transmit the virus but for efficient transmission three aphids are required (Aglave *et al.*, 2007). Other aphid vectors are: *Rhopalosiphum maidis, R. prunifoliae* and *Myzus persicae*. In association with Banana bunchy top virus, CMV was also transmitted by *Pentalonia nigronervosa*. Seed transmission has also been reported (Gold, 1972). Infected propagating material is more responsible for the outbreak of the disease than aphids (Avgelis, 2008).

A different type of banana mosaic disease was observed in Andhra Pradesh and Tamil Nadu which is transmitted to tomato, tobacco, *Nicotiana glutinosa, N. rustica, Cucumis melo, Gynura aurantiaca, Chenopodium amaranticolor* and *C. quinoa* in addition to cucumber. The virus was designated a Filaree strain of banana mosaic (Rao, 2006).

HOST RANGE

The associated CMV has many alternative host plants and has the broadest host range among the plant viruses (Roosinck, 2002). Cucumber mosaic virus has been reported infecting banana and causing banana chlorosis disease from various parts of the world including India (Mali and Deshpande, 1976; Singh *et al.*, 1995; Bouhida and Lockhart, 1990; Pappu *et al.*, 1991; Hu *et al.*, 1995). In India, forty-four cultivated banana varieties including *Barsai* and *Velchi* are susceptible to mosaic infection. However, *Musa balbisiana, M. chillicarpa, M. coccinia* and *M. acuminata* were found immune (Capoor and Varma, 1968). Banana mosaic virus was transmitted mechanically to *Cajanus cajan, Capsicum annuum, Chenopodium amaranticolor, Cicer arietinum, Cucumis sativus, Cucurbita pepo, Glycine max, Tomato, Nicotiana glutinosa, N. tabacum, Phaseolus vulgaris, Vigna mungo, Vigna radiata,* and *Vigna unguiculata,* and *Vigna sinensis.* The virus produced characteristic symptoms of Chlorosis, chlrorotic streaks followed by necrotic streaks and developed local lesions in *C. amaranticolor*. The Indian isolates are transmitted by sap inoculation to tobacco and the virus from inoculated plants was confirmed as cucumber mosaic virus (CMV) by DAC-ELISA. Transmission of CMV also causes severe mosaic and leaf deformations in cucumber. The virus is propagated on *N. glutinosa* plants by mechanical inoculation.

THE PATHOGEN

The virus was purified by the method of Tomlinson *et al.* (1973). Electron microscopy of negatively stained partially purified virus preparation revealed the presence of isometric particles of about 28-29 nm in size (Aglave *et al.*, 2007; Cherian *et al.*, 2004). Virus is the type member of the genus *Cucumovirus* in the family *Bromoviridae.* Virus preparations contain more than one particle components. Cucumber mosaic virus (CMV), a positive sense, single stranded RNA plant virus with a tripartite genome, is the type member of the genus *Cucumovirus.* CMV has a world -wide distribution and exists as a variety of isolates that differ in host range and pathogenicity (Palukaitis *et al.*, 1992). A number of CMV isolates have been described previously and classified into two subgroups, I and II, according to serological relationships, peptide mapping of the coat protein (CP), and nucleotide sequence identity (Palukaitis *et al.*, 1992). More recently, phylogenetic analysis of a number of CMV isolates led to a further subdivision of subgroup I into subgroups IA and IB (Roossinck, 2002; Roossinck *et al.*, 1999). Cucumber mosaic virus (CMV) causing mosaic and chlorosis of banana in India was characterized on the basis of biological, serological and RT-PCR amplification of Coat protein (Aglave *et al.*, 2007). The coat protein gene of cucumovirus causing mosaic in banana was cloned and sequenced in India (Cherian *et al.*, 2004) and elsewhere.

DIAGNOSIS AND MANAGEMENT

Cucumber mosaic virus can be reliably detected by mechanical inoculation to a range of diagnostic hosts like *Chenopodium amaranticolor, C. quinoa* and *Vigna unguiculata* (Francki *et al.*, 1979). The virus can be detected in ELISA from leaf, sucker of infected plants (Aglave *et al.*, 2007). This method is very efficient even in detecting the virus in symptomless plant. The virus can be successfully detected by electron microscopy, and by serological techniques like ELISA. CMV distribution in plants is uneven, hence serological and molecular techniques are more reliable than biological (Kirmani *et al.*, 1996). Reverse transcription-polymerase chain reaction (RT-PCR) assays, restriction enzyme analysis, and comparison of nucleotide sequences are used to identify isolates of cucumber mosaic cucumovirus (CMV) that belong to subgroup I from infectious chlorosis-affected banana plants.

An RT-PCR assay was done with primers designed in a conserved region of the 3' end of the CMV coat protein gene amplified a 486- to 488-bp DNA fragment from infected banana samples. CMV was detected in leaf blades, midrib, and pseudostem of infected plants, but not in asymptomatic plants. Restriction enzyme analysis of PCR products using EcoRI and MspI showed that the samples tested from diseased banana were infected with CMV subgroup I isolates. This result was confirmed by sequencing PCR-amplified products, which showed 98% homology to sequences of CMV subgroup I isolates but only 76% homology to sequences of CMV subgroup II isolates. The RT-PCR is a reliable assay to detect CMV in banana plants and, when combined with restriction enzyme digestion, provides a simple means of identifying the CMV subgroup present (Singh *et al.*, 1995; Raj *et al.*, 1998; Cherian *et al.*, 2004). For comparative detection of cucumber mosaic virus in banana, PCR is more sensitive than dot-blot and ELISA. Dot blot is more sensitive than ELISA. (Hu *et al.*, 1995).

Banana chlorosis disease detection based on serological biological and CP gene detection would help in development of diagnostic tool for identifying and certifying planting material to check spread of disease. Most of the farmers are now using tissue culture plants for cultivation. This diagnostic tool will also help for the indexing of explants before they are used for micropropagation in commercial tissue culture industry (Aglave *et al.*, 2007). Virus-free plantlets have been obtained by meristem culture of heat treated suckers (Gupta, 1986). Among the banana genotypes, the clones with ABB genomes showed complete field resistance (Mohan and Lakshmanan, 1988).

ABACA MOSAIC

Abaca mosaic disease affects fiber yield as well as fiber quality of *Musa textilis* (abaca, Manila hemp) in Philippines and is a major constraint in abaca production. Leaves of affected plants show yellowish or light green streaks. Petioles and midribs are mottled with dark green and yellowish streaks, even when no symptoms appear on the leaves (Frison and Putter, 1989). *Musa textilis, Marantha arundinacea,* and *Canna indica* are natural hosts of abaca mosaic virus. Experimentally, the virus was transmitted to several experimental hosts, including banana. The virus is transmitted by vegetative propagation and tissue culture, as well as by aphids (mainly *Rhopalosiphum maidis* and *Aphis gossypii*) in a nonpersistent manner. The virus is also Mechanically transmitted (Eloja *et al.*, 1962). A potyvirus with flexuous rod shaped particles measuring about 680 nm was associated with the disease. Sequence data, host reactions and serological relationship indicate that abaca mosaic virus should be considered a distinct strain of sugarcane mosaic virus (SCMV) and strain designation as SCMV-Ab was suggested (Gambley *et al.*, 2004). The virus can be detected by ELISA using antibodies for sugarcane mosaic virus (Eloja and Tinsley, 1963). No information is available on management of the disease.

C. VIRUS AND VIRUS- LIKE DISEASES OF CITRUS

Citrus belongs to family *Rutaceae* which is represented by approximately 150 genera and 1600 species of plants found in tropical and warm regions of the world. The genus *Citrus* is the most popular genus in subfamily *Aurantioideae* (Swingle and Reece, 1967). Among the group True citrus, following genera and species are included: *Fortunella* (4 spp.), *Clymenia* (2 spp.), *Microcitrus* (7 spp.), Citrus subgenus, *Citrus* (10 spp.), and Citrus subgenus *Papeda* (6 spp.). *Eremocitrus* and *Microcitrus* are other genera in this family. Important species and hybrids are: *Citrus aurantifolia*

(Key lime) *Citrus maxima* (Pummelo), *Citrus medica* (Citron), *Citrus reticulata* (Mandarin orange), *Citrus aurantium* (Bitter or sour orange), *Citrus latifolia* (Persian lime), *Citrus limon* (Lemon), *Citrus limonia* (Rangpur lime), *Citrus paradisi* (Grapefruit), *Citrus sinensis* (Sweet orange) and *Citrus tangerine* (Tangerine).

Citrus fruits are produced all around the world. About 140 countries produce various kinds of citrus fruits. Major Citrus producing countries are situated in both Northern and Southern Hemispheres. World citrus production for the year 2005 was estimated as 94.8 million MT (FAOSTAT-FAO statistical Database, 2005). The estimated production of citrus fruits in India is 3.79 million tons from an area of 0.45 million ha.

In citrus, more than 40 virus and virus-like diseases have been described from most citrus growing countries (Ahlawat, 2012). Other important pathogens infecting citrus are phytoplasmas earlier known as Mycoplasma-like organisms (MLOs) and viroids (250–400 nucleotides). So far only one member of class Mollicute, the citrus stubborn pathogen has been cultured in synthetic medium. Based on its helical morphology, this organism was named as *Spiroplasma citri*. Important virus and virus-like diseases will be described on following pages.

CITRUS PSOROSIS

The name "psorosis" was coined for a local lesion forming disease in Florida (Swingley and Webber, 1896). Two types of psorosis have been proposed, psorosis -A and Psorosis-B (Fawcett, 1932, 1933; Wallace, 1968). Although both develop similar bark lesion symptoms, but they produced different symptoms on leaves and fruits. Psorosis-A was considered to be mild strain, whereas, psorosis-B is a severe strain of psorosis disease complex. In India, psorosis-A disease has been reported on Mosambi sweet orange but without bark lesions (Ahlawat, 1989). In ultrathin sections, inclusions resembling P_2 like protein were observed in phloem cells of infected plants. Later studies revealed association of filamentous virus particles in mature leaves of Indian isolate (Ahlawat and Chakraborty, 1990). It has now been confirmed that this disease is distinct from psorosis and was designated as Indian citrus ringspot.

SYMPTOMS

Characteristic symptoms of psorosis-A disease in field trees are bark scaling and wood staining of stem (Plate-23). As the lesions grow, gum may exude from the margin of the lesions. The bark is break away as these lesions get bigger in size. Leaf symptoms may develop on infected field trees as clearing and flecking of veins in the whole or part of the young leaves in new growth flush. These symptoms can be clearly seen when the leaf is shaded from direct sun and viewed against the light. Symptoms on young leaves usually disappear as soon as such leaves mature or even earlier, but in psorosis-B infection, leaf symptoms persist in mature leaves and can also occur on fruits. In addition of vein clearing symptoms a few leaves also develop oak leaf pattern in field trees if these are infected with concave gum. These effects will disappear as the leaves become mature.

Axillary shoots of experimentally inoculated sweet orange seedlings curve downward with smalling of leaves. These leaves show necrosis and fall down. This reaction is known as "shock reaction". Infected plants are not killed by shock reaction and in due course normal new leaves will appear but with typical vein flecking symptoms. Following are the major symptoms of psorosis disease which can be useful for field and laboratory identification.

1. Shock reaction of young flushes in inoculated Sweet orange and some other *Citrus* spp.
2. Vein clearing or flecking or oak leaf pattern on young leaves which persists for a short period.
3. Blotches or spots on mature leaves.
4. Bark scaling on stem and branches in field affected trees.
5. Wood impregnated with gum, which forms an irregular circular pattern when viewed in the cross section of the trunk.

TRANSMISSION

Citrus psorosis virus (CPV) is readily transmitted by grafting. Some isolates of CPV are transmitted by mechanical inoculation on herbaceous hosts like *Chenopodium quinoa* and *C. amaranticolor* where it produce local lesions. Mechanical transmission from citrus to citrus is done by knife or razor slashing into stem. Psorosis virus is suspected to be seed transmitted in Carrizo citrange (Bridges *et al.*, 1965) but authentic evidence is lacking (Roistacher, 1993). Transmission of psorosis virus by dodder (*Cuscuta compacta*) has also been reported (Price, 1963).

HOST RANGE

Grapefruit, mandarins, sweet oranges, lemons, limes and Dweet tangar are the natural host of psorosis virus. Citranges, Lemons, Mandarins, *P. trifoliata*, Rough lemon, Sour orange, *Citrus daidai*, and *Clausena* sp. are symptomless hosts of psorosis-A virus in field. Under experimental conditions, psorosis virus has been transmitted to grapefruits, sweet oranges, Orlando tangelo, sour orange, Dweet tangor, West Indian lime, lemons, mandarins, pummelo, rough lemon, tangelo, Etrog citron, *C. excelsa*, Citranges (Carrizo, Rusk, Troyer), *Citrus daidai*, *Clausena* sp., and Sweet lime. Several herbaceous plant species including *Chenopodium amaranticolor* are susceptible (Storm and Streets, 1963; Timmer *et al.*, 1978).

Sweet orange seedlings are used as diagnostic and indicator host for psorosis virus. Inoculated seedlings will show symptoms of shock reaction, young leaf flecking along veinlets and oak-leaf pattern in young leaves and ring like pattern on mature leaves within a few weeks at cool temperature. Such plants will show bark scaling and flaking within six years or more. Seedlings of mandarin, Dweet tangor, citron and Mexican lime may also be used as indicator hosts of psorosis-A virus. *Chenopodium amaranticolor* may be used local lesion hosts of psorosis virus.

THE PATHOGEN

The virus associated with psorosis disease is *Citrus psorosis virus* (CPsV), the type member of the genus *Ophiovirus* (Milne *et al.*, 2000). Virus Particles have a length of 1500-2500 nm; 760 nm and a width of 3 nm, or 9 nm. Psorosis virus (CPsV) is a bipartite single stranded, negative strand RNA virus classified as genus *Ophiovirus* in family *Ophioviridae* (Martin *et al.*, 2005). CPsV virions are highly kinked filaments 3–4 nm in diameter of at least two sizes (García *et al.*, 1994), which contain three single-stranded RNAs of negative polarity and a 48-50-kDa coat protein (Sánchez de la Torre *et al.*, 2000; Milne *et al.*, 2000).

Genome of psorosis virus is segmented and consists of three segments of circular, negative-sense, single-stranded and double-stranded RNAs; that form a panhandle, a structure formed of complementary sequences. RNA-2 and RNA-3 are fully or partially sequenced. Viral genome encodes structural proteins of 48700 kDa which has been sequenced and a function assigned. Variation in populations of citrus psorosis virus has been observed (Martin *et al.*, 2006).

DIAGNOSIS AND MANAGEMENT

Field diagnosis of CPsV is done on the basis of bark scaling symptoms on tree trunk or limbs. Commonly, laboratory diagnosis is done by grafting samples on indicator seedlings (sweet orange or mandarin). Inoculated plants will show a shock reaction, vein clearing/vein banding of young leaves and mottling or ringspot on mature leaves. Mechanically inoculated *C. quinoa* leaves will show local lesions (Roistacher *et al.*, 2000; Garnsey, 1993). Both polyclonal and monoclonal antibodies are available for serodiagnosis in DAS-ELISA, TAS-ELISA, direct tissue blot immunoassay (DTBIA) (Djelouah *et al.*, 2000; D'Onghia *et al.*, 2001; Maria *et al.*, 2006). Diagnosis of CPsV is more reliably done in RT-PCR (Garcia *et al.*, 1997), nucleic acid hybridization and immunosorbent electron microscopy (Martin *et al*, 2004).

The first budwood certification programme was started against citrus psorosis in California and is still a very successful method to eradicate citrus psorosis. Another option to get virus free planting material is the use of nucellar clones. Psorosis virus is also eliminated through heat treatment of infected citrus plants by keeping them for 2-3 months in hot air at alternating temperature of 40°C for 16h during day and 30°C for 8h during night (Calavan *et al.*, 1972). Psorosis disease is now being eliminated by the use of virus-free healthy planting material which may also be obtained through shoot tip grafting (Navarro *et al.*, 1980).

In Brazil, a bark scaling disease of citrus (BBSC) is found in Bahia state commonly known as Bahia psorosis. It affects most of citrus species and varieties, but especally severe on grapefruit. BBSC symptoms include a darkening and thickening of the bark leading to scaly lesions on trunk and branches; young branches show dieback; and significant gum extrusion. Symptoms resemble to psorosis-A but data revealed that BBSC is distinct disease, although its aetiology remains unknown (deSimone *et al.*, 2010). A combined strategy of strong sanitation effort, lower planting densities and rouging has been recommended to manage Bahia bark scaling disease.

CITRUS TRISTEZA

There are very few diseases that caused huge crop losses as tristeza did in citrus. Phytophthora epidemic caused worldwide destruction of citrus in the 19th century. Sour orange was then identified a rootstock which was resistant to *Phytophthora* and was used as acceptable rootstock to raise sweet orange and mandarin trees in many countries. However, the trees raised on sour orange rootstock declined/died within two to three years. This condition was noticed in USA, Gulf States, Java, Argentina, Brazil and several other countries where sour orange was used as root stock. The term "tristeza" was coined by Moreira (1942) to this decline malady which means "sadness" in Portuguese language. Reason to this decline was attributed as an incompatibility between sweet orange and mandarin scions and sour orange rootstock (Bioletti *et al.*, 1904). But it was later described as a disease condition (Webber, 1925). First evidence that this disease was graft transmissible and a virus was involved in this malady was published in 1946 (Fawcett and Wallace, 1946). At the same time, it was reported from Brazil that tristeza disease was transmissible by brown citrus aphid, *Toxoptera citricidus* Kirk (Meneghini, 1946). Finally, electron micrographs of thread like particles were obtained from purified preparations from tristeza infected lime trees under electron microscope (Kitazima *et al.*, 1964).

Citrus tristeza virus (CTV) had ruined citrus industry in many countries where sour orange was used as rootstock. Tristeza destroyed 15 million trees in Argentina, 6 million or more trees in Tucuman, 6 million trees in Brazil, 3 million trees in California and an unestimated number in many citrus growing countries. However, in countries where sour orange was not a root stock like Japan, India, the Philippines, Taiwan etc. the trees were not so severely damaged (Reuther *et al.*, 1978). In India, decline symptoms were observed on either sweet orange grafted on rough lemon or sweet orange rootstocks (Vasudeva *et al.*, 1959). In India, Tristeza was first observed in Bombay state (Nagpal, 1959) and later reported from several parts of India (Capoor, 1963). Tristeza is widely distributed in most citrus growing regions of the India (Capoor and Rao, 1967; Nariani and Raychaudhuri, 1968; Nariani *et al.*, 1965, 1966, 1970; Raychaudhuri *et al.*, 1977; Reddy, 1965; Reddy and Rao, 1961). Estimates indicate that tristeza destroyed about a million trees in India (Ahlawat, 1997a). CTV is more damaging to Indian sweet oranges (Mosambi) than to other citrus cultivars (Ahlawat *et al.*, 1995) and about 70% incidence of tristeza (stem pitting) has been recorded in Mosambi sweet orange trees on different rootstocks (Mali and Chaudhuri, 1993). In places where aphid vector, *Toxoptera citricida* is present, severe strains of CTV are commonly available (Ahlawat and Raychaudhuri, 1988). In certain citrus growing regions of India CTV strains that cause seedling yellows were also found (Ahlawat and Raychaudhuri, 1988).

Tristeza and its vector, *Toxoptera citricida* might be originated in Asia which is also the original home of citrus. CTV along with its vector, *T. citricida* is widely distributed in seedling trees of Darjeeling orange (Mandarin) in Darjeeling and Sikkim hills of India suggesting the spread of the disease by vector in this region of citrus origin (Ahlawat and Sardar, 1976). Tristeza disseminated to most parts of the world through the movement of infected planting material probably through traders and tourists and within countries by aphid vectors. Tristeza disease is worldwide in distribution wherever citrus is grown (Price, 1970, CABI, 2010). In India, a mixed infection of tristeza and greening bacterium is common in declining sweet oranges in Maharashtra (Ghosh *et al.*, 2003). Synonym(s) of tristeza virus are citrus quick decline virus (Fawcett and Wallace, 1946; Capoor, 1975), grapevine A virus, hassaku dwarf virus, and lime dieback virus.

SYMPTOMS

The most important symptom of tristeza virus disease (CTV) is decline, sudden wilting and drying of leaves in certain stock-scion combinations (Plate-24). In fruiting citrus trees, leaves of wilting trees fall down but the fruits remain hanging till they become dried or mummified. Rotting of roots and absence of functional feeder roots is a common phenomenon in CTV infected trees. Absence of new growth and bronzing of leaves in CTV infected field trees is also common. CTV infected trees flower heavily with comparatively more fruit setting. In Asia, various strains of CTV, generally referred to as seedling yellows (CTV-SY), tristeza (CTV-T), stem pitting (CTV-SP), and a mild type that have been widespread for many years. Any of these strains may exist in a citrus plant, or they may occur together, as a complex. In highly susceptible hosts like Mexican lime or Kagzi lime and *Citrus excelsa*, CTV produces vein clearing of leaves (Plate-25) and stem pitting. In India, Stem pitting in Coorg mandarin is known to occur due to CTV infection (Manjunath *et al.*, 1989). Tristeza infected trees also show some microscopic symptoms like necrosis of sieve tubes in phloem tissues of sweet orange trees on sour orange rootstock (Schneider, 1959). After infection, the amount of phloem is reduced resulting in a girdling effect. Girdling prevents movement of photosynthetic food to the roots resulting utilization of reserve starch and subsequent decay of roots and appearance of top symptoms (Schneider, 1959). Decline (Plate 24), stem pitting (Plate 26) and seedling yellows (Wallace and Drake, 1972) are also the reaction of tristeza. In India, seedling yellows strain of tristeza had been observed earlier but was restricted in its distribution (Vasudeva *et al.*, 1959, Capoor, 1965).

TRANSMISSION

CTV is readily transmitted by bud or scion grafts. It is also transmitted by dodder, *Cuscuta americana, C. subinclusa and C. reflexa.* CTV is transmitted mechanically by bark wounding or by slashing stem or leaf abrasion (Garnsey and Muller, 1988). In India and many other countries, CTV is transmitted by seven species of aphids colonizing on citrus plants. These aphid species are: *Toxoptera citricida* Kirk., *Aphis gossypii* Glover, A. *citricola* van der Goot (= *A. spiraecola* Patch), *A. craccivora* Koch, *M. persicae* Sulz., *Toxoptera aurantii* Bsy and *Dactynotus jaceae* L. (Vasudeva *et al.*, 1959; Varma *et al.*, 1960, 1965; Meneghini, 1946; Dickson *et al.*, 1951; Norman and Grant, 1956). *T. citricida* (Plate 27) is the most effecient vector of CTV throughout the world. *A. gossypii* is an important vector, especially in places where *T. citricida* is not established as yet. *Aphis citricola* and *Toxoptera aurantii* (Capoor and Rao, 1967) have also been shown to be potential vectors of CTV but their efficiency is very low. All these aphid vectors transmitted CTV in semi-persistent manner. In places where *T. citricida* is not present, *Aphis gossypii, A. craccivora, Myzus persicae* and *Dactynotus jaceae* play the role of vectors, but they are less efficient (Ahlawat and Raychaudhuri, 1988).

HOST RANGE

Most of the citrus species/cultivars and citrus relatives are host of citrus tristeza virus. CTV develops visible symptoms only in a few hosts and in most of the species of citrus it remains symptomless. Trifoliate orange has been reported to be immune to tristeza (Costa *et al.*, 1949) but later it was shown to be infected by some strains of seedling yellows (Wallace and Drake, 1972). Rootstock species, *Poncirus trifoliata* is resistant to most isolates of CTV, but not to members of the CTV resistance-breaking (RB) strain presently found in New Zealand (Harper *et al.*, 2010). Hosts like Rusk citrange, Grapefruit, Hassaku. Kumquats, Limequats, Mandarins, Mediterranean Pera, Tahiti lime and Tangor show stem pitting symptoms. Hosts like Limes (*Citrus aurantifolia*) cultivars Kagzi, Key, West Indian, Mexican lime, and sweet oranges show vein clearing and stem pitting symptoms respectively. Pummelo shows both stem pitting and seedling yellows symptoms. Following hosts show variable symptoms: Uvalve, Citron, Etrog citron, Citrumelo 1452, *Citrus amblicarpa, Citrus daidai, Citrus excels, Citrus hystrix, Citrus junos, Citrus macrophylla,* Sweet lime, *Aegle marmelos, Aeglopsis chevalieri, Afraegle paniculata, Citropsis gelatiana, Citrus aurantium, Citrus excelsa, Citrus limon, Citrus medica, Citrus paradisi, Citrus reticulata, Citrus sinensis, Microcitrus australis, Pamburus missionis, Passiflora gracilis, P. caerulea, P. incense* and *P. incarnata.*

In North-eastern region of India, natural infection of CTV was observed in seedling trees of Khasi mandarin, Rough lemon, Assam lemon, and kapura tenga. However, wildly distributed *Citrus indica* neither showed any CTV symptoms in field nor showed CTV symptoms when inoculated to indicator host, Kagzi lime (Bhagabati *et al.* 1989). Symptoms on some important hosts are given below as described by Wallace and Drake (1951) and Lee and Bar-Joseph (2000).

- *Citrus aurantifolia* cvs West Indian lime, Key lime, Mexican lime, Galego lime, Kagzi lime—Chlorotic vein flecks (Plate 25), vein corking (Plate 28), leaves cupped or canoe-shaped; stem pitting (Plate 26).
- *C. sinensis* on *C. aurantium* roorstock — Sudden wilting, decline and death, overgrowth on bud union and 'honeycombing' just below union.
- *C. sinensis, C. paradisi* — Stem pitting.

Best assay host for CTV is small fruited acid lime seedlings (*Citrus aurantifolia)* West Indian lime, Key lime, Mexican lime, Galego lime, Kagzi lime (Capoor, 1961). Symptoms on inoculated plants develop within a few weeks under cool climate. Main symptoms are vein clearing, leaf cupping, stem pitting and stunting of plants. Other indicator hosts are *Citrus macrophylla* and *C. excelsa*. Symptoms in these hosts will develop as veinlet clearing in young leaves and pitting in stem. *Citrus amblycarpa* (*Nasnaran* Mandarin) is also a good indicator host for CTV which develop symptoms similar to Kagzi lime upon inoculation (Pandey *et al.*, 1987).

THE PATHOGEN

Citrus tristeza virus particles are highly flexuous and filamentous measuring 10-12 nm in width and 2000 nm in length (Fig. 19). CTV is a member of the genus *Closterovirus* within the family *Closteroviridae*, is one of the most complex single stranded RNA plant viruses (Bar-Joseph *et al.* 1970, 1979; Bar-Joseph and Lee, 1989).

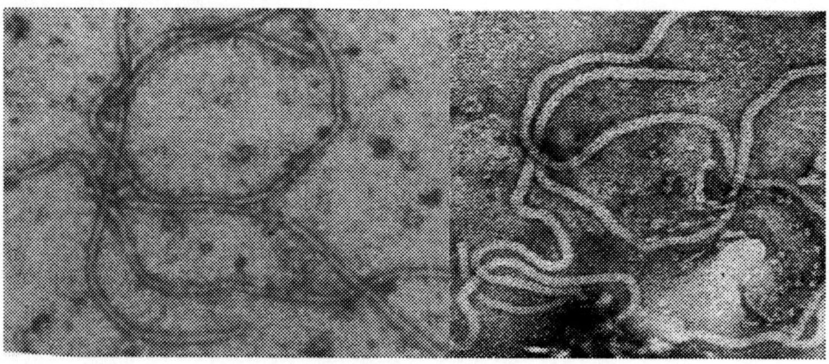

Fig. 19: Virus particles of CTV

Tristeza virus genome is monopartite. The nucleic acid is linear, positive-sense and single-stranded RNA of approximately 19.3 kb (Bar-Joseph and Lee, 1989) with two proteins (CP) 25 and 27 kDa covering 95 and 5% of particle length respectively. Complete CTV genome has been sequenced and is 19,226 to 19302 nt long. Several workers have reported complete sequences of CTV genome of several isolates including the coat protein gene and deposited in GenBank (Bar-Joseph and Lee, 1989; Harper *et al.*, 2010; Karasev *et al.*, 1995, Pappu *et al.*, 1993; Roy and Brlansky, 2010). Partial genome of CTV has been sequenced from a CTV isolate in a mixed infection (Manjunath *et al.*, 2005). Coat protein genes of several Indian isolates of CTV have also been cloned and sequenced (Manjunath *et al.*, 1993; Ghosh *et al.*, 2009; Roy *et al.*, 2003). Based on genomic sequences, four different clusters *viz* Kpg1, Kpg2, Kpg3 and CTK 27 have been identified from Northeast Himalayan region which is considered as home of citrus. These isolates were found to be genetically related to exotic isolates like T-30 (mild isolate from Florida), T-385 (severe strain from Spain), SY 568 (Severe isolate from California), VT (Israel Severe) and PB 61 from Australia indicating that diversified CTV population is present in the NE Himalayan region of India from where CTV originated (Biswas *et al.*, 2008). The sequence variation of the Indian CTV-B165 isolate was 19,247 nt in length with 12 open reading frames (ORFs) and was structurally identical to other CTV isolates. Molecular characterization of CTV isolates has been done by several workers throughout world and coat protein gene sequences selected from genbank were used in sequence and phylogenetic analysis to determine mild or severe isolates (Barbarossa and Vovlas, 2010). It has been suggested that there is extreme genome stability and low mutation rates in CTV. These studies have significant implication in evolution and emergence of new CTV strains, and in the use of CTV as a stable expression vector (Weng *et al.*, 2010). CTV isolates have been grouped into six genotypes; T3, T30, T36, VT, trifoliate resistance breaking (RB) and B165 (Harpez *et al.*, 2010; Roy *et al.*, 2010).

DIAGNOSIS AND MANAGEMENT

CTV strains have been detected by way of differential transmission, differential hosts, serological or through molecular techniques (Balaraman and Ramakrishnan, 1977; Chakraborty *et al.*, 1993; Lee *et al.*, 1997). CTV spread in nature is either by infected planting material or through insect vectors. CTV is easily detected by graft inoculation of samples to Mexican lime (Kagzi lime) indicator host where the leaves of inoculated plants will develop vein clearing or flecking symptoms within a few weeks under low temperature conditions (18-25°C). Traditionally, CTV has been detected and categorized by different biological indexing methods inoculating the appropriate indicator plants and observing the symptoms (Garnsey *et al.*, 2005) but, the major limitations of this technique are time consuming to confirm the detection, expensive and laborious and often imprecise. Although polyclonal antibodies have been developed against several isolates of CTV world wide, a monoclonal antibody (MCA-13) has been produced and is being used to discriminate mild and severe strains (Permar *et al.*, 1990; Chakraborty and Ahlawat, 2001; Chakraborty *et al.* 1993; Cambra *et al.*, 1991). Polyclonal and monoclonal antibodies against CTV have been prepared against various isolates/strains in several laboratories world over including from the recombinant coat protein gene of citrus tristeza virus. In practice, CTV is detected by electron microscopy, enzyme linked immunosorbent assay (ELISA), dot immunobinding assay (DIBA), Immuno electron microscopy (IEM) with polyclonal or monoclonal antibodies, cDNA probes, light microscopy after azure-A staining of sections (aggregates of CTV are seen in light microscope) and by PAGE observing dsRNAs (Dodds *et al.*, 1984). Antibodies raised by the expression of coat protein gene of CTV are efficient at trapping virus particles in double antibody sandwich ELISA assays. Serological detection techniques like ELISA, immunoprinting ELISA or tissue blot assay are more commonly used for CTV diagnosis in indexing programmes all over the world (Ahlawat and Raychaudhuri, 1988; Ahlawat, 1997b; Ahlawat *et al.*, 1992; Ahlawat and Pant, 2003; Roch Pena and Lee, 1991; Bar-Joseph *et al.*, 1979; Garnsey *et al.*, 1993; Cambra *et al.*, 2000; Biswas *et al.*, 2009; Ghosh *et al.*, 2009).

Use of polymerase chain reaction (PCR), not yet a competitor of immunological methods of CTV in field diagnosis and indexing because of their reliability, rapidity and low relative cost, offers a quick detection and alternative technique (Karasev *et al.*, 1995; Pappu *et al.*, 1993; Biswas, 2008). A multiplex real-time PCR assay has also been developed to detect and quantify CTV genotypic mixture present in infected plants (Ananthakrishnan *et al.*, 2010). Recently developed genotype specific primers for 5 known CTV genotypes revealed the presence of all CTV genotype in India by RT-PCR (Roy *et al.*, 2010).

Sometimes, citrus trees are also infected with a mixed population of citrus yellow mosaic badnavirus-a dsDNA virus and tristeza virus (Fig. 20). Both these RNA and DNA viruses are detected by duplex PCR technique (Ghosh *et al.*, 2008; Roy *et al.*, 2005). Quantitative detection of CTV in citrus and aphids can be done by real-time reverse transcription-PCR (Saponari *et al.*, 2008). RT-PCR, multiple molecular marker assay, coat protein (CP) gene analysis and Single strand conformation polymorphism analysis are also used for CTV detection in field trees (Yokomi *et al.*, 2010). Identification of diversified citrus tristeza virus population isolates from eastern Himalayan region of India has been determined by heteroduplex mobility assay and nucleotide sequence analysis (Biswas *et al.*, 2004). Several methods have been suggested to contain CTV which are briefly described below:

- Introduction and prevention of CTV infected material by quarantine. Eradication of infected trees in the field must continue as far as possible to avoid spread of CTV through insect vectors (Yokomi *et al.*, 2010).

- To identify CTV or vector- free areas (Ahlawat and Sardar, 1976).

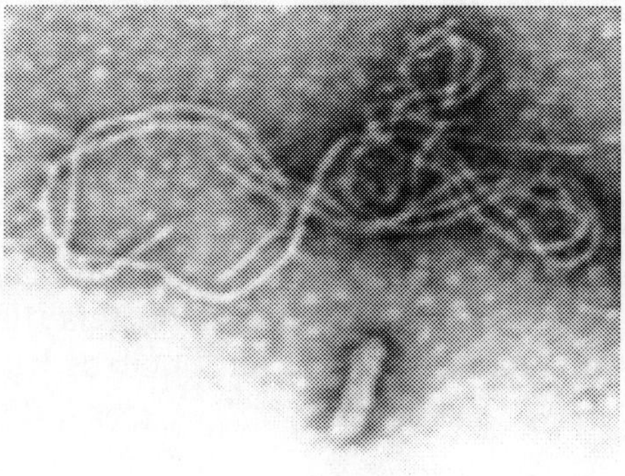

Fig. 20: Mixed infection of CTV with badnavirus

- To develop and supply healthy planting material through budwood certification programmes (Reuther, 1978; Calavan, 1974). Tristeza virus-free planting material is now being obtained by shoot tip grafting (Vijayakumari *et al.*, 2006a and 2006b). CTV tolerant rootstocks should be used.
- In places where severe strains of CTV and its effective aphid vector are endemic, cross protection technology using mild strain is the only practical method to reduce the severe effect of the virus. Cross protection has been successfully used in Brazil, South Africa, Australia and in several other countries. In India, cross protection was also suggested (Balaraman and Ramakrishnan, 1980). However, this technology could not be as successful in India as elsewhere (Chakraborty and Ahlawat, 2001). Similar results of failure of cross protection have been reported from California (Roistacher and Dodds, 1993). Protective CTV isolates can be obtained by attenuating severe isolates by passage through *Passiflora gracilis* or *P. caerulea* via aphid transmission.
- Aphid vector(s) control is an important method to minimize CTV incidence either by spraying insecticides or by biological methods.
- A long term solution to CTV management is to produce transgenic citrus genetically resistant or immune to CTV. Pathogen-drived transgenic *Citrus* for resistance against Citrus tristeza virus has been developed (Febres *et al.*, 2005). Genes used for this purpose were native and untranslatable versions of CP genes, replicase genes, antisense sequences and RNA interference ((Roman *et al.*, 2005; Lee *et al.*, 2007). Genetic transformation of

Citrus paradisi was obtained with antisense and untranslatable RNA-dependent RNA polymerase genes (Cevik *et al.*, 2006; Moore *et al.*, 2000). Sour orange (*C. aurantium*), Mexican lime/Kagzi lime (*C. aurantifolia*), and grapefruit (*C. paradisi*) were transformed with CTV genes (Gutierrez *et al.*, 1997; Dominguez *et al.*, 2000; Febres *et al.*, 2002; Roman *et al.*, 2005; Moore *et al.*, 2000; Pérez and Ochoa, 1998; Tarafdar *et al.*, 2009). Transgenic citrus has been reviewed by Peña and Navarro. (1999).

Note: Hassaku (*Citrus hassaku*) is a popular variety in Japan. It is monoembryonic and most of Hassaku trees are infected with a dwarfing disease known as Hassaku dwarf. Studies on this disease revealed that this disease in Japan is caused by a strain of citrus tristeza virus (Tanaka *et al.*, 1969).

CITRUS CRINKLY LEAF

Crinkly leaf disease of citrus was first described by Fawcett in 1936 and detailed description of the crinkly leaf disease was given by Wallace, 1959. In India, crinkly leaf disease was reported on Lisbon and Nepali round lemons from Darjeeling hills (Ahlawat and Sardar, 1976). Crinkly leaf disease has also been reported from Florida (Garnsey, 1968) and also from Australia (Fraser, 1961). Lemon is the main host of crinkly leaf virus and the disease popularly known as lemon crinkly leaf disease.

SYMPTOMS

Symptoms of crinkly leaf disease appear both on young leaf and mature leaf. The young leaves of infected lemon plants show vein flecking symptoms but sometimes Eureka and Lisbon lemon show numerous necrotic pin-point spots on newly developing leaves (Fraser, 1961). The numbers of such spots were greatest on the growth which appears immediately following inoculation and decreased in later growth. As the spotted leaves grew and reached maturity, they develop crinkling. Mature leaves of infected lemon show leaf puckering or crinkling and blistering (Fig. 21). Affected leaves also show warping and pocketing appearance due to irregularities of growth in different parts of the leaf blade (Fawcett,

1936, Klotz and Fawcett, 1941; Fawcett and Bitancourt, 1943). Leaf flecking symptoms of crinkly leaf disease resemble to feeding by citrus red mite (*Panonychus citri*). Affected trees show an upright, restricted type of growth. Fruits on infected lemon trees remain small, coarse and bumpy. When sweet orange trees are inoculated with crinkly leaf infected lemon scions, typical veinal flecking symptoms of psorosis complex are developed (Fawcett and Bitancourt, 1943; Reuther *et al.*, 1978). Hence, at one time this disease was considered to be a component of psorosis complex. Pin point spotting on young leaves of inoculated Eureka lemon plants was observed in Australia and India (Fraser, 1961; Ahlawat and Sardar, 1976). However, Dweet tangor consistently developed psorosis young leaf symptoms when inoculated with pure culture of crinkly leaf virus.

Fig. 21: Symptoms of crinkly leaf disease on lemon

TRANSMISSION

Citrus crinkly leaf virus is transmitted by grafting including tissue grafting to Mexican or Kagzi lime (*Citrus aurantifolia)*, Eureka lemon (*C. limon*), Lisbon lemon, and Dunken grapefruit (*C. paradisi*). Crinkly leaf virus is also transmitted by mechanical inoculation from citrus to citrus (seedlings of Dunken grapefruit and Eureka lemon) and herbaceous hosts like beans, cucumber, petunia, tobacco and tomato. Local lesion symptoms are developed on mechanically inoculated plants of cowpea (*Vigna sinensis*) and bean (*Phaseolus vulgaris*) after 2-4 days of inoculation. No vector of this virus is known but natural spread is sometimes observed probably by contamination of operational tools. Seed transmission of the virus has also been observed. Seedlings of sweet lime, sour orange, rough lemon, *Poncirus trifoliata,* Emperor and Cleopatra mandarin, and Orlando tangelo showed no symptoms. Strains in Crinkly Leaf Virus have also been suggested (Fraser, 1961).

HOST RANGE

Susceptible host species are found in the Family Apocynaceae, Leguminosae-Papilionoideae, and Rutaceae. Following Citrus species are susceptible to experimental inoculation: Citron, Etrog citron, Clementine, Grapefruit, Lemons, Lime (kagzi, Key, Mexican and West Indian), Cleopatra mandarin, Pummelo, Satsuma, Sour orange, Sweet lime and sweet orange. Following species of herbaceous hosts are susceptible by mechanical inoculation: *Chenopodium amaranticolor, C. quinoa, Crotalaria spectabilis, Cucumis sativus, Gomphrena globosa, Momordica balsamina, Phaseolus vulgaris cv. Red Kidney, P. aureus, Vigna unguiculata* and *Catharanthus roseus. Citrus aurantifolia, C. limon, C. paradisi, C. reticulata, C. sinensis,* Lemons and limes (Maxican lime) are highly susceptible. Puckering symptoms in Mexican lime are comparatively severe. Puckering has also been observed on leaves of Etrog citron (*C. medica*) and *C. excelsa*. However, symptoms on sweet orange and mandarins are mild and may be visible to some young leaves only. Indian isolates of crinkly leaf virus showed symptoms on inoculated Washington navel, Rangpur lime and Kagzi lime in addition to Eureka and Lisbon lemons. The virus is transmitted mechanically to *Sesamum indicum* in addition of some other hosts listed above. Symptoms on some important hosts upon inoculation are given below.

- *Lisbon lemon*: Spots produced on early formed leaves and were in some cases so numerous that they became confluent, and some distortion of the young leaves resulted.
- *Eureka lemon*: Asteroid local lesions developed.
- *Meyer lemon*: Few or more spots and slight crinkling developed.
- *Lemon seedlings, Etrog citron*: Leaf crinkling develops after a few weeks of inoculation.
- *Pear-shaped and knight citrons*: The spots were few and not well defined. In a few cases very slight crinkling occurred in one or two leaves.
- *Sweet orange*: Few spots and slight crinkling developed.
- *Vigna unguiculata:* Vein clearing and malformation developed.
- *Crotalaria spectabilis:* Systemic chlorotic veinbanding developed.
- *Cowpea cultivar Blackeye*: Chlorotic lesions develop on mechanically infected leaves within a few days.

Better local lesion assay host for crinkly leaf virus is *Vigna unguiculata* cv. 'Blackeye'. Among *Citrus* spp. indicator hosts are: lemons, limes, Etrog citron and Sour orange.

THE PATHOGEN

A polyhedral virus approximately 27 nm has been reported from California isolates of crinkly leaf (Yot-Dauthy and Bove, 1968). Crinkly leaf virus is now classified as *ilarvirus* of subgroup 2 of *ilarviruses* with four nucleoprotein components. The size of virus particles ranges from 26 to 32 nm. Crinkly leaf virus is serologically related to other citrus ilarviruses causing citrus variegation and citrus leaf rugose diseases. Crinkly leaf virus is considered to be a strain of infectious variegation ilarvirus. Genome of ilarviruses will be discussed in the chapter of leaf rugose and infectious variegation.

DIAGNOSIS AND MANAGEMENT

Citrus crinkly leaf ilarvirus is detected by inoculation to indicator host(s) as described above. The virus is also detected by ELISA using polyclonal and monoclonal antibodies. Since citrus ilarviruses have been sequenced (Scott and Ge, 1995 a and b), crinkly leaf virus is successfully detected by molecular techniques like Polymerase chain reaction (PCR) and nucleic acid hybridization. Crinkly leaf virus can be managed by the use of healthy planting material. *Poncirus trifoliata* is tolerant to crinkly leaf virus and hence can be used as rootstock in endemic areas.

CITRUS INFECTIOUS VARIEGATION

Infectious variegation in citrus was described as a possible virus disease in 1931 by Petri. His description of symptoms and his illustrations are similar to those reported by Fawcett and Klotz (1939), who transmitted the causal agent by bud-grafting from variegated lemon to sour orange. They noticed that some of the symptoms were similar to those associated with psorosis and suggested that the diseases might possibly be related to psorosis. Klotz and Fawcett (1941) noted that crinkly leaf of lemon is sometimes associated with infectious variegation and it was not certain whether infectious variegation differed sufficiently from crinkly leaf to be considered a separate disease. From cross-protection studies, it was concluded that infectious variegation is caused by a strain of the psorosis virus.

Initially, infectious variegation disease was reported to infect lemons and Sour orange in California but now the disease is probably distributed worldwide. This disease has been observed in the United States of America, Cyprus, Turkey, Australia and Mediterranean countries, Italy and Portugal (Grant and Smith, 1960; Bahia Bennoni *et al.*, 2004; Gerola *et al.*, 2005). In India, infectious variegation was recorded on *Citrus decumana* and lemon plantations in Darjeeling and Sikkim hills by the senior author (Yora *et al.*, 1977). This disease was initially studied as a strain of citrus psorosis (Wallace, 1957) but later investigations revealed it to be a distinct disease. Strains in infectious variegation virus do exist. For example, California and Florida strains of infectious variegation virus differ in reaction on *Crotalaria spectabilis* (Desjardins and Wallace, 1966).

SYMPTOMS

Common symptoms in lemon (*Citrus limon)* and Sour orange (*C. aurantifolia)* trees are malformation of mature leaves and psorosis-like flecking of young-leaves. Infectious variegation virus causes mosaic-like symptoms including crinkling, distortion, flecking of young leaves, pin point spotting and shoe-string effect on leaves. Affected trees remain stunted.

TRANSMISSION

Infectious variegation virus is transmitted by mechanical inoculation (Grant and Corbett, 1960); transmitted by grafting (Fawcett and Klotz, 1939); and possibly transmitted by seeds.

HOST RANGE

Grant and Smith (1960) reported infectious variegation in Florida and transmitted the causal agent by bud and leaf-piece grafts to several different species of citrus. Citrus cultivars susceptible to infectious variegation are; Bergamot, Citrons, Etrog citron, Clementine, Grapefruit, lemons, Lime (Kagzi, Key, Mexican, West Indian), Mandarin (Cleopatra), Pummelo, Satsuma, Sour orange, Sweet lime and Sweet orange. Willow leaf mandarin and *P. trifoliata* are symptomless hosts of the virus. *Poncirus trifoliata* appears to be tolerant to infectious variegation virus. Symptoms on common citrus hosts are given below:

- Eureka lemon, Etrog citron, *C. decumana* and Grapefruit: Chlorotic mottle and Puckering.
- *Lemon, Sweet and Sour Orange*: Leaf distortion, epinasty, chlorotic mottle and Leaf Pattern (Plate-29).

Following plant species are susceptible to experimental virus inoculation: *Capsicum frutescens, Capsicum frutescens* var. *grossum, Chenopodium quinoa, C. amaranticolor, Cucumis sativus, Crotalaria spectabilis, Gomphrena globosa, Momordica balsamina, Phaseolus aureus, P. vulgaris, Sesamum indicum, Vigna sinensis* (cowpea) and *Vigna unguiculata*. Symptoms produced by infectious variegation virus upon mechanical inoculation in some major hosts are described below (Desjardins, 1969; Desjardins and Wallace, 1966, Garnsey, 1968):

- Cucumber (*Cucumis sativus*): Diffuse chlorotic rings on inoculated cotyledons.
- Kidney bean (*Phaseolus vulgaris*): On variety Bountiful, the symptoms appear as distinct yellow vein banding in systemic manner.
- Gomphrena (*Gomphrena globosa*): Diffuse spots on inoculated leaves and interveinal chlorosis on systemically infected leaves.
- Cowpea cv Black eye: Chlorotic local lesions on inoculated leaves followed by systemic infection.

Lemon seedlings and cucumber variety National Pickling was found to be the best indicator host but *Vigna unguiculata* (W) and *Capsicum frutescens* var. grossum (W) are also used as assay hosts.

THE PATHOGEN

The virus particles associated with infectious variegation disease were of approximately 30 nm (Corbett and Grant, 1967; Garnsey, 1974; Rana *et al.*, 1974; Desjardins and French, 1970). Spherical virus particles of 30 nm were also observed from tissues of sour orange infected with Indian isolates of infectious variegation in leaf dip preprations (Fig. 22) (Yora *et al.*, 1977). Infectious variegation virus is an ICTV approved species of the genus *Ilarvirus* in the family *Bromoviridae*. In the mature sieve elements of virus-infected *Phaseolus vulgaris* leaves when viewed in electron microscope contain systems of parallel membranes along which spherical particles, of about 30 nm diameter which were aligned in simple rows (Gerola *et al.*, 2005).

The genome of infectious variegation ilarvirus is segmented; tripartite with positive-sense, single-stranded RNA. Minor species of non-genomic nucleic acid are also found in virions. The genome was fully sequenced (Calvert *et al.*, 1988). RNA-3 is fully sequenced and encodes movement protein and coat protein genes (Scott and Ge, 1995a). Coat protein gene appeared to be highly conserved (lowest similarity among all CIVV sequences 93%), especially at the N-terminal, indicating at the molecular level that both types of symptoms are indeed produced by the same virus species. Sequence analysis showed low genetic variability of CIVV isolates in the CPg confirming the high degree of coat protein gene conservation (Kubaa *et al.*, 2010). Surprisingly, at the nucleotide and amino acid levels and in the secondary features of the intergenic region, CVV appeared closer to the other subgroup 2 Ilarviruses than to Citrus leaf rugose virus to which it is serologically related and has been considered to be evolutionary related (Bahia Bennani *et al.*, 2004; Gonsalves and Garnsey, 1975 a,b). Viral genome encodes structural proteins which are involved in viral encapsidation and in protein movement. The virus is serologically related to citrus leaf rugose virus but distantly.

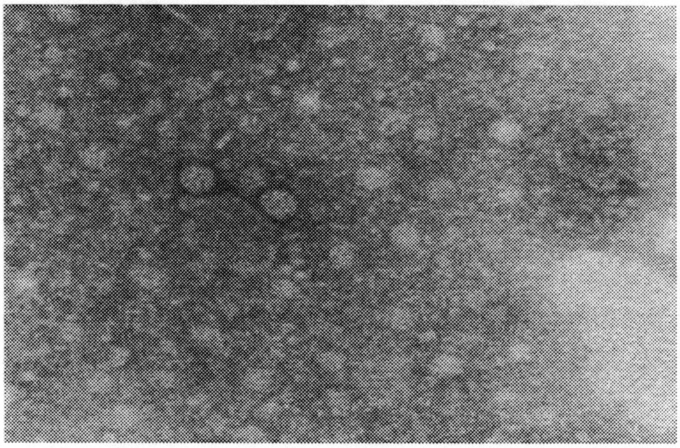

Fig. 22: Virus particles associated with Infectious variegation in dip Preparation

DIAGNOSIS AND MANAGEMENT

Citrus infectious variegation ilarvirus can be detected by graft inoculation to lemon seedlings and by sap inoculation to susceptible herbaceous hosts as described above. Polyclonal and monoclonal antibodies have been produced against this virus (Davino and Garnsey, 1984) and used for serological detection in techniques like ELISA. Other techniques like nucleic acid hybridization assays and PCR are also being used to detect the infectious variegation virus (Bahia Bennani *et al.*, 2004). Using primers designed for the conserved region, the virus was detected by RT-PCR and PCR positives were confirmed by ELISA and biological indexing (Bahia Bennani, 2004).

Since infectious variegation virus is not vector transmitted, chances of field spread are very rare. Therefore, This disease can be effectively controlled by propagating young plants from virus-free mother trees. Infectious variegation virus can also be inactivated by keeping infected plants for two-three months for hot air treatment at alternating temperature of 40°C for 16h/day and 30°C for 8hr/night (Calavan *et al.*, 1972). An antiviral principle obtained from seeds of *Celosia cristata* inhibited tobacco streak ilarvirus and may be effective to infectious variegation ilarvirus (Sumiya *et al.* 2004).

CITRUS LEAF RUGOSE

Citrus leaf rugose virus disease has earlier been reported by the name of Citrus psorosis virus complex and Citrus crinkly-leaf-type virus. The name "citrus-leaf-rugose virus" (CLRV) was proposed for a mechanically transmitted citrus virus which was discovered in Florida and has been described earlier as "crinkly-leaf-type virus"(Garnsey, 1974, 1975).

SYMPTOMS

Trees of Indian Key lime or Mexican lime (*Citrus aurantifolia),* Lemon (*C. limon),* and grapefruit (*C. paradisi)* are found to be naturally infected with leaf rugose virus. In Eureka lemon trees and on these hosts symptoms appear as flecking of young leaves and malformation or puckering of old leaves. Fruits on affected trees are small and lumpy. Citrus leaf rugose virus (CLRV) causes leaf rugosity in Mexican lime, and severe stunting of young Duncan grapefruit (*C. paradisi*) plants (Garnsey, 1968).

TRANSMISSION

Leaf rugose is not transmitted by a vector but field spread is suspected (Garnsey, 1968). CLRV is experimentally transmitted through contamination of cutting tools. It is transmitted by mechanical inoculation and by grafting. It is possibly transmitted by seeds but there is no authentic report.

HOST RANGE

Host range of citrus leaf rugose virus includes both citrus and herbaceous hosts. Among citrus species, Lime (Kagzi, Key, Mexican, West Indian), Dunken grapefruit, Eureka lemon, sweet orange are important hosts. Herbaceous hosts like *Chenopodium amaranticolor, C. quinoa, Crotalaria spectabilis, Cucumis sativus, Gomphrena globosa, Momordica balsamina, Nicotiana tabacum, Petunia hybrida, Phaseolus vulgaris cv.* Red Kidney, *Catharanthus roseus,* and *Vigna unguiculata* are susceptible to leaf rugose infection (Garnsey, 1975). Symptoms develop upon inoculation on some important hosts are given below:

- *Mexican lime or Indian key lime*: Variable rugosity of leaves
- *C. limon* (Eureka lemon): Pinpoint chlorotic flecks on expanding leaves, persistent symptoms without leaf distortion
- *C. paradisi* (Duncan grapefruit): Severe stunting and chlorosis in some isolates.
- *Citrus sinensis, C. limon, C. reticulata* — Crinkling and flecking
- *Phaseolus vulgaris* (bean) cv. Red Kidney: Small, necrotic local lesions develop on inoculated leaves. No systemic infection.
- *Vigna unguiculata*: Vein clearing and malformation of leaves.
- *Crotalaria spectabilis*: Systemic chlorotic veinbanding.

Phaseolus vulgaris cv. 'Red Kidney', *Vigna unguiculata* (Early Ramshorn cowpea), and *Crotalaria spectabilis* are local lesion hosts (Garnsey, 1975) and can be used as assay host of the virus. In addition Mexican lime seedlings and Red Kidney bush bean are also used as indicator hosts of virus.

THE PATHOGEN

Citrus leaf rugose virus (CLRV) particles are round with icosahedral symmetry. The capsid is quasi-isometric and has a diameter of 25-32 nm (Yot-Dauthy and Bove, 1968, Garnsey, 1974, 1975). CLRV belongs to group IV positive sense, single stranded RNA viruses.The virus is an ICTV approved species of the genus *Ilarvirus* in family *Bromoviridae.*

The genome of CLRV is tripartite, consists of four segments of linear positive-sense, single-stranded RNA. Based on A_{260}/A_{280} ratios amount of RNA in the particles is estimated to be similar to that of Tulare apple mosaic and tobacco streak ilarviruses (12-14%). The RNAs in NPs 1, 2, 3 and 4 are designated as RNAs 1, 2, 3 and 4, respectively (Gonsalves & Garnsey, 1974). Their Molecular Weights are: *c.* 1.1×10^6, *c.* 1.0×10^6, *c.* 0.7×10^6, and *c.* 0.3×10^6, respectively. The RNAs are not infective singly. However, non-infective mixture of RNAs 1+2+3 became highly infective with the addition of either RNA 4 or coat protein (Gonsalves and Garney, 1975). The complete genome of CLRV has been sequenced and was 7888 nucleotides long. RNA-1 has 3099 nucleotides, RNA-2 was about 2817 nucleotides long. RNA-3 has also been sequenced, but only an estimate is given, complete sequence is 1972 nucleotides long. Subgenomic RNA-4 is an mRNA derived from RNA-3 negative strand template. The multipartite genome is divided among more than one type of particle and the segments are distributed between 3 different types of particles. The genomic organization of ilarviruses is identical to that of other genera in the family *Bromoviridae* (Scot and Ge, 1995c).

Two strains have been described in leaf rugose virus (Garnsey, 1975). This virus is serologically related but not identical to citrus variegation virus, citrus crinkly leaf virus and Tulare apple mosaic virus (Garnsey, 1974; 1975). CLRV has properties similar to Tulare apple mosaic virus.

DIAGNOSIS AND MANAGEMENT

Citrus leaf rugose virus is detected by inoculation to citrus and herbaceous indicator hosts. The virus can be detected by virus apecific antisera in gel diffusion test (Garnsey and Purcifull, 1969). Monoclonal antibodies have also been prepared against CLRV and have been used to detect CLRV in ELISA and gel diffusion tests. The virus can also be detected in PCR and nucleic acid hybridization tests (Ge and Scott, 1994; Scott and Ge, 1995 a and c). Citrus leaf rugose disease can be managed by similar methods as have been mentioned earlier with other ilarviruses.

CITRUS LEPROSIS AND ZONATE CHLOROSIS

Leprosis disease was first reported from Florida on sweet orange as scaly bark and neil-head rust disease (Knorr, 1968; Humes, 1901; Fawcett, 1911). This is probably the most important virus disease of citrus in Brazil at the present time (Rodrigues, 2000). The Brazilian citrus industry spends over US $100 million annually (about 21% of citrus production cost) on miticides to control the *Brevipalpus* mite vector of leprosis virus (Omoto, 1998; Rodrigues *et al.*, 2003). Without control of the mite vector and reduction of inoculum in the field by pruning, the disease can kill citrus trees in three years. Presently, this disease is present in countries of South America, Central America, Mexico (Rodrigues *et al.*, 2003; Bastianel *et al.*, 2006). It is a major problem of sweet oranges in Brazil and Argentina (Rossetti, 1996). Zonate chlorosis is a similar disease to leprosis but is not as harmful as leprosis. Zonate chlorosis is found in coastal regions of Brazil. It is currently known that two different virus types cause leprosis disease, the prevalent cytoplasmic type (CiLV-C) and the rare, nuclear type (CiLV-N).

SYMPTOMS

Leprosis affected trees show heavy defoliation, fruit drop and dieback followed by death of infected tree. On leaves of infected trees, symptoms appear as chlorotic patches with necrotic centre and chlorotic transluscent halo occasionally impregnated with gum (Plate 30). These lesions are either dark brown with raised surface or brilliant yellow areas (Bitancourt, 1955). Foliar symptoms also develop on tangerines and mandarins. Disease symptoms also appear on twigs and branches as yellowish or chlorotic flat patches or lesions. These patches later developed into raised large brown or reddish areas which subsequently turn greyish and may be confused with citrus canker (Plate 31). Lesions may fuse togather and kill the stem by girdling which will not allow the photosynthate to travel to roots *via* phloem resulting death of tree (Rossetti, 1980). Zonate chlorosis produces chlorotic areas on leaves of infected plants. Symptoms in fruits first appear as flat yellow patches which become larger, black and brown with somewhat depressed areas when fruits mature. Fruits in severely affected trees may fall prematurely (Rossetti, 1996).

TRANSMISSION

Leprosis disease is transmitted by grafting (Knorr, 1968; Chagas *et al.*, 1984 a). Both forms of leprosis are transmitted by mite vector(s). Vectors of leprosis disease are different species of mites. Mite, *Brevipalpus phoenicis* is a vector of leprosis in Brazil. Leprosis disease is more efficiently transmitted by larvae of *B. phoenicis* than by adults (Chagas *et al.*, 1984 b; Rosetti, 1996). In Brazil, both leprosis and zonate chlorosis are transmitted by *B. phoenicis* vector. In Florida, leprosis is transmitted by another mite species, *B. californicus* and in Argentina and Venezuela, the vector is *B. obovatus*. This disease is also transmitted by mechanical inoculation from citrus to citrus and herbaceous hosts (Colariccio *et al.*, 1995; Lovisolo *et al.* 1996).

HOST RANGE

Sweet oranges and mandarins are highly susceptible to leprosis. Kumquats, *Poncirus trifoliata* (trifoliate orange), and sour orange are susceptible but grapefruits, Tahiti lime, and rough lemon are comparatively less susceptible. Lemons and limes appear to be immune to leprosis (Rodrigues *et al.*, 2003; Bastianel *et al.*, 2006). Symptoms of leprosis on some citrus and noncitrus hosts are given below:

- *Bahianinha Sweet orange*: Leaves show large chlorotic lesions without necrotic centre.
- *Pera Sweet orange*: Round or elongated lesions measuring 5-12 mm with chlorotic or necrotic centre develop on inoculated plants (Plate 30).
- *Caipira sweet orange* and *Tahiti lime*: Show typical leaf symptoms of leprosis.
- *Cleopatra mandarin*: Large chlorotic areas sometimes covering the whole leaf

CiLV-C type leprosis was transmitted by mechanical inoculation to herbaceous hosts, *Chenopodium quinoa, C. amaranticolor* and *Gomphrena globosa* (Colaricio *et al.*, 1995). CiLV-C was also experimentally mite-transmitted to *Solanum violaefolium* (Rodrigues *et al.*, 2005), *Phaseolus vulgaris* (Groot *et al.*, 2006), *Malvaviscus arboreus, Hibiscus rosa-sinensis, Grevilea robusta, Bixa orellana, Commelina benghalensis* (Nunes *et al.*, 2006) and *Arabidopsis thaliana* (Kubo *et al.*, 2010). However, CiLV-N infects only sweet oranges and mandarins. Symptoms on some herbaceous hosts are described below:

- *Chenopodium amaranticolor, C. quinoa, C. album, C. murale*: Conspicuous chlorotic lesions with necrotic centre and transluscent halo developed on inoculated leaves.
- *C. capitatum, C. foliosum*: Necrotic local lesions.
- *C. polysspermum*: Brown spots with chlorotic halo
- *Gomphrena globosa*: Local lesions of brown colour develop on inoculated leaves.

THE PATHOGEN

Bacilliform particles measuring 100 – 110 X 40-50 nm similar to rhabdoviruses as associated with coffee ring spot virus and with some orchid viruses (orchid fleck virus) were also observed in nucleus and cytoplasm of leprosis infected cells (Kitajima *et al.*, 1972). There are two known genera of plant-infecting rhabdoviruses, *Cytorhabdovirus* and *Nucleorhabdovirus*. Cytorhabdoviruses replicate in the cytoplasm of infected cells in association with masses of thread-like viroplasm structure, and virus morphogenesis occurs in association with vesicles of the ER (Van Regenmortel *et al.*, 2000). Nucleorhabdoviruses accumulate in the perinuclear space with some particles scattered in the cytoplasm (Hull and Matthews 2002). Both types of particles (CiLV-C and CiLV-N) have been reported in citrus leprosies (Guerra-Moreno *et al.* 2005). In ultrathin sections of lesions caused by CiLV-C, virions measuring 120-130 x 50-70 are present within membrane-bound vesicles and/or electron dense, circular to elliptical, vacuolated viroplasms, usually single, are present in the cytoplasm (Colariccio *et al.*, 1995). CiLV-C genome has been completely sequenced and it has been proposed as the type member of a new genus of plant viruses, *Cilevirus* (Kubo *et al.*, 2010). Leprosis virus (CiLV-N) belongs to the genus *Nucleorhabdovirus* in the family *Rhabdoviridae*; order *Mononegavirales.*

Complete sequence of the cytoplasmic Citrus leprosis virus (CiLV-C) has been done and its genome organization determined (Guerra-Moreno *et al.* 2007). The genome consists of two segments (5 and 9 kb) of positive sense ssRNA with poly-A tail (Pascon *et al.*, 2006; Locali-Fabris *et al.*, 2006). CiLV-C has been shown to have two RNAs (RNA 1 and RNA 2). RNA 1 and RNA 2 have been sequenced (Guerra-Moreno *et al.*, 2007). Its unique genomic sequence, structure, and organization led Locali-Fabris *et al.*, 2006 to propose CiLV-C as the prototype of the *Cilevirus,* a new virus genus. Genome assembly of CiLV-N reveals a close association with orchid fleck virus (Roy *et al.*, 2013)-an unclassified bipartite, negative sense RNA plant virus (Peng *et al.*, 2013). Recently a new virus closely related to CiLV was reported and named as citrus leprosis virus cytoplasmic type-2 (CiLV-C2) and was classified as a member of the genus *Cilevirus* (Roy *et al.*, 2013).

DIAGNOSIS AND MANAGEMENT

Typical symptoms of lesions in sweet oranges and mandarins are of satisfactory diagnosis of leprosies in field trees. Diagnosis of leprosis disease is done by sap inoculation to *Chenopodium* spp. or to *Gomphrena globosa.* Electron microscopy is routinely used for leprosies virus detection (Kitajima *et al.*, 2003). Leprosis can be detected in Western blotting (Manjunath *et al.*, 2006), PTA-ELISA, RT-PCR (Locali *et al,* 2003) and by immunofluorescence (Calegario *et al.* 2010). Control of leprosis is based on elimination of inoculum by pruning of affected trees and to reduce mite population by applying acaricides (Bastianel *et al.*, 2006). The disease is extremely destructive if mite control is not applied.

SATSUMA DWARF

Satsuma orange is a variety of mandarin orange grown extensively in Japan where it has a dominating satsuma orange market. Satsuma trees are preferably grown in the cool subtropical regions of Japan, Spain, central China, Turkey, Russia, South Africa, South America and in United States of America. A disease similar to Satsuma dwarf has been reported from Cheju Island in Korea, Mainland China and Turkey (Azeri, 1973, Cui *et al.*, 1991). Satsuma dwarf is a common disease of satsumas and was first shown to be a graft transmissible disease in 1952 in Japan (Yawada and Sawamura, 1952). Incidence of satsuma dwarf disease is increasing due to widespread use of top grafting in addition to its natural transmission.

SYMPTOMS

Boat and spoon-shaped malformed leaves, leaf crinkling, downward curling, leaf deformation short internodes and a witches'-broom appearance of the twigs are the most characteristic symptoms of Satsuma dwarf disease. Boat-shaped leaves develop under temperatures ranging from 13/8°C to 23/18°C (day/night). Affected trees of Satsuma mandarin are stunted. Severely affected trees bear small and malformed fruits. In Japan, citrus mosaic virus, Navel orange infectious mottling virus and natsudaidai dwarf virus are serologically related and possibly are the strains of Satsuma dwarf virus (SDV) (Tanaka and Imada, 1976; Iwanami and Koizumi, 2000). Major symptoms of these diseases are enations, multiple flushing, stunting or dwarfing, reduction in number and size of leaves and shoots, shortened internodes, and small-sized fruits with thick peel. Fruit yield is much reduced and depend on infection of severe or mild strain (Iwanami *et al.*, 1993). Navel orange infectious mottling caused by one of Satsuma dwarf group viruses is characterized by formation of yellow mottling on leaves. Several new varieties of tangelo and tangor become dwarfed if affected by Satsuma dwarf (Shimizu and Miyoshi, 2003).

TRANSMISSION

Satsuma dwarf virus (SDV) is transmitted by grafting (Ushiyama, 1981; Yamaguchi, 1984). Mechanical inoculation by rubbing or knife slashing is successful not only to various herbaceous but also citrus plants (Kishi and Tanaka, 1964; Tanaka, 1972; Tanaka and Kishi, 1963, Tanaka and Imada, 1974). Soil-borne transmission has been confirmed by many experiments and observations in Japan (Izawa, 1966; Ushiyama and Ohgaki, 1970) but vector remains unknown. SDV is transmitted upto 8.6% in *Phaseolus vulgaris* seed.

HOST RANGE

Principal host of Satsuma dwarf virus (SDV) is satsumas (*Citrus unshiu*) but most species of citrus and citrus relatives are susceptible to SDV

(Iwanami *et al.*, 1993; Miyakawa, 1968). Common hosts of SDV are orange, lemon and Pummelo. Among citrus relatives, SDV infects *Aegle marmelos, Aeglopsis chevalieri, Atalantia monophylla, Clymenia polyandra, Fortunella polyandra, Poncirus trifoliata, Swinglea glutinoa* (Iwanami *et al.*, 1993). China laurestine (*Viburnum odoratissimum)* is a plant commonly used as windbreak for citrus in Japan is also susceptible to SDV and thought that it may act as a reservoir for Satsuma dwarf infection (Koizumi *et al.*, 1988). Following plant species are susceptible by sap inoculation: cowpeas (*Vigna unguiculata*), *Nicotiana tabacum, Petunia hybrida, Phaseolus vulgaris, Phaseolus aureus*, sesame (*Sesamum indicum*), *Chenopodium quinoa, Chenopodium capitatum, Vigna sinensis, Citrus unshiu* and *Physalis floridana* (Tanaka and Nakanishi, 1972; Usugi and Saito, 1977). Diagnostic symptoms developed on some mechanically inoculated host species are given below.

- *Chenopodium quinoa:* Chlorotic and necrotic local lesions develop on inoculated leaves.
- *Physalis floridana*: Systemic mottling and malformation of leaves.
- *Sesamum indicum:* Local lesions; systemic vein clearing, necrosis and leaf deformation.
- *Vigna unguiculata:* Local lesions; vein clearing and mottling, necrotic streaking of petioles and stems.
- *Citrus unshiu:* Stunting, leaf malformation and rosetting.

Satsuma mandarin, White sesame, satisfaction kidney bean, *Chenopodium quinoa, Physalis floridana* and black eye cowpea are used as indicator hosts. *Chenopodium quinoa* is a good local lesion host for virus assay (Tanaka and Imada, 1974).

THE PATHOGEN

Virus preparation contains few virions and 3 sedimenting component(s) in purified preparations (Usugi and Saito, 1979). Top component (T) consists of empty capsids. Middle component (M) and bottom component (B) contain genomic RNAs of buoyant densities in CsCl of about 1.43 and 1.46 g/cm^3, respectively. Virus particles are round with icosahedral symmetry and have a diameter of 26-28 nm (Saito and Hibino, 1962). Virions are found in the cytoplasm. Infection with SDV results in formation of characteristic tubules containing single rows of virions in the infected cells (Hibino *et al.*, 1977). Intercellular inclusions and tubules containing arrays of spherical virus particles are found in leaf cells of satsuma and some herbaceous plants systemically infected with Satsuma dwarf virus (Saito and Hibino, 1962). SDV has been classified by ICTV as the type member of the genus *Sadwavirus* in the family *Comoviridae*. SDV genome is segmented and bipartite. Genome segements are distributed among 2 particle types of linear, positive-sense, single-stranded RNA, RNA1 and RNA2 (Iwanami *et al.*, 1996, 1999). All virions have two species of coat proteins with molecular weights of about 42,000 and 22,000 (Tanaka and Imada, 1974). Virions encapsulate two species of RNAs of about 7.0 kb and 5.4 kb both of which have a poly-A sequence at the 3' end (Usugi and Saito 1977, 1979, Iwanami *et al.*, 2001). Complete genome is 10100 nucleotides long and proteins constitute about 63% of the particle weight (B), or 67% of the particle weight (M), or 100% of the particle weight (T). Viral genome encodes both structural and non-structural proteins (Usugi and Saito, 1979).

DIAGNOSIS AND MANAGEMENT

In field affected citrus trees, diagnosis is done by observing spoon shaped leaves. The virus can be detected by inoculating indicator hosts such as white sesame (*Sesamum indicum*) Blackeye cowpea and Satisfaction kidney (Tanaka and Kishi, 1963). ELISA is widely used in Japan for SDV, CiMV, and NDV indexing. ELISA with monoclonal antibodies is used to distinguish citrus mosaic strain from other SDV (Nozu *et al.*, 1986). DAS-ELISA, SDS-PAGE, and electroblot immune assay diversty are also used to detect Satsuma dwarf and related viruses (Iwanami *et al.*, 1993). A simple and rapid detection kit using an immunochromatographic assay (ICA) has been developed in Japan (Nario Kusano, 2006). Satsuma dwarf and related viruses can be eliminated from citrus by heat-treatment ((Leki and Yamada, 1984) or by shoot-tip grafting. Certification of budstock is the primary counter measures to control the disease. Citrus planting material should be certified free ffrom SDV by methods such as those of the EPPO certification scheme for citrus (OEPP/EPPO, 1995). Propagation of budwood by topgrafting should be avoided. If affected trees are found, they should be immediately removed from the orchard. Soil should be fumigated with fumigated pesticides like chloropicrin and others (Isoda *et al.*, 1991).

NAVEL ORANGE INFECTIOUS MOTTLING

Navel orange infectious mottling virus (NIMV) is known to infect citrus in Japan for a long time (Tanaka *et al.*, 1971). Symptoms of this disease consist of large diffused chlorotic blotches and mottling of leaves with brown spots on the under surface. The virus is easily transmitted by graft inoculations from infected trees to citrus seedlings (Tanaka and Yamada, 1972). NIMV is also transmitted by mechanical inoculations to herbaceous hosts (Tanaka and Imada, 1974). Host range of citrus mosaic (Japan) and navel infectious mottling is similar to Satsuma dwarf virus which has been described in earlier chapter. NIMV is considered as a strain of Satsuma dwarf virus (Tanaka and Imada 1976). The virus particles are isometric in shape measuring 23 nm in diameter which are smaller in size than that of citrus mosaic virus (28 nm). The genome of the virus is positive sense ssRNA. Citrus mosaic virus (CiMV), Natsudaidai dwarf virus (NDV) and Navel orange infectious mottling virus (NIMV) are considered to be distantly related strains of Satsuma dwarf virus (SDV).

CITRUS MOSAIC

A mosaic disease of citrus has been reported from Japan (Ishigai and Jinno, 1958). Mosaic disease in Japan is commonly occurring in Satsuma mandarin in most citrus growing areas of Japan (Yamamoto and Yamaguchi, 1980; Usugi *et al.*, 1986). Mosaic diseases have also been reported from various parts of India (Ahlawat *et al.*, 1985; Bhagabati *et al.*, 1989; Pandey *et al.*, 1986; Dakshinamurty and Reddy, 1975; Naidu and Reddy, 1985). However, limited information is available on mosaic diseases reported from India and it is not known whether these diseases are similar to mosaic disease of citrus as has been reported from Japan.

SYMPTOMS

In Japan, infected Satsuma trees show large diffuse blotches and mottling on the leaves, with brown spots on the under surface of the leaf (Tanaka *et al.*, 1971). Similar symptoms also appear on the rind of fruit at the time of developing colour, near maturity or in matured fruits (Usugi *et al.*, 1986). However, Indian scientists have reported different symptoms of mosaic in different cultivars and locations. In Maharashtra, affected mosambi sweet orange trees show irregular green/yellow patches on leaves (Pandey *et al.*, 1986). In Darjeeling hills, field infected Darjeeling orange

(*Citrus reticulata*) trees show mosaic symptoms as severe mottling of leaves which is sometimes confused with greening symptoms (Plate 32). Affected trees are comparatively stunted (Ahlawat *et al.*, 1985).

TRANSMISSION

Mosaic disease of citrus in Japan is transmitted by graft and mechanical inoculations (Tanaka and Yamada, 1972; Tanaka and Imada, 1974). Field observations in Japan suggested soil transmission of mosaic disease but no vector has been identified (Iwanami and Leki, 1996). Seed borne nature of the Indian citrus mosaic has been suggested (Ahlawat *et al.*, 1985). Mosaic disease of citrus from South India has been transmitted by dodder laurel (Reddy *et al.*, 1985). In India, some isolates are transmitted by aphid vector, *Toxoptera citricidus* and also by sap inoculation (Pandey *et al.*, 1986, Dakshinamurty and Subbayya, 1981). The North-eastern region isolates of mosaic are transmitted by aphids, *Myzus persicae* and *Aphis craccivora* but not by *T. citricidus* (Ahlawat *et al.*, 1985).

HOST RANGE

Japanese citrus mosaic has a very wide host range in citrus and in herbaceous plants as well (Usugi *et al.*, 1986). Hosts susceptible to Satsuma dwarf virus are also susceptible to mosaic virus in Japan (Kishi and Tanaka, 1964). In India, citrus species/cultivars susceptible to mosaic infection are: Sweet orange (*Citrus sinensis*) cultivars mosambi, sathgudi, Washington navel; Indian acid lime (*C. aurantifolia*) cultivars kagzi nimbu; Rangpur lime (*C. limonica*), grapefruit (*C. paradisi*), mandarin (*C. reticulata*) cvs. Cleopatra mandarin, Nagpur santra, Sikkim or Darjeeling orange, Kinnow mandarin; lemon (*C. limon*) cv. Lisbon, Pummelo (*C. grandis*) and bale (*Aegle marmelos*). Symptoms in these hosts appear as chlorosis of the leaves with clear mosaic pattern, extreme reduction of leaf lamina and wavy margin of the leaves (Ahlawat *et al.*, 1985).

THE PATHOGEN

Virus particles of Japanese citrus mosaic virus are polyhedrons measuring 28 nm in diameter. Purified preparation had three centrifugal components in sucrose density gradients (Tanaka and Imada, 1976; Usugi *et al.*, 1986). Although the virus is unclassified as yet but the properties of the virus indicate that the virus is similar to Comoviruses and the nepoviruses (Bruening, 1977; Harrison and Murrent, 1977).

Genome of Japanese citrus mosaic virus consists of two RNA species (RNA1 and RNA2) 7.0 kb and 5.4 kb (Iwanami *et al.*, 1993). Mosaic virus sequences showed maximum similarity (78%) with satsuma dwarf virus suggesting that citrus mosaic and Satsuma dwarf viruses are related to the viruses of the genera comoviruses and the nepoviruses (Iwanami and Leki, 1996). However, low similarity in sequences suggests that these two viruses are distinct. The Darjeeling orange mosaic virus in India did not react with antiserum of Japanese citrus mosaic virus (Ahlawat and Chenulu, 1994) suggesting that mosaic diseases in India and Japan are caused by different viruses.

DIAGNOSIS AND MANAGEMENT

Mosaic disease is detected by grafting on to susceptible hosts or by sap inoculation to herbaceous hosts like sesamum. Japanese mosaic virus can be detected by serological tests like ELISA (Usugi and Tsuchizaki, 1982). Mosaic virus in Japan can be managed by same approaches as have been described for management of Satsuma dwarf.

VEIN ENATION AND WOODY GALLS

Woody galls were first described on rough lemon in Australia (Fraser, 1958) but vein enation was first observed in California on leaves of sour orange, rough lemon and Mexican lime (Wallace and Drake, 1953). Vein enation and woody galls were later shown to be different symptoms of the same disease (Wallace and Drake, 1960, 1962). Now vein enation/ woody gall disease is known to occur in many citrus growing countries, United States (California), South America (Peru, Brazil), Africa (South Africa, Kenya, Libya, Reunion, Tanzania), Oceania ((New South Wales, Australia), New Zealand, EPPO region (Spain, Libya and Turkey), Asia (China (Zhejiang), India, Iran and Japan. This disease is also present in European Union countries wherever citrus is grown. Cooler coastal regions are more favorable to gall formation. In India, galls have been reported on Rangpur lime in Darjeeling and Sikkim hills, on Kagzi lime in Rajasthan (Chenulu and Ahlawat, 1994) and on rough lemon rootstock of nucellar Mosambi sweet orange in Maharashtra (Mali *et al.*, 1975, 1976; Mali and Chaudhuri, 1993) and in South India (Manjunath, 1987).

SYMPTOMS

Symptoms that gave the disease its name are the small enations (upto 1 mm) on the veins on under surface of leaves with corresponding depressions on the upper surface of leaves and galls on stem and branches of affected trees (Plate 33) (Garnsey, 1988).

TRANSMISSION

Vein enation-woody gall disease is bud and graft transmitted. It is also transmitted by dodder and by aphids, *Toxoptera citricida* (McLean, 1954), *Myzus persicae* (Wallace and Drake, 1959), *Aphis gossypii* (Laird and Weathers, 1961), *Toxoptera aurantii* (Hermoso de Mendoza *et al.*, 1993) and *T. citricidus* (Maharaj and da Graca, 1989) in persistent manner. It is not seed transmitted.

HOST RANGE

Host range of vein enation/woody gall virus is restricted to citrus species only. Vein enations on leaves of several citrus species have been observed upon inoculation but most severe vein enation symptoms are developed on rough lemon, West Indian lime and sour orange. Galls are formed on stems in several citrus hosts but severity of galls develops on *C. Volkameriana,* Rough lemon (*Citrus jambhiri*), Sour oranges and bitter sweet orange (*C. aurantium*). In India, enations have been observed on grapefruit, *C. amblycarpa, C. macroptera, C. latipes* and *C. pennivesculata* (Mali *et al.,* 1976). Grapefruit appears to be a tolerant host. Kagzi lime or Small-fruited acid lime seedlings are used as indicator host for vein enaion and *Citrus volkameriana* and rough lemon seedlings are good indicator hosts for woody gall symptoms. Injury near thorns can initiate early gall formation.

THE PATHOGEN

Isometric virus-like particle measuring 27 nm have been observed in electron microscope in the phloem cells of galls or vein enations (Iwanami *et al.,* 1992). Citrus vein enation virus is probably a member of Luteoviridae (da Graca and Maharaj, 1991). Serological relationship of this virus to barley yellow dwarf (PAV and MAV strains) further suggested that this virus is a member of family Luteoviridae (Clark and da Graca, 2000) most closely related to cereal yellow dwarf-RMV and RPV strains now classified as *Poleroviruses* (Pringle, 1998).

DIAGNOSIS AND MANAGEMENT

Woody galls on stems and trunks and branches are diagnostic symptoms in field trees. Citrus vein enation-woody gall luteovirus is detected by grafting bark chips of suspected material on woody indicators. Enations are seen in 5-8 weeks after inoculation on underside of leaves of limes (*Citrus aurantiifolia*), rough lemons (*Citrus jambhiri*) or sour oranges (*Citrus aurantium*) at a temperature 24-27°C max. day/18-21°C min. night. Swellings or galls appear more slowly on stems of *C. jambhiri* or *C. volkameriana*. The virus can also be detected by electron microscopy. Citrus vein enation virus (CVEV) can also be detected by PAGE analysis of infected samples by observing double-stranded RNAs. CVEV has been detected in ELISA and ISEM using some commercial Barley yellow dwarf virus kits (Clark and da Graça, 2000) but attempts to amplify cDNA using luteovirus primers have not been successful, thus the luteovirus proposal remains unconfirmed.This disease can be managed by adopting practices such as to avoid the use of susceptible rootstocks and use clean planting material. Shoot-tip grafting and thermotherapy are effective to manage the disease (Roistacher and Kitto, 1977; Roistacher, 1979).

TATTER LEAF = CITRANGE STUNT

Tatter leaf disease of citrus was first discovered in 1962 on Meyer lemon trees (Wallace and Drake, 1962). Affected Meyer lemon trees were symptomless but when tissue grafts made from Meyer lemon to *Citrus excelsa* plants, leaves of inoculated plants developed yellow blotches and edge malformation. Further investigations revealed that two viruses were involved in original Meyer lemon, tatter leaf causing symptom in *C. excelsa* and citrange stunt causing symptoms on citrange and citremon (Wallace and drake, 1968) but separation of two viruses could not be achieved by sequential passaging through a variety of citrus hosts (Roistacher *et al.*, 2000). Tatter leaf virus was therefore also known as Citrange stunt virus (Roistacher, 1988). Tatter leaf disease is wide spread in several citrus growing countries in Asia, South Africa and Australia including USA and Brazil. Yield on infected mandarin (*Citrus reticulata*) trees on trifoliate orange (*P. trifoliata)* rootstock is 75% less than those of virus-free trees (Takahara *et al.*, 1988).

SYMPTOMS

Trees on susceptible rootstocks (Trifoliate orange and its hybrids) are stunted with chlorotic foliage. Seedlings of Mexican lime (*Citrus *aurantifolia*) and Kalpi lime (*C. excelsa*) showed young leaf symptoms as chlorotic mottle and tearing-like or tattering of the margin of leaves. Affected plants bloom heavily and fruits on such trees mature early but often die. Citrus tatter leaf virus (CTLV) produces striking leaf symptoms and stunting in citrimon (*Poncirus trifoliata* × *C. limon*) and Rusk and Troyer citranges (*P. trifoliata* × *C. sinensis*) (Wallace and Drake, 1963). Ringspot symptoms have also been observed on infected citrumelo leaves. Citrange leaves develop chlorotic spots, mottling and distortion. Stem of citrange plants may be deformed and have a zigzag growth pattern and develop chlorotic areas on stem.

TRANSMISSION

Tatter leaf virus is transmitted from citrus to citrus by grafting and mechanical inoculation (Garnsey, 1974). CTLV is seed transmitted in *Chenopodium quinoa,* cowpea, lily and soybean seeds (Nishio *et al.*, 1982). It also appears to be seed transmitted in Eureka lemon (Tanner *et al.*, 2011). No natural vector is known.

HOSTRANGE

Susceptible hosts of Tatterleaf-Citrange stunt virus are: *Citrus excelsa,* Citranges (Carrizo, Rusk, Troyer), Citremons (1448, 1449), Satsuma or sweet orange on *P. trifoliata* roots. Clearly visible and severe symptoms develop on citranges. *Aeglopsis chevalieri,* Lemons, Rough lemon, *Severinia buxifolia,* Sour orange, Sweet orange and West Indian lime are symptomless hosts of tatter leaf virus. CTLV also infects herbaceous hosts like *Chenopodium quinoa, Phaseolus vulgaris, Vigna sinensis (cowpea)* cv. Blackeye and Masterpiece bean. Citrange, citremon and citrumelo stems develop a 'zig-zag' growth pattern. Host range of CTLV is wide and susceptible hosts are both in citrus and herbaceous plant species: *Citrus aurantifolia, C. excelsa, C. limon, C. medica, C. sinensis, P. trifoliate, C. sinensis* × *Poncirus trifoliata, P. trifoliata* × *C. paradisi cv. Swingle, P. trifoliata* × *C.limon* (Calavan *et al.*, 1963; Garnsey and drake, 1968; Miyakawa, 1980b, Miyakawa and Matsui, 1976; Wallace and Drake, 1968; Marais and Lee (1986).

CTLV is also mechanically transmitted to herbaceous hosts. These hosts are: *Amaranthus tricolor, Catharanthus roseus, Chenopodium amaranticolor, Cucurbita pepo, Dianthus barbatus, D. chinensis, faba beans, Gomphrena globosa, N. debneyi, Petunia hybrida, Tetragonia tetragonioides, Vigna unguiculata,* peas, tomatoes, and soybeans. (Nishio *et al.*, 1982; Tanaka and Imada, 1974). 19 plant species in nine families are susceptible to CTLV (Semancik and Weathers, 1965). Indicator hosts are: *Citrus excelsa* seedlings, Rusk citrange seedlings and local lesion host is *Chenopodium quinoa.*

THE PATHOGEN

Virus particles of tatter leaf are filamentous and flexuous with a clear modal length of 650 nm and 19 nm wide with cross banding pattern and with a helical construction of 3.4 nm pitch (Semancik and Weathers, 1965, Miyakawa and Matsui, 1976). This virus was considered to be closely related strain of apple stem grooving *Capillovirus* rather than a distinct virus (Ohira *et al.*, 1994) but sequences of CTLV genomic RNA has shown it to be indistinguishable from apple stem grooving virus, type member of genus *Capillovirus* in family *Flexiviridae* (Ohira *et al.*, 1995; Yoshikawa *et al.*, 1993). Genome of CTLV is monopartite with only one particle size of linear, positive-sense, single-stranded RNA. Genomic RNA is partially sequenced and sequenced region is 2956 nucleotides long. RNA species has a molecular weight of 2.83 x 106 Da and produces a single protein band of molecular weight 27 x 103 Da in SDS-PAGE (Nishio *et al.*, 1989). Complete genome sequence of CTLV from Meyer lemon has also been determined to be 6,495 nucleotides (nts), with two overlapping open reading frames (ORFs) and a poly (A) tail at the 3' end. (Satyanarayana *et al.*, 2009).

DIAGNOSIS AND MANAGEMENT

Bud union crease in trees on trifoliate orange and trifoliate hybrid rootstocks are the symptoms useful for field diagnosis of CTLV. Indicator plants can detect the virus in laboratory. Sap inoculation to *Chenopodium quinoa* is also used for detection and assay of virus which develops chlorotic or necrotic spots on the inoculated leaves, and its upper leaves show vein clearing, twisting and stunting. ELISA is used for detection of CTLV using polyclonal and monoclonal antibodies (Kawai and Nishio, 1990; Kawai *et al.*, 1996). In PAGE analysis, two typical dsRNA bands were observed in infected samples (Frison and Taher, 1991). PCR-based techniques have also been used for CTLV diagnosis (Hailstones *et al.*, 2000) and the virus can be detected in multiplex PCR with several viroids (Ito *et al.*, 2002). A real time RT-PCR assay using SYBR green has also been used to detect and quantify CTLV RNA-targets (Liu *et al.*, 2011).

Tatter leaf virus is managed by using virus-free budlines for propagation. CTLV cannot be eliminated by shoot-tip grafting alone (Roistacher and Kitto, 1977). Heat treatment for 30 days at 35-40°C/30°C (day/night) followed by shoot-tip grafting (STG) can be an effective therapy (Koizumi, 1984). Incubation of budsticks on synthetic medium *in vitro* for 10-14 days at 32°C, followed by shoot-tip grafting can produce CTLV-free plants with 30-50% success (Navarro *et al.*, 1989). Long-term heat treatment of affected plants for 90 or more days at 40°C/30°C (day/night) can also eliminate CTLV (Miyakawa, 1980a).

INDIAN CITRUS RING SPOT

Indian citrus ringspot disease was originally recorded on kinnow mandarin trees at Abohar in India. Based on symptoms on inoculated Mosambi sweet orange plants, this disease was tentatively considered to be a strain of psorosis-A disease (Ahlawat, 1989). But further studies revealed that the ringspot disease in India is not related to psorosis (Byadgi and Ahlawat, 1995). This disease is widely distributed in citrus orchards in India. Interestingly, this disease was also recorded on lemon and Khasi mandarin in Northeast India-the home of citrus (Bhagabati *et al.* 1989). In trees on Sohsarkar (*Citrus karna*) and Karna khatta, incidence of this disease was around 55%, but in trees on Troyer citrange roots, incidence was only 7% (Ahlawat and Pant, 2003). There is no report of this disease outside India so far but could be present. Field affected trees show typical bright yellow rings in most of the leaves and such trees look as ornamental trees (Plate 34). Although the incidence of the disease in Kinnow mandarin trees –a hybrid (Willow × king mandarins) was observed upto 90% in several orchards in or near Abohar in Punjab but as an average the incidence of ringspot disease ranged from 5-83.8% and yield loss was estimated from 20.5 to 98.38 % (Byadgi and Ahlawat, 1995; Thind and Arora, 1997; Hoa and Ahlawat, 2004). Fruit quality on affected trees is reduced with less juice and TSS (Thind *et al.*, 1998). Based on detailed studies on ringspot disease and the virus associated with it, this disease was named as *Indian citrus ringspot disease* (Byadgi and Ahlawat, 1995) and the causal virus as *Indian citrus ringspot virus* (ICRSV) (Pant and Ahlawat, 1998).

SYMPTOMS

Natural infection of Indian citrus ringspt virus (CRSV) has been observed in sweet oranges (Malta, Mosambi, Satgudi (Chini), mandarins (Nagpur, Sikkim, Coorg and Kinnow), Kagzi lime and Kagzi Kalan (*Citrus aurantifolia*). Leaves of affected trees showed typical symptoms as complete or broken rings with yellow margins and green centre (Plate 34) ranging from mere dots to 4-18 mm in diameter (Byadgi *et al.*, 1993). Rings may be one to several on a leaf and their size may be variable. Fruits on affected trees remain smaller with less juice. Experimentally inoculated sweet orange plants show vein clearing or flecking symptoms during the first spring flush, 2-4 months after graft inoculation followed by rings on matured leaves (Plate 35). (Ahlawat, 1989; Byadgi *et al.*, 1993; Hoa and Ahlawat, 2004). Symptoms of ring spot disease develop only at low temperature during winter season in North India.

TRANSMISSION

Indian citrus ringspot disease is transmitted by wedge, side and leaf patch grafting. Ring spot disease could not be transmitted by several aphid species (*Aphis gossypii, A. craccivora, A. citricola* and *Myzus persicae*). However, natural transmission of the disease has been observed on kinnow mandarin seedlings in the vicinity of infected trees (Ahlawat, 2007) suggesting that a vector could be involved. This disease was transmitted by dodder, *Cuscuta reflexa* from citrus to citrus (Byadgi and Ahlawat, 1995) and also by mechanical transmition from citrus to some herbaceous hosts (Pant *et al.*, 2000). It is not seed or soil transmitted (Byadgi and Ahlawat, 1995; Hoa *et al.*, 2004; Rutici *et al.*, 2000c). However, virus particles of this virus were detected in pollens collected from ringspot infected trees suggesting the possibility of pollen transmission of this virus (Pant and Ahlawat, 1998).

HOST RANGE

Citrus and several of its relative and hybrids are susceptible to Indian citrus ring spot virus (ICRSV) by graft inoculation. Typical symptoms of ringspot develop on inoculated plants of Kinnow mandarin, Sweet oranges (Mosambi, Pineapple, Malta and Satgudi), Nagpur mandarin and Rough lemon (Plate 36). Sour orange, khasi or Darjeeling mandarin, Rangpur lime, Kagzi lime and Kagzi kalan (*C. aurantifolia* (Christen) Swingle) and Karna khatta (*C. karna* Raf.) are also susceptible and are easily infected by graft-inoculation (Hoa and Ahlawat, 2004; Hoa *et al.*, 2004). Symptoms in these hosts varied from shock reaction to leaf flecking, vein banding, chlorotic spots and ring spots. Diagnostic symptoms developed on citrus and herbaceous hosts under experimental inoculation are described below (Byadgi *et al.*, 1993; Pant *et al.*, 1997, 2000, Hoa and Ahlawat, 2004). Assay hosts are *C. amaranticolor, C. quinoa* and saxa bean.

Herbaceous Hosts Susceptible to ICRSV

- *Chenopodium amaranticolor* and *C. quinoa, Glycine max* (Soybean cv Hodgson), *Vigna unguiculata* (Cowpea cv Black): Necrotic local lesions are developed on inoculated leaves (Plate 37).
- *Phaseolus vulgaris* cv Saxa, Singtami, gheusemi, alapatri, *Vicia faba major*: Local lesions with dark brown margins, followed by vein clearing, browning and necrosis of leaf veins and tissues, mottling of leaves (Plate 37).

THE PATHOGEN

A filamentous virus associated with Indian citrus ringspot virus was initially reported by Ahlawat and Chakraborty (1990). Purified preparations showed three types of particles in electron microscope. The major filamentous and flexuous particles with a modal length of 650 nm with clearly visible cross banding were tentatively identified as capillovirus (Byadgi *et al.*, 1993; Byadgi and Ahlawat, 1995; Pant and Ahlawat, 1998; Ahlawat *et al.*, 1997). However, other components were later found to be the host components. Main virions associated with ICRSD are of 650 nm modal length and about 13 nm in diameter, are flexuous rods with a distinct zipper-like structure (Fig. 23). Indian citrus ringspot disease is now known to be caused by *Indian citrus ringspot virus*, the only current member of the genus *Mandarivirus* in the family *Flexiviridae* (Adams *et al.*, 2004, 2005). Further, based on serological and molecular studies, Indian cirus ringspot showed some affinities to potex, Carla, fovea and allexiviruses but does not fall into any known genus and hence was proposed to be a new species (Rutici *et al.*, 2000a). The virus particles are relatively robust, easily purified, and highly immunogenic. ICTV classified Indian citrus ringspot virus in family *Flexiviridae*, genus *Mandrivirus*, and type species *Indian citrus ringspot virus* (Fauquet *et al.*, 2005). ICRSV genome is positive sense single stranded RNA of about 7.5 kb without poly A tail and with a coat protein (CP) of 34 kDa. ICRSV particles have 6% nucleic acid (Byadgi and Ahlawat, 1995). The viral genome was fully sequenced which consists of 7560 nucleotides. It contains six open reading frames (ORFs) which encode putative proteins of 187.3, 25, 12, 6.4, 34 and 23kDa respectively. Function of ORF6 is unknown. Phylogenetic analysis of the complete genome and each ORF separately, and database searches indicate that ICRSV, though showing some similarities to potexviruses, is significantly different, as in the presence of ORF6, the genome and CP sizes, and particle morphology (Rutici *et al.*, 2000b). Based on sequence

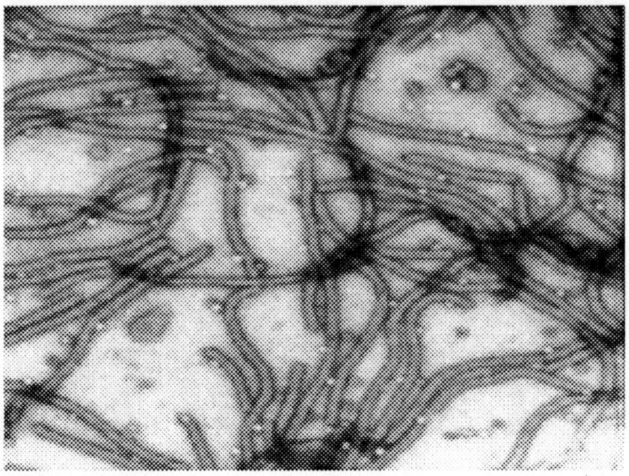

Fig. 23: Purified preparations of ICRSV particles.

analysis and inoculation on a panel of Citrus species/cultivars several isolates of ICRSV have been identified (Hoa and Ahlawat, 2004). Coat protein genes of these isolates have been cloned and sequenced (Accession No. AY255006, AY255007, AY255008, AY255009).

DIAGNOSIS AND MANAGEMENT

Diagnosis and management of ICRSV has been reviewed from time to time (Ahlawat *et al*, 1995; Ahlawat and Pant, 2008). In Field trees, ICRSV is easily detected by yellow rings on leaves especially during winter period. A diagnostic panel of citrus species comprising of Kinnow mandarin, Mosambi sweet orange, Malta sweet orange, Rough lemon and King mandarin showed variable symptoms upon graft inoculation and can be used for isolate detection (Hoa and Ahlawat, 2004). ICRSV is easily detected by sap inoculation on indicator plants such as *Chenopodium amaranticolor* and *C. quinoa* which react with local lesions. Systemic hosts like Kinnow mandarin and Mosambi sweet orange can also be used for disease diagnosis by grafting with test scions. Virus particles can be seen in electron microscope. Virus detection is also done in ELISA, ISEM (Fig. 24), and dot immunobinding assay (DIBA) (Fig. 25) (Pant and Ahlawat, 1998; Rutici *et al.*, 2000c; Hoa and Ahlawat, 2004). Nucleic acid hybridization (Fig. 26) and western blotting are also used to detect ICRSV (Byadgi and Ahlawat, 1995; Hoa and Ahlawat, 2004; Hoa *et al.*, 2004; Ahlawat and Pant 2002). Digoxigenin-labeled cDNA and RNA probes both complementary to the sequences of ICRSV detected it in infected plants (Rutici *et al.*, 2000c). Virus specific primers have been designed and used for virus diagnosis including its strains (Dl, Pu, Ab, Ah) in RT-PCR (Fig. 27, 28) (Hoa and Ahlawat, 2004; Hoa *et al.*, 2004).

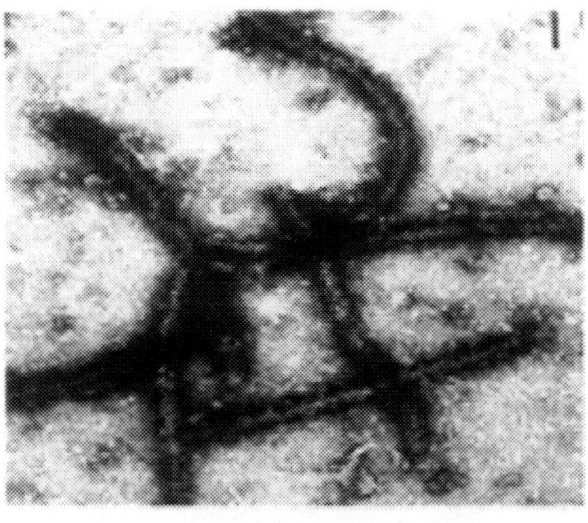

Fig. 24: ISEM decoration test of ICRSV particles with homologous antiredism.

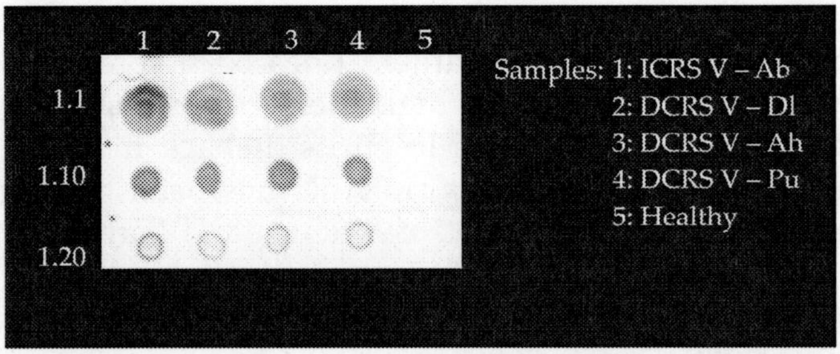

Fig. 25: All the four isolates of ICRSV detected by DIBA using antibodies from Ab isolate.

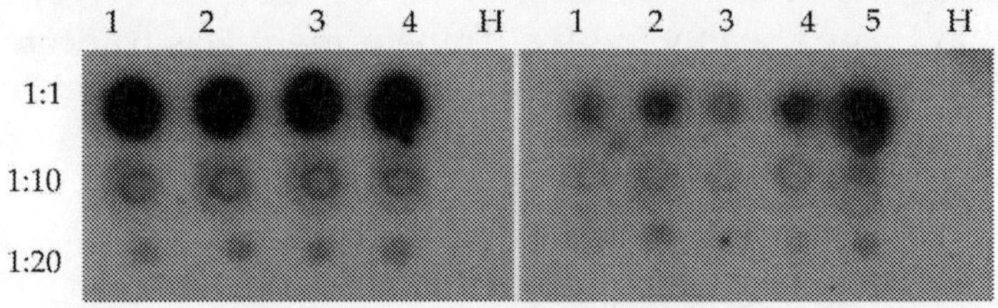

Fig. 26: Detection of ICRSV isolates by NASH using probes marked from 350bp (left) and 450bp (right) amplicons of ICRSV-Ab.

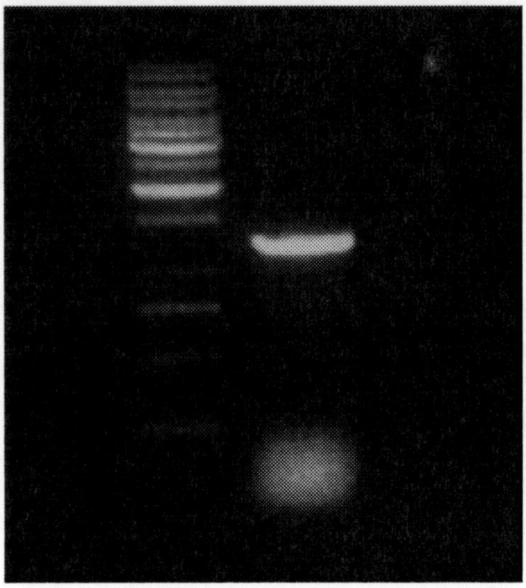

Fig. 27: PCR detection of ICRSV with virus specific primers

Incidence of ICRSV was very high in Kinnow mandarin and sweet oranges in India. Masking of symptoms of the disease during hot climate is a common phenomenon and hence identification of ICRSV-free trees is difficult at the time of propagation. This is one of the reasons that the disease has not been identified earlier. It has been spreading for a long time inadvertently approaching high incidence of the disease and is believed to be one of the factors in citrus dieback in India (Ahlawat, 1997a). Use of virus-free planting material has been recommended to contain this disease (Ahlawat, 1997b; Ahlawat and Pant, 2003; Byadgi *et al.*, 1994). Virus-free propagation material in Kinnow mandarin and Mosambi sweet orange has been developed through shoot-tip grafting and regeneration of plants through somatic embryogenesis and by embryo culture (Hoa *et al.*, 2004). Another strategy developed for management of ICRSV was the use of CP gene construction. Gene constructed in pB1121 binary vector was mobilized to *Agrobacterium* strain LBA 4404 and co-cultivated in French bean var. saxa. Transformant was truncated version of CP gene of 900bp. The presence of gene in transformed plants was confirmed by PCR, southern hybridization and nucleic acid spot hybridization (Hoa and Ahlawat, 2004). This technology will be helpful to develop virus resistant plants in near future.

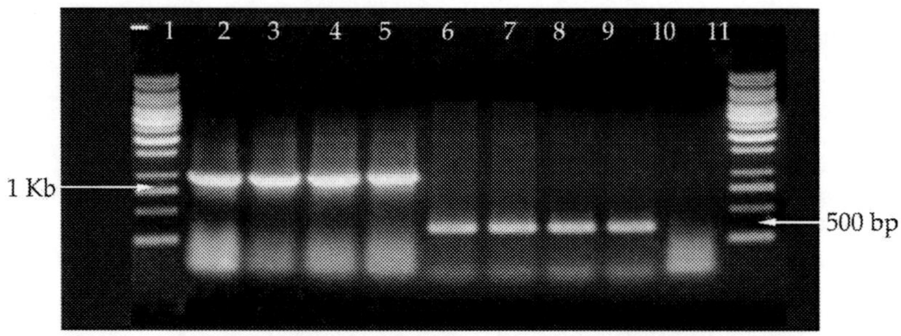

Fig. 28: RT-PCR amplification of four isolates using two sets of primers

Lane 1 and 11: 1Kb DNA ladder, Lane 2 and 6: ICRSV-Ab
Lane 3 and 7: ICRSV-DlLane 4 and 8: ICRSV-Ah
Lane 5 and 9: ICRSV-Pu Lane 10 : Healthy control

CITRUS RING SPOT

A ring spot disease in citrus was observed in California as early as 1949 (Reuther *et al.*, 1978; wallace and Drake, 1968). Symptoms of psorosis-like leaf pattern on young leaves followed by rings on mature leaves were observed in some samples when grafted on sweet or sour orange, Eureka and Lisbon lemons and Brazilian Sweet Lime. Citrus ring spot disease is now known to occur in Spain, Algeria, France, Greece, Italy (including Sicily), Turkey, Iran, Algeria, South Africa, Argentina, and Uruguay. This is an economically important disease in many areas of Argentina and other countries but distinct from Indian citrus ringspot.

SYMPTOMS

First symptom of the disease on tissue graft inoculated plants develops as vein flecking and chlorotic spotting similar to psorosis disease but later symptoms develop as bright yellow spots or distinct circular or irregular rings which may be gum impregnated. Necrotic lesions may also develop on young stem which sometimes encircle the stem resulting death of the upper part of the tree. Although symptoms of ring spot disease were similar to psorosis-A virus infection but no bark lesions were developed in inoculated plants for several years after inoculation (Reuther *et al.*, 1978). Fruits may also show ringspot symptoms. Most *Citrus* spp. and their hybrids, especially grapefruits (*C. paradisi*) and oranges (*C. sinensis*) are susceptible but some ringspot isolates do not infect tangelos (*Citrus paradisi* x *C. reticulata*) or *Poncirus trifoliata* (Timmer *et al.*, 1978).

TRANSMISSION

Citrus ring spot disease is easily transmitted by tissue grafting. It is also transmitted from citrus to citrus, petunia and periwinkle and back through dodder, *Cuscuta subinclusa* (Desjardins *et al.*, 1969). No insect vecor is known but evidence of natural spread is there.

HOST RANGE

Citrus ringspot virus infects both citrus and herbaceous hosts. Citrus hosts susceptible to ringspot are: Citrange Troyer, Citremon (1448-1449), *Citrus excelsa*, Clementine, Grapefruit, Lemons, Lime (Kagzi, Key, Mexican, West Indian), Mandarin (Parson's Special, Szinkom), *Poncirus trifoliata* (Pomeroy), Rangpur lime, Rough lemon, *Severinia buxifolia*, Sour orange, Sweet oranges and Dweet tangor. Grapefruit is highly susceptible host to ringspot virus. Herbaceous hosts of the virus are: cowpea, bean, *Chenopodium quinoa, Gomphrena globosa, Crotalaria spectabilis, Petunia hybrida* and *Zinnia elegans*. An isolate of citrus ringspot (CRSV-6) from Florida showed typical symptoms of psorosis including bark lesions and was mechanically transmitted to herbaceous hosts and back to citrus (Garnsey and Timmer, 1988). Upon mechanical transmission, Etrog citron and Mexican lime developed typical symptoms of ringspot, necrotic etch, leaf mottle and shock reaction. *Gomphrena globosa* developed chlorotic blotches and spots. Necrotic local lesions developed after 7-10 days of inoculation on mature leaves of *Chenopodium quinoa*. Sweet orange seedlings and *Chenopodium quinoa* are used as indicator hosts of the virus.

THE PATHOGEN

The virus particles associated with ring spot disease were identified as novel spiral filamentous type of two different length and the virus name was proposed as 'Spirovirus" (Derrick *et al.*, 1988). Virus particles are in an open circular form and a linear form, plus intermediates. The linear form may appear as a sinuous structure about 10 nm diameter and 1500-2500 nm long; it is a two stranded structure with loops at the ends, formed from a collapsed circle. Open circular form is a highly kinked filament 3-4 nm in diameter and 300-500 nm long. Virion morphology suggests that ring spot virus represents a new genus of viruses similar to the *Tenuiviruses* in family *Bunyaviridae* (Milne *et al.*, 1996; Garcia *et al.*, 1994; Navas-Castillo *et al.*, 1993). They proposed the name of the virus as "Ophiovirus". Ringspot virus has a 48kDa protein. Genome of *Tenuiviruses* is an ambisense divided ssRNA, each RNA is circularized by means of a panhandle structure of complementary base sequences at 5′ and 3′ termini.

DIAGNOSIS AND MANAGEMENT

Ringspot virus is easily detected by sap inoculation to *Chenopodium quinoa* which develops local lesions upon inoculation. The virus can be detected serologically in immunosorbent electron microscopy (ISEM), DAS-ELISA and RT-PCR (Derrick *et al.*, 1988; da Graca *et al.*, 1991; Garcia *et al.*, 1997). Normal methods for controlling graft-transmissible viruses of citrus (sanitation and certification) have partly failed in endemic areas (Frison and Taher, 1991). Budwood certification programme established against Citrus ringspot virus in some countries also failed to control the disease.

CITRUS YELLOW MOSAIC

In 1992, a Pummelo tree was observed with bright yellow mosaic and vein banding symptoms in an orchard near Indian Institute of

Horticultural Research, Bangalore, India. Later such symptoms were observed in several Sathgudi and Chini orange trees in Karnataka and Andhra Pradesh. Budwood collected from these sources were grafted on Mosambi sweet orange seedlings which developed typical yellow mosaic symptoms. Bacilliform virions were observed in electron microscope from affected samples. Initially the disease was reported as *citrus mosaic disease* (Ahlawat *et al.*, 1996) but it was suggested in 13th Conference of International Organization of citrus Virologists (IOCV) held in China in 1995 that the name of *'citrus mosaic disease'* (Ahlawat *et al.*, 1996) should be renamed as *'citrus yellow mosaic'* to distinguish it from other mosaic diseases reported in India and elsewhere. Yellow mosaic disease is widely distributed in India. The incidence of yellow mosaic disease (10-70%) was observed in Sathgudi and Chini sweet orange trees in citrus orchards and nurseries in Andhra Pradesh (Ahlawat *et al.*, 1999). Citrus yellow mosaic disease is restricted in its distribution only to India so far.

SYMPTOMS

Symptoms of yellow mosaic disease in sweet orange and Pummelo trees consist of pale yellow to bright yellow mosaic pattern irregularly distributed on leaf surface (Plate 38). Yellow flecking along the leaf veins is also seen quite often. Yellow mosaic virus is systemic in nature and almost all the leaves of an infected tree will show symptoms. Because of bright yellow mosaic symptoms, affected trees sometimes appear like ornamental trees. Photosynthetic activities are reduced as most of the leaf surface becomes chlorotic resulting in reduced vigor and yield. Fruits on affected trees are comparatively smaller, have mosaic symptoms (Plate 38) and often with bitter in taste.

TRANSMISSION

Yellow mosaic disease is transmitted by wedge grafting and T-budding from citrus to citrus. It is also transmitted by a mealybug vector, *Planococcus citri* (Reddy, 1997; Ahlawat and Pant, 2003). Yellow mosaic virus is not transmitted by aphids, *Myzus persicae, Aphis gossypii, A. spiraecola, A. citricola, A. craccivora,* mealybugs, *Brevennia rehi* and *Sacchricoccus sacchari.* This disease is also transmitted by mechanical inoculation (stem slashing) to Pummelo, *Citrus decumana,* and Sathgudi sweetorange. Inoculated plants develop typical yellow mosaic symptoms of the disease after 90 days post-inoculation. (Ahlawat *et al.,* 1996).

HOST RANGE

Citrus yellow mosaic virus (CYMV) has been recorded mainly on sweet oranges (*Citrus sinensis*), but was experimentally transmitted on *Aegle marmelos*, *C. limonia*, *Fortunella* species, grapefruits (*C. paradisi*), lemons (*C.limon*), mandarins (*C. reticulata*), pummelos (*C. grandis*), *C. volkameriana*, *C. jambhiri*, *C. aurantium* and *C. mitis*. Mider symptoms develop in *C. limettioides*, *C. aurantiifolia* and *C. medica* (Ahlawat *et al.*, 1996). Symptoms of the disease equally develop during winter and summer suggesting that temperature variation does not affect development of symptoms. The reaction of yellow mosaic on some susceptible cultivars is given below:

- *Sweet orange (Mosambi, Satgudi, Chini, Malta)*: Typical mosaic with scattered green and chlorotoc areas (Plate 39).
- *Rangpur lime*: Chlorotic spots, vein banding, mild mosaic, malformation.
- *Volkamer lemon*: Yellow spot, yellow blotches, yellow mosaic (Plate 39).
- *Pummelo*: Yellow mosaic, smalling of leaves, golden yellow spot (Plate 39).
- *Grapefruit cv. Dunken*: Vein flecking, yellow mosaic.
- *Sour orange*: Chlorotic and yellow spots, yellow mosaic.
- *Citrus decumana*: Yellow mosaic, smalling of leaves, chlorotic spots, stunting.
- *Citrus mitis*: Yellow rings and smalling of leaves (Plate 39).
- *Rough lemon*: Yellow flecking, mosaic, yellow spot.
- *Mandarin cv. Nagpur*: Vein banding.
- *Sweet lime*: Mild mosaic.
- *Citron*: Vein yellowing.

THE PATHOGEN

A previously unreported badnavirus is associated with citrus yellow mosaic disease (Ahlawat *et al.*, 1996). Citrus yellow mosaic virus (CYMV) has non-enveloped bacilliform particles, 30 × 150 nm (Fig. 29), located in the cytoplasm, typical of badnaviruses. In ultrathin sections, large numbers of bacilliform particles were observed in the cytoplasm of tissue from CYMV infected plants (Fig. 30). Yellow mosaic bacilliform virus reacted specifically in ISEM test to antisera of seven badnaviruses, sugarcane mosaic badnavirus, commelina yellow mottle virus, banana streak virus, cacao swollen shoot virus, diascorea bacilliform virus, canna yellow mottle virus and kalanchoe top spotting virus but not to rice tungro bacilliform particles (Ahlawat *et al.*, 1996; Pant and Ahlawat, 1997). Based on its bacilliform morphology and serological reactivity, yellow mosaic virus was identified as a member of genus *Badnavirus* in family *Caulimoviridae*. Its badnavirus nature was further confirmed by amplifying its genomic nucleic acid in PCR using degenerated primers located on consenses sequences of badnaviruses (Fig. 33) (Ahlawat *et al.*, 1996).

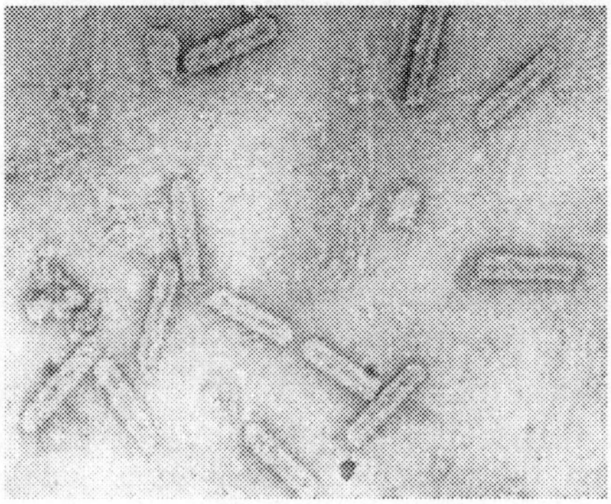

Fig. 29: Badnavirus particles associated with citrus yellow mosaic disease (CYMD)

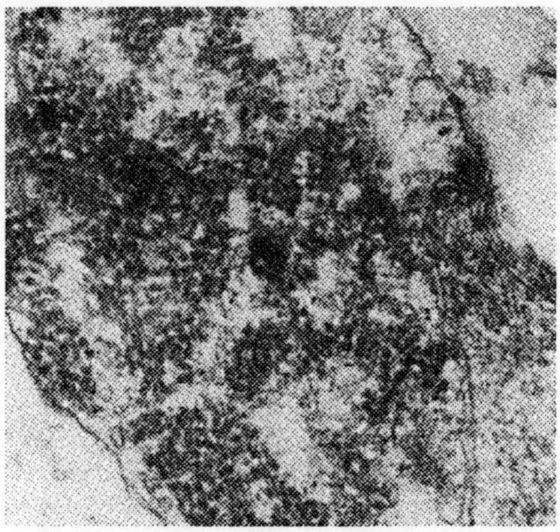

Fig. 30: Badnavirus particles in cytoplasm of CYMD infected cell

Genome of yellow mosaic virus is double stranded DNA. A full length genome from Rangpur lime has been sequenced which consists of 7522 nucleotides with 6 ORFs. Intergenic region of 724 nucleotides between ORF6 and ORF1 consists of putative promoter elements (Singh *et al.*, 2009). ORF 3, the largest ORF, encodes a putative polyprotein for functions involved in virus movement, assembly and replication. The other ORFs encode proteins whose exact functions are not completely understood. SDS-PAGE and western blot analysis of purified virions showed one major band of 32 kDa and two minor bands of 22 and 19 kDa. 32 kDa is the major capsid protein and other two proteins are degraded products of 35 kDa. Nucleic acid from purified virions showed two bands of 21kb and 6.6kb in non-denaturing agarose gel. The 6.6 kb is the linear form of the viral DNA and 21kb is the circular form of DNA (Ahlawat *et al.*, 1997). Phylogenetic analysis of the badnaviruses revealed that citrus yellow mosaic badnavirus (CYMBV) is most closely related to Cacao swollen shoot virus. It was demonstrated that a construct containing 1.4 copies of the cloned CYMBV genome could infect sweet orange *via* Agrobacterium-mediated inoculation (Huang and Hartung, 2001). Nucleotide sequences obtained during variability studies have been submitted to the NCBI GenBank nucleotide sequence database and have been assigned Accession numbers, EU489744 and EU489745 (Borah *et al.*, 2009). In India, partial polyprotein gene of ORF III, partial ORF 6 and intergenic region, intergenic region of Tirupati isolates on Rangpur lime, acid lime and Sathgudi sweet orange have been sequenced (Accession No. AY950594, AY945935, AY944237, AY953373, AF347695).

DIAGNOSIS AND MANAGEMENT

Virus specific antibodies against CYMBV have been prepared and used for virus trapping (Fig. 31) and decoration tests ((Fig. 32). CYMBV is also detected in DAS-ELISA (Ahlawat *et al.*, 1996). Viral DNA was cloned at Hind III site in P^{UC18} vector and transformed into competent *E. coli* strain NM 522 cells. Recombinant colonies were screened and finally two virus specific clones of 2.0 and 4.0 kb were identified. These clones were radiolabeled and used as probe to detect the virus in field trees and mealybug vector. 2.0 kb clone is more efficient probe and detects virus in DNA-DNA hybridization in even 25ng of the total DNA or 95pg of viral DNA or 25 ng of tissue per spot (Plate 40) (Reddy, 1997; Ahlawat *et al.*, 1997). CYMBV is also detected in PCR using degenerate oligonucleotide primers (Fig. 33). A PCR product of approximately 1.5 kb and 1.0 kb in size was obtained (Ahlawat *et al.*, 1996; Baranwal *et al.*, 2003). By using two different sets of primer pairs specific for partial polyprotein gene of CYMBV genome, amplified product of 726bp and 245bp were obtained (Ghosh *et al.*, 2007). CYMBV was also detected in vector, *Planococcus citri* by PCR (Singh *et al.*, 2004). Citrus is a perennial crop and is subjected to infection by more than one virus in the field trees, hence mixed infections are not uncommon (Fig. 34). Simultaneous detection of CYMBV with other viruses or virus-like pathogens from mixed infection could be possible in multiplex PCR system (Fig. 35) (Baranwal *et al.*, 2005; Ghosh *et al.*, 2008). *Citrus decumana* and grapefruit are good indicator hosts of the virus which develop bright yellow mosaic symptoms and can be used for disease diagnosis by graft inoculation. Virus diagnosis is also possible by electron microscopy either by thin sectioning or by leaf dip preparations. Mealybug vector of CYMBV is not an efficient vector and hardly plays any role in natural spread. Therefore, this virus can be eliminated by using healthy planting material which may be obtained by shoot tip grafting and proper indexing.

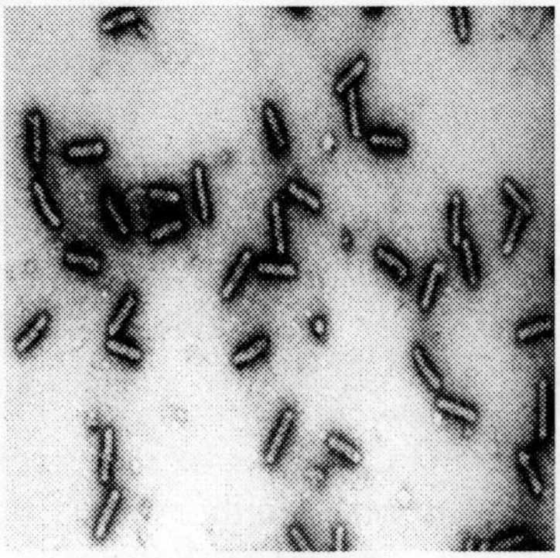

Fig. 31: Trapping of virus particles in ISEM

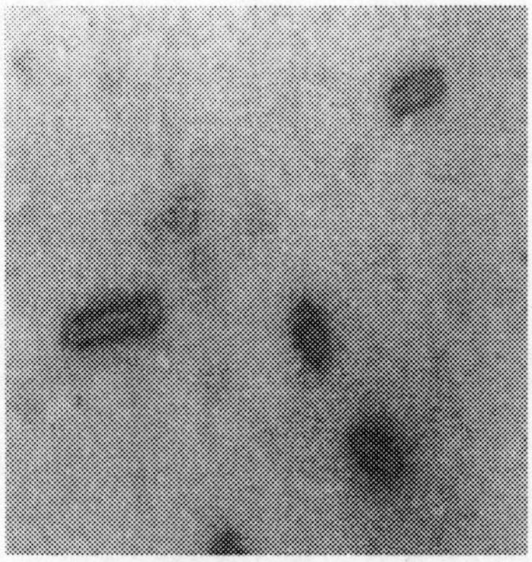

Fig. 32: Decoration of virus particles with specific antiserum

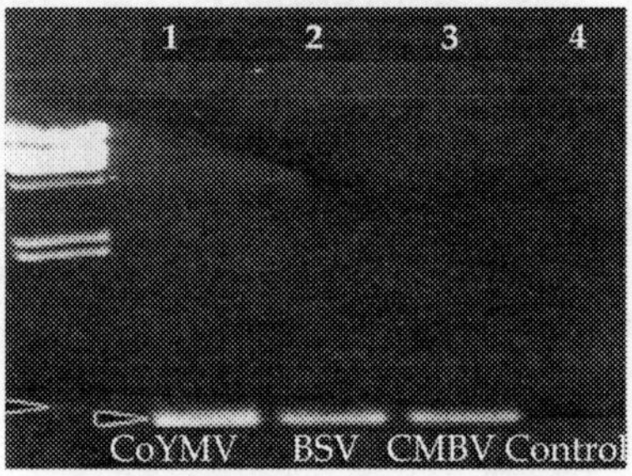

Fig. 33: Detection of CYMBV in PCR along with COYMV and BSV badnaviruses

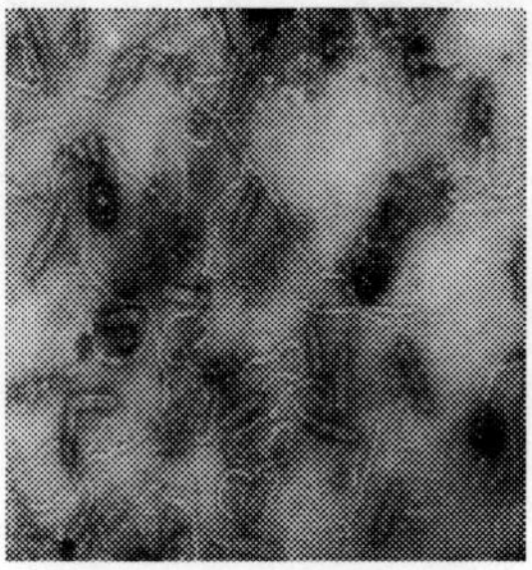

Fig. 34: Mixed infection of CTV, ICRSV and CYMBV isolated from naturally infected tree.

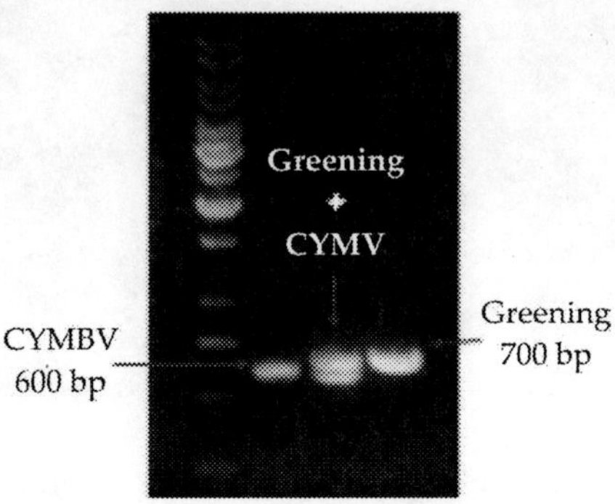

Fig. 35: Detection of CYMBV and *Candidatus* Liberibacter asiaticus in duplex PCR

PUMMELO YELLOW VEIN MOSAIC

Pummelo yellow vein mosaic disease was initially observed on a Shaddock tree at Bangalore, India (Ahlawat *et al.*, 1993). Young leaves of naturally infected Pummelo trees showed yellow flecks or dots which later developed in to mottling and yellow flecking along the veins. Some of the flecks fused together forming either rings or large chlorotic/necrotic raised spots (Plate 41). This disease is transmitted by wedge and leaf patch grafting but not by sap inoculation.

Natural and graft transmitted samples of Pummelo were screened under electron microscope in leaf dip preparations. Bacilliform particles with an outer membrane measuring 166 × 36 nm, typical of rhabdoviruses were observed both in natural and glasshouse inoculated plants (Fig. 36). So far in citrus, rhabdovirus is known to be associated only with leprosis disease of grapefruit and sweet oranges and found only in Brazil (Kitajima *et al.*, 1972). However, the symptoms of yellow vein mosaic are entirely different from leprosies and hence was considered as an unreported disease. The name of this disease is tentatively proposed as 'Pummelo yellow vein mosaic' and the associated virus as 'Pummelo yellow vein mosaic rhabdovirus which is a tentative member of *Rhabdoviridae* (Ahlawat *et al.*, 1993). Further studies are required to establish the relationship of the two rhabdoviruses infecting citrus in India and Brazil.

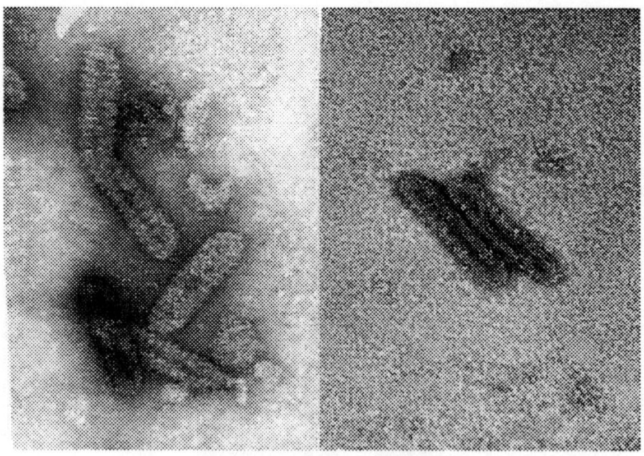

Fig. 36: Rhabdovirus particles associated with Pumello yellow mosaic virus

CITRUS YELLOW VEIN CLEARING

Citrus yellow vein clearing disease appears to be widely distributed in Indian subcontinent including Pakistan. This disease has been reported from India (Ahlawat *et al.*, 2001; Alshami *et al.*, 2003), Pakistan (Bove, 1989; Catara *et al.*, 1993; Ashfaq *et al.*, 2005), and recently from Turkey (Onelge, 2003; Onelge *et al.*, 2010). In India, symptoms of yellow vein clearing were observed by the senior author in 1996 on Etrog citron (*Citrus medica*), Sour orange, Lisbon lemon, *Citrus pectinifera* and Citranges in the germplasm collection at Punjab Agricultural University (PAU) campus at Abohar and at Ludhiana (Plate 42) and on Lisbon lemon and Rangpur lime at Anand in Gujarat suggesting wider distribution of yellow vein clearing disease in India.

SYMPTOMS

Symptoms of yellow vein clearing disease appear as clearing of leaf veins which are clearly visible against transmitted light (Plate 42). Water soaking along veins on underside of the leaf is the most characteristic symptom of the disease (Plate 43). Clearing of lateral veins starts as flecks of varying length not involving the entire vein but subsequently most veins are involved. The yellow flecks may extend laterally into the tissue adjacent to the veins. Midveins show little clearing even on leaves with severely affected lateral veins. Symptoms remain constant even after the leaf becomes mature and vein clearing symptoms on affected trees can be seen year around. Variable symptoms may develop on different *Citrus* species but clearing of veins is the major symptom. In Pakistan, mean disease incidence ranged from 68.75 to 98.75 percent (Yasir Iftikhar *et al.*, 2010).

TRANSMISSION

Yellow vein clearing disease is easily transmitted by wedge and bud grafting from Etrog citron to various citrus species causing symptoms as described above. Indian isolates are mechanically transmitted to *Phaseolus vulgaris* cvs. Gheusami, Saxa and Singtamy causing veinal necrosis, necrotic spots, blotching (Plate 44 and 45) followed by blight symptoms and on *Chenopodium quinoa* causing necrotic local lesions. Indian isolates of yellow vein clearing virus were not transmitted by dodder, *Cuscuta reflexa*, aphids, *Aphis gossypii*, *A. citricola*, *Myzus persicae*, whitefly, *Bemisia tabaci* and mealybug, *Planococcus citri*. It is also not seed transmitted. However, Turkey isolate of CYVCV has been reported to be transmitted by *A. craccivora* from infected lemon to bean plants upto 69% and with *A. spiraecola* upto 42% (Onelge *et al.*, 2010).

HOSTRANGE

Following citrus species/varieties are susceptible by experimental transmission in India causing typical symptoms of the disease: Etrog citron, Sour orange, Rough lemon, Rangpur lime, Mosambi and Malta sweet orange, Grapefruit, Kinnow mandarin, Nagpur orange, Galgal (*C.limon*), Pummelo, Decumana, Calamondin, Karna Khatta and *C. pectinifera* (Alshami *et al.*, 2003). Yellow vein clearing has also been reported on several Citrus species/varieties or selections in Pakistan and Turkey (Catara *et al.*, 1993; Onelge *et al.*, 2010).

THE PATHOGEN

Initial results of immunosorbent electron microscopy (ISEM) tests with antiserum of Indian citrus ringspot virus (ICRSV) showed decorated and undecorated flexuous particles in samples of yellow vein clearing disease (Fig. 37). These results suggested that a filamentous virus with similar morphology to ICRSV was present in field affected samples. However, through experiments yellow vein clearing virus (CYVCV) was separated from mixed infection and brought to pure culture (Ahlawat *et al.*, 2001). Purified preparations of the virus showed filamentous virions measuring 685 nm in length and 14 nm in width in electron microscope (Fig. 38) (Alshami *et al.*, 2003). OD values of yellow vein clearing virus preparations at 260 and 280 nm were 0.6442 and 0.5735 respectively. A260/A280 ratio of purified virus prepation was 1 : 123. Filamentous virus particles with a modal length of 600-700 nm and diameter of 13-14 nm have also been observed in leaf dip preparation electron microscopy with Pakistani isolates (Grimaldi and Catara, 1996). They also reported aggregation of filamentous particles in phloem cells of thin sections from lateral symptomatic veins. Genome of the virus is ssRNA with a coat protein of 32 kDa (Alshami *et al.*, 2003). Studies conducted on this virus so far indicate that this is an unreported virus of citrus species and needs more studies before its final classification by ICTV.

DIAGNOSIS AND MANAGEMENT

Polyclonal antibodies of Indian isolates of yellow vein clearing virus have been prepared and used for serodiagnosis and serological relationship. This virus is not serologically related to nine filamentous viruses including Indian citrus ringspot mandarivirus in ISEM tests (Byadgi and Ahlawat, 1995). Yellow vein clearing virus specific antibodies decorated all particles in purified sample from glasshouse raised cultures (Fig. 39). Antibodies of yellow vein clearing virus (CYVCV) have been effectively used for virus detection from citrus and non citrus hosts both by DAS

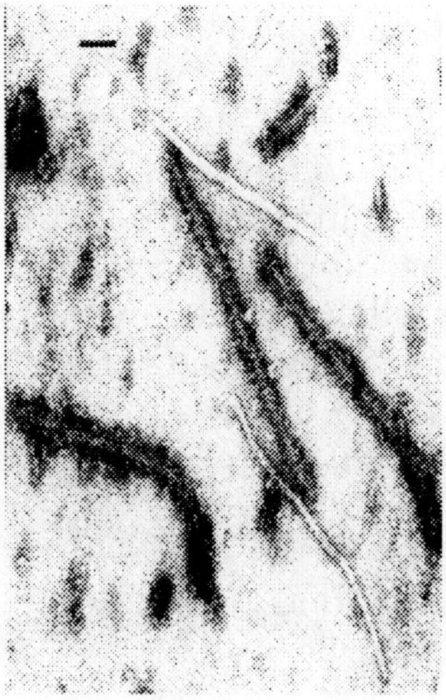

Fig. 37: Electron micrograph of yellow vein clearing virus (CYVCV) in ISEM with antiserum of ICRSV. Only ICRSV particles were decorated but not yellow vein clearing suggesting that two virues are different.

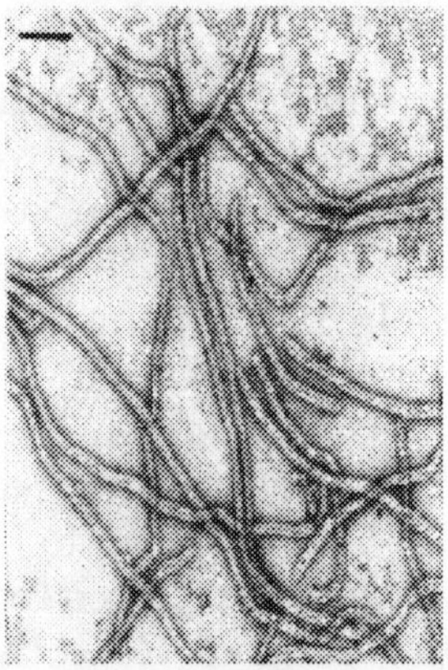

Fig. 38: Purified virus particles of yellow vein clearing virus

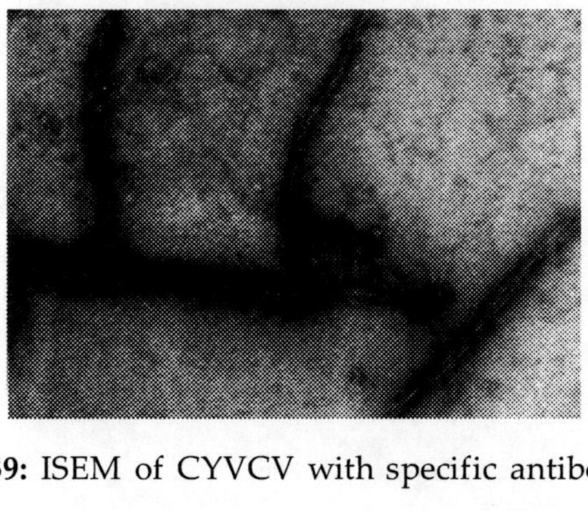

Fig. 39: ISEM of CYVCV with specific antibodies.

and DAC-ELISA. Homologous antibodies successfully detected the virus in decoration tests of ISEM. CYVCV could not be amplified with ICRSV specific primers in PCR. Since these two viruses have similar morphology and are always found in mixed infection, there could be a risk of CYVCV movement through planting material if both the viruses are not indexed. CYVCV can also be detected by sap inoculation to *C. quinoa*. Yellow vein clearing disease can be easily identified by the characteristic symptom of water soaking of veins on the lower side of leaves. This typical symptom is not found with ICRSV. Use of virus-free planting material is the only recommended method to contain yellow vein clearing disease.

DWEET MOTTLE = LEAF BLOTCH VIRUS

Dweet mottle symptoms were first observed in California in 1960s when graft inoculations were made from a symptomless but sick looking Cleopatra mandarin tree from Florida to dweet tangor. The pathogen associated with dweet mottle remains symptomless in several hosts but symptoms appear only on Dweet tangor. Therefore, the name of the disease was given "Dweet mottle". In 1984, Spanish Scientists discovered a new disease showing bud union incompatibilities in certain scion/rootstock combinations and this disease was given the name Citrus Leaf Blotch disease (Navarro *et al.*, 1984). Based upon reaction in indicator plants, citrus leaf blotch virus (CLBV) was found to be closely related to Dweet mottle virus (DMV). CLBV has been reported from different citrus varieties in Japan, Australia, Florida and Spain. Recently, CLBV has been found in Nagami kumquat and Calamondin (*Citrus mitis* or *madureusis*) in Italy (Guardo *et al.*, 2007 a and b), and in several citrus varieties in New Zealand (Harper *et al.*, 2008).

SYMPTOMS

Dweet tangor disease is found in trees showing bud union crease on trifoliate rootstocks and stem pitting in Etrog citron. However, major symptom of the disease is chlorotic blotching on leaves of infected dweet tangor. Several trees grafted on citrange rootstock from Valencia (Spain) showed decline and bud union crease (Galipienso *et al.*, 2001, 2004; Vives *et al.*, 2002).

TRANSMISSION

Citrus leaf blotch virus (CLBV) is graft transmitted from citrus to citrus and showed a bud union disorder of Nagami kumquat and Calamondin scions grafted on trifoliate rootstocks. CLBV can also be transmitted by contamination of knife blade (Roistacher *et al.*, 1980). In Spain, CLBV is seed-transmitted causing concern in the nursery industry (Krueger *et al.*, 2005; Robert *et al.*, 2005). It is experimentally transmitted from citrus to citrus by stem-slash inoculation and from citrus to *Nicotiana occidentalis* and *N. benthamiana* by mechanical inoculation with partially purified virus preparations.

HOST RANGE

Host range of the virus is within *Citrus* species but some herbaceous hosts are also susceptible to CLBV. Susceptible hosts to CLBV are: Cleopatra mandarin, Dweet tangor, Nagami Kumquat, Etrog citron, Pineapple sweet orange, Nules Clementine and Troyer citrange. Dweet tangor and Etrog citron are more sensitive and other hosts are tolerant. The virus remains symptomless in inoculated plants of *Nicotiana occidentalis, N. cavicola* and *N. benthamiana* (Guardo *et al.*, 2009). Dweet tangor seedlings are used as indicator hosts in which chlorotic blotches are developed upon inoculation.

THE PATHOGEN

Citrus leaf blotch virions are filamentous measuring 960 × 14 nm. Citrus leaf blotch virus (CLBV) is type species of the putative new genus *Citrivirus* in family *Flexiviridae* (Mayo and Haenni, 2006). Genome of CLBV is a single-stranded positive sense RNA (ssRNA) of about 9 kb (Mr 3×10^6) containing a 42 kDa protein. CLBV genomic RNA has been completely sequenced (Galipienso *et al.*, 2001, 2004; Vives *et al.*, 2001). The CLBV genomic RNA contains three open reading frames (ORFs) encoding a polyprotein involved in replication, a potential movement protein and the capsid protein. Dweet mottle virus (DMV) genome has also been sequenced which has 8,747 nucleotides (nt) excluding the 32 poly-A tail. DMV genomic RNA contains three putative open reading frames (ORFs) and untranslated regions of 73 nt at the 52 and 541 nt at 32 termini. Nucleotide sequences of DMV were compared with members of the families *Alphaflexiviridae* and *Betaflexiviridae* (Hajeri *et al.*, 2010). Protein database searches showed 98–99% matches of DMV ORFs with CLBV sequences. Phylogenetic analysis based on the RdRP core domain revealed that DMV is closely related to CLBV as a member of the genus *Citrivirus*. DMV did not satisfy the molecular criteria for demarcation of an independent species within the genus *Citrivirus*, family *Betaflexiviridae*, and hence, DMV can be considered a CLBV isolate (Hajeri *et al.*, 2010).

DIAGNOSIS AND MANAGEMENT

Dweet tangor, Etrog citron and pineapple sweet orange are appropriate diagnostic indicators when propagated in Troyer citrange. Inoculated plants should be decapitated and preferably incubated in a temperature controlled glasshouse (26°C/18°C day/ night). The symptoms of leaf blotching usually appear in young shoots (Galipienso *et al.*, 2000). Citrus leaf blotch virus is also detected by ELISA, dot-blot hybridization with digoxigenin-labeled cDNA probes, tissue print hybridization and by RT-PCR (Galipienso *et al.*, 2004). Dweet mottle and CLBV are difficult to eliminate by shoot-tip grafting or thermotherapy (Vives *et al.*, 2005). CLBV is seed transmitted hence virus-free buds and rootstocks should be used.

CITRUS BLIGHT

Blight disease of citrus kills about five lakh trees in Florida and 10 million in Brazil every year. Blight is now known to occur in several citrus growing states of United States of America, Brazil, Caribbean Basin, Central America, Colombia, Argentina, Venezuela, Uruguay, South Africa, Costa Rica and Australia (Timmer *et al.*, 1984, 2000; Childs, 1979; Lima, 1982; Villalobos *et al.*, 2005). It has been commonly observed that trees near the infected tree may get affected more quickly than trees at a distance (Yokomi *et al.*, 1984). Etiology of blight disease has not been determined although several attempts have been made and are being made to determine the etiology of blight disease (R.H.Brlansky-persnal communication, 2010).

SYMPTOMS

The symptoms of the disease are normally observed on bearing trees (8-10 yr old). Blight causes a general decline of tree canopy with wilt, leaf loss, twig dieback, and poor growth flushes. Symptoms may be confined to one part of a tree and can be confused with those of other diseases like huanglongbing, tristeza or by burrowing nematodes (*Radophilus citrophilus*). Foliage on affected trees become dull green as observed by the senior author in Florida and Brazil. Affected trees produce numerous water sprouts on the trunk and the lower branches followed by leaf drop and subsequent dieback (Plate 46a). Wilting of tree may be of a few branches or the entire canopy of a tree show wilting. Delayed flush, thin foliage and delayed flowering are also symptoms of blight. Fewer but normal size fruits develop on affected trees but they have high acidity and soluble solids. Stress symptoms of citrus trees are seen in successive stages of decline due to blight (Albrigo *et al.*, 1986).

TRANSMISSION

Blight disease is transmitted by natural root grafting from affected trees to healthy trees suggesting the involvement of a graft transmitted pathogen (Tucker *et al.*, 1984; Albrigo *et al.*, 1992; Timmer *et al.*, 1991; Beretta *et al.*, 1991; Marais and Lee, 1991). It is believed that a soil borne pathogen or a pathogen of very limited aerial movement is involved with blight disease (Yokomi *et al.*, 1984).

HOST RANGE

Citrus blight is common in all citrus cultivars on all rootstocks and seedling trees. Sweetorange and grapefruit varieties on Rough lemon, Rangpur lime, Trifoliate orange and Carrizo citrange rootstocks are more susceptible than on other rootstocks. Sweet oranges are the most affected but mandarins are generally not affected. Trees on *Citrus jambhiri*, *C. nobilis* and *Poncirus trifoliata* rootstocks are highly susceptible. Trees on *Citrus macrophylla*, *C. volkameriana* and *Citroncirus* x *webberi* rootstocks are comparatively less susceptible but trees on *Citrus aurantium* and *C. reshni* rootstocks are tolerant, and those on *C. sinensis* are highly resistant (Young *et al.*, 1980). Although it is difficult to establish an indicator host for such diseases but Valencia sweet orange on rough lemon may be used as indicator host. Blight syndrome may develop in inoculated plants after several years of inoculation.

THE PATHOGEN

Infectious nature of the disease has been established (Timmer *et al.*, 1992). Rod shaped virus particles were observed in extracts from blighted trees (Brlansky and Howd, 2001). Recently, an I*daeovirus*, similar to raspberry bushy dwarf virus has been shown to be associated with citrus blight (Derrick *et al.*, 2006). Several proteins have been observed in vacuum extracts of root and stem pieces of blighted trees in Florida and Brazil (Beretta *et al.*, 1993) which are absent or present in much lower concentration in healthy trees. These extracts showed filamentous structures when viewed in electron microscope (Derrick *et al.*, 1991). However, the exact nature of causal agent for blight remains undetermined.

DIAGNOSIS AND MANAGEMENT

Bilght disease is detected by visual symptoms and water uptake by affected trees through syringe injection and percentage of amorphous plugs obstracting xylem vessels. Best diagnosis of blight disease in field is syringe test for water uptake (Plate 46b) (Lee *et al.*, 1984). For syringe test, a small hole is drilled into trunk of tree and a syringe is used to inject water into trunk. Blighted trees will not take up water, whereas healthy trees usually absorb water readily (Cohen and Wutscher, 1979). Zinc concentration and water soluble phenolics are high in blighted trees and may be used for diagnosis (Wutscher *et al.*, 1977). Serological tests for the 12 kDa and other proteins that are unique to blighted trees are commonly used to diagnose blighted trees (Derrick *et al.*, 1992). Disease diagnosis in young trees can be done by immunospot and Western blot techniques and other immunological techniques (Bausher and Sweeney, 1991). For management of blight disease, blight-affected trees should be replaced with trees on tolerant rootstocks. Heavy pruning of infected trees helps in good growth of trees and normal bearing for a few years (Rossetti *et al.*, 1988).

CONCAVE GUM – BLIND POCKET

Based on psorosis young-leaf symptoms- a common characteristic of psorosis-A disease, concave gum-blind pocket were thought to be members of the diseases of "psorosis family" (Wallace, 1959). However, concave gum was unrelated to psorosis as evidenced by cross-protection experiments (Roistacher and Calavan, 1965). Concave gum disease is largely present in the Mediterranean region although it is now found worldwide.

SYMPTOMS

Field affected trees showed mild to severe stunting. Trunk and large limbs are distorted. Concavities normally develop on the scion trunk close to bud union. Trunk symptoms are seen on the sensitive scion.

Concave Gum

Symptoms of concave gum disease are formation of concavities of 10-20 cm or more in length and half as wide and numerous in stem and branches in addition to young leaf symptoms of vein flecking and oak-leaf pattern (Plate 47). Concave gum disease has following characteristics:

1. It has broad concavities or depressions of various sizes on the trunk and limbs.
2. There is presence of gum beneath the concavities, in between wood and bark.
3. There is gum exudation through bark-cracks in the centre or around the margins of the depression at certain seasons.

Blind Pocket

In blind pocket, concavities similar to that of concave gum are also developed on stem and branches but in this case concavities are more in number, deeper and narrower (Plate 48). Major identification of concave gum and blind pocket is based on symptoms on field trees and on the development of oak-leaf patterns on graft inoculated sweet orange, Dweet tangor or mandarin indicators which show hard gum like substance below the blind pockets. On the basis of trunk and leaf symptoms they are thought to be closely related and will be considered here as one single disease.

TRANSMISSION

Agent of concave gum-blind pocket is transmitted by graft inoculation of bark patches, buds or shoots. There is no report on mechanical or vector transmission.

HOST RANGE

Host range is restricted to *Citrus* species. Typical symptoms of the disease develop on mandarin, tangelo, grapefruit and a pummelo-like hyuganastsu (*Citrus tamurana*) in Japan (Leki and Ito, 1996). Infected Sour orange, lemon, and Citron show only oak leaf pattern but not trunk or branch depressions. Symptomatic hosts of concave gum-blind pocket are: Clementines, Grapefruits, Mandarins, Siamelo, Sweet oranges, Tangelos (Orlando, Sampon, Sunshine, Waft) and Dweet tangor. Mandarins and sweet oranges are highly susceptible. Sweet orange, Dweet tangor and mandarin seedlings are used as indicator hosts where oak-leaf pattern symptoms develop on leaves within a few weeks after inoculation.

THE PATHOGEN

Causal agent of this disease has not been established as yet. Based on its graft transmission, this disease is considered to be caused by a virus-like pathogen.

DIAGNOSIS AND MANAGEMENT

Diagnosis of the disease can be done by observing oak leaf pattern symptom by Indexing on mandarin and Dweet tangor. In field trees, disease can be identified by oak leaf pattern of young leaves during spring and typical depressions on trunk and branches. For management of disease budwood free of concave gum-blind pocket (CG-BP) should be used. The agent of CG-BP is readily eliminated by shoot tip grafting (STG). Thermotherapy can also be used to eliminate the agent of CG-BP (Calavan *et al.*, 1972; Roistacher, 1977).

CITRUS CHLOROTIC DWARF

A new virus-like disorder named citrus chlorotic dwarf has been reported from Eastern Mediterranian region of Turkey with an incidence of 20.3% (Korkmaz *et al.*, 1994). Chlorotic dwarf disease spreaded fast and reached to an epidemic level in Mersin and Adana regions. Now chlorotic dwarf disorder is being considered as one of the serious diseases of citrus in that region (Cinar *et al.*, 1993, 1994).

SYMPTOMS

Field symptoms consist of a V-shaped notch and chlorotic flecking of young leaves, warping, crinkling, inverted cupping and variegation of mature leaves (Kersting *et al.*, 1996). Leaves on affected trees were smaller and have a notch near the apex (Korkmaz *et al.*, 1995).

TRANSMISSION

The causal agent is graft-transmitted. The vector of the causal agent is a whitefly, *Parabemisia myricae* (Kuwana) (Korkmaz *et al.*, 1994). It is transmitted from rough lemon to rough lemon by whitefly vector in a semi-persistent to persistent manner (Cinar *et al.*, 1994; Önelge and Çýnar, 2010). Chlorotic dwarf is also transmitted mechanically by stem slash inoculation from rough lemon to rough lemon.

HOST RANGE

Chlorotic dwarf agent can affect most of the citrus species and cultivars. Rough lemon is highly susceptible and symptoms on rough lemon seedlings appeared 5-8 weeks after graft inoculation. Better symptoms develop at a temperature of 20-25°C but can also be developed at 30-35^{0}C (Korkmaz *et al.*, 1994). Following varieties of citrus are commonly found affected in field: Aydin, Interdonata, Lamas and Kutdiken lemons, Minneola Tangelo, Fremont and Satsuma mandarins, Marsh seedless, Star Ruby and Redblush grapefruit, Valencia, Washington Navel and other sweet oranges, Rough lemon, Sour orange and Alemow. Rough lemon, Sour orange and Alemow are the best diagnostic and indicator hosts.

THE PATHOGEN

Preliminary purification attempts revealed that the pathogen is probably phloem restricted and has isometric particles. The ultrastructure of CCD infected bark sections showed unknown filamentous virus-like particles (Howd *et al.*, 2002). However, the precise nature of these virus-like particles remains unclear. Recently, an apparently new geminivirus-like agent were found associated with the disease (Loconsole *et al.*, 2012). The newly identified virus has a circular single-stranded DNA genome comprising five open reading frames (ORFs) with sequence homologies with those encoded by geminiviruses. The larger genome size (3.64 vs. 2.5-3.0 kb) and a number of differences in its structural organization, identified this virus as a highly divergent member of the family *Geminiviridae*, to which the provisional name of Citrus chlorotic dwarf-associated virus (CCDaV) is assigned (Loconsole *et al.*, 2012). CCD did not react in ELISA to antisera against citrus variegation, citrus leaf rugose, and satsuma dwarf viruses (Kersting *et al.*, 1996).

DIAGNOSIS AND MANAGEMENT

CCD is detected by graft inoculation to diagnostic hosts such as Rough lemon, Alemow and Sour orange. PCR and qPCR assays were successfully used for determining its presence in the CCDD-affected plants obtained by graft propagation (Loconsole *et al.*, 2012). Management of CCD is possible by using healthy planting material and to control or minimize the whitefly population. Biological control of whitefly vector population is possible by releasing the aphelinid wasp (*Eretmocerus debachi* Rose and Rosen) (Kersting *et al.*, 1996).

IMPIETRATURA DISEASE

Impietratura disease of citrus is prevalent in all countries of the Mediterranean basin and also in Iran, Venezuela, South Africa and Cyprus. The name 'impietratura' has been used to designate a disease that cause skin hardening and albedo gumming of some citrus fruits (Ruggieri, 1955). In India, this disease has been reported from Kalimpong (Ahlawat *et al.*, 1984) on a local shaddock-type tree identified as *Citrus decumana*. Typical symptoms of the disease were also observed on *Soh-Shyrkhoit* and *Citrus latipes* in Northeastern region of India-the home of citrus (Bhagabati *et al.*, 1986).

SYMPTOMS

Affected trees bear abnormally small fruits which never develop to normal size and remain small, hard like a stone and also become misshapen or lop sided (Plate 49). Such fruits fell prematurely during June-July. Fruits bearing branches had depressed brownish spots. There are gum deposits under such spots on the wood (Ahlawat *et al.*, 1984). Fruits from affected trees show symptoms of skin hardening and albedo gumming even when they are ¼th of size. Distinct discoloured spots are usually seen as raised bumps or protuberances on the rind. If such fruits are sectioned, gum spots may appear throughout the rind and albedo and a brownish gum impregnated tissue can be seen below hard area of fruit (Plate 49 and Fig. 40). Hardening, malformation and albedo gumming of fruits are typical symptoms of impietratura disease in field trees. This disease can be confirmed by cutting hard fruits and see the layer of gum-impregated tissues below hard areas (Caruso *et al.*, 1993). Affected fruits rarely grow to full size and fell down prematurely. Gum impregated tissues have also been observed in young green twigs of infected shaddock trees in Kalimpong region of India (Ahlawat *et al.*, 1984). Leaf flecking symptoms were also observed on Volkamer lemon, Rough lemon, Grapefruit, Pummelo, Sanguinello sweet orange and *Citrus decumana*.

TRANSMISSION

Impietratura disease is transmitted by grafting and no other method of transmission is known. Symptoms of the disease appear when grafted plants flower and fruit. It is apparently not seed-transmissible (Bar-Joseph, 1976). Impietratura disease has been transmitted by insertion of pollen from infected trees under the bark of healthy Orlando tangelo test trees (Vogel and Bové, 1980).

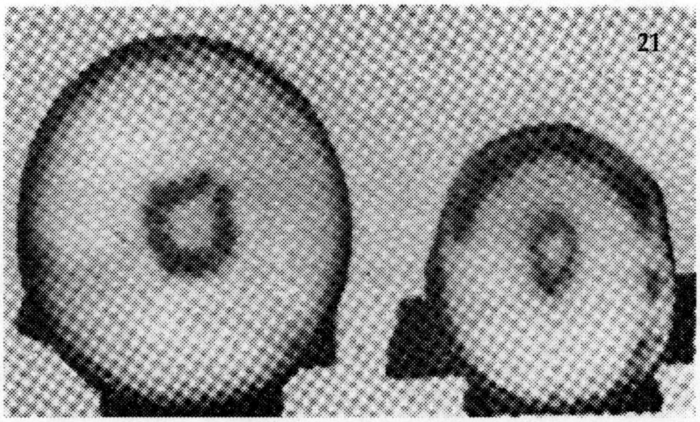

Fig. 40: Sections of healthy and impietratura infected fruits. Albedo gumming can be seen in infected fruit (right).

HOST RANGE

Fruits of several Citrus species are susceptible to impietratura disease. In India, symptoms of impietratura were observed on fruits of lemon, mandarin, grapefruit, sweet orange and tangelo (Mali and Chaudhuri, 1993). Volkamer lemon and rough lemon are good indicator hosts of impietratura disease. Rough lemon is a better indicator for fruit symptoms than volkamer lemon, grapefruit and sweet orange (Caruso *et al.*, 1993). Sweet orange seedlings develop vein flecking and oak-leaf patterns on young leaves after a few weeks of inoculation at low temperature was considered to be diagnostic for impietratura disease (Bar-Joseph and Loebenstein, 1970).

THE PATHOGEN

To date, etiology of the disease has not been proved. Symptoms of Impietratura disease are similar to boron deficiency but analysis of leaves from affected trees showed adequate boron contents (Ruggieri, 1955; Papasolomontos, 1965). Evidence of transmissibility further suggests that the disease is caused by a virus.

DIAGNOSIS AND MANAGEMENT

Diagnosis of this disease is possible by inoculating indicator plants and observing the symptoms on field trees as described above. Appearance of oak leaf pattern symptoms in young emerging leaves during first growth flush during cooler spring months in association with fruit symptoms can also be used for diagnosis of the disease in the field. Impietratura disease can be eliminated by certification programmes involving thermotherapy followed by shoot-tip grafting. Legislative measures and institutions are involved in the production of healthy propagating material in Greece and some other Mediterranean countries and certified material is indexed in glasshouse and also in the field (Caruso *et al.*, 1993; Bar-Joseph and Ben-Shalom, 1993).

CITRUS YELLOW VEIN DISEASE

Citrus yellow vein disease was initially observed on limequat (*Citrus aurantifolia* X *Fortunella japonica*) in California (Weathers, 1957). This disease showed synergistic reaction with some other citrus viruses like psorosis and vein enation virus (Weather, 1960). Symptoms on West Indian lime, Rough lemon, Etrog citron and *Citrus macrophylla* appear as bright yellow veins of leaves and yellow blotches on stem. Yellowing may extend to petiole and young twigs upto a few centimetres. Occasionally vein banding effect also develops. Discoloration is sometimes restricted to half of the leaf but usually whole leaf is affected. Yellow vein disease is transmitted only by grafting and no other means of disease spread is known. Following hosts are susceptible to yellow vein disease: Citron (Etrog), *Citrus macrophylla, C. volkameriana*, Kumquats, Lemons, Limequats, Rough lemon, Tangelo (Orlando) and West Indian lime. Trifoliate orange seems to be immune (Weather, 1960).

There is no information on etiology of the disease. Plant infected by yellow vein virus and citrus vein enation virus will show highly synergistic effect in Lime, Sweet orange, lemon and rough lemon as shown individually by these viruses. After shotgun cloning and sequencing, several clones showed high degree of homology with *umbraviruses* and *luteoviruses* in the case of yellow vein whereas the same *luteovirus* species was identified in vein enation inoculated plants. It has been proposed that the *luteovirus* is a major component of both diseases and the "enation" evolves to "yellow vein" as a result of the synergism between the umbra and luteo viruses (Kanrar *et al.*, 2010). Yellow vein disease is detected by graft transmission to susceptible hosts. Eradication of infected trees and to avoid propagation from infected trees is the only recommended methods to contain this disease.

LEAF CURL DISEASE

Leaf curl disease of citrus has been reported from Brazil and India (Salibe, 1959, 1965; Mali *et al.*, 1976). Curling and distortion of leaves similar to the one caused by heavy infestation of aphids are the major foliar symptoms. Affected trees though bloom heavily yet set only a few fruits which remain smaller in size. Shoots become weak and can be easily broken under slight pressure. Leaf curl disease is transmitted by graft inoculations. No other mode of disease spread is known. Hosts susceptible to leaf curl disease are: Citrons, Grapefruits, Lemons, Mandarins, Pummelos, Sweet lime, Sweet oranges and Lime (Kagzi, Key, Mexican, West Indian). The virus has been experimentally transmitted to the following hosts: Baianinha sweet orange, Hamlin and Pera sweet oranges, Maxirica Tangerine, Marsh seedless grapefruit, Doce citron (*Citrus medica*), Comprida citron, and Siciliana lemon (Salibe, 1965). There is no information on the etiology of leaf curl disease of citrus. In Brazil, Caipira seedlings are used as indicator host. Bud inoculated seedlings of Caipira sweet orange show yellowing of leaves which are dropped prematurely. Leaves on newly developed shoots are severely curled and seedlings may die within one year of inoculation. No control measures are known but as a precaution, infected trees should be destroyed and propagation must be avoided from symptomatic trees.

CRISTACORTIS

Cristacortis was first described on Tarocco sweet orange trees grafted on sour orange in Corsica ((Vogel and Bove, 1964, 1968). Affected trees showed severe stem pitting in Sour orange and Tarocco orange. This disease is wide spread in Corsica and affects trees of Clementine on Sour orange rootsock (Vogel and Bove, 1972). Cristacortis disease affects many species/varieties of citrus in Corsica, Sardinia, Sicily, Spain, Morocco, Algeria, Egypt, Turkey, Mauritius, North Africa, Mediterranean Basin,

South America, Yemen and Vietnam (Scaramuzzi and Catara, 1968; Timmer *et al.*, 2000; Reuther *et al.*, 1978).

Trunk, limbs and shoots of naturally infected trees show vertical depressions due to pits in the wood with corresponding pegs on cambial side of bark. Most conspicuous symptoms are numerous pits on trunk and main shoots. With certain species like tangelos, gum-like substance stains at top of peg and bottom of pit. Cristacortis infected trees also show weaker branches which may bend towards ground. Verna sweet orange tree on sour orange graft inoculated with the cristacortis agent showed stem pitting on both scion and stock. Flecking and oak-leaf pattern symptoms on young leaves also develop on affected trees with cristacortis (Plate 47) (Vogel and Bove, 1974). Infected trees develop comparatively larger fruits than normal but fewer in number. Cristacortis disease is transmitted by graft and bud inoculation only to *Citrus* species. No natural spread of cristacortis pathogen has been observed (Vogel and Bove, 1976, 1980). Susceptible hosts to cristacortis are: Orlando tangelo, Dweet tangor, Sour orange, Sweet orange, Mandarin, Clementine, Grapefruit, Rough lemon, Satsuma, Siamelo, Sweet lime and *Citrus pectinifera*. Sweet orange seedlings are used as indicator host which develop Vein flecking and oak-leaf patterns on young leaves within a few weeks after inoculation at cool temperature. Tangelo seedlings can also be used as test plants.

Nature of pathogen associated with cristacortis has not been established as yet. Infected plants show leaf flecking and oak leaf pattern similar to psorosis group of viruses and therefore this disease was placed within the so called psorosis group of diseases (Vogel and Bove, 1974). Cristacortis disease can be diagnosed by inoculating indicator hosts for disease diagnosis. The presence of stem pitting on the sour orange stock in addition to pits on the scion is valuable for diagnosis. Use of cristacortis free budwood is only method of disease management so far.

CITRUS LEATHERY LEAF

Leathery leaf disease of citrus was reported in Sikkim orange (*Citrus reticulata*) trees in Darjeeling and Sikkim hills in 1972 (Ahlawat, 1975).

Leaves on affected trees become leathery and thick in texture. Citrus leathery leaf virus (CLLV) is graft transmissible from Sikkim orange to several citrus species/cultivars. CLLV is readily transmitted mechanically from citrus to citrus and herbaceous hosts. It is also transmitted by aphid, *Aphis gossypii* in non-persistent manner (Ahlawat *et al.*, 1979). Susceptible hosts of leathery leaf are: mandarin cultivars Sikkim Orange, Khasi Orange and Coorg Orange; Sweet orange cvs. Mosambi, Sathgudi and Washington navel; Eureka, Lisbon and Hill lemons; Sour orange, Rangpur lime, Grapefruit, Pummelo, Kagzi lime, Rough lemon and *C. decumana* but not trifoliate orange. In sweet orange plants, psorosis like leaf flecking symptom appears which persisted for about a week and later disappeared. In other hosts, thickening of the uppermost few leaves are the major symptoms (Fig. 41A). Leaves of inoculated plants show downward curling and death of apical bud (Ahlawat *et al.*, 1979).

In mechanically inoculated plants systemic mottle of leaves develops in *Gomphrena globosa* (Fig. 41b), *Phaseolus mungo, P. vulgaris* cultivars Red Kidney and Singtam local, *P. aureus* cultivars NP-23, PM-1 and ML-8. In *Sesamum indicum*, symptoms appear as chlorotic spots on inoculated leaves followed by systemic mosaic and twisting of leaves. In *Cucumis sativus*, mild mottling of leaves and in periwinkle downward curling of the leaves were the major symptoms. *Petunia hybrida, Nicotiana glutinosa* and *Physalis floridana* are symptomless hosts of the virus. This virus is not mechanically transmitted to 17 herbaceous hosts. Leathery leaf virus has thermal death point between 65° and 70°C, dilution end-point between 10^{-6} and 10^{-7} and longevity at room temperature (22-25°C) 96 h and at 5°C 19 days.

(a) (b)

Fig. 41: (a) Symptom of leathery leaf on Darjeeling mandarin, (b) Symptoms of leathery leaf on *Gomphrena globosa* showing mosaic symptoms

Nature of pathogen associated with citrus leathery leaf has not been established as yet. Based on psorosis-like young leaf vein flecking symptoms, it is believed that the virus belongs to psorosis group of viruses. leathery leaf disease can be easily diagnosed by sap inoculation to herbaceous hosts like *Gomphrena globosa* and *Sesamum indicum* which produce clear diagnostic symptoms. Use of virus-free planting material and control of aphid vector are the measures effective for disease management.

MULTIPLE SPROUTING

A disease with multiple bud sprouts in sweet orange trees was described from South Africa (Schwartz, 1970). A similar disease was also observed in kagzi lime trees (*Citrus aurantifolia)* from India (Ahlawat and Chenulu, 1985b). A bud transmissible disease with almost similar symptoms has also been reported from India with the name of "Blastomania" (Mali *et al.*, 1975). However, Blastomania disease has been shown to be associated with phytoplasma-like bodies (Sharma and Singh, 1988). On affected trees, many shoots develop from a single bud giving a look of witches' broom symptoms (Fig. 42). Branches with multiple sprouts may show negatively geotropic habit (Searle, 1969). The growth of field infected trees is bushy with multiple branching having deformed leaves. Key lime trees in India develop extremely reduced size of leaves on multiple sprouts. Multiple branches remained green in colour for quite a long time or until they are removed.

Multiple sprouting disease agent is graft transmitted to several citrus hosts. It is also transmitted by mechanical inoculations. Multiple sprout disease is also sap transmitted in India (Ahlawat and Chenulu, 1985b). The Indian isolate was transmitted by mechanical inoculation from citrus to *Chenopodium quinoa*. Multiple sprouting has been shown to be sap transmitted from citrus to *Chenopodium quinoa, C. amaranticolor, Nicotiana glutinosa* and *Vigna sinensis* showing virus-like symptoms (Majorana and Schwarz, 1972). *Celosia argentea* and *Gomphrena globosa,* Bergamots and Grapefruits are symptomless hosts. Citranges (Carrizo, Rusk, Troyer), Citrumelos and Trifoliate orange are tolerant hosts of multiple sprouting. Sweet orange plants showed milder symptoms than sour orange, lemon and grapefruit. Causal agent of multiple sprouting in India is also easily sap transmitted from affected leaves of multiple sprouts to *C. quinoa* causing local lesions. However, the multiple sprouting agent could not be transmitted back to citrus or any other host from local lesions of *C. quinoa*.

Fig. 42: Twigs of key lime showing multiple sprouting-Photo by author from IARI, Regional Station Pune, India

Repeated attempts to identify virus-like particles from citrus and sap inoculated *C. quinoa* were unsuccessful (Ahlawat and Chenulu, 1985b). The cause of this disease is yet to be ascertained from Indian citrus and also from South Africa. With the growth habit of affected plants, they can be easily identified in the field. Multiple sprout agent can also be detected by sap transmission to *C. quinoa* where it produces local lesions on inoculated leaves. Removal of affected branches and propagation from MS-free trees are the best ways to manage this disease effectively.

CITRUS DIEBACK DISEASE IN INDIA

Dieback is a serious disorder of citrus species and cultivars in India and is responsible for decline of citrus industry in the country. This disease complex can be identified from distances as it causes defoliation of young branches and dieback of twigs from tips. Affected trees normally do not die but vigour, general health and production is adversely affected (Raychaudhuri *et al.*, 1969, 1977). Dieback or decline of citrus trees is also common in Pakistan (Akhtar and Ahmad, 1999).

In dieback syndrome, leaves of affected trees fell down from twigs and branches resulting symptoms of defoliation and dieback of twigs (Plate 50). This is followed by secondary growth of axillary buds which consist of short, upright, small weak shoots and smaller leaves, sometime showing bushy appearance. Leaves on these shoots may show typical symptoms of zinc deficiency such as green blotches or small green spots on yellow tissue or greening of veins with chlorosis of interveinal tissue. These branches may develop unusual and excessive flowering. Dieback of twigs moves downward and affected twigs may be attacked by weak wood rotting fungi. Initially, symptom starts on one or two branches but subsequently the whole tree is involved. Affected trees bear excessive flowering which produces sun blotched fruits that seldom mature. Feeder roots become dark and show rotting.

Citrus dieback syndrome has been studied by several scientists of different disciplines in India. Hence, different causes of dieback have been suggested such as soil types and conditions, high water table, nutritional deficiencies and infection by fungi, viruses, phytoplasmas, viroids and nematode infestation (Raychaudhuri *et al.*, 1969, 1977). Citrus dieback symptoms have also been attributed by citrus tristeza virus (Capoor, 1963; Vasudeva and Capoor, 1958). Greening disease now known as Huanglongbing was also suggested to be the cause of dieback in India (Fraser and Daljit Singh, 1966, 1968; Fraser *et al.*, 1966; Nariani *et al.*, 1967). Further investigations revealed that the disease is caused by a complex of the Huanglongbing organism and certain fungi (Raychaudhuri *et al.*, 1967). With the development of diagnostic technology and increasing knowledge of new pathogens, cause of dieback syndrome is now known due to infection of certain viruses (Tristeza, Indian citrus ringspot and citrus yellow vein badnavirus), Huanglongbing bacterium and exocortis and hop stunt viroid in two or more combinations in addition of huanglongbing (Ahlawat, 1997a). In Pakistan also, dieback disease has been attributed to both biological (Huaglongbing, CTV, Exocortis viroid, Phytophthora gummosis, Bud-union crease) and abiological (Over-irrigation, poor drainage, excess salt and/or heavy ploughing for clean cultivation) factors. Huanglongbing is probably one of the major causes of citrus decline in Pakistan. Fungi similar to isolated from dieback-affected citrus trees in India have also been isolated from Pakistan (Raychaudhari *et al.*, 1977; Ahlawat and Raychaudhuri, 1988, Girmaldi and Catara 1989). The role of these fungi has probably been overemphasized, since huanglongbing in itself is enough to induce decline and dieback. Dieback disease has also been described in Australian citrus which is probably caused by Huanglongbing bacterium (Broadbent *et al.*, 1976).

Development of new diagnostic methodology and technology for freeing viruses and virus-like pathogens through techniques like shoot tip grafting (STG) became extremely helpful to minimize dieback incidence. STG and recent diagnostics like ELISA and PCR are of great advantage to raise new plantings free from pathogens and minimize the century old dieback in India. Diagnostic methods and management practices for individual pathogen involved in dieback syndrome have been described with their description in this book.

BUD-UNION CREASE

Bud-union abnormalities became apparent as soon as the practice of growing seedling trees was shifted to grow grafted trees. Bud union crease was first described in Palestine on Shamaouti sweet orange trees grafted on rough lemon rootstock (Yedidyah, 1937). Tatter leaf which induces bud union crease in different citrus species propagated on trifoliate rootstocks is a good example of bud union crease. Bud union crease has been reported from several countries on different rootstock-scion combination (McClean, 1974; Navarro *et al.*, 1993). Usually bud union disorder appears to be a non-destructive disease. In Bloodred and Mosambi sweet orange old trees, symptoms of bud union crease have been described in India (Bhutani *et al.*, 1972). Symptoms of bud union crease appear at a point of bud union as continuous depression in the wood with corresponding projections from the inner face of the bark. Sometimes this depressed line is continuous (Plate 51), but in most cases indentations with gum impregnation were observed. In some cases bark appear normal but if the bark is removed at the bud union, clear symptoms of the disease are visible. In India, bud-union crease has been frequently observed on Mosambi sweet orange and Nagpur mandarin trees grafted on rough lemon (Plate 51). There are no symptoms on leaves and fruits. Transmission is normally through infected budwood or bark patches.

The causal agent is graft-transmissible, but its nature is unknown. Bud-union crease disorder has been considered to be due to a virus or a genetic factor (Reuther *et al.*, 1978). Bove (1995) has provided evidence that the bud-union crease of sweet orange trees on rough lemon rootstock is caused by a graft-transmissible pathogen. Trees indexed positively for gummy bark viroid were also positive for bud-union constriction (Nour-Eldin, 1968). Nucellar sweet orange clones on rough lemon roots grew healthy upto eight years of age but develop symptoms within four years after inoculation (Reuther *et al.*, 1978). Selection of disease-free scion budwood and avoidable susceptible stionic combinations are the advisable approaches to contain the disease. The disease is controlled by using new-line sweet orange on rough lemon where old lines are found to make abnormal unions The disease can also be eliminated by shoot tip grafting *in vitro* (Navarro *et al.*, 1993).

CITRUS MEASLES

Measles in citrus is a disorder that occasionally affects a few trees in a grove. The disease is rarely found in Florida and Brazil for a long time but it has also been recorded in California (near Farmersville) for the first time. Symptoms of measles appear as numerous pale yellow spots on leaves of affected trees. These spots are seen as yellow from upper surface and tan to brown and raised on the lower surface of leaves. Once a tree is affected there is no masking of symptoms and they reappear every year throughout the tree canopy. Varieties of sweet orange such as Hamlin, Pineapple, Valencia, and Navel orange trees and several other citrus species develop symptoms of measles. Cause of measles disorder is not known and has been considered as a genetic disorder (Whiteside *et al.*, 1988). However, this disorder has been shown to be graft transmitted to sweet orange using blind buds from symptomatic field trees (Lee *et al.*, 1993). Faint translucent spots occurred on young leaves which then developed into typical measles spots as the leaves matured. Thus graft transmissible nature of citrus measles was confirmed. Symptoms of measles disorder develop at day temperature of 38°C and night temperature of 18°C. There is no information on the causal agent. Graft transmission of measles disorder to sweet orange indicator plants suggests that the disease is caused by an infectious agent similar to viruses (Lee *et al.*, 1993). Budwood from contaminated trees should be avoided to manage the disease.

KASSALA DISEASE

Kassala disease was first observed in 1987 from a place known as Kassala in Eastern Sudan (Bove, 1995). Symptoms of Kassala disease have also been observed in Southern Yemen. Gum impregnation in bark and mild stem pitting on affected grapefruit trees are the main symptom of the disease. Mild stem pitting of kassala disorder resembles to those of cachexia in mandarin and gummy bark in sweet orange. Symptoms of cassala disease can be easily observed after removing the outer bark of grapefruit scion. Whitish bark tissues below the bark show impregated streak of reddish-brown gum like material near the bud union. Pits in xylem of grapefruit trees are also seen after removal of outer bark. Bark gumming can stand upto 2 m above the bud union. The nature of the agent is unknown. Leaves and fruits on affected trees do not show any symptom. Graft-transmissibility of cassala disease agent has not been studied. Natural transmission through budwood is possible but not demonstrated. Host range of this agent is restricted to *Citrus* species only. This disease is diagnosed by Scraping the bark of grapefruit trunk above the bud union which reveals gum deposits. In field trees, the disease is diagnosed by scraping the bark of the grapefruit trunk above the bud union which will show gum deposits. Bark removal across the bud-union shows stem pitting on the grapefruit scion. Cassala disease free budwood should be used for disease management.

RIO GRANDE GUMMOSIS (RGG)

A disease with the name of ferment gum disease (Rio Grande gummosis) was first reported on a grapefruit tree in California (Calavan, 1961). This disease was further studied and reviewed by Childs (1978). It occurs in the Caribbean Basin, Somalia and Morocco. Grapefruit and lemon are the most commonly affected species. Narrow cracks (lesions) exuding yellow gum are seen in bark of affected trees. Severely affected trees produce more gum and their branches and trunk become black upto 100 cm of lesions. The cambial tissues below the bark become pink or orange in colour which may be seen after removal of bark. Gum is water soluble and hence washed away during rains. As the symptoms advance, lesions become darker and wood tissue become impregnated with gum followed the death of bark and decay of wood. Transmission and cause of the disease are not known. In Florida, the disease has been associated with an uncharacterized, graft-transmissible agent, unrelated to psorosis (Powell *et al.*, 1998). The disease is aggravated by application of chlorides and other fertilizer salts and occurs more commonly on calcareous soils or those with calcareous subsurface layers. Symptoms are the only method of disease diagnosis. No control measures are known for the disease except to avoid making large pruning wounds near the trunk (Timmer and Powell, 2006).

CITRUS SUDDEN DEATH DISEASE

Citrus Sudden Death (CSD) is a new disease that was first reported in 1999 at the Triangulo mineiro and Sao Paulo in Brazil and was considered to be a new threat to Brazilian citrus. Citrus Sudden Death (CSD) was a new epidemic disease of alarming proportions that already has caused the death of more than 1 million sweet orange trees grafted on Rangpur lime in Brazil (Maccheroni *et al.*, 2005). Graft-transmission tests reproduced the symptoms of the disease (Barros *et al.*, 2005). The death of the plants was due to disorganization and death of phloem vessels at the graft union (Kitajima, 2010). Recently, sudden death virus has been transmitted to *Chenopodium quinoa, C. amaranticolor, Nicotiana clevelandii* and *Nicotiana* sp. (Santos *et al.*, 2010). However, a newly isolated and identified tymo-like virus (TlV) was detected from mainly CSD affected orange trees (Barros *et al.*, 2005; Maccheroni *et al.*, 2005). They named this new virus as Citrus sudden death-associated virus (CSDaV). EM preparations showed the presence of two types of isometric particles, typical of *Tymoviridae* (Maccheroni *et al.*, 2005). A tymo-like virus isolated from CSD-trees was closely related to the tymovirus group that possesses a poly-A tail, but not to a tymovirus with a t-RNA like structure in the 3′ end of genome (Barros *et al.*, 2005). Further studies are necessary to verify if the isolated tymo-like virus, or more virulent CTV variants or a synergistic interaction of both is the cause of CSD (de Jesus Junior *et al.*, 2005). The complete genomic sequence of this virus was obtained and phylogenetic analysis showed a close relationship with viruses of the *Marafivirus* genus. In all affected trees, *Marafivirus* was constantly detected in association with citrus tristeza virus. However, citrus sudden death virus was detected in all symptomatic leaves of inoculated *Chenopodium quinoa, C. amaranticolor, Nicotiana clevelandii* and *Nicotiana* sp., while CTV was never detected (Santos *et al.*, 2010) suggesting that the disease is caused by *Marafivirus* genus.

CITRUS DISEASES CAUSED BY PROKARYOTES

In citrus, prokaryotes cause two serious and economically important diseases, huanglongbing, commonly called as citrus greening disease and citrus variegated chlorosis. Huanglongbing or greening disease is caused by an insect-borne bacterium which is restricted to sieve tubes of phloem tissues. Although the name 'greening' was commonly used name but it was changed to 'huanglongbing' in 1995 and shall be used in this chapter. Citrus variegated chlorosis is caused by a xylem-limited bacterium, *Xylella fastidiosa* and is found in Brazil. These diseases will be discussed in detail in following pages.

HUANGLONGBING (GREENING) DISEASE

A disease with the name of yellow shoot was reported from China as early as 1919 (Reinking, 1919). Later, this disease was shown to be graft transmitted (Lin, 1956). A transmissible disease with similar symptoms was described from South Africa by the name of greening disease (McClean and Oberholzer, 1965). Transmission of the greening agent by the African citrus psyllid vector, *Trioza erytreae*, was also published in 1965 (McClean and Oberholzer, 1965). Transmission of Huanglongbing (HLB) agent by the Asian citrus psyllid vector, *Diaphorina citri*, was reported simultaneously in 1967 in India (Capoor *et al.*, 1967) and in the Philippines (Martinez and Wallace, 1967). HLB causes 30-100% loss and about 100 million trees are affected with HLB throughout the world each year. Since HLB is a vector borne disease, it spreads very fast in citrus groves. For example, in Brazil, HLB was first recorded in 2004 and since then about three million trees have been eradicated under HLB control programme.

Dieback in India had been a baffling problem since 1920 and was considered to be a serious disease (Asana, 1958). Dieback was attributed by several factors, including soil disorders, nutritional deficiencies, twig fungi, and viruses such as citrus tristeza virus (Capoor, 1963; Cheema and Bhat, 1929). However, after the visit to major citrus growing areas of the country Dr. L. Fraser concluded that the dieback affected citrus trees in India carry the symptoms of citrus greening (huanglongbing) disease similar to the one known to occur in South Africa and suggested that greening appears to be the cause of citrus dieback disease in India (Fraser and Daljit Singh, 1966; Fraser *et al.*, 1966). The experimental proof for the presence of huanglongbing disease came in 1967 when successful transmission of HLB agent was obtained by the psyllid vector (Capoor *et al.*, 1967). However, it was established that dieback in India is associated with HLB and some fungi (Raychaudhuri *et al.*, 1972, 1977). Further studies revealed that dieback symptoms are caused by viruses like citrus tristeza, Indian citrus ringspot Mandrivirus, Citrus yellow mosaic badnavirus, HLB and citrus yellow corky vein viroid (Exocortis+hop stunt) in two or more combinations but HLB is a major component in citrus dieback syndrome in India (Ahlawat, 1997a, 2007). In India, HLB is widely distributed in all citrus growing states (Nariani and Raychaudhuri, 1968, Nariani *et al.*, 1967,1970; Gupta *et al.*, 1972; Raychaudhuri *et al.*, 1972; Bhagabati *et al.*, 1989; Ahlawat and Sardar, 1976). In addition to Asia and Africa where HLB is present for several decades, huaglongbing has also been reported in the Western hemisphere from Brazil in 2004 (Coletta-Filho *et al.*, 2004; Teixeira *et al.*, 2005a), Florida in 2005 (Halbert, 2005) and in Cuba in 2006 (Cordero *et al.*, 2010).

SYMPTOMS

Citrus trees can be infected by HLB bacterium at any stage of the tree right from nursery plants to orchard trees. Generally, young trees are more susceptible than older ones. Affected trees are stunted, have spreaded growth and sparse yellow foliage and show twig dieback. HLB symptoms first appear on one or two branches but finally the whole tree is affected (Plate 52). Leaf symptoms of HLB disease remain confusing with zinc deficiency but through experiments, it has been established that the most characteristic symptom of greening disease is mottling of leaves (Plate 53a) (Varma *et al.*, 1993). Defoliation followed by dieback of twigs is also a common symptom of HLB disease in India. Mottled leaf symptoms are conspicuous on sweetorange leaves but less prominent in mandarin leaves. Mottled symptoms are characterized as discolouration of inter-veinal areas, alternating with green corresponding islands similar to mineral deficiency symptoms (zinc deficiency). Midribs and lateral veins of mature leaves of affected trees often show yellowing. Secondary growth from axillary buds having short, upright, small weak shoots with bushy appearance often develop on affected trees. New leaves develop on these shoots are small with upright growth. These leaves show different types of discolouration resembling zinc, manganese or iron deficiency. Initially, symptoms develop on one or two branches but ultimately the whole tree is involved. Seedling mandarin trees in Sikkim, Darjeeling hills, Nepal, Bhutan show yellowing of leaves on affected trees. African form of HLB develops leaf symptoms under cool conditions (20-25^{0}C) but Asian form of greening shows symptoms under both cool and warm temperatures (up to 35^{0}C). Symptoms of zinc deficiency are often confused with greening symptoms.

Affected trees show excessive and early flowering. Fruit set is very poor and off-season blooming is common on affected trees. Fruits on HLB affected trees are smaller, lop sided, drop easily and poorly coloured and remain dull green (Plate 53b). Such fruits contain brown aborted seeds (Plate 53c). There is no symptom on trunk or twigs.

TRANSMISSION

Huanglongbing (HLB) is transmitted by bud, shoot, bark or leaf patch grafting. HLB disease is also transmitted experimentally by dodder (*Cuscuta campestris* or other species) to periwinkle (*Catharanthus roseus)* and tobacco (Garnier and Bove, 1983), and tomato (Duan *et al.*, 2008). HLB is seed transmitted in periwinkle and dodder seeds but does not appear to be seed transmitted in citrus seeds. Natural transmission of HLBs is through psyllid vectors in persistent manner. South African HLB is transmitted by psyllid, *Trioza erytreae* Del Guercio (McLean and Oberholzer, 1965). Asian HLB is transmitted by another psyllid, *Diaphorina citri* Kuw (Capoor *et al.*, 1967, 1974; Martinez and Wallace, 1967; Salibe and Cortez, 1966; Celino *et al.*, 1966). Indian form of HLB is also transmitted by *Trioza erytreae,* the vector of South African greening and African form of HLB is transmitted by *Diaphorina citri,* vector of Asian greening. Nymphs of *D. citri* are more efficient vectors than adults. Different biotypes are known to be present in *D. citri* which differ in their vectoral capacity of HLB transmission (Ahlawat *et al.*, 1995). American form of the HLB is also transmitted by *D. citri*. *T. erytreae* transmits HLB pathogen in South Africa, Madagascar Island and Yemen. *D. citri* transmit HLB pathogen in Asia, Southeast Asia, North, central, South America, Western Soudi Arabia, and Iran. HLB pathogen multiplies in psyllid vector but transovarial transmission is not known. Preferred hosts of *D. citri* are: *Murraya paniculata, M. exotica, C. aurantifolia, Bergera koinigii, Clausena* spp. and for *T. erytreae,* lemon and most citrus species and *Vepris undulate, Clausena anisata, Zanthoxylum capense, Toddalia lanceolata* are preferred hosts.

HOST RANGE

HLB pathogen infects many citrus cultivars, species, hybrids and citrus relatives. Cultivars of following citrus spp. irrespective of the root-scion combination are susceptible: Citrons, Grapefruits, Kumquats, Lemons, Mandarins, Pummelos, Rough lemon, Rangpur lime, Sweet lime, Sweet oranges, Tangelos and Limes (Kagzi, Key, Mexican, West Indian lime). Mandarins, Sweet oranges, grapefruits and Tangelos are highly susceptible to HLB pathogens followed by lemon, Rough lemon and sour orange. Citrus relatives are also hosts of HLB. Among these *Murraya paniculata* (Lopes *et al.*, 2005, 2006, Zhou *et al.*, 2007), *Clausena lansium,* (Ding *et al.*, 2006), *Severinia buxifolia* (Hung *et al.*, 2000) and *Calodendron capense* (Garnier *et al.*, 2000) are important hosts of HLB pathogen. HLB pathogen infects periwinkle and tobacco experimentally through dodder (Garnier and bove, 1993). Tomato has also been infected with HLB pathogen through dodder (Duan *et al.*, 2008). All the three *Liberibacters* are transmitted to periwinkle through dodder, *Cuscuta compestris*. Dodder, *Cuscuta indecora* is also host of *Ca* Liberobacter spp. and these pathogens multiply well in *Cuscuta indecora* (Hartung *et al.*, 2010). *D. citri* transmits Asian form of HLB (*Candidatus* Liberibacter asiaticus) into *M. paniculata* and *M. exotica* but not in *Bergera* (*Murraya*) *koenigii*. Disease symptoms rarely develop in *Murraya* plants but pathogen can be determined by back-inoculations from *M. paniculata* to Madam Vinous sweet orange suggesting that *Murraya* is host of HLB (Damsteegt *et al.*, 2010). *Clausena lansium* has also been shown to be susceptible host to HLB bacterium (Deng *et al.*, 2007). Sweet orange and tangelo are used as indicator hosts of HLB bacterium. Typical symptoms of the disease develop in these hosts upon graft inoculation.

THE PATHOGEN

Huanglongbing disease (HLB) was considered to be a virus disease till 1970 but it is now known that a bacterium, '*Candidatus* Liberibacter species' is associated with it. At present, three liberibacter species, *Candidatus* Liberibacter asiaticus, *Candidatus* Liberibacter africanus and

Candidatus Liberibacter americanus infecting citrus are known. Based on the peptiglycan in the cell envelop of HLB organism, it was established that HLB disease is associated with a bacterium of gram negative type (Garnier and Bove, 1993 Garnier *et al.*, 1984, 1993). Characterization of the HLB pathogen was done through sequencing of its 16S ribosomal DNA and ribosomal protein gene. The term '*Candidatus*' has been proposed for organisms in "waiting position' for putative taxa of bacteria like organisms causing Huanglongbing (HLB) disease that could not be described in enough detail to warrant establishment of novel taxa (Murray and Stackebrandt, 1995). HLB bacterium is now a new genus in the *Alphaproteobacteria* division (gram negative bacteria) and given the name of new genus as "*Candidatus* Liberibacter". Penicillin treatment of HLB affected trees showed remission of symptoms further suggesting the bacterial nature of the pathogen. In Citrus, three species of *Candidatus* Liberibacter have been identified to be associated with HLB agents of Asian, South African and American. These species for Asian, African and American HLBs are: "*Candidatus* Liberibacter asiaticus" (Ca.L.a), "*Candidatus* Liberibacter africanus" (Ca.L.af.) and "*Candidatus* Liberibacter americanus" (Ca.L.am.) respectively (Jagouiex *et al.*, 1996; Coletta-Filho *et al.*, 2004; Teixeira *et al.*, 2005b; Halbert, 2005). To determine the taxonomy of the causal bacterium,16S rDNA of HLB of Indian isolates on sweet orange from pune has been amplified using PCR, cloned and sequenced (Garnier *et al.*, 1993). These genes have also been sequenced from HLB infected Nagpur mandarin (Das, 2009). Based on these studies, HLB bacterium has been shown to be á class of *Protobacteria*. Recent studies on Indian isolates of HLB revealed that Indian isolates belong to the *Candidatus* Liberibacter asiaticus strain (Ghosh *et al.*, 2013).

A complete circular '*Candidatus* Liberibacter asiaticus' (LAS) genome has been obtained by metagenomics, using the DNA extracted from a single LAS infected psyllid. The 1.23-Mb genome has an average 36.5% GC content. Multiprotein phylogenetic analysis confirmed '*Ca.* L. asiaticus' as an early-branching and highly divergent member of the family *Rhizobiaceae* (Duan *et al.*, 2010). This is the first genome sequence of an α-proteobacteria that is both an intracellular plant pathogen and insect symbiont. (Duan *et al.*, 2009).

DIAGNOSIS AND MANAGEMENT

Huanglongbing can be easily detected in hosts like sweet oranges just by observing mottling of leaves-a characteistic symptom of HLB disease. Sparse foliage and 'open' type growth in affected trees, lopsided fruits with colour inversion and aborted brown black seeds are other diagnostic symptoms in the field. Since HLB remains symptomless in some hosts, it is essential to index field samples in glasshouse on indicator seedlings of sweet orange (Madam Vinous, Pineapple, Hamlin, mosambi and tangelo (Orlando, Seminole) by graft or vector inoculation. HLB Inoculated plants will show typical symptoms of yellowing and mottling after several months. Best way of biological diagnosis of HLB disease is by transmitting the disease to periwinkle through dodder. In periwinkle, symptoms of smalling of leaves, chlorosis and witches' broom are developed in much shorter time than in *Citrus* spp. Electron microscopy is a good technique for detection of liberibacter species. *Ca.* Liberibacter species are observed in sieve tubes of phloem tissues when thin sections of infected tissues are observed in electron microscopy (Garnier and bove, 1996). Pleomorphic bodies are seen in sieve tubes of phloem tissue of affected plants (Fig. 43). Enzyme-linked immunosorbent assay (ELISA) with monoclonal antibodies (MA) was also used for detection of HLB bacterium. Citrus HLB pathogens are easily detected by immunofluorescence (IF) both in sections and extracts using greening specific MAbs (Plate 54) (Garnier and bove, 1993; Ahlawat, 1993; Ahlawat, 2010; Ahlawat and Pant, 2010). HLB bacterium is also detected by immunoelectron microscopy (Ke *et al.*, 1993).

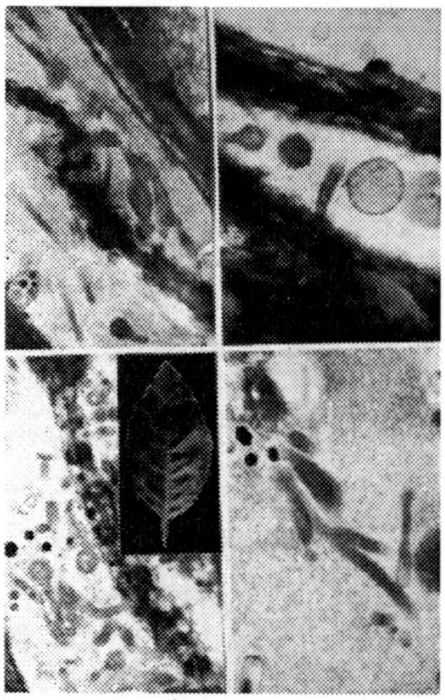

Fig. 43: Electron micrograph of thin secions showing pleomorphic HLB bacteria in sievetubes of infected tissue.

Asian form of HLB can be detected by DNA-DNA hybridization. A fragment of bacterial genome (In 2.6) from Pune strain of Ca. L. asiaticus has been used as a probe for the detection of HLB bacterium in DNA-DNA hybridization in field trees and in individual *Diaphorina citri* collected on infected trees in Delhi and Bangalore in India (Fig. 44) (Bove *et al.*, 1993). Direct plant tissue-print and insect squash-blot methods have also been used for large scale detection of *Candidatus* Liberibacter spp. by real-time PCR with new primers (Bertolini *et al.*, 2010). For rapid and early detection of HLB bacterium non-radioactive tissue blot technique was found to be very useful (Nageswara-Rao *et al.*, 2013). Detection of HLB bacterium through PCR, real time PCR and nested PCR is now commonly used in India and elsewhere (Jagouiex *et al.*, 1996; Ahlawat *et al.*, 2003; Baranwal *et al.*, 2003, 2004; Gouda *et al.*, 2006, Das, 2004; Cordero *et al.*, 2010, Bertolini *et al.*, 2010, Golmohammadi *et al.*, 2010). A membrane based DNA template preparation protocol has been recently developed for simultaneous detection of *Candidatus* Liberibacter asiaticus and citrus yellow mosaic badnavirus (Baranwal and Arya, 2009; Baranwal *et al.*, 2005).

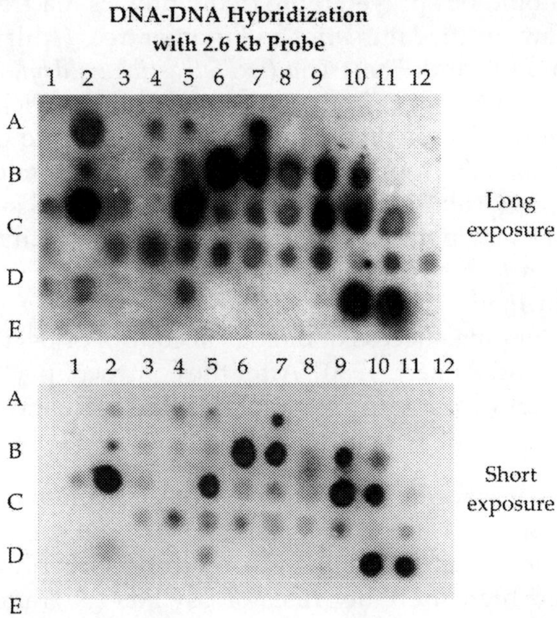

Fig. 44: Detection of HLB bacteria in DNA-DNA hybridization using In 2.6kb probe. Upper one is with long exposure and lower one is with short exposure.

Immediately after the discovery that HLB pathogen is a gram negative bacterium, tetracycline and chemotherapeutants like B.P.-101 and CPAN No.MJN 1891 were injected to the trunk of infected trees in several countries including India (Capoor *et al.*, 1973; Capoor and Thirumalachar, 1973; Ahlawat *et al.*, 1989; Bove *et al.*, 1980). Chemotherapy however provided remission of disease symptoms but the symptoms reappeared as soon as the therapy was discontinued. Preventing citrus tree from becoming infected with HLB-associated agents is the only way to manage HLB most effectively. Following preventive measures are suggested:

1. Symptomatic trees must be removed to eliminate Liberibacter inoculum.
2. To reduce psyllid vector population by insecticide treatments or by biological control approaches.
3. Production of healthy citrus planting material and to maintain insect-proof nurseries for developing new orchards and to replace infected trees.

It has been commonly observed that more trees at the border are infected than in the centre of a grove. It is therefore advisable that either a suitable barrier crop/trees should be identified at regional basis or trees at the border should be sprayed more frequently. Psylla-free areas may be identified to grow certified nurseries and mother trees (Ahlawat and sardar, 1976). In Reunion Island, *Tamarixia* (ex-*Tetrastichus*) *dryi*, an ectoparasite of *T. erytreae* was imported from Transvaal, South Africa, in December 1974. *Tamarixia (ex-tetrastichus) radiata*, an ectoparasite of *D. citri*, was imported in April 1978 from Punjab, India with a view to control psyllids by these parasites (Aubert and Quilici, 1984). These two imported psyllid parasites were absent in this Island and obviously these imported parasites were not present there. Absence of *Tamirixia* parasites (or Psyllid hyperparasites) made it possible to eliminate psyllid vectors from this Island and explain the success of biological control of psyllids in Reunion Island (Etienne and Aubert, 1980). After the control of psyllids in Reunion Island there is significant improvement in citrus plantations as compared prior to introduction of parasites (Aubert *et al.*, 1984).

It has been observed in Mekong Delta of Vietnam that planting guava (*Psidium guajava*) in citrus orchards can increase commercial life of citrus plantings as compared to monoculture citrus. Greenhouse studies in Florida revealed high mortality rates of psyllids (95% mortality) within 6-9 days when confined to guava in no-choice situation. The effect may be due to volatile compounds produced by guava that are deleterious to psyllids (Gottwald *et al.*, 2010). It has recently been shown that the volatiles of guava repel *D. citri*, and this explains the guava-tree effect (Noronha Jr. *et al.*, 2008).

In addition to eradication of sources of inoculum and control of psyllid vectors, growing disease-free plants is an important factor to manage HLB. HLB-free healthy citrus plants can now be obtained by shoot-tip grafting (Vijyakumari *et al.*, 2006a). Use of tolerant rootstocks is another option to manage the disease. *Citrus macroptera* and *C. indica* were found free from greening infection in Northeastern region of India and may be used as sources of resistance (Bhagabati *et al.*, 1989).

CITRUS VARIEGATED CHLOROSIS

Citrus variegated chlorosis disease (CVC) caused by a bacterium, *Xylella fastidiosa* is widely distributed in Brazil (Rossetti *et al.*, 1990; lee *et al.*, 1991). This disease is also reported from Paraguay and Argentina (Brlansky *et al.*, 1991). In Brazil, variegated chlorosis causes approximately 20% loss to the growers and is now estimated to exceed several million dollars loss per year in Brazil (Rossetti, 1993, 2000).

SYMPTOMS

Initially chlorosis begins on younger leaves but by the time of maturity of leaves, intensity of chlorosis is greatly increased which is visible only on upper surface of leaves. In young infected trees, severe leaf chlorosis develops within leaf veins. Gummy lesions develop underside of the leaves corresponding yellow areas on the upper surface of leaves. These lesions appear as dark brown, necrotic and with slightly raised surface due to gum deposit. Fruits on affected trees remain small, hard and have higher contents of sugar. Fruits ripe earlier on affected trees. Initially, symptoms appear on one or two branches but subsequently whole tree is infected. Two years after infection, infected trees become uneconomical. Symptoms of this disease are more severe under conditions of high temperature and drought (Beretta *et al.*, 1993). Affected trees are stunted with protruding branches.

TRANSMISSION

Variegated chlorosis is transmitted by grafting from field infected trees to healthy citrus plants. CVC is also transmitted by natural root grafting (He *et al.*, 2000). Variegated chlorosis associated bacterium, *Xylella fastidious* is transmitted through xylem feeding leafhoppers and sharpshooters from even non-symptomatic alternate hosts to the pathogen. At least 11 species of sharpshooters are known to vector CVC in Brazil. Main vectors are, *Dilobopterua costalimai, Acrogonia terminalis, Oncometopia facialis, Bucephalogonia xantopis* and *Plesiommata corniculata* (Grevena *et al.*, 1998). In Argentina, sharpshooters, *Bucephalogonia xanthophis; Scopogonalia subolivacea; Hortensia similis; Rotigonalia limbatula; Sonesimia grossa; Chlorotettix latocintus* and *Curtara samera* from cicadellidae family and two species of embracidae family; *Ceresa ustulata* and *Entylia carinata* are main vector of *Xylella fastidious* (de Coll *et al.*, 2000). The bacterium is also seed transmitted (Rodrigues Neto *et al.*, 1992).

HOST RANGE

Most of the sweet orange cultivars are susceptible to variegated chlorosis (Beretta *et al.*, 1993). Tangerins and Tahiti lime are symptomless carriers of the pathogen. Plant species in at least 28 families of monocotyledonous or dicotyledonous are natural hosts of the bacterium associated with variegated chlorosis disease and could be the natural source of pathogen survival. Some of the important weed hosts of *X. fastidiosa* are: *Conium maculatum; Chloris halophila; Digitaria* sp.; *Paspalum regnelli; Axonopus compresus; Blainvillea biaristata; Commelina erecta; Medicago polymorpha; Phalaris angusta; Lepidium aetes; Stachys arvensis; Talinum paniculata* and *Sida rhombifolia* (de Coll *et al.*, 2000). CVC does not kill trees, but trees become less productive within a few years following infection

THE PATHOGEN

A gram-negative xylem limited bacterium, *Xylella fastidiosa* is associated with Citrus variegated chlorosis (Beretta *et al.*, 1994; Chagas *et al.*, 1992). The bacterium, *X. fastidiosa* is found in leaf midribs, fruit columella and petioles (Rossetti, 1993). CVC bacterium has rippled cell wall which composed of three layers and bacterial cells are 0.3-0.5 μm in diameter and 1 to 3.5 μm in length (Chagas *et al.*, 1992). Causal bacterium of variegated chlorosis has been cultured on a modified PW medium. Koch's postulate has been fulfilled by inoculating cultured bacterium to sweet orange and re-isolated the same bacterium from inoculated plants (Chang *et al.*, 1993). Coffee leaf scorch is also caused by a strain of *Xylella fastidiosa* from citrus (Li *et al.*, 2001) suggesting that the CVC strain is apparently closely related to the bacterium causing coffee leaf scorch.

DIAGNOSIS AND MANAGEMENT

Variegated chlorosis bacterium is detected by light and electron microscopy in xylem tissues. CVC bacterium can be detected by specific antibodies and in DAS-ELISA and by immunoprinting (Garnier *et al.*, 1993). Dot-immunobinding assay (DIBA) (Lee *et al.*, 1993), Western blot assay and PCR are also used for bacterium detection (Beretta *et al.*, 1993, 1997). Molecular assays based on DNA hybridization or PCR diagnostics are also available for CVC diagnosis (Pooler and Hartung, 1995).

Remission of disease symptoms is possible by treatment of infected trees with oxytetracycline. In Brazil, if 30% of trees in a grove are infected, the recommendation is to remove the entire grove. In Brazil, control strategies to manage CVC include nursery certification, protection of nursery stock from infection, and pruning of symptomatic parts of trees to remove inoculum sources. Control of sharpshooters is very difficult. Long term approaches involve the identification and development of CVC resistant sweet orange cultivars (Chung and Brlansky, 2009). High level of resistance has been reported in Rangpur lime against *X. fastidiosa* (Lopes *et al.*, 2010). For effective management of the disease, following practices are recommended

1. Infected trees up to three years of age should be removed.
2. Effected branches of trees 3-5 years of age should be removed.
3. Infected trees older than 5 years should be pruned 50-10 cm below any symptomatic leaf (Beretta *et al.*, 1996).
4. Propagation material must be obtained from registered nurseries (Rossetti, 2000).

CITRUS DISEASES CAUSED BY MOLLICUTES

Mollicutes are a class of bacteria which are distinguished by the absence of a cell wall. Mollicutes are parasites of various animals and plants, living on or in the host's cells. These organisms are very small, measuring only 0.2-0.3 µm in size and have a very small genome size. Cells of these organisms are pleomorphic in nature and require sterols that make the cell membrane somewhat more rigid. Many are able to move about through gliding, but members of the genus *Spiroplasma* are helical and move by twisting. The best-known genus in Mollicutes is *Mycoplasma*. Mollicutes, *Phytoplasma* and *Spiroplasma* are plant pathogens and some may survive in their insect vectors.

CITRUS STUBBORN DISEASE

Citrus stubborn disease (CSD) was first reported in California State of USA and was shown to be transmitted by grafting (Fawcett *et al.*, 1944;

Fawcett, 1946; Haas *et al.*, 1944). CSD occurs in citrus grown in the Coachella Valley and interior valleys of central and southern California and Arizona. It is present in the Western USA, Eastern Mediterranean region including Israel, Syria, Turkey, Morocco, etc., Middle East and North America. The disease is prevalent in temperate regions with arid or semi-arid climates where citrus is grown with irrigation. CSD is not known to occur in tropical or sub-tropical regions. It is economically important disease as the yield and quality of fruits are greatly reduced (Calavan and Christiansen, 1966). Stubborn disease is now known to be caused by a Mollicute, *Spiroplasma citri.*

SYMPTOMS

Leaves on stubborn affected trees are small, abnormally upright, stiff and pinched at the tip or cup or heart- shaped. Growth of leafy branches from multiple axillary buds is common. Leaves become chlorotic resembling zinc, iron or manganese deficiency. Premature leaf and fruit drop is common. Twigs show dieback symptoms and affected trees remain stunted and bushy in growth (Plate 55a). There is normal but unseasonal flowering resulting fruits of various ages on the same tree. Fruits are acorn-shaped, malformed, lop sided with curved columella (Plate 55b). There is excessive abortion of seeds in fruits on affected trees and fruits have insipid sour or bitter flavor. Fruits are smaller with blue, dense or cheesy albedo having prominent vascular network (Carpenter *et al.*, 1965; Calavan, 1969).

TRANSMISSION

The pathogen associated with stubborn disease (*Spiroplasma citri*) is transmitted by tissue grafting (Fawcett *et al.*, 1944). Though the disease is not seed transmitted but stubborn pathogen is known to be present in the coats of aborted seeds from fruits of affected trees (Calavan *et al.*, 1974). *Spiroplasma citri* is naturally transmitted by leafhopper vectors: *Circulifer (Neoaliturus) tenellus, Scaphytopius nitrides* and *S. acutus* in USA (Oldfield *et al.*, 1976; Oldfield, 1988), *Circulifer (Neoaliturus) haematoceps* is the vector in the Mediterranean area (Bové, 1986). *S. citri* is transmitted in a persistent and circulative manner and multiplies in its vector. Both male and female can transmit but no transovarial transmission was obtained (Oldfield *et al.*, 1984). Successful transmission of *S. citri* by the leafhopper depends on the ability of the *Spiroplasmas* to pass through the insect gut cells, to multiply in various tissues, and to finally cross the salivary gland cells. It is solely from the salivary glands that *Spiroplasmas* are inoculated into the sieve tubes during insect feeding (Labroussaa *et al.*, 2010).

HOST RANGE

Spiroplasma citri infects plant species in 18 dicotyledonous families and one in monocotyledonous family and can infect both citrus and non-citrus hosts. Following are susceptible citrus hosts of stubborn pathogen: Calamondin, Carrizo, Rusk, Troyer Citranges, Grapefruit, Kumquat, Lemons, Mandarin, limes, Pummelo (Chandler), Rough lemon, Satsuma, Sour orange, Sweet lime, sweet orange, and Limes (Kagzi, Key, Mexican, West Indian). In addition to citrus, *Spiroplasma citri* also infects several other hosts and cause diseases in them (Gumpf, 1988). For example, it is the causal agent of brittle root disease of horseradish in Illinois (Sullivan *et al.*, 1987) and sesame yellows in Turkey and Iran. White clover, chrysanthemum and periwinkle are hosts of stubborn pathogen and can be infected by direct inoculation by leafhopper vectors (Reuther *et al.*, 1978). London rocket, many *Brassica* spp., wild radish and marigold are some of the natural/experimental hosts of *S. citri*. *Catharanthus roseus, Crepis echiodes, Echium* sp. and *Sesamum indicum* are found to be naturally infected in eastern Mediterranean region of Turkey (Kersting *et al.*, 1993). Carrot purple leaf disease is also known to be associated with *Spiroplasma citri*. Recent studies confirmed that carrot is a host of *S. citri* which got infected by beet leafhopper, *circulifer tenellus* (Mello *et al.*, 2009). Sweet orange seedlings and tangelo are used as indicator hosts in *Citrus* species. Symptoms in inoculated plants appear as leaf tip chlorosis, leaf mottle, leaf cupping, stunting and short internodes.

THE PATHOGEN

The stubborn pathogen was established as a Mycoplasma-like organism based on electron microscopic studies of infected tissues and tetracycline therapy of infected plants (Igwegbe and Calavan, 1970; Lafleche and Bove, 1970). The pathogen was finally cultured in artificial media simultaneously from France (Saglio *et al.*, 1971) and from California (Fudl-Allah *et al.*, 1971). *Spiroplasma citri* cells are wall-less, pleomorphic with characteristic spiral morphology. The minimum viable length of a helix is 2.0 × 0.1–0.2 µm. The helices are motile by flexing or rotation. Some strains are non-motile and non-helical. Based on helical morphology of this cultured mollicute, name "Spiroplasma" was coined to the causal organism of stubborn disease (Saglio *et al.*, 1973). Taxonomic position of *S.citri* is given below:

- *Kingdom*: Bacteria
- *Phylum*: Firmicutes or tenericutes
- *Class*: Mollicutes
- *Order*: Mycoplasmatales

- *Family*: Spiroplasmataceae
- *Genus*: Spiroplasma
- *Species*: Citri

DIAGNOSIS AND MANAGEMENT

For diagnosis stubborn disease, graft inoculations are made on sensitive indicator plants such as Madame Vinous sweet orange. *Spiroplasma citri* is easily detected by culturing on synthetic media (Fig. 45) (Lee and Davis, 1984) and by electron microscopy of infected plants and vector (Fig. 46) (Fudl-Allah *et al.*, 1971; Lafleche and Bove, 1970). It is also detected in ELISA (Saillard and Bové, 1983; Clark *et al.*, 1989) and by immunodiffusion test. DAPI staining technique can also detect *S. citri* (Dale, 1988). DNA extracted from organisms in liquid medium facilitated assay for *S. citri* by PCR (Rangel *et al.*, 2005). *S. citri* is also detected by IC-PCR (Saillard *et al.*, 1996). Best tissue for detection of *S. citri* in real time PCR is fruit columella because pathogen titer remains highest in tissues of columella (Yokomi *et al.*, 2010).

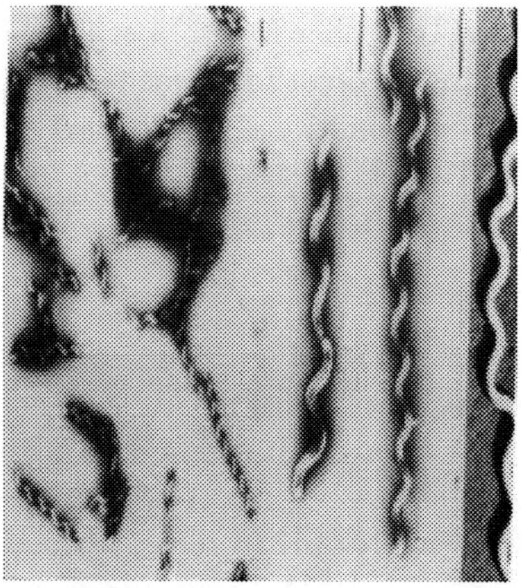

Fig. 45: Spiral form of *S. citri* in culture.

Stubborn disease is difficult to manage because the pathogen, *S. citri* has several natural weed hosts and is transmitted by several leafhopper vectors. Production of healthy budwood is therefore a practical means of control. Trees showing symptoms should be rogued and replaced. Healthy trees can be obtained by eliminating the pathogen through shoot tip grafting. Control of weed hosts and vectors are important for effective management of stubborn disease. It is suggested that *S. citri* has to be covered by virus-free certification schemes for citrus, such as the one now under development by EPPO. It has been suggested that trap plants should be planted in the vicinity of orchards to manage vectors (Gumpf, 1988).

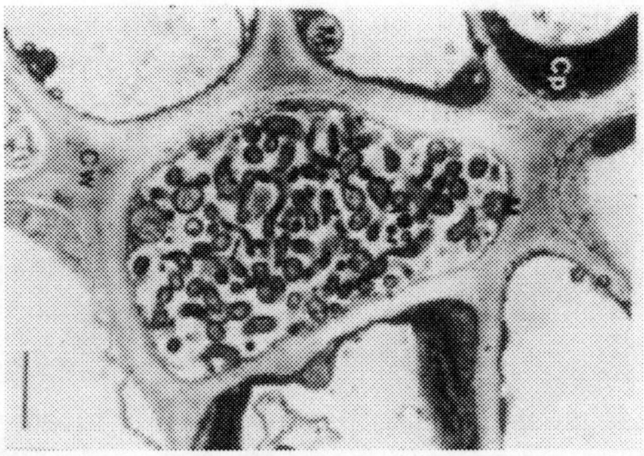

Fig. 46: *Spiroplasma citri* in a sieve tube cell showing pleomorphic bodies.

WITCHES' BROOM DISEASE OF LIME

Witch's broom disease of lime (WBDL) was first observed in coastal plain of the Sultanate of Oman near the border of United Arab Emirates (UAE) around 1970 (Bove, 1986). This disease has now spread to UAE, Muscat and is spreading fast in Iran (Bove, 1995; Salehi *et al.*, 2001). It was estimated that over 98% of limes currently grown in Oman are infected with WBDL. In India, Witches' broom disease has also been found on acid lime (*Citrus aurantifolia*) trees in some citrus growing areas of Maharashtra, Andhra Pradesh, Tamil Nadu and Karnataka States (Ghosh *et al.*, 1999a). Blastomania is a somewhat similar to witches' broom disease reported on *C. aurantiifolia* and *C. limonia* in India (Mali *et al.*, 1975). Blatomania has been shown to be associated with phytoplasmas (Sharma and Singh, 1988).

SYMPTOMS

Symptoms of witches' broom disease (WBDL) in infected trees appear as small pale leaves, highly proliferating shoots, short internodes and witches' broom (Plate 56a). Leaves drop prematurely and infected branches have distorted twigs characteristic of witches' broom symptoms. In advanced stages, infected branches show die-back symptoms (Plate 56a). Leaves on older witches' brooms die but remain attached to them. Fruits on healthy part of the tree are normal but on affected branches, fruit size is reduced and such fruits often dried. In final stages of infection, most infected shoots and twigs die and only a few witches' brooms are left. At this stage, there is no yield and growers have no option except to remove such trees (Bove *et al.*, 1988). Trunk and branches of tree having witches' broom symptoms are excessively ramified.

TRANSMISSION

Witches' broom disease is transmitted by grafting affected twigs to healthy lime and Troyer Citrange seedlings (Garnier *et al.*, 1991). WBDL is also transmitted by dodder to periwinkle (Plate 56b) (Garnier *et al.*, 1991; Bove, 1986; Bove *et al.*, 1988; Ghosh *et al.*, 1999 a). A leafhopper, *Hishimonas phycitis* is widely present and reproduces on infected lime orchard trees in Oman and UAE and is suspected to be the vector of the disease (Bove *et al.*, 1993).

HOST RANGE

Witch's broom disease agent naturally infects lime (kagzi, Key, Mexican, acid), citron, Palestine sweet lime and sweet limetta. Experimentally, this disease is graft transmissible to lemon, Rough lemon, Trifoliate orange, Troyer citrange, *Citrus excelsa, C. macrophylla, Etrog citron, C. ichangensis, C. karna,* Rangpur lime, Meyer lemon and *C. hystrix* (Bove *et al.*, 1996). Causal agent of WBDL is transmitted to eggplant, ornamental eggplant, two species of tobacco, Jimson weed, black nightshade and tomato by grafting and dodder. Disease symptoms on these hosts are yellowing, smalling of leaves, short internodes, virescence, phyllody, witches' broom, stunting, wilting and death of infected plants (Salehi *et al.*, 2001). Acid lime or Mexican lime seedling are used as indicator host. The best indicator host is periwinkle. Dodder inoculated periwinkle plants show yellowing of mid vein, smalling of leaves and witches' broom symptoms.

THE PATHOGEN

Witches' broom disease is caused by a phytoplasma both in Oman and in India (Bove *et al.*, 1988; Ghosh *et al.*, 1999b). The size of the WBDL Phytoplasma genome is 720 kbp. Genomic similarities with Phytoplasmas of sunhemp, sesame, and alfalfa phyllodies were demonstrated, and it was established that the WBDL phytoplasma belongs to the sunhemp phyllody phylogenetic subgroup (Zreik *et al.*, 1995). Phytoplasma associated with WBD has been classified as *Candidatus* Phytoplasma aurantifolia (Zreik *et al.*, 1995).

DIAGNOSIS AND MANAGEMENT

Witches' broom disease can be easily identified in field trees by its specific symptoms. WBDL is easily detected by inoculated indicator plants by grafting and periwinkle plants by dodder. In thin section electron microscopy pleomorphic phytoplasma bodies can be viewed in sieve tubes of phloem tissues in sections of midrib of diseased periwinkle or lime leaves (Bove *et al.*, 1988; Ghosh *et al.*, 1999b). Presence of *Candidatus* phytoplasma aurantifolia (Ca.P. a) is also detected by DAPI test (Ghosh *et al.*, 1999b). Eighteen Monoclonal antibodies have been prepared against witches' broom disease of lime by Professor Bove and his group and are being used in DAS-ELISA or immunofluorescence to detect *Ca.* P.a. in plants and insect, *H. phycitis* (Garnier *et al.*, 1991). DNA probes specific to Ca.P.a. have been prepared and used to detect witches' broom Phytoplasma in DNA-DNA hybridization tests from plants as well as from insects (Bove *et al.*, 1993). PCR amplification of "*Candidatus* Phytoplasma aurantifolia" 16S rDNA was obtained with universal primer P1 (forward) and specific primer WB3 (reverse) (Zreik *et al.*, 1995; Bove *et al.*, 2000). The Indian isolates were also detected by nested PCR (Ghosh *et al.*, 2013).

Strict regulations are required for movement of propagation material from countries or regions where witches' broom is present to avoid the introduction of this disease in places where it is not found. In endemic areas, eradication of infected plants should be done as quickly as possible. Resistant lines of lime, if available, should be planted. Attemps are being made to regenerate somatic hybrid plants of *Citrus sinensis* (sweet orange) and *Citrus micrantha,* a progenitor species of lime (Khan and Grosser, 2004). Healthy planting material can also be obtained by shoot tip grafting.

CITRUS RUBBERY WOOD DISEASE

Rubbery wood (RW) disease of citrus is known to occur in India for a long time (Ahlawat and Chenulu, 1985a and b). RW is quite prevalent in North eastern region of India, Kota region of Rajasthan and Nagpur region of Maharashtra state in India (Chenulu and Ahlawat, 1993). In some orchards of lime and lemon, incidence of RW disease was recorded upto 30% in North eastern states of India. In Akola and Tirupati regions of India, a disease of Nagpur mandarin locally known as 'Valia' appears similar to rubbery wood disease. RW is a disease of minor importance in India. A disease with similar symptoms also occurs in Brazil which is now considered to be a reaction of exocortis viroid.

Rubberywood affected trees are characterized with development of thin and linear shoots which bend towards ground owing to unusual flexibility or elasticity. Such trees are stunted and bushy, and become totally unproductive. Leaves are slightly chlorotic and smaller in size. Glasshouse inoculated plants of kagzi lime (*Citrus aurantifolia*), lemon and mandarin develop willowing habit and their shoots bend to an extent that they can touch the ground (Plate 57). Usually there is no leaf fall of lemon and mandarin seedlings in the glasshouse but RW infected plants show heavy leaf fall. Severely affected glasshouse inoculated plants show dieback and die following 4-5 year of infection. After the death of main shoot, numerous thin branches come out at the base of the plant. These branches have short internodes and smaller leaves giving the plant a bushy appearance. Swelling of buds was also noted in Eureka lemon plants (Ahlawat and Chenulu, 1985a).

Rubberywood disease is transmitted by grafting when branches showing rubberywood are grafted to test plants. RW is also transmitted by dodder, *Cuscuta reflexa* from lemon to lemon. In Electron miceoscope, Phytoplasma like bodies were seen in sieve tubes of phloem tissues of ultrathin sections of diseased plant but not in tissues of healthy plant (Ahlawat, 1987, 1989; Ahlawat *et al.*, 1986). Rubbery wood phytoplasma did not show any reaction in immunofluorescence tests with monoclonal antibodies of *Candidatus* Phytoplasma aurantifolia, the causal pathogen of witches' broom disease of lime. RW phytoplasma is detected either by electron Microscope or through biological indexing on lemon seedlings (Ahlawat and Pant, 2003). Propagation from infected trees should be avoided for better management of RW phytoplasma disease.

CITRUS DISEASES CAUSED BY VIROIDS

Viroids are small, naked, single stranded RNA molecules that infect plant cells and cause diseases. Viroids are not enclosed in a protein coat like viruses. They generally consist of less than 400 nucleotides and do not contain any gene. Viroids can cause many commercially important diseases such as coconut cadang-cadang, citrus exocortis, coconut root wilt and potato spindle tuber and several others. Based on the symptoms expression in *Etrog citron* (the indicator host for most citrus viroids), five citrus viroids have been identified. The molecules of citrus viroids consist of ribonucleotides ranging in size from 284 to 371, with the exception of one CEVd isolate variant of 463 nucleotides.

CITRUS EXOCORTIS DISEASE

Exocortis disease of citrus is also known as "scaly butt" in Australia (Benton *et al.*, 1950; Bitters, 1952). Exocortis disease is mainly a disease of trifoliate orange rootstock and resulting in failure of several scion varieties on trifoliate orange. Citrus exocortis viroid is one of the most ancient viroids among pospiviroids. Exocotis affected trees remain stunted and this dwarfing factor of exocortis had been practically exploited to develop dwarfing trees for density planting in Israel and some other countries (Cohen, 1968; Ashkenazi and Oren, 1988). In India, exocortis disease is commonly present in Nagpur mandarin and Mosambi sweet orange on Rangpur lime and rough lemon rootstocks (Nariani *et al.*, 1968; Patil and Warke, 1968; Ghosh *et al.*, 2002; Kapur *et al.*, 1974). Hop stunt viroid has also been isolated from Nagpur mandarin and Mosambi sweet orange on rough lemon and Rangpur lime rootsocks in India (Ramachandran *et al.*, 2005).

SYMPTOMS

Canopy of exocortis affected tree is normally symptomless. Trees raised on susceptible rootstocks develop symptoms below the bud union on the rootstock at or below ground level. Characteristic symptoms of exocortis disease on susceptible rootstocks are shelling, scaling or splitting of rootstock bark above the ground level (Plate 58a). Affected trees remain stunted. Small elongated areas of dead bark may have slight amount of gum beneath the bark flakes. Bark-scaling may extend below ground up to the main root. Bark-scaling takes longer time to develop ranging from 4-8 years after infection (Bitters, 1952).

TRANSMISSION

Citrus Exocortis disease is transmitted by grafting and natural spread may take place through contaminated budwood. In nurseries and orchards, exocortis disease spreads through contaminated operational tools (budding knives and pruning shears) or even by propagating tools and worker's contaminated hands (Garnsey and Jones, 1967; Garnsey, 1968). Citrus exocortis is also transmitted mechanically to some herbaceous hosts. Exocortis pathogen is not transmitted by seed in citrus species but it is seed-borne in seeds of *Impatiens walleriana* to the extent of 66% and in *Verbena* species upto 28% (Singh *et al.*, 2009). Exocortis is also transmitted by dodder, *Cuscuta subinclusa* from citrus to citrus and Petunia (Weathers, 1965 a,b). No insect vector is known for exocortis disease or any other viroid disease.

HOST-RANGE

Citrus exocortis viroid (CEVd) can infect large number of citrus species, its hybrids and citrus relatives. Exocortis affects large number of rootstock varieties including the ones commonly used as root-stocks like Trifoliate orange, Citron, Lemons and Rangpur lime. Citranges, Etrog citron, Harvey lemon, Mandarin, limes, Sweet lime and Tahiti lime are susceptible hosts to exocortis. In India, exocortis has been isolated from mandarin and sweet orange budded on *Citrus karna* and Rangpur lime (Nariani *et al.*, 1968; Raychaudhuri *et al.*, 1977). Herbaceous hosts, *Gynura aurantiaca* (Flores and Rodriguez, 1981), tomato, *Chrysanthemum morifolium, Datura stramonium* (Weathers *et al.*, 1967; Duran-Vila *et al.*, 1988; Schwinghamer and Broadbent, 1986), petunia (weather, 1965b), potato and various species of *Scopolia* (Singh and Clark, 1973) are also susceptible to exocortis. Arizona 861-S1 Etrog citron is used as indicator host for exocortis viroid. The symptoms on this indicator host develop as leaf epinasty and/or stunting, midvein necrosis, petiole necrosis, leaf tip browning within weeks to months after inoculation at warm temperature (Plate 58b). Best symptoms of exocortis disease develop at a temperature of 28–32°C.

THE PATHOGEN

A circular and linear form of infectious RNA molecules is associated with exocortis disease. Complete genome of exocortis viroid has been sequenced and contains 371 to 375 nucleotides which is the largest of citrus viroids (Gross *et al.*, 1982; Viswader *et al.*, 1982). Citrus exocortis viroid (CEVd) is a species in the genus *Pospiviroid* with a central conserved region similar to the type member, potato spindle tuber viroid (Duran-Vila *et al.*, 1988). In exocortis affected trees in India, two viroids have been identified, one Indian tomato bunchy top viroid (Mishra *et al.*,

1991) and the other Citrus B (CVd-11a) group of viroids (Ramachandran *et al.*, 1993).

DIAGNOSIS AND MANAGEMENT

Citrus exocortis viroid (CEVd) can be easily identified in field trees if raised on sensitive rootstocks like Trifoliate orange, Citron and Rangpur lime. Indexing on Arizona 861-S1 Etrog citron will detect the presence of exocortis viroid by producing characteristic symptoms as described above. Exocortis viroid is detected by sequential gel electrophoresis (sPAGE) under native and denaturing (dPAGE) conditions and silver staining. dPAGE increases the resolution of infectious linear and circular forms of viroids (Baksh *et al.*, 1984; Flores, 1988; Duran-Vila *et al.*, 1993). Reverse PAGE (R-PAGE) is also commonly used to detect viroid infections. CEVd is also detected by molecular hybridization of tissue imprints on a nitrocellulose membrane using radioactive probes (dot blots) and by hybridization of CEVd-cDNA probes to RNA dot blots and Northern transfers (Gillings *et al.*, 1988; Palacio-Bielsa *et al.*, 2000). In India, exocortis viroid is detected by R-PAGE (Rustem and Ahlawat, 1999; Ghosh *et al.*, 2002). CEVd can also be detected by molecular techniques like nucleic acid hybridization and RT-PCR (Bernad and Duran-Vila, 2006; Murcia *et al.*, 2008).

To manage citrus exocortis disease, viroid sensitive rootstocks should not be used for propagation. In Australia, desert lime (*Eremocitrus glauca*) has been suggested to be a good rootstock for citrus in which inoculated viroids did not reach detectable titers (Bani Hashemian *et al.*, 2010). Budwood for propagation should be obtained only from indexed viroid-free trees. CEVd contamination should be avoided by separation of healthy and diseased plants in nurseries. Operational tools like budding knives, shears, clippers etc. must be sterilized. Sterilization of tools can be done by treatment with dilute solutions of sodium hypochlorite or formaldehyde or a mixture of the two.

CACHEXIA (XYLOPOROSIS)

The term "Cachexia" was coined to a condition of Orlando tangelo showing gumming and browning of phloem tissue, wood pitting and bark cankers (Childs, 1950). Major cachexia disease problem occurs when infected budwood are grafted on sensitive rootstocks or infected trees are top worked with sensitive cultivars. Cachexia is found in most citrus growing regions of the world. A disease with similar symptoms termed as xyloporosis was first reported on citrus trees grafted on Palestine sweet lime in Israel (Reichert and Periberger, 1934). A possible relationship between cachexia (Childs, 1950) and the xyloporosis, a disease found in the Mediterranean region has been speculated (Childs, 1952; Roistacher, 1988). However, it now appears that cachexia and xyloporosis are not synonymous and that the disease in sweet lime (xyloporosis) is probably not caused by the cachexia viroid (Roistacher, 1988).

SYMPTOMS

Cachexia disease induces symptoms of discolouration and gum impregnation of the phloem and cambial pitting of the wood with corresponding pegging of the bark on mandarins, tangelos, tangors, *Citrus macrophylla, C. volkameriana* and some other varieties. Inner bark surface of contaminated trees contains numerous pegs that fit in to corresponding depressions in the wood. Cachexia disease affects Alemow, Clementine, kumquats, Mandarins, Mandarin hybrids, Rangpur lime, Satsuma, Sweet lime Tangelos (Nocatee, Orlando, Seminola, Wekiwa) and Ellendale tangor (*Citus reticulata X C. sinensis*). Cachexia also infects cucumber (*Cucumis sativus* cv Suyo) causing mild leaf rugocity in inoculated leaves.

Symptoms of xyloporosis appear as small conoid pits with corresponding small pegs beneath the bud union and brown discoloration of bark and wood of affected trees. Swelling sometimes occurs at the bud union. Leaves on infected trees become small and yellow and fruit set is heavy. Sweet lime is highly susceptible to xyloporosis. Phloem discoloration in the tangelo scion is distinct, and diagnostic. Symptoms of sticky exudation on lower leaf surface and dieback of twigs were also developed (Terranova *et al.*, 1991).

Parson's Special mandarin on rough lemon rootstock is used an indicator host for cachexia (Calavan and Christiansen, 1985). Cachexia induces a strong reaction in Parson's special mandarin showing gum near bud union and point of cut back, pits in wood and gum in bark develop in inoculated indicator plants within one year at warm temperature.

TRANSMISSION

Cachexia disease is transmitted by grafting (Childs, 1952). Cachexia is mechanically transmitted when healthy plants were inoculated by knife cuts from a positive cachexia-infected citron. A quick dip of the knife blade in a 1% sodium hypochlorite solution completely prevented transmission.

THE PATHOGEN

Based on transmission similarities with exocortis viroid (CEVd), viroid aetiology of cachexia was proposed (Roistacher *et al.*, 1983; Semancik *et al.*, 1988). Genome of citrus cachexia viroid (CCaVd) has been sequenced (Semancik *et al.*, 1988; Levy and Hadidi, 1993) which contains 300 nucleotides. Both circular and linear forms of RNAs are present in CCaVd. The variable domain V was extremely conserved in all the cachexia isolates. Five nucleotide differences in both the upper (three nucleotides) and the lower (two nucleotides) strands of the V domain have been identified as the most characteristic feature discriminating cachexia from non-cachexia isolates (Palacio and Duran-Vila, 2000).

DIAGNOSIS AND MANAGEMENT

Cachexia viroid can be detected by its specific symptoms it produces in trees on susceptible rootstocks. Indexing is done by graft inoculation on indicator, Parson's special mandarin. Cachexia viroid can be detected in direct PAGE (dPAGE) analysis of citron or cucumber tissue 6 to 8 weeks after inoculation (Semancik *et al.*, 1988). Cachexia viroid is also detected by sequential gel electrophoresis (sPAGE) and silver staining (Duran-Vila *et al.*, 1991), dot blot hybridization with labeled nucleic acid probes and RT-PCR (Levy and Hadidi, 1993).

For management of cachexia disease, viroid-free cultivars should be propagated, especially if cachexia viroid-susceptible rootstocks such as Rangpur lime or tangelos are to be used. Healthy cachexia-free planting material can be developed by shoot tip grafting of infected bud lines. The operational tools like budding knives, shears, clippers etc. must be sterilized to avoid spread of the disease. Sterilization of tools can be done by treatment with sterilant like sodium hypochlorite (1%) or formaldehyde or a mixture of the two.

CITRUS YELLOW CORKY VEIN

Citrus yellow corky vein disease is known to occur on *Citrus* species in India and has been reported by several names such as yellow corky vein, yellow mid vein and yellow vein banding (Reddy *et al.*, 1974; Reddy and Naidu, 1989; Sharma and Pandey, 1983; Ahlawat and Chakraborty, 1993). Yellow corky vein disease is now most accepted name of this disease. Yellow corky vein disease is widely distributed in citrus plantations in India. Recently, a similar disease has been reported on Washington navel orange in Fars province of Iran (Bagherian and Izadpanah, 2010).

SYMPTOMS

In naturally infected Kagzi lime (*Citrus aurantifolia*), symptoms of yellow corky vein disease develop as yellowing of midrib and lateral veins followed by formation of corky tissues on lower surface of leaf veins (Plate 59). Yellowing and corking may cover all the veins or part of them. Corking is also developed on petioles, both young and old branches of yellow corky vein affected trees (Plate 60E). Shrinking of midrib and crinkling of leaves have also been observed in advanced cases of infection (Plate 60D). Glasshouse inoculated plants showed dieback symptoms within 3-4 years of inoculation (Rustem and Ahlawat, 1999).

TRANSMISSION

Yellow corky vein disease (CYCVD) is transmitted by wedge grafting and by bud grafting. CYCVD is also transmitted by mechanical inoculation from kagzi lime to kagzi lime using glycine buffer. The agent of yellow corky vein is not transmitted through seed and no insect vector is known. Maximum transmission of the disease was obtained at a temperature of $41.7^{0}/22.7°C$ (maximum/minimum).

HOST RANGE

Most *Citrus* species and cultivars are susceptible to yellow corky vein infection. Susceptible *Citrus* species are: Citranges (Carrizo, Savage, Troyer), Etrog Citron, *Citrus decumana*, Mandarins (Cleopatra, Darjeeling/Khasi, Nagpur, Kinnow), Lemons (Galgal, hill, Volkamer), Limes (Kagzi and kagzi kalan), Nasnaran, Rough lemon (Jatti khatti), Sour orange, Sweet orange (Folkan and Mosambi), and Trifoliate orange. These hosts showed yellow vein banding, vein corking,veinal necrosis and crinkling of leaves. Epinasty of leaves was also observed in some hosts. Symptoms developed on some susceptible hosts are given in Plate 60. Kagzi lime (*Citrus aurantifolia*) has been established as indicator host of yellow corky vein disease.

THE PATHOGEN

Viroid aetiology of yellow corky vein disease was established by polyacrlamide gel electrophoresis (PAGE) especially return PAGE (R-PAGE) (Plate 61) and by nucleic acid hybridization studies. Yellow corky vein pathogen reacted positively with specific probes of exocortis and hop stunt viroids suggesting that two viroids, exocortis and hop stunt were associated with yellow corky vein disease (Rustem and Ahlawat, 1999; Ramachandran *et al.*, 2003). Sequencing studies revealed that CEVd component consists of 370 nucleotides. Hop stunt component was also subsequently sequenced. It consisted of 295 nucleotides (Roy and Ramachandran, 2003).

DIAGNOSIS AND MANAGEMENT

Yellow corky vein disease is easily detected in field by specific yellow corky vein symptoms (Plate 59,60). It can be easily detected by graft inoculation on indicator host, Kagzi lime seedlings which will develop typical symptoms of the disease one week after inoculation. Yellow corcky vein viroid is also detected in R-PAGE (Plate 61) and dot-blot hybridization using CEVd and hop stunt viroid probes (Fig. 47 A, B). It is also possible to detect both the viroids associated with yellow corky vein disease in RT-PCR (Roy and Ramachandran, 2006). Yellow corky vein disease can be managed by use of resistant rootstocks like Karna khatta and Soh Sarkar. These viroids can also be eliminated by shoot tip grafting. Kagzi lime has high degree of polyembryony and hence nucellar plants can be used as planting material.

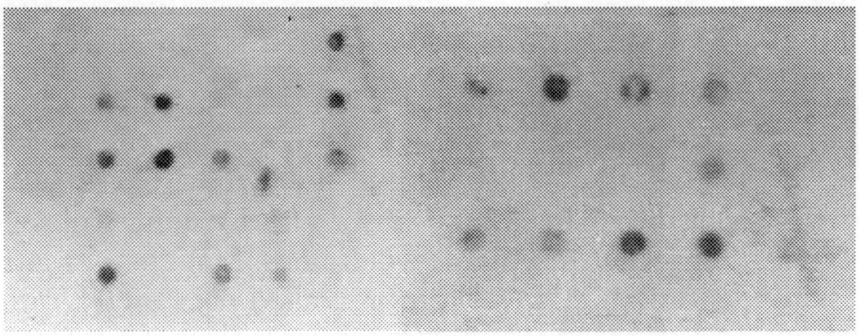

(a) (b)

Fig. 47: A. Dot blot hybridization of CYCV with exocortis specific probes. B. Dot blot hybridization of CYCV with hop stunt specific probes

CITRUS GUMMY BARK

Gummy bark disease was first reported as phloem discolouration of Baladi and other sweet orange trees grafted on sour orange rootstock in Egypt in 1954 (Nour-Eldin, 1956, 1959). This disease is now known to occur in a number of eastern Mediterranean, North African, near East countries and Middle East countries (Roistacher, 1991; Bové, 1995). Infected trees show stunting with streaks of reddish–brown gum impregnated tissue near bud union when bark of trunk is scrapped. Stem pitting is also developed in affected trees which can be easily seen after removal of bark. In addition to these symptoms bud-union crease can also be seen on such trees (Bové, 1995). Gummy bark disorder is transmitted by grafting contaminated buds, shoots and bark. Host-range of this disorder is restricted to citrus species and cultivars. Sensitive species are Sweet oranges and rough lemon (Nour-Eldin, 1968). citrus gummy bark viroid is a sub-species of the hop stunt viroid, and thus is a member of the genus *Hostuviroid*. (Sofy *et al.*, 2010). Diagnosis of the disease is done by bark patch grafting on sweet orange on sour orange and rough lemon rootstocks. The gummy bark disease can also be identified by sPAGE and slot-blot hybridization using viroid-specific probes and RT-PCR (Bernard *et al.*, 2005; El-Dougdoug *et al.*, 2010). The disease is managed by use of gummy bark-free budwood.

GUM POCKET AND GUMMY PITTING

The terms 'Gum pocket', 'Wood pitting' and 'Gummy pitting' have been used to describe a characteristic disorder of trifoliate orange. These diseases commonly occur in old budwood lines. Main symptoms of these disorders are the formation of gum impregnated pegs on the cambial face which fit in the pits of wood with an indented ring at the bud-union, flattened and longitudinal fissures and laminate shelling of the bark (Du Charme and Knorr, 1954; Fernandez-Valiela *et al.*, 1965; Foguet and Oste, 1968).Gummy pitting disease is transmitted by grafting and also by direct propagation from contaminated trees (Fraser *et al.*, 1976). Sweet orange, Marsh grapefruit, Clementine mandarin and Eureka lemon trees grafted on trifoliate orange rootstock develop better symptoms of the disease. Sweet orange on *Poncirus trifoliata* is used as indicator host. Naturally occurring viroid infections contain a mixture of sequence variants (Elleuch *et al.*, 2006). The disease can be identified in field trees by observing symptoms of gum impregnated pits in the surface of the wood of trifoliate orange rootstock and also by inoculating indicator host (sweet orange on trifoliate roots) under experimental condition. RT-PCR and multiplex PCR systems have also been suggested to detect the associated viroid (Almeyda-Leon *et al.*, 2002). Use of budwood derived from shoot tip grafting *in vitro* or nucellar embryony have been suggested to manage the disease.

SHELL BARK

Shell bark disease was reported in old-line lemon trees in California (Calavan and Weathers, 1959). Symptoms of this disease appear in trees older than 10 years or more. Outer bark of affected trees cracks into vertical strips and dies. Lemon trees develop symptoms above the bud union. Shell bark symptoms are more severe in Eureka lemon than in Lisbon lemon. Shell bark is a transmissible factor (Calavan and Weathers, 1959). The etiology of shell bark is yet to be determined but trees with shell bark symptoms usually carry a number of citrus viroids, including citrus exocortis viroid. Shell bark is identified in field affected trees as cracking of outer bark into vertical strips (Calavan and Weathers, 1959). Shell bark disease can be eliminated by nucellar embryony or shoot tip grafting *in vitro* and thus healthy planting material is obtained.

DWARFING CITRUS TREES BY VIROIDS

Dwarfing of citrus trees on trifoliate orange rootstock infected with a graft transmissible factor (GTF) without showing scaling symptoms on rootstock was reported from Australia (Fraser *et al.*, 1961). The yield was enhanced per ha through close planting of such dwarfed orange trees (Long *et al.*, 1972). The dwarfing factor was graft transmissible and specific viroids were found associated with dwarfing trees in Australia and Israel (Schwinghamer and Broadbent, 1987; Gilling *et al.*, 1991; Bar-Joseph, 1993). Israel and Australia are the major countries taking advantage of dwarfing technology and 115ooo dwarfing buds were sold between 2003-2006 but it has now been realized that the use of graft transmissible dwarfing (GTD) is incompatible with the operations of a high health status budwood scheme. It has become obvious that GTD has benefits while conditions for tree growth and management are optimal, but any stress on trees (*e.g.*, drought, salinity) will have a greater adverse effect on GTD-inoculated trees than their viroid-free counterparts (Barkley and Hutton, 2007). This dwarfing factor is known as viroid dwarfing and graft transmissible dwarfing (GTD). GTD is bud, graft and mechanically transmitted but not by vector or seed. Trees in field should be inoculated during the first year (Hutton *et al.*, 2000). All *Citrus* spp. are susceptible to citrus dwarfing, but only *Poncirus trifoliata* and its hybrids (*e.g.*, citranges) and Rangpur lime provide a dwarfing response when inoculated with dwarfing factor. Natural isolates of *Citrus viroid III*- CVd-III (variant CVd-IIIb) +/- *Hop stunt viroid*- HSVd (variant CVd-IIa)] are associated with dwarfing.

CITRUS BENT LEAF VIROID

Several citrus viroids were obtained when Etrog citron was inoculated with graft transmissible citrus dwarfing agent (GTCDA) in Israel. GTCDA isolate 225 # T showed dwarfing in some grapefruit trees in Israel. Citrus bent leaf viroid (CBLVd) consists of 318 nts which can be arranged in the viroid specific rod like structure (Kisse and Simons, 1985). CBLVd is the first member of the apple scar skin viroid subgroup infecting citrus plants. Citrus bent leaf viroid has been classified in the genus *Apscaviroid* in family *Pospiviroidae* and strain *Citrus bend leaf viroid* (Index of viruses, 2006).

D. VIRUS DISEASES OF PAPAYA

Papaya (*Carica papaya)* is cultivated in most countries with a tropical climate. Green papaya fruit and the tree's latex are both rich in an enzyme called papain, a protease. Papaya is a rich source of antioxidants like carotenes, vitamin C and flavonoids. Global papaya production averaged 67,50,40 tons grown in area of 3,83, 430 ha (FAO STAT, 2007). In india, papaya is grown in about 80,000 ha with an annual production of 7,00,000 tons (FAO STAT 2007). Brazil, Mexico, Nigeria, India and Indonesia have consistently been the top producers, contributing more than 71% of the total world production (FAOSTAT, 2004). In India, many varieties of papaya are under production. Among them important ones are: Ranchi, Coorg, Honey Dew, Co1, Co2, Co3, Co4, Co5, Co6, CP81, Pant 1,2,3, Pusa Majestic, Pusa Dwarf, Pusa Delicious, Pusa Giant, Pusa Nanha, and Pink Flesh Sweet. Species related to *Carica papaya* include *C. pentagona, C. cauliflora, C. pubescens,* and *C. stipulata*. Strong incompatibility between *C. papaya* and these other species has hampered the production of interspecific hybrids. Papaya crops suffer from several diseases and pests but the most ubiquitous and widespread of which are virus diseases of papaya. These virus diseases will be discussed in this chapter.

PAPAYA RINGSPOT VIRUS

Papaya ringspot virus (PRSV) affects production and productivity of papaya in every region of the world by decreasing photosynthetic capacity of the plants, which subsequently display stunted growth, deformed and inedible fruit, and eventually, plant mortality. Ringspot disease of Papaya is the biggest constraint to papaya production in India and elsewhere. In India, ringspot disease is recorded from all the states where papaya is commercially grown. The incidence of the disease is sometimes 100% in some plantations (Capoor and Varma, 1958; Khurana, 1974; Yemewar and Mali, 1980). The disease reported by Capoor and Varma, 1958 as mosaic disease of papaya was probably papaya ringspot disease. This disease has been characterize in India (Jain *et al.*, 1998)

SYMPTOMS

Papaya ringspot disease affects plants of all ages. It produces typical mosaic symptoms showing chlorosis of the leaves with dark green blisters. The lamina is extremely reduced and becomes malformed (Plate 62A). Larger mosaic areas are seen on stem. In extreme cases the leaves show shoe string effect (Plate 62A). The fruits are elongated and reduced in size. The stem, petioles and fruits develop elongated water-soaked areas showing circular rings or lesions. Distinctive ringspot patterns on fruit develop consisting of concentric rings and spots, or c-shaped markings - these markings persist during ripening and can darken to become orange-brown rings as the fruit matures (Plate 62B and C). After the symptoms are fully developed, the plants show decline with marked reduction in growth. As the disease advances, older leaves fell down and small tuft of leaves left at the top which show vein clearing or spots or mosaic symptoms. Papaya ringspot virus type-W causes mosaic on leaves of affected plants of cucurbits, fruits are deformed and yield is reduced in *Cucurbita maxima*.

TRANSMISSION

The disease is transmissible by mechanical inoculations and also by grafting. The virus is naturally transmitted by aphids in a non-persistent manner. Both the coat protein (CP) and helper component (HC-Pro) are required for vector transmission. This virus is not soil or seed-transmitted. The aphid vectors are: *Myzus persicae, Aphis malvae, A. gossypii, A. medicaginis* and *Macrosiphum sonchi* (Capoor and Varma, 1958).

HOST RANGE

Hostrange of PRSV is restricted to the members of families *Caricaceae, Chenopodiaceae* and *Cucurbitaceae*. The local lesion hosts of PRSV are: *Chenopodium quinoa* and *Chenopodium amaranticolor* where local lesions develop on inoculated leaves. Systemic hosts are: *Lagenaria siceraria, Trichosanthes anguina, Cucurbita maxima, C. moschata, C.pepo var. Medullosa, C. metuliferus, C. anguria var. Anguria, Citrullus vulgaris, C. fistulosus, C. lunatus, Cucumis sativus, C. meeusii, C. dispaceous, C. dinteri, C. hardwickii* (Chenulu and Ahlawat, 1993, Yeh *et al.*, 1984). A few alternate weed hosts, *Cucumis melo, Xanthium indicum, Cassia tora* and *Physalis minima* have been identified in India. These hosts showed symptoms of leaf reduction, mosaic, leaf yellowing, yellow vein and stunting. PRSV in these weeds was confirmed by biological test, ELISA, and EM (Tomar *et al.*, 2009). PRSV isolates belong to either one of two major strains, P or W. The P strains infect both papaya and cucurbits whereas the W strains infect only cucurbits but not to papaya (Tripathi *et al.*, 2008). Its P biotype is a devastating pathogen of papaya crop and its W biotype of cucurbits. The virus can be assayed on *Chenopodium quinoa* and *Chenopodium amaranticolor.* The virus has a thermal inactivation point between 53°C and 55°C. and dilution end-point between 10^{-3} and 10^{-4}. The virus survives aging *in vitro* at room temperature for only 26h.

THE PATHOGEN

Papaya ringspot virus (PRSV) belongs to the genus *Potyvirus,* family *Potyviridae.* Virions are filamentous, non-enveloped and flexuous measuring 760–800 × 12 nm. Virus particles contain 94.5% protein and 5.5% nucleic acid. The PRSV genome consists of a unipartite linear single-stranded positive sense RNA of 10 326 nucleotides with a 5' terminus, genome-linked protein, VPg. The genomic RNA contains one large open reading frame that starts at nucleotide positions 86 to 88 and ends at positions 10118 to 10120, encoding a polyprotein of 3344 amino acids. In India, complete nucleotide sequence of the PRSV pathotype P (10317 nt) and W (10335 nt) has been done. The Indian PRSV sequences when compared with 15 available sequences showed similarity suggesting no or very little variability (Jain *et al.*, 2004; Parmeshwari *et al.*, 2009).

DIAGNOSIS AND MANAGEMENT

Papaya ring spot disease can be easily identified in the field by visual inspections. Serological studies were made with antibodies raised in rabbit (Roy *et al.*, 1999). During recent years, recombinant CP of *Papaya ringspot virus* (PRSV) was used for the production of polyclonal antibodies which was expressed as a fusion protein (<"72 kD) containing a fragment of *E. coli* maltose binding protein. Polyclonal antibodies from rabbits immunized with the fusion protein, successfully detected natural infection of PRSV in papaya and cucurbits samples in direct antigen-coated enzyme-linked immunosorbent assay (Agarwal *et al.*, 2009). Other techniques like immunosorbent electron microscopy (ISEM) and PCR have been used for PRSV detection (Agarwal *et al.*, 2009; Saxena *et al.*, 1994, 1995; Sharma *et al.*, 2004a). PRSV can also be detected by dot-blot hybridization using Biotin-labeled RNA probes (Vidya *et al.*, 2010). The recombinant coat protein specifically detected PRSV in ELISA (Maheshwari *et al.*, 2012).

Papaya ringspot is a serious disease of papaya *(Carica papaya* L.) that has only been partially controlled by conventional methods. The development of commercial virus-resistant transgenic papaya provides a tangible approach to control PSRV in Hawaii and other laboratories (Gonsalves, 1998; Yeh *et al.*, 1988). Agrobacterium mediated transformation protocol to develop transgenic papaya resistant to papaya ringspot virus has been developed in India (Sharma *et al.*, 2004b). Two varieties of papaya, SunUp and Rainbow, that had been genetically-modified to be resistant to the papaya ring spot virus (Maureen *et al.*, 1992) are commercially grown in Hawaii since 1998. The other methods to manage PRSV are : burning all infected plants, using PRSV-tolerant varieties, pesticides, and anti-aphid netting for the first three months, when available and introducing mild strains of the PRSV virus into plants to prevent the symptoms of more severe strains (cross-protection) (Wang *et al.*, 1987).

PAPAYA LEAF CURL

Papaya leaf curl disease was first reported from India (Thomas and Krishnaswamy, 1939; Nariani, 1956). Now the disease is known to occur in other South Asian countries like China, Taiwan etc. (Wang *et al.*, 2004; Lee *et al.*, 2003). The disease has also been reported from Nigeria and other African countries (Noe-Becerra *et al.*, 1999).

SYMPTOMS

The disease is characterized by severe curling, crinkling and distortion of the leaves accompanied by vein clearing and reduction of leaf lamina (Fig. 48). The leaf margins are rolled downward and inward in the form of inverted cap. The veins get thickened and turn dark green. The leaves become leathery and brittle, and petioles are twisted. Vein enations also develop in advanced stages of infection. Affected plants bear only a few flowers and fruits. In advanced stages, defoliation takes place and growth is stunted. Finally, there is no fruit develops on affected trees. Papaya leaf curl virus has a wide host range and infects sunnhemp, chilli, petunia, zinnia, *Datura stramonium* and several weeds and ornamental plants. The virus is experimentally transmitted to tomato and tobacco where it produces the symptoms of leaf curl.

Fig. 48: Leaf curl disease of papaya

TRANSMISSION

In nature, the disease spread by white fly, *Bemisia tabaci* (Nariani, 1956; Saxena *et al.*, 1998a). The disease is not mechanically transmissible.

THE PATHOGEN

Papaya leaf curl is caused by a *begamovirus* in family *Geminiviridae.* Typical twin virions are observed in the diseased papaya cells by electron microscopy. Papaya leaf curl china virus is a single stranded DNA virus (Lee *et al.*, 2003). The Indian isolates have bipartite genome. The genome of several isolates of papaya leaf curl virus has been sequenced. However, the CP-ORF of Taiwan isolates (PaLCV-Tw) was somewhat different from papaya leaf curl virus in India (PaLCV-Id). The DNA sequences of CP-ORF of PaLCV-Tw were 80% identical to those of PaLCV-Id, and their translated amino acid sequences were 77% identical. This indicates that PaLCV-Tw and PaLCV-Id are two different species or strains (Lee *et al.*, 2003). In India, isolates of Papaya leaf curl virus have also been sequenced (Saxena *et al.*, 1998b).

DIAGNOSIS AND MANAGEMENT

The symptoms of the disease are so conspicuous that the disease can be identified in the field without any difficulty. Other methods of virus diagnosis such as electron microscopy, nucleic acid hybridization and PCR are also available (Saxena *et al.*, 1998b, 2004). The best way for the disease management is to remove infected plants as soon as the symptoms are visible to avoid further spread of the disease. Papaya leaf curl and its vector, whitefly have several weed hosts which play great role in epidemiology of this virus. Therefore, as far as possible clean plantation should be done. Transgenic plants resistant to papaya leaf curl shall be a good approach in the near future (Saxena and Singh, 2005).

OTHER VIRUS DISEASES OF PAPAYA

PAPAYA MOSAIC

Papaya plants infected with mosaic virus are characterized by stunted growth, chlorotic mottling or mosaic, vein clearing, downward curling of leaves, and irregular leaf margins (Breman, 1997). The virus is sap-transmitted, but it is not transmitted by seed. Virions are filamentous and about 530 nm in length. The virus has been reported occurring in the Americas (Brunt *et al.*, 1996). *Papaya mosaic virus* caused by *a potexvirus* does not occur in India.

PAPAYA DROOPY NECROSIS AND PAPAYA APICAL NECROSIS

Both these diseases are characterized with apical necrosis and plant death. Initial symptoms of papaya droopy necrosis (PDNV) are drooping and curvature of leaves. Petioles become shorter, the crown will have a distinct bunchy appearance with the abscising of leaves. Stem tip becomes necrotic and the plant eventually dies. Initial symptoms of papaya apical necrosis (PANV) are leaf yellowing followed by rapid wilting of leaves. Apical portions of plants then become necrotic and the entire plant usually dies. *Papaya droopy necrosis virus* (PDNV) and *Papaya apical necrosis virus* (PANV) are the papaya *rhabdoviruses*. Both viruses have similar particle morphology (Noe-Beceria *et al.*, 1999). The virus has been found infecting commercial papaya plantations in Mexico and Venezuela (Noe-Beceria *et al.*, 1999). Phytoplasma bodies have also been observed from papaya plants showing yellows symptoms in Mexico through scanning electron microscope (Lebsky *et al.*, 2009). The symptoms of apical necrosis have also been observed in India (Plate 63).

PAPAYA LETHAL YELLOWING

The virus associated with papaya lethal yellowing is isometric, transmitted by mechanical means only and causes serious chlorosis in papaya. The virus is suggested to be of the *Tombusvirus* group (Silva *et al.*, 1997) and known to occur in Brazil.

PAPAYA DIEBACK

A dieback disease of papaya has been reported from Australia (Harding and Teakle, 1988). No information is available on its aetiology and other properties.

E. VIRUS DISEASES OF OLIVE

The olive tree is often called as the Tree of Life. The low incidence of cardiovascular disease in many Mediterranean populations has been attributed to the benefits of an olive-oil rich diet. Studies demonstrated that the components of olive leaf extract are also toxic to a wide range of bacteria, protozoa, yeasts, parasites and fungi. Seven viruses infecting olive (*Olea europaea* L.) have been reported, four of which belong to nepovirus group. Others are strawberry latent ringspot virus (SLRV), cherry leaf roll virus (CLRV), arabis mosaic virus (AraMV) and olive latent ringspot virus (OLRV). One virus belonging to the cucumovirus group: cucumber mosaic virus (CMV) and two, olive latent virus-1 (OLV1) and olive latent virus-2 (OLV-2) have also been reported on olive.

Most of these viruses remain symptomless but in Portugal and Italy olive trees showed severely distorted leaves and misshapen fruit due to SLRV infection. All these viruses are mechanically transmitted except SLRV (da Clara Henriques, 1994; Gallitelli and Savino, 1985). Experimentally transmitted viruses have very narrow host range comprising seven species in five dicotyledonous families (Savino *et al.*, 1983). OLRV is mechanically transmitted to *Chenopodium amaranticolor, C. quinoa* and *Gomphrena globosa. C. quinoa* and *C. amaranticolor* produce small chlorotic/necrotic local lesions followed by systemic mosaic and, sometimes, apical necrosis upon inoculation. In *G. globosa* inoculated leaves develop red-rimmed necrotic local lesions but upper non-inoculated leaves develop distortion and mottling. *G. globosa* is a satisfactory local lesion host.

In expressed sap of *C. quinoa,* infectivity is lost after dilution to 10^{-3}, heating for 10 min at 60°C or storing for 15 days at 20°C (Savino *et al.*, 1983). Olive latent ringspot virus has isometric particles *c.* 28 nm in diameter which sediment as three components and contain two functional species of single-stranded RNA (Savino *et al.*, 1983). Olive latent virus-2 has quasispherical to bacilliform virus particles (Savino *et al.*, 1983; Martelli, 2008). However, no other information is available on these viruses. Virus particles occur in the cytoplasm, scattered or in discrete paracrystalline or crystalline arrays or within tubules and in the central vacuole (Di Franco *et al.*, 1989). The nucleic acid of the virus is single-stranded RNA. The RNA migrates as two species (RNA-1 and RNA-2). Both RNA species are necessary for infectivity. The coat protein is not glycosylated (Savino *et al.*, 1983). Immunosorbent electron microscopy can be used for virus identification.

Although the virus has no known vector but classified as a member of the nepovirus group on the basis of particle size and morphology, physico-chemical properties and intracellular behaviour. The virus particles resemble those of tobacco ringspot virus (Chu and Francki, 1979) and artichoke vein banding virus (Gallitelli *et al.*, 1984). Other olive viruses can be differentiated from olive latent ringspot virus by serology, differences in experimental host range and, in part, by cytopathological features. Olive latent virus 2, another sap-transmissible virus infecting olive, is distinguished by having quasi-spherical to bacilliform particles (Savino *et al.*, 1984).

SLRV, CMV and CLRV can be detected in DAS-ELISA test (da Clara Henriques, 1994). Production of virus-free propagative material to be used in new plantations is important to manage these diseases. Disease-free selection, sanitation and implementation of certification schemes are essential to manage the olive diseases.

PHYTOPLASMA DISEASES OF OLIVE

Phytoplasma diseases in olive have been reported from Italy and Spain. Symptoms appear very similar, but it was not known whether these diseases are caused by the same or related phytoplasmas (Del Serrone and Barba, 1997; Font *et al.*, 1998). Infected olive trees (*Olea europaea*) in Spain showed abnormal shoots, shortened internodes and proliferation (witches' broom). In Italy, diseased plants showed dwarfed branches with shortened internodes, leaf deformation and chlorosis, proliferation of axillary buds with witches' broom. It is suspected that the cicadellid, *Hyalesthes obsoletus* can transmit the phytoplasma in the field. In both cases, PCR analysis revealed the presence of phytoplasmas in diseased olive trees. In Spain, it is noted that it belongs to the stolbur group. In Italy, the name: olive witches' broom disease has been given.

F. DISEASES OF COCONUT PALM, DATE PALM AND OIL PALM

F-1. DISEASES OF COCONUT PALM (*Cocos nucifera*)

The coconut palm (*Cocos nucifera* L.) is one of the most important plantation crops. This is extensively grown in many tropical countries and the population in several tropical countries depends on coconut palm cultivation as their livelihood. Coconut is used in several ways like beverages, oil for cooking etc. Palm is affected by a virus, few phytoplasmas and viroid diseases which will be briefly discussed in this chapter.

VIRUS DISEASES

FOLIAR DECAY

Coconut foliar decay disease was first reported in *Cocos nucifera* from Saraoutou, Vanuatu where the disease is endemic (Calvez *et al.*, 1980). Natural infection is restricted to *Cocos nucifera* in which necrosis of leaflets is the main symptom. Chlorosis first appears in leaflets between fronds at position of 5 and 11 below the spear leaf. Subsequently the whole frond becomes chlorotic. The central whorl in the crown become yellow. Marginal necrosis followed by collapse of fronds which hang down. Affected plants of susceptible cultivars die within six months to two years after infection. Local cultivars 'Vanuatu' and 'Vanuatu dwarf' are symptomless but introduced 'Malayan red dwarf' shows severe disease symptoms. The virus was not transmitted by mechanical inoculation but was transmitted by a planthopper vector; *Myndus taffni*. Experimentally infected plants by vector mostly show progressive yellowing and necrosis. The assay hosts is *Cocos nucifera* cv. Malayan Red Dwarf (W).

The virus particles associated with the disease are isometric; 20 nm in diameter; angular in profile; without a conspicuous capsomere arrangement (Randle and Hanold, 1989). Genome consists of single-stranded; circular DNA (Randle *et al.*, 1986). Genome possibly unipartite. A 1203 nucleotide segment of the sequenced encapsidated circular single-stranded DNA is of 1291 nucleotides (Rohde *et al.*, 1990; Randles *et al.*, 1992).

The virus can be detected in dot-blot hybridization with radioactive and non-radioactive probes. Virus was also detected in the body of the insect vector by these probes (Hanold *et al.*, 1988; Randles *et al.*, 1992). *In situ* hybridization of rachis tissue showed localisation of DNA within the phloem, but its distribution in the phloem was uneven.

VIROID DISEASES OF COCONUT

COCONUT CADANG-CADANG DISEASE

Coconut cadang-cadang viroid (CCCVd) is responsible for a lethal disease of coconut plant first reported in the early 1930 in the Philippines and widely known as Cadang-cadang disease. The disease causes heavy losses to the coconut palm industry in the Philippines. The name cadang-cadang comes from the word gadang-gadang that means dying in the local dialect. Coconut cadang-cadang viroid (CCCVd) is a lethal viroid of Coconut (*Cocos nucifera*), Anahaw (*Livistona rotundifolia*) Buri (*Curypha elata*), and African Oil Palm (*Elaeis guineensis*). Every year one million coconut palms are killed by CCCVd and over 30 million coconut palms have been killed since Cadang-cadang has been discovered and the loss of production is valued at about \$80 – \$100 for each planting site (Hanold and Randle, 1991).

SYMPTOMS

The first symptoms in the early stage develop within two to four years of infection as rounding of nut shape; development of equatorial nut scarifications; and appearance of fine, translucent, bright yellow spots on leaves. About two years later, known as medium stage, the inflorescence become stunted then become necrotic and eventually killed, nut production decline, and finally stopped. General chlorosis and death of the crown are the advanced symptoms. During the final stage, roughly 6 years after the first symptoms are recorded, the yellow/bronze fronds start to decrease in size and number. Finally, all the leaves coalesce, leaving just the trunk of the palm "standing like a telephone pole". Eight to ten years are taken between the appearance of first symptoms and death of the palm. African oil palm has similar symptoms as coconut but also have orange spotting on palms.

TRANSMISSION

The mode of natural spread of the viroid is unknown. A low rate of seed transmission was observed and pollen transmission is suspected. The pathogen was experimentally transmitted by high pressure injection into the shoot of germinating nuts. CCCVd can spread through mechanical inoculation primarily through contaminated farm tools.The inoculum may be spreading by contaminated seed or operational tools.

HOST RANGE

The natural hosts of cadang-cadang identified are *Cocos nucifera, Corypha elata, Elaeis guineensis* and *Oredoxa regia*. The experimental hosts are*: Adonidia merrillii, Areca catechu, Caryota cumingii, Chrysalidocarpus lutescens, Livistona rotundifolia, Phoenix dactylifera, Ptychosperma macarthurii* and *Roystonea regia* (Batugal *et al*., 2005).

THE PATHOGEN

Cadang cadang viroid (CCCVd) is small, circular, single-stranded RNA, highly base pared, forming a stable rod-like structure. There are four different RNAs found in CCCVd, two of them fast and the other two slow (the difference between fast and slow RNA was based on their mobilities in electrophoresis gel). In the early stages of the disease, RNA fast 1 and RNA fast 2 appear, while in the late stages RNA slow 1 and RNA slow 2 are detected. In coconut palm, the minimal 246-247 nucleotides form are detected after inoculation but it is progressively replaced by larger nucleotides form of 287-302 nucleotides when symptoms of disease develop. CCCVd belongs to potato spindle tuber viroid subgroup (Randles, 1975; Hanold and Randles, 1990).

DIAGNOSIS AND MANAGEMENT

The viroid can be detected by polyacrylamide gel electrophoresis (PAGE) and by dot blot hybridization with specific radioactive RNA probes (Hanold and Randle, 1991). The infection can be avoided by using sterile operational tools. No resistance has been found.

TINANGAJA DISEASE

Tinangaja is a lethal disease of coconut palm which was first described in 1917 in Guam. This disease resembles coconut cadang-cadang which occurs in the Philippines and is caused by coconut cadang-cadang viroid (CCCVd). Symptoms of the disease appear as fine, yellow spot on leaves. Nuts may be round and scarified. Affected palms may decline or die like cadang-cadang disease. Tinangaja disease is caused by coconut trinangaja viroid (CTiVd), having 64% nucleotide sequence homology with CCCVd. CTiVd is a single stranded circular RNA with 254 nucleotides (Keese *et al.*, 1988; Boccardo *et al.*, 1981). Coconut tinangaja viroid (CTiVd) is associated with tinangaja disease, although it has not been verified that inoculation of this viroid causes the disease. Detection can be done by polyacrylamide gel electrophoresis, by using molecular techniques like hybridization using radioactive probes and RT-PCR (Hanold and Randles, 1991). This procedure can also be used for CCCVd, and oligoprobes designed to be specific for either CCCVd and CTiVd can distinguish between the two viroids in coconut leaf extracts.These two viroids although similar but are distinct. Coconuts from Asia and South Pacific have been found to have viroids with similar nucleic acid sequences of CCCVd. The pathogenicity of CTiVd is uncertain (Hodgson *et al.*, 1998).

OTHER VIROID-LIKE SEQUENCES

It has been observed in countries of Pacific region, South asia and French Polynesia that coconut and other monocotyledons have viroid-like sequences similar, but not identical, to coconut cadang-cadang viroid. However, these sequences need to be establish as viroid species (Hanold and Randles, 1991).

PHYTOPLASMA DISEASES OF COCONUT PALM

Four important diseases of phytoplasmal origin, coconut root (wilt), lethal yellowing, blast and tatipaka have been repoted on coconut palm. In India, rootwilt and tatipaka are commonly found.

ROOT (WILT) OR KERALA WILT

Kerala Wilt Disease (KWD) of coconut palms, formerly named as root wilt disease is the most important single threat of coconut production in Kerala, South India. The disease causes heavy losses to coconut plantation (Solomon *et al.* 1989). This disease was first reported from Kerala (Butler, 1908; Varghese, 1934). The disease is found in eight districts of Kerala, in some districts of Tamil Nadu, Karnataka and Goa with an incidence ranging from 1.52 to 75.63 %. It is so far restricted to India only.

SYMPTOMS

The symptoms of the disease normally appear when the plants are about 30 months old. Most characteristic symptom is bending of leaflets called as 'flaccidity'. In older palms, yellowing and marginal necrosis of leaflets of the leaves of central and outer whorls are the typical symptoms of the disease.The leaves become smaller and the crown is reduced. Roots of affected palms show degeneration of phloem, disorganized tracheal elements and tylosis in the metaxylem. They eventually rot. The disease is not lethal but productivity is significantly reduced. Rotting of root system, drying of spathe and necrosis of spikelets are observed in certain cases (Menon and Pandalai, 1958). The nuts from diseased palms have thinner husk and fibers are weak and less firm. The kernel is thinner and remains soft and flexible. The oil content is reduced and oil loses its quality as well.

TRANSMISSION

The root wilt phytoplasma is transmitted by insect vectors, *Stephanitis typica*-a lace-bug (Shanta *et al.*, 1964; Mathen *et al.*, 1990) but planthopper, *Proutista moesta* and a leafhopper, *Sophonia greeni* are also reported to be the vector of the disease (Rajan and Mathen, 1984, 1985). There is no evidence of seed transmission. Symptoms on vector inoculated plants develop 9-24 months after inoculation. Experimental transmission of the disease to phytoplasmal indicator host, periwinkle (*Catharanthus roseus*), was successful through the dodder, laurel, *Cassytha filifonnis* (Sasikala *et al.*, 1989).

THE PATHOGEN

A phytoplasma is known to be associated with Kerala wilt disease. In electron microscopy pleomorphic phytoplasma bodies measuring 250-400 nm were observed (Solomon *et al.*, 1983; Parthasarathi, 1974). These organisms are limited to the sieve tubes of the phloem tissues. Phylogenetic analysis revealed that 16S rRNA gene of coconut root (wilt) phytoplasma is closely related to sugarcane white leaf phytoplasma and member of 16Sr XI group (Manimekalai *et al.*, 2009, 2010)

DIAGNOSIS AND MANAGEMENT

The disease can be identified with specific diagnostic symptoms on diseased palm (Radha and Lal, 1972). EM and ELISA systems have been standardized for pathogen diagnosis (Chandra Mohan, 2009; Sasikala *et al.*, 2005). PCR and nested PCR using phytoplasma specific primers have been designed to amplify the 16S rRNA gene. Expected products of 480 bp and 1342 bp were amplified in PCR and nested PCR respectively (Manimekalai *et al.*, 2009, 2010). For disease management, removal of infected plants is recommended in mildly affected areas. Movement of planting material from infested areas to new areas should be avoided. Malayan green dwarf (Kalparaksha) has been released as a resistant variety with high field tolerance for planting in heavily infested areas (Chandra Mohan 2009). Tetracycline therapy gives temporary remission of disease symptoms (Solomon *et al.*, 1989).

LETHAL YELLOWING

Earliest record of lethal yellowing disease was in 1834 in Cayman Islands. Later it was recorded from Belize, Bahamas (1946), Cuba (1920s), Dominican Republic (1915), Haiti (1920s), Honduras, Jamaica (1955). Lethal yellowing disease is now found in many countries of Africa, North America, Mexico - Yucatan Peninsula, South America, Guyana (unconfirmed). Lethal yellowing destroyed about 300,000 coconut palms in Miami (USA) in less than five years. By 1979, an estimated 4 million coconut palms had been killed by palm lethal yellowing disease in Jamaica. The host list of palm species attacked by lethal yellowing is large and includes *Phoenix dactylifera* L.; *P. canariensis* Hort., and *P. reclinata* Jacq. (Cronje *et al.*, 2000). The disease is known by several names such as Awka disease (Nigeria), Cape St Paul wilt (Ghana), Kaincope disease (Togo), Kribi disease (Cameroon), unknown disease (Jamaica) (English) Jaunisse létale des palmiers, pourriture du bourgeon terminal (Haiti) (French) Amarilles letal de las palmeras, pudrición del cogollo (Cuba) (Spanish) (Howard, 1983). The term leathal yellowing was first used to a disease of coconut palm in Jamaica (Nutnam and Robert, 1955).

SYMPTOMS

Leathal yellowing (LY) refers to the progressive discolouration of fronds. Developing fruits of the coconut start dropping from the palm followed by the formation of new inflorescences which rapidly become necrotic. 'Shelling' or premature dropping of most coconut is the early diagnostic symptom of the disease which is followed by a rapid and generalised yellowing, leading to the death of the palm. In date palm, the fronds become desiccated and grey-brown instead of becoming yellow. Presence of yellow leaf in mid-crown is highly indicative of LY. Bright yellow fronds are characteristic of the disease in 'Jamaica tall' and 'West Africa tall' but in other varieties fronds may turn bronze or brown. A soft rot of the growing point occurs, converting the meristematic area into a putrid, slimy mass. The crown topples from the palm, leaving a naked trunk like a 'telephone pole'. Necrosis of younger tissues and a soft rot within the crown of affected palms are the characteristic symptoms of the disease (Tsai and Maramorosch, 1987). Infected trees may die within three to six months after infection.

TRANSMISSION

A planthopper, *Haplaxius crudus* (former name *Myndus crudus*) was suggested the vector of lethal yellowing and was later confirmed in Florida (Tsai and Maramorosch, 1987; Howard, 1983).

HOST RANGE

The main host of lethal yellowing is coconut (*Cocos nucifera*) but the disease has also been found on dates (*Phoenix dactylifera*), *P. canariensis* and *Pandanus utilis*. About 30 species of palms are susceptible to LY infection. The following genera are susceptible: *Allagoptera, Arenga, Arikuryroba, Borassus, Caryota, Chrysalidocarpus, Cocos, Corypha, Dictyosperma, Gaussia, Hyophorbe, Latania, Livistona, Mascarena, Nannorrhops, Phoenix, Pritchardia, Trachycarpus, Veitchia* (Harris, 1977; Howard *et al.*, 1979).

THE PATHOGEN

The causal agent of lethal yellowing disease is a Phytoplasma (Beakbane *et al.*, 1972; Al-Awadhi *et al.*, 2002; Parthasarthy, 1974). Typical phytoplasma bodies were found in sieve tubes of infected plants. They were ovoid, elongated and filamentous in shape and were bounded by a triple-layered structure comprising two electron-dense layers with a transparent layer in between (Plavsic-Banjac *et al.*, 1972). Lethal yellowing disease has been reported to be associated with phytoplasma subgroup 16SrIV (Myrie *et al.*, 2012).

DIAGNOSIS AND MANAGEMENT

Lethal yellowing phytoplasma can be detected by electron microscopy and fluorescent staining (DAPI). Control is best achieved by planting of resistant cultivars like Panama Tall and Malayan Dwarf. Control of the vector *H. crudus* with insecticides can avoid the spread of the disease. Oxytetracycline-HCl treatment of infected plants could cause remission of disease symptoms (McCoy, 1972). Management of the grass populations in coconut plantations has also been suggested as a means of control, since some grass species are relatively poor hosts for the nymphs of the vector (Howard, 1990).

TATIPAKA DISEASE

Tatipaka disease first appeared in Tatipaka village of Razole Taluk in the east Godavary District of Andhra Pradesh, India following a cyclone in 1949 (Rethinam *et al.*, 1989). Palms in the age group of 25 to 60 years are more susceptible. The disease is now found in east and west Godavary, Srikakulam and Nellore districts of Andhra Pradesh, India. Development of an abnormally large crown with dark green inner leaves and higher yield is the precursor of disease incidence. Subsequently the crown becomes smaller in size producing progressively shorter leaves. The stem begins to taper. The leaves give a fascinated appearance due to improper unfolding of leaflets. The affected tree produces smaller bunches with atrophied barren nuts (Ramapandu and Rajamannar, 1983). The disease is caused by a phytoplasma but no vector or other means of spread are known. The disease can be controlled as mentioned under root wilt.

COCONUT BLAST

Coconut blast disease has been reported from Africa and Indonesia. Blast is a disease of nursery plants. Based on its transmission by a cicadellid, *Recilia mica* Kramer (Julia, 1979), earlier hypothesis of association of fungus or rickettsia-like organisms (RLO) were ruled out. Blast disease affects both oil and coconut palms. Tissues at the base of spear leaf become black. Necrosis then affect the meristem giving a soft rot like smelling. In transmission tests by vector, *R. mica* to *Catharanthus roseus* plants, symptoms of wilting, progressive dwarfing of flowers, crinkling and yellowing of the leaves, and eventual complete inhibition of growth and flower production were obtained. Examination of affected periwinkle plants under the electron microscope revealed the presence of phytoplasma in the phloem. Experiments with treatment of diseased plants by means of tetracycline suggest that blast is a disease of Phytoplasma origin (Dollet, 1980).

F-2. VIRUS DISEASES OF DATE PALM

Although there is no authentic report on virus diseases of date palm. A report appeared in May, 2009 that "Date-palm orchards come under viral attack" from Khairpur. Date growers of different areas of Khairpur were worried due to a viral attack on date-palm orchards. The disease has affected roots and trunks of date palms and production of dates. For the last three years, this virus attacks date-palm orchards during the season as a result of which entire orchards get destroyed.

Brittle leaf disease, also called "Maladie des Feuilles Cassantes" in French, was first observed in Nefta, Tozeur and Degache date plantations (Tunisia) and in Adrar, M'zab and Biskra (Algeria). This is a new lethal disease and has been reported from Tunisia since 1960s, but is now taking alarming proportions. 36,000 trees are now affected by this disease of unknown aetiology and many other trees have already been killed and removed. Symptoms resemble those of manganese deficiency, but sprays or injections of manganese did not solve the problem (at least there is a delay in symptom expression), and mineral soil analysis could not reveal differences between diseased and healthy plots. Patterns of diseased trees observed in the field suggest a biotic origin, as affected trees seem to cluster into foci. A small RNA band in diseased palms was observed. This RNA was only detectable at a low level in old leaves collected from symptomatic plants. Properties (Size and mobility in S-PAGE analysis) of this RNA were similar to those of viroids (Saadi *et al.*, 2006).

Quarantine measures seem to be the only means of limiting the spread of the disease. Since manganese is deficient in unhealthy palms, this nutrient could be brought to these palms either by spraying or by injection.

PHYTOPLASMA DISEASES OF DATE PALM

AL WIJAM DISEASE

This disease has been reported from Al Hassa (Saudi Arabia). In Arabic, Al Wijam means poor or unfruitful. The disease is characterised by a retardation in terminal bud growth and rosetting symptoms. Newly formed leaves are reduced in size and marked by a faint narrow, yellow longitudinal line on the midribs. Leaves become chloritic and their life span is reduced. Death of leaves starts from the distal end and extends towards the base. Diseased spathes split open before their complete emergence and are reduced in size. The number and size of the bunches produced are also reduced year after year till the diseased palm fails to produce and dies (Elarosi *et al.*, 1983). Positive amplification bands were obtained from DNA templates extracted from diseased tissue of date palm using the Polymerase Chain Reaction (PCR). These DNA tests offer basic support to the hypothesis that the cause of Al-Wijam disease is a Phytoplasma.

YELLOW STREAK DISEASE OF DATE PALM

Symptoms of yellows disease were observed in Egypt for the first time on a few date palm trees. A wide range of symptoms developed on date palm. The first main symptom is pale or bright yellow streak on the midrib of the leaf. This streak sometimes appears as longitudinal line on the midrib or as a lesion varies in size from small spots to patches. Lesions tend to be more extensive on the pinna of leaves. Sometimes, the lesions becomes more prominent as yellowish lines running parallel to veins and complete bright yellow discoloration of leaves can be observed in severe infection. Also, yellowing streak could be seen on bunches of date palm. Diseased spathes split open before their complete emergence and reduced in size and numbers year after year. Fruits on diseased tree do not attain the normal size.

The disease agent is a phytoplasma. It can be detected by fluorescence microscopy. Polymerase chain reaction (PCR) assays using primer pairs designed for identification of phytoplasma using universal (R16mF2/R16mR1) and specific nested primer (R16F2n/R16R2) provide a sensitive means of detecting this pathogen. Nested PCR assays using the specific primer confirmed the presence of phytoplasma (Ammar *et al.*, 2005). This is the first record of a phytoplasma associated with streak yellows on date palm in Egypt.

FATAL YELLOWING OF OIL PALM

Fatal yellowing is a serious disease of oil palm (*Elais guineensis* Jacq.) in central and South America. Fatal yellowing is also known by several names such as lethal spear rot, Pudricion del cagollo and Amar elecimento fatal. The symptoms of the disease are: patches of chlorotic or necrotic tissue of the spear leaf affected by a wet rot and dying back of palms

(Beuther *et al.*, 1992). Viroid like nucleic acid was associated with fatal yellowing disease (Singh *et al.*, 1988).

REFERENCES

Adams, M.J., Antoniw, J.F., Bar-Joseph, M., Brunt, A.A., Candresse, T., Foster, G.D., Martelli, G.P., Milne, R.G. and Fauquet, C.M. 2004. The new plant virus family *Flexiviridae* and assessment of molecular criteria for species demarcation. *Arch. Virol.* 149: 1045–1060.

Adams, M.J., Accotto, G.P., Agranovsky, A.A., Bar-Joseph, M. and Boscia, M. 2005. Family *Flexiviridae*. *In*: Fauquet, C.M., Mayo, M.A., Maniloff, J., Desselberger U. and Ball, L.A. (*Eds*) Virus Taxonomy, Eighth Report of the International Committee on Taxonomy of Viruses, Elsevier Academic Press, London. pp. 1089–1124.

Agarwal, S., Reddy, M.K. and Jain, R.K. 2009. Production of polyclonal antibodies. Using recombinant coat protein of papaya ringspot virus and their use in immunodiagnosis. *J. plant Biochem. and Biotech.* 18: 109–111.

Aglave, B., Krishnareddy, A., Patil, M. and Andhale, M.S. 2007. Molecular identification of a virus causing banana chlorosis disease from Marathwada region. *Intl. J. Biotechnol. Biochem.*, 3: 13–23.

Ahlawat, Y.S. 1975. Leathery leaf – a new citrus disease isolated from greening affected mandarin trees. *Indian Phytopath.*, 28: 146.

Ahlawat, Y.S.1987. Association of mycoplasma-like organisms with rubbery wood disease of citrus. *In*: 3rd Reg. Workshop on Plant Mycoplasma, INSA, New Delhi, p. 12.

Ahlawat, Y. S. 1989. Psorosis 'A' disease of citrus in India, *Indian Phytopath,* 42: 21–25.

Ahlawat, Y.S. 1989. Association of mycoplasma-like organisms with rubbery wood disease of citrus. *In*: Plant Diseases caused by Fastidious prokaryotes (*Eds.*) Raychaudhuri, S.P. and Varma, A. Today and Tomorrow Printers and Publishers, New Delhi. pp. 105–108.

Ahlawat, Y.S. 1993. Newer techniques for quick detection of greening disease of citrus. *Indian Phytopath,* 46: 276.

Ahlawat, Y.S. 1997a. Viruses, greening bacterium and viroids associated with citrus (*Citrus* spp.). *Indian J. Agric. Sci.,* 67: 51–57.

Ahlawat, Y.S. 1997b. Status of virus and virus-like pathogens infecting citrus in India and future strategies for citrus improvement. *Indian Phytopath.,* 50: 192–193.

Ahlawat, Y.S. 2007. Citrus decline: Problems and Preventions. *Indian Phytopath.* 60: 1–12.

Ahlawat, Y.S. 2010. Diagnosis of plant viruses and allied pathogens. Studium Press (India) Pvt. Ltd., New Delhi.

Ahlawat, Y.S. 2012. Virus diseases of citrus and management. Studium Press (India) Pvt. Ltd., New Delhi.

Ahlawat, Y.S. and Sardar, K.K. 1976. Distribution of greening and tristeza diseases of citrus in various blocks of Darjeeling district. *Sci. and Cult.* 42: 175–177.

Ahlawat, Y.S. and Sardar, K.K. 1976. Lemon crinkly leaf virus disease in India. *Indian J. Hort.,* 33: 168–171.

Ahlawat, Y.S. and Chenulu, V.V. 1985a. Rubberywood - a hetherto unrecorded disease of citrus. *Curr. Sci.,* 54: 580–581.

Ahlawat, Y. S. and Chenulu, V.V. 1985b. Rubberywood and multiple sprouting-two newly discovered diseases of citrus. *Indian Phytopath.*, 38: 625.

Ahlawat, Y.S. and Chenulu, V.V. (*ed.*) 1994. Virus and Mycoplasma diseases of fruit crops in India. ICAR, New Delhi, p. 73.

Ahlawat,Y.S. and Raychaudhary, S.P. 1988. Status of citrus tristeza and dieback disease in India and their detection. *In*: Proc. 6th Internatl. Citrus Congress, (Eds.) R. Goren and K. Mendel, Tel Aviv, Isarel pp. 871–879.

Ahlawat, Y.S. and Chakraborty, N.K. 1990. Etiology of psorosis 'A' disease of citrus. *Indian Phytopath*, 42: 239.

Ahlawat, Y.S. and Chakraborty, N.K. 1993. Yellow vein banding disease of citrus in India. *Indian Phytopath*, 46: 274.

Ahlawat, Y.S. and Pant, R.P. 2002. Status of virus and virus-like diseases of citrus and production of diagnostics in India. *In*: 15th Conf. IOCV, Nov. 11th-16th 2001, Paphos, Cyprus. p. 108.

Ahlawat, Y.S. and Pant, R.P. 2003. Major virus and virus-like diseases of citrus in India, their diagnosis and management. *Ann. Rev. Plant Path.* 2: 447–474

Ahlawat, Y.S. and Pant, R.P. 2008. Indian citrus ringspot virus. *In*: Characterization, Diagnosis and management of Plant Viruses (*Eds.*) Rao, G.P. Myrta, Arben and Kai Shu Ling, Studium press Houston, USA pp. 97–107.

Ahlawat, Y.S. and Pant, R.P. 2010. *In*: IPM system in Agriculture. Virus and virus-like diseases of plants Vol. 10. (*Eds*) Upadhyay, R.K., Ahlawat, Y.S., Dubey, O.P., Baranwal, V.K. and Srivastava, K.P. Aditya Books Pvt. Ltd. New Delhi. p. 381. pp. 71–94.

Ahlawat, Y. S., Nariani, T.K. and Sardar, K.K. 1979. Leathery leaf: a new virus disease of citrus. *Indian Phytopath.* 32: 198–201.

Ahlawat, Y.S., Chenulu, V.V., Gumpf, D.J. and Pandey, P.K. 1986. Some newly discovered virus and mycoplasma diseases of citrus in India. In Abs. 10th Conf: IOCV, IOCV, Riverside, CA.

Ahlawat, Y. S., Payak, M.M. and Thirumalachar, M.J. 1989. Control of citrus greening disease in India. *Curr: Sci.* 58: 921–922.

Ahlawat, Y.S., Chenulu, V.V., Chakraborty, N.K. and Viswanath, S.M. 1984. Occurrence of impietratura disease of citrus in India. *Curr. Sci.*, 53: 384–385.

Ahlawat, Y.S., Chenulu, V.V. ,Viswanath, S.M., Pandey, P.K. and Bhagabati, K.N. 1985. Mosaic disease of citrus in India. *Curr. Sci.*, 54: 873–874.

Ahlawat, Y.S., Chakraborty, N.K., Varma, A. and Raychaudhuri, S.P. 1992. Diagnosis of diseases caused by viruses and fastidious Prokaryotes in citrus in India. *In*: Proc. Intl. Soc. Citriculture, VII, Intl. Citrus Cong. pp. 749–751.

Ahlawat, Y.S., Chakraborty, N.K., Jagdishchandra, K., Srivastava, M. and Varma, A. 1993. Association of a Rhabdovirus with yellow vein mosaic, a new disease of citrus. pp. 455–457. *In*: Proc. 12th Conf. IOCV, IOCV, Riverside.

Ahlawat, Y.S., Varma, A., Chakraborty, N.K. and Srivastava, K.P. 1995. Biotypes and vectoral capacity of *Diaphorina citri* in transmitting citrus greening BLO. *In*: Proc. 2nd Agric. Sci. Cong. (*Eds.*) M.S. Swaminathan, N.S. Randhawa, T.N. Ananthakrishanan and P. Narain. Nat. Acad. Agric.Sci., IARI, New Delhi-110012 pp. 222–232.

Ahlawat, Y.S., Byadgi, A.S., Varma, A. and Chakraborty, N.K. 1995. Detection of ringspot virus and greening BLO in citrus and their management, *In*: Detection of plant pathogens and their management (*Eds.*) Verma, J.P., Varma A. and Kumar, D. Angkor Publishers, New Delhi.

Ahlawat, Y.S., Pant, R.P., Lockhart, B.E.L., Srivastava, M., Chakraborty, N.K. and Varma, A. 1996. Association of a badnavirus with citrus mosaic disease in India. *Plant Disease,* 80: 590–592.

Ahlawat, Y.S., Varma, A., Pant, R.P., Shukla, A. and Lockhart, B.E.L. 1996. Partial characterisation of a badnavirus associated with citrus yellow mosaic disease in India. *In*: Proc. 13th Conf. IOCV, IOCV, Riverside, CA. pp. 208–217.

Ahlawat, Y.S., Byadgi, A.S., Pant, R.P. and Varma, A. 1997. Citrus ringspot capillovirus and Citrus mosaic badnavirus: Two emerging pathogens infecting citrus in India. *In*: Management of threatening Plant diseases of National importance (*Eds.*). V.P. Agnihotri, A.K. Sarbhoy and D.V. Singh, Malhotra Pub. House, New Delhi. pp. 211–225.

Ahlawat, Y.S., Reddy, B.V.B., Pant, R.P. and Varma, A. 1997. Diagnosis of virus and virus-like diseases of citrus in India. *In*: Proc. Reg. Expert Consult. Applica. Biotech. Pl. Pest Management. RAP publication: 1997/40, 25-28 Feb. 1997, IARI, New Delhi, India. pp. 298–306.

Ahlawat, Y.S., Reddy, D.R., Moses, J.E., Pant, R.P., and Singh, S.P. 1999. Distribution of yellow mosaic disease of citrus in South India. *Citrograph Magazine,* p. 7.

Ahlawat, Y.S., Al-Shami, A.A.A., Pant, R.P. and Prasad, Ramji. 2001. Association of a filamentous virus causing yellow vein symptoms in citrus in India. In: 15th Conf. IOCV, Paphos, Cyprus, p. 86.

Ahlawat, Y.S., Baranwal, V.K. Thinlay, Doe Doe, and Majumder, S. 2003. First report of citrus greening disease and associated bacterium *Candidatus* liberobacter asiaticus from Bhutan. *Plant Dis.,* 87: 448.

Akhtar M.A. and Ahmad, I. 1999. Incidence of citrus greening disease in Pakistan. *Pakistan J. Phytopath.,* 11: 1–5.

Al-Awadhi, H.A., Hanif, Asma, Suleman, P. and Montaser, M. 2002. Molecular and microscopical detection of phytoplasma associated with yellowing disease of date palm *Phoenix dactylifera* L. in Kuwait. *Kuwait J. Sci. Eng.,* 29: 87–109.

Albrigo, L.G., Syvertsen, J.P. and Young, R.H. 1986. Stress symptom of citrus trees in successive stages of decline due to blight. *J. American Soc. Hort. Sci.* 111: 465–470.

Albrigo, L.G., derrick, K.S., Timmer, L.W., Tucker, D.P.H. and Graham, J.H. 1992. Inability to transmit citrus blight by limb grafts. *In*: Proc. 12th Conf. IOCV, IOCV, Riverside, CA. pp. 131–138.

Allen, R.N. and Dale, J.L. 1981. Application of rapid biochemical methods for detecting avocado sunblotch disease. *Ann. Appl. Biol.,* 98: 451.

Allen, R.N., Palukaitis, P. and Symons, R.H. 1981. Purified avocado sunblotch viroid causes disease in avocado seedlings. *Aust. Plant Path.* 10: 31–32.

Almeyda-Leon, I.H., Iracheta-cardenas, M.M., Jasso-argumedo, J., Kurti-Diaz, S.A., Ruiz-Beltran, P. and Roch-Pena, M.A. 2002. Re-examination of citrus viroids in Tahiti lime in Mexico. *Revista Mexicana de Fytopatologia* 20: 152–160.

Alper, M., Bar-Joseph, M., Salomon, R. and Loebenstein, G. 1978. Some characteristic of an avocado strain of tobacco mosaic virus. *Phytoparasitica,* 6: 15–20.

Alshami, A.A.A., Ahlawat, Y.S. and Pant, R.P. 2003. A hitherto unreported yellow vein clearing disease of citrus in India and its viral etiology. *Indian Phytopath.* 56: 422–427.

Ammer, M.I., Amer, M.A. and Rashid, M.F. 2005. Detection of Phytoplasma associated with yellow streak disease (*Phoenix dactylifera*) in Egypt. *Egyptian J. Virol,* 2: 74–86.

Ananthakrishnan, G. Venkataprasanna, T., Roy A. and Brlansky, R.H. 2010. Characterization of the mixture of genotypes of a Citrus tristeza virus isolate by reverse transcription quantitative real-time PCR. *J. Virol. Meth.,* 164: 75–82.

Anonymus. 1992. International network for the improvement of banana and plantain. *In*: Annual report 1992. International network for the improvement of banana and plantain, Montpellier, France, pp. 29–33.

Asana, R.D. 1958. The citrus dieback problem in relation to cultivation of citrus fruits in India. *Indian J. Hort.*, 15: 283–286.

Ashfaq, M., Iftikhar,Y., Mughal, S.Y. and Khan, M.A. 2005. Occurrence, distribution and transmission of yellow vein clearing virus (YVCV) of lemon in Faisalabad. *Pak. J. Phytopath.*, 17: 101–104.

Ashkenazi, S. and Oren, Y. 1988. The use of citrus exocortis virus (CEV) for tree size control in Israel – a practical aspect. *Proc. Interl. Soc. Citriculture*, 2: 917–919.

Aubert, B. and Quilici, S. 1984. Biological control of the African and asian citrus psyllids (Homoptera: Psylloidea), through eulophid and Encyrtid parasites (Hymenoptera: Chalcidoidea) in Reunion Island. pp. 100–108.

Aubert, B., Sabine, A. and Picard, P. 1984. Epidemiology of the greening disease in Reunion Island before and after the biological control of the African and Asian citrus psyllas. *Proc. Inernl. Soc. Citriculture* 1: 440–442.

Avgelis, A. 2008. Cucumber mosaic virus on banana in crete. *J. Phytopath.*, 120: 20–24.

Azeri, T. 1973. First report of satsuma dwarf virus disease on satsuma mandarins in Turkey. *Plant Dis. Reptr.*, 57: 149–153.

Bagherian, S.A.A. and Izadpanah, K. 2010. Two novel variants of hop stunt viroid associated with yellow corky vein disease of sweet orange and split bark disorder of sweet lime. 21st International Conference on Virus and other GraftTransmissible Diseases of Fruit Crops. *Julius-Kühn-Archiv*, 427: 105–113.

Bahia Bennani, Celso Mendes, Mustapha Zemzami, Houssine Azeddoug and Gustavo Nolasco. 2004. Citrus variegation virus: Molecular variability of a portion of the RNA 3 containing the coat protein gene and design of primers for RT-PCR detection. *European J. Plant Pathol.*, 108: 155–162.

Baksh, N., Lee, R.F. and Garnsey, S.M. 1984. Detection of citrus exocortis viroid by polyacrylamide gel electrophoresis. *In*: Proc. 9th Conf. IOCV. IOCV, Riverside, CA. pp. 343–352.

Balaraman, K. and Ramakrishnan, K. 1977. Studies on strains and strain interaction of citrus tristeza virus. U.S.A. Tech. Univ. Agric. Sci., Hebbal. Bangalore. Series No. 19: 62.

Balaraman, K. and Ramakrishnan, K. 1980. Strain variation and cross protection in citrus tristeza virus on acid lime. *In*: Proc. 8th Conf: IOCV, IOCV, Riverside, CA pp. 60–68.

Bani Hashemian, S.M., Barbosa, C.J., Pina, J.A. and Duran-Vila, N. 2010. Is desert lime (*Eremocitrus glauca*) Resistant to Viroid Infection or only a Poor Viroid Host? Proc. 17th Conf. IOCV, IOCV, Riverside, CA.

Baranwal, V.K., Majumder, S., Ahlawat, Y.S. and Singh, R.P. 2003. Sodium Sulphite yields improved DNA of higher stability for PCR detection of Citrus Yellow Mosaic Virus from citrus leaves in *J. Virol. Method*, 112: 153–156.

Baranwal, V.K., Mazumder, S., Jitendra Singh, Suryanarayana, V., Ghosh, D.K. and Ahlawat, Y.S. 2004. PCR detection of *Candidatus* Liberibacter asiaticus, the agent of Huanglongbing disease in citrus. *Indian Phytopath*, 57: 164–168.

Baranwal, V.K., Mazumder S., Ahlawat, Y.S. and Singh, R.P. 2005. A novel approach for simultaneous detection of citrus yellow mosaic virus and citrus greening bacterium by multiplex polymerase chain reaction. *Indian J. Biochem.*, 4: 528–533.

Baranwal, V.K. and Arya, M. 2009. Novel approaches of nucleic acid template preparation for the detection of RNA and DNA viruses and greening bacterium in citrus by RT-PCR and PCR. *In*: 5th Internl. Conf. on Plant Path. in Globalized era. Nov. 10-13, 2009, New Delhi, India. p. 47.

Barbarossa, L. and Vovlas, C. 2010. Molecular characterization of Citrus tristeza virus Isolates from Epirus (Greece). *In*: Proc. 17th Conf. IOCV, IOCV, Riverside, CA.

Bar-Joseph, M. 1976. Some effects of temperature on symptom appearance and therapy of citrus impietratura disease. *In*: Proc. 7th Conf. IOCV, IOCV, Riverside, CA. pp. 105–108.

Bar-Joseph, M. 1993. Citrus viroids and citrus dwarfing in Israel. *Acta Hort.*, 349: 271–276.

Bar-Joseph, M. and Loebenstein, G. 1970. Leaf flecking on indicator seedlings with citrus in Israel. A possible indexing method. *Plant Dis. Rep.*, 54: 643–646.

Bar-Joseph, M. and Ben-Shalom, J. 1993. A combined indexing and budwood propagation method for citrus. *Phytoparasitica,* 21: 269–272.

Bar Joseph, M., Loebenstein, G. and Cohen, J. 1970. Partial purification of virus-like particles associated with the citrus tristeza disease. *Phytopathology,* 60: 75–78.

Bar Joseph, M., Garnsey, S.M. and Gonsalves, D. 1979. The closterovirus: A distinct group of elongated plant viruses. *Adv. Virus Res.*, 25: 93–168.

Bar-Joseph, M. and Lee, R.F. 1989. Citrus tristeza virus; Description of plant viruses No. 353 (No. 33 revised). Common. Mycol. Inst./Assoc. *Appl. Biol.* Kew, Surrey, England.

Barkley, Particia (Broadbent) and Hutton, R.J. 2007. Dwarfing citrus trees by viroids (IOCV slide collection at www.ivia.es/iocv)

Barros, C.C.P., de Jesus Junior, W.C., Teixeira, D.C., Gomes, C.C., Dutra, L.S. and Nagata, T. 2005. Molecular characterization of tymo-like virus isolated from orange trees with sudden death symptoms. Proc. 16th Conf.IOCV, IOCV, Riverside, CA. (Abs).

Bastianel, M., Freitas-Astua, J., Kitajima, E.W. and Machado, M.A. 2006. The citrus leprosis pathosystem. *Summa Phytopathologica,* 32: 211–220.

Bateson M.F. and Dale J.L. 1995. Banana bract mosaic virus: Characterization using potyvirus specific degenerate PCR primers. *Arch. Virol.* 140: 515–527.

Batugal, P., Ramanatha Rao, V. and Oliver, J. (*eds.*) 2005. Coconut genetic resources. international plant genetic resources institute- regional office for Asia, the Pacific and Oceania (IPGRI-APO), Serdang, Selangor DE, Malaysia

Baulcombe, D.C. 1996. Mechanisms of pathogen-derived resistance to viruses in transgenic plants. *Plant Cell,* 8: 1833–1844.

Bausher, M.G. and Sweeney, M.J. 1991. Field detection of citrus blight using immunological techniques. *Plant Dis.*, 74: 168–170.

Beakbane, A.B., Slater, C.W., and Posnette, A.F. 1972. Mycoplasma in the phloem of coconut, *Cocos nucifera* L. With lethal yellowing disease. *J. Hort. Sci.*, 47: 265.

Benton, R.J., Bowman, F.T., Fraser, L. and Kebby, R.G. 1950. Stunting and scaly butt of citrus associated with *Poncirus trifoliata* rootstock. *N.S.W. dept. agric. Sci. Bull.,* 70. 20.

Bernad, L., and Duran-Vila, N. 2006. A novel RT-PCR approach for detection and characterization of citrus viroids. *Mol. Cell. Probes* 20: 105–113.

Berreta, M.J.G., Rossetti, V., Teixeira, A.R.R. and Sempionato, O. 1991. Positive diagnostic tests for declinio on plants root-graft inoculated in Brazil. *In*: Proc. 11th Conf. IOCV. IOCV, Riverside, CA. pp. 256–260.

Beretta, M.J.G., Lee, R.F., Derrick, K.S., Davis, C.L., and Barthe, G.A. 1994. Culture and serology of a *Xylella fastidiosa* associated with citrus variegated chlorosis in Brazil. *In*: Proc. 7th Citrus Cong, Italy. pp. 830–31.

Beretta, M.J., Derrick, K.S., Lee, R.F., Barthe, G. and Rosetti, V. 1993. Citrus declinio in Brazil: Rate of spread and serological comparison with other declines. *In*: Proc. 12th Conf. IOCV, IOCV, Riverside, CA. pp. 113–115.

Beretta, M.J.G., Lee, R.F., Barthe, G.A., Thome Neto, J., Derrick, K.S and Davis, C.L.1993. Citrus variegated chlorosis: Detection of *Xylella fastidiosa* in symptomless trees. In: Proc. 12th Conf. IOCV, IOCV, Riverside, CA. pp. 306–311.

Beretta, M.J.G., Rodas, V., Garcia Jr. A., and derrick, K.S. 1996. Control of citrus variegated Chorosis by pruning. *In*: 13th Conf. IOCV, IOCV, Riverside, CA. pp. 378–379.

Beretta, M. J. G., Barthe, G.A. Ceccardi, T.L., Lee, R.F. and Derrick, K.S. 1997. A survey for strains of *Xylella fastidiosa* in citrus affected by citrus variegated chlorosis and citrus blight in Brazil. *Plant Dis.* 81: 1196–1198.

Bernard, L., Moreno, P., Bove, J.M. and Duran-Vila, N. 2005. Viroids in gummy bark sources from the Sultanate of Oman. In: Proc. 16th Conf. IOCV, IOCV, Riverside, pp. 272–279.

Bertolini, E., Lopes, S.A., Miranda, M.P., Arilla, A., Ayres, A.J., Lopez, M.M., Bove, J. M. and Cambra, M. 2010. Direct plant tissue-print and insect squash-blot methods for large scale detection of *Candidatus* Liberibacter spp. by real-time PCR with new primers. *In*: Abs. 18th Conf. IOCV, IOCV, Riverside, CA p. 45.

Beuther, E., Wiese, U., Lukacs, N., Van Slobbe, W.G. and Riesner, D. 1992. Fatal yellowing of oil palms: Search for viroids and dsRNA. *In*: Plant diseases of viral, viroid, mycoplasma and uncertain etiology (*Ed*) Karl Maramorosch, Publisher Mohan Prilamani for Oxford and IBH Pub. Co. Pvt. Ltd. New Delhi pp. 19–43.

Bhagabati, K.N., Ahlawat, Y.S., Chakraborty, N.K. and Borthakur, B.C. 1989. Distribution of greening, tristeza and mosaic diseases of citrus in North-Eastern States of India. *Indian Phytopath.* 42: 552–555.

Bhagabati, N.K., Borthakur, B.C. and Ahlawat, Y.S. 1986. Occurrence of impietratura disease of citrus in North-Eastern India. *Indian Phytopath.* 39: 110–111.

Bhutani, V.P., Bakhshi, J.C. and Knorr, L.C. 1972. Biochemical changes in healthy and decline sweet orange trees associated with bud union crease. *In*: Proc. 5th Conf. IOCV, IOCV, Riverside, CA. pp. 229–233.

Bioletti, F.T., Gowrie, W. and Cillie, P.J. 1904. Citrus culture in Cape colony. Report of Commission of inquiry into the causes of the failure of citrus trees in Cape Colony. *Agr. J., Cape of Good Hope*, 21: 1–20.

Biswas, K.K., Manjunath, K.L., Lee, R.F. and Ahlawat, Y.S. 2004. Identification of diversified citrus tristeza virus population isolates from eastern Himalayan region of india by heteroduplex mobility assay and nucleotide sequence analysis. *In*; Nat. symp. Molecular diag. for management of viral Dis. Div. Pl. Path. IARI, New Delhi. p. 9.

Biswas, K.K. 2008. Molecular diagnosis of citrus tristeza virus in mandarin (*Citrus reticulata*) orchards of Darjeeling hills of West Bengal. *Indian J. Virol.* 19: 26–31.

Biswas, K.K., Sunita Kumari, Tarafdar, A., and Ramachandran, P. 2008. Biological and genetic diversity of citrus tristeza virus population from North-Eastern Himalayan region of India. *Indian J. Virol.*, 19: 59.

Biswas, K.K., Tarafdar, B,K,, Jayakumar, B.K. and Pun, K.B. 2009. Strategy for the production of citrus tristeza virus-free mandarin (*Citrus reticulata*) planting material in the Darjeeling hills of India. *Indian Phytopath*, 62: 376–380.

Bitancourt, A.A. 1955. Estudos sobre a leprose dos citrus. I-Distribuicao geografica e sintomatologia. Ii-transmissao natural as folhas. III. transmissao natural as frutas. IV-Experiencias de tretamento. *Arq. Inst. Biol.*, 22: 161–231.

Bitters, W.P. 1952. Exocortis on trifoliate. *Citrus Leaves* 32: 14–16.

Boccardo, G., Beaver, R.G., Randles, J.W., and Imperial, J.S. 1981. Tinangaja and bristletop, coconut diseases of uncertain etiology in Guam and their relationship to cadang-cadang disease of coconut in Philippines. *Phytopathology,* 71: 1104.

Borah, B.K., Anthony Johnson, A.M., Sai Gopal, D.V.R. and Dasgupta, I. 2009. Sequencing and computational analysis of complete genome sequences of Citrus yellow mosaic badna virus from acid lime and Pummelo. *Virus Genes,* 39: 137–140.

Bouhida, M. and Lockhart, B.E.L. 1990. Increase in importance of cucumber mosaic virus infection in greenhouse grown bananas in Morocco. *Phytopathology,* 80: 981.

Bouhida, M., Lockhart, B.E.L. and Olszewski, N.E. 1993. An analysis of the complete nucleotide sequence of a sugarcane bacilliform virus genome infectious to banana and rice. *J. Gen. Virol.,* 74: 15–22.

Bové, J.M. 1986. Stubborn and its natural transmission in the Mediterranean area and in the Near East. *FAO Plant Prot. Bull.,* 34: 15–23.

Bove, J.M. 1986. Witches' broom disease of lime. *FAO Plant Prot. Bull.,* 34: 217–218.

Bové, J.M. 1989. Virus and virus-like diseases of citrus in Pakistan. *FAO Report,* p. 56.

Bové, J.M. 1995. Virus and Virus-like diseases of citrus in the near east region. *In*: Virus and virus-like diseases of citrus in the Near East region. Chapter 20, The Sudan. FAO, Rome. pp. 303–318.

Bové, J.M. 1995. Oman-Witches' broom disease of lime. *In*: Virus and virus-like. Diseases of citrus in the Near East region, FAO, Rome. pp. 217–229.

Bove, J.M., Garnier, M., Mjeni, A.M. and Khayrallah. 1988. Witches' broom disease of small fruited acid lime trees in Oman: First MLO disease of citrus. *In*: Proc. 10th Conf. IOCV, IOCV, Riverside, CA. pp. 307–309.

Bové, J.M., Bonnet P., Garnier M., Aubert B. 1980. Penicillin and tetracycline treatments of greening disease-affected citrus plants in the glasshouse, and the bacterial nature of the Procaryote associated with greening. *In*: Proc. 8th Conf. IOCV, IOCV, Univ. Calif., Riverside, CA. pp. 91–102.

Bove, J.M., Zreik, L., Danet, J.L., Bonfils, J., Majeni, A.M.M. and Garnier, M. 1993. Witches' broom disease of lime trees: monoclonal antibody and DNA probes for the detection of the associated MLO and the identification of a possible vector. *In*: Proc. 12th Conf. IOCV, IOCV, Riverside, CA. pp. 342–348.

Bove, J.M., Garnier, M., Ahlawat, Y.S., Chakraborty, N.K. and Varma, A. 1993. Detection of Asian strains of the greening BLO by DNA-DNA hybridization in Indian orchard trees and Malaysian *Diaphorina citri* psyllids. *In*: Proc. 12th Conf. IOCV, IOCV, Riverside, CA. pp. 258–263.

Bove, J.M., Navarro, L., Bonnet, P., Zreik, L. and Garnier, M. 1996. Reaction of citrus cultivars to graft inoculation of *Phytoplasma aurantifolia*-infected lime shoots. *In*: Proc. 13th Conf. IOCV, IOCV, Riverside, CA. pp. 249–251.

Bove, J.M., Danet, J.L., Bananej, K., Hassanzadeh, N., Taghizadeh, M., Salehi, M. and Garnier, M. 2000. Witches' broom disease of lime (WBDL) in Iran. *In*: Proc. 14th Conf. IOCV, IOCV, Riverside, CA. pp. 207–212.

Braithwaite, K.S., Egeskov, N.M. and Smith, G.R. 1995. Detection of sugarcane bacilliform virus using polymerase chain reaction. *Plant Dis.,* 79: 792–796.

Breman, L.L. 1997. *Papaya mosaic potexvirus* in *Portulaca* spp. Plant Pathology circular, Gainesville No. 382.

Bridges, G.D., Youtset, C.O. and Nixon, R.R. 1965. Observations indicating psorosis transmission by seed of Carrizo citrange. *Proc. Fla. Sta. Hort. Soc.,* 78: 48–50.

Brlansky, R.H. and Howd, D.S. 2001. Purification of virus-like particles from blight-affected citrus trees. *In*: 15th Conf. IOCV, IOCV, Riverside, CA pp. 297–305.

Brlansky, R.H., Davis, C.L.Timmer, L.W., Howd, D.S. and Contreras, J. 1991. Xylem limited bacteria in citrus from Argentina with symptoms of citrus variegated chlorosis. *Phytopathology* 81: 1210.

Broadbent, P., Fraser, L.R. and McGechan, R. 1976. Australian citrus dieback. Proc. 7th Conf. IOCV, IOCV, Riverside, CA pp. 141–146.

Bruening, G. 1977. Comovirus group. CMIAAB Desc. Plant Viruses, No. 199.

Brunt, A.A., Crabtree, K., Dallwitz, M., Gibbs, A. and Watson, L. 1996. Viruses of plants. CAB International, Wallingford, UK. 1484.

Brunt, A. K. Crabtree, M. Dallwitz, A. Gibbs, and Watson, L. 1996. Viruses of plants. CABInternational, Wallingford, UK. 1484.

Butler, E.J. 1908. Report on coconut palm disease in Travancore. *Agric. Res. Instt. Pusa Bull.*, 9: 23.

Byadgi, A.S. and Y. S. Ahlawat. 1995. A new viral ringspot disease of citrus *(Citrus* spp.) in India. *Indian* J. *Agric. Sci.*, 65: 763–770.

Byadgi, A.S., Ahlawat, Y.S. Chakraborty, N.K., Varma, A., Srivastava, M. and Milne, R.G. 1993. Characterisation of a filamentous virus associated with citrus ringspot in. India. *In*: Proc. 12th Conf. IOCV, IOCV, Riverside, USA. pp. 155–162.

Byadgi, A.S., Ahlawat, Y.S., Chakraborty, N.K. and Varma, A. 1994. Management of citrus ringspot disease. In*dian Phytopath,* 47: 326.

CABI. 2010. Citrus tristeza virus. *[Distribution map].* Distribution maps of plant diseases, April, Map 289 (Edition 7).

Calavan, E.C. 1961. Ferment gum disease (Rio Grande gummosis) of grapefruit. *Calif. Citrogr.* 46: 231–232.

Calavan, E.C. 1969. Investigations of stubborn disease in California: Indexing, effects on growth and production, and evidence for virus strains. *Proc. First Internl. Citrus Symp.*, 3: 1403–1412.

Calavan, E.C. 1974. Registration and certification of citrus propagative material in California. *In:* Jackson, L.K., Krezdorn, A.H. and Soule, J. (*Eds.*). Proc. 1st inernl. Citrus short course – citrus rootstocks. Fla Coop. Exr. Serv. Inst. Food and Agric Sci.., Univ. Florida, Gainesville, Fla. pp. 137–145.

Calavan, E.C. and Christiansen, D.W. 1966. Effects of stubborn disease on various varieties of citrus trees. *Israel J. Bot.*, 15: 121–132.

Calavan, E.C. and Christiansen, D.W. 1985. Variability of cachexia reaction among varieties of rootstocks within clonal propagations of citrus. *In*: Proc. 3rd Conf. IOCV. IOCV, Riverside, CA. pp. 76–85.

Calavan, E.C. and weathers, L.G. 1959. Transmission of growth retarding factor in Eureka lemon trees. *In*: Citrus virus diseases. J.M. Wallace ed. Univ. California, Div. Agric. Sci. Berkeley. pp. 167–177.

Calavan, E.C., Christiansen, D.W. and Roistacher. C.N. 1963. Symptoms associated with tatter-leaf virus infection of Troyer citrange rootstocks. *Plant Dis. Reptr.*, 47: 971–975.

Calavan, E.C., Roistacher, C.N. and Nauer, E.M. 1972. Thermotherapy of citrus for inactivation of certain viruses. *Plant Dis. Reptr.*, 56: 976–980.

Calavan, E.C., Harjung, M.K., Fudl-Allah, A.E.-S.A. and Bowyer, J.W. 1974. Natural Incidence of stubborn in field grown citrus seedlings and budlings. *In*: Proc. 6th Conf. IOCV, IOCV, Riverside, CA. pp. 16–19

Calegario, R.F., Kitajima, E.W., Salaroli, R.B., Rezende, J.A.M. and Locali-Fabris, E.C. 2010. Immunofluorescence studies on the distribution of citrus leprosis virus (CiLV-C) antigen in citrus leaf lesions. *In*: Abs. 18th Conf. IOCV held in Caminas/SP, p. 55.

Calvert, L.A., Lee, R.F. and Hiebert, E. 1988. Characterization of the RNA species of citrus variegation virus and complementary DNA clones. *In*: Proc. 10th Conf. IOCV, IOCV, Riverside, CA. pp. 327–333.

Calvez, C., Renard, J.L. and Marty, G. 1980. Tolerance of the hybrid coconut local × rennell to new hybrids disease. *Ol~agineux,* 35: 443–451.

Cambra, M., Gorris, M.T., Roman, M.P., Terrada, E., Garnsey, S.M., Camarasa, E., Olmos, A. and Colomer, M. 2000. Routine detection of citrus tristeza virus by direct immunoprinting ELISA method using specific monoclonal and recombinant antibodies. pp. 34–41.

Cambra, M., Camarasa, E., Gorris, M.T., Garnsey, S.M. and Carbonell, E. 1991. Comparison of different immunosorbent assays for citrus tristeza virus (CTV) using CTV specific monoclonal and Polyclonal antibodies. pp. 38–45.

Capoor, S.P. 1959. *In*: Bunchy top disease of banana I, Banana mosaic II. Proc. Pro. Sem. Dis. Hort. Plants, ICAR, New Delhi pp. 163–172.

Capoor, S.P. 1961. Kazilime: An indicator plant of citrus decline virus in India. *Indian Phytopath,* 14: 109–112.

Capoor, S.P. 1963. Decline of citrus trees in India. *Bull. Nat. Inst. Sci. India,* 24: 48–64.

Capoor, S.P. 1965. Presence of seedling yellows complex in the citrus of South India. *In*: Proc. 3rd Conf. IOCV, IOCV, Riverside, California. pp. 30–35.

Capoor, S.P. 1975. Role of the tristeza virus in citrus dieback complex. *Indian J. Hort.* 32: 1–6.

Capoor, S.P. and Rao, D.G. 1967. Tristeza virus infection of citrus in India. *In*: Proc. Intl. Symp. on subtropical and tropical Horticulture. Hort. Soc. of India, Bangalore. pp. 723–736.

Capoor, S.P. and Varma, P.M. 1958. A mosaic disease of papaya in Bombay. *Indian J. Agric. Sci.,* 28: 225–233.

Capoor, S.P. and Varma, P.M. 1968. Investigations on a mosaic disease of banana in the Deccan. *Indian Phytopath.,* 21: 135.

Capoor, S.P. and Thirumalachar, M.J. 1973. Cure of greening-affected citrus plants by Chemotherapeutic agents. *Plant Dis. Reptr.,* 57: 160–163.

Capoor, S.P., Rao, D.G. and Viswanath, S.M. 1967. *Diaphorina citri* Kuway., a vector of greening disease of citrus in India. *Indian J. Agric. Sci.,* 37: 572–576.

Capoor, S.P., Thirumalachar, M.J., Pandey, P.K. and Chakraborty, N.K. 1973. Control of the Greening Disease of Citrus by B.P.-101: A New chemotherapeutant. *In*: Proc. 6th IOCV, IOCV, Riverside, CA. pp. 50–52.

Capoor, S.P., Rao, D.G. and Viswanath, S.M. 1974. Greening disease of citrus in the Deccan Trap country and its relationship with the vector, *Diaphorina citri* Kuwayama. *In*: Proc. 6th Conf: IOCV, IOCV, Riverside, CA, pp. 43–49.

Carpenter, J.B., Calavan, E.C. and Christiansen, D.W. 1965. Occurrence of excessive seed abortion in citrus fruits affected with stubborn disease. *Plant Dis. Reptr.* 49: 668–672.

Caruso, A., Davino, M. and Terranova, G. 1993. Observations on impietratura disease symptoms in four citrus species. *In*: Proc. 12th Conf. IOCV, IOCV, Riverside, CA. pp. 379–382.

Catara, A., Azzaro, A., Davino, M. and Polizzi, G. 1993. Yellow vein clearing of lemon in Pakistan. *In:* Proc. 12th Conf. IOCV, IOCV, Riverside, CA. pp. 364–367.

Celino, C.S., Salibe, A.A. and Cortez, R.E. 1966. *Diaphorina citri* Kuway., the insect vector for the leaf mottle virus of citrus in the Philippines. Manila, Bureau of Plant Industry.

Cevik, B., Lee, R.F. and Niblett, C.L. 2006. Genetic transformation of Citrus with antisense and untranslatable RNA-dependent RNA polymerase genes of Citrus tristeza closterovirus. *Turkish J. Agric. and Forestry,* 30: 173–182.

Chagas, C.M., Rosetti, V. and Chiavegato, L.G. 1984a. Transmission of leprosis symptoms by grafting infected tissue. PP. 215-217. In: Proc. 9th Conf. IOCV, IOCV, Riverside, CA.

Chagas, C.M., Rosetti, V. and Chiavegato, L.G. 1984b. Effectiveness of the different life cycle stages of *Brevipalus phoenicis* Geijskes in leprosis transmission, *In*: Proc. 9th Conf. IOCV, IOCV, Riverside, CA. pp. 211–214.

Chagas, C.M., Rossetti, V. and Beretta, M.J.G. 1992. Electron microscopy studies of xylem limited bacterium in sweet orange affected with citrus variegated chlorosis disease in Brazil. *J. Phytopathology* 134: 306–312.

Chakraborty, N.K., Ahlawat, Y.S., Varma, A., Jagdish Chandra, K., Ramapandu, S. and Kapur, S.P. 1993. Serological reactivity in citrus tristeza virus strains in India. *In*: 12th Proc. IOCV, IOCV, Riverside, CA. pp. 108–112.

Chakraborty, N.K. and Ahlawat, Y.S. 2001. Indexing of cross protected and non-cross protected citrus cultivars / species for citrus tristerza virus by enzyme linked immunosorbent assays. *Plant Dis. Res.* 16: 10–16.

Chandra Mohan, R. 2009. State of art of coconut root (wilt) disease in India. *In*: 5th Internl. Conf. on Plant Pathology in the globalized era, Nov. 10-13, 2009, New Delhi, india (Abst.) p. 43.

Chang, C.J., Garnier, M., Zreik, L., Rossetti, V. and Bove, J.M. 1993. Culture and serological detection of the xylem limited bacterium causing citrus variegated chlorosis and its identification as a strain of *Xylella fastidiosa*. *Curr. Microbiol.* 27: 137–142.

Cheema, G.S. and Bhat, S.C. 1929. The dieback disease of citrus trees and its relation to the soils of western india. *Bull. Bombay Dept. Agric.*, 155: 1–48.

Chenulu, V.V. and Ahlawat, Y.S. 1993. *In*: Virus and Mycoplasma diseases of fruit crops in India (*Eds.*) Chenulu, V.V. and Ahlawat, Y.S. ICAR, New Delhi, India p. 34.

Cherian, K.A., Menon, R., Suma, A., Nair, S. and Sudeesh, M.V. 2002. Effect of banana bract mosaic disease on yield of commercial varieties in Kerala. *In*: Global Conf. on Banana and Plantain, Bangalore, India. p. 155.

Cherian, A.K., Baranwal, V.K., Malathi, V.G., Pant, R.P. and Ahlawat, Y.S. 2004. Banana streak virus from India and its detection by polymerase chain reaction. *Indian J. Biotech.*, 3: 409–413.

Cherian A.K., Baranwal, V.K., Ahlawat, Y.S. and V.G. Malathi, V.G. 2004. Standardizing a polymerase chain reaction test to detect banana streak virus. Session 1. Genetic resources and improvement-Posters. Internl. Cong. On Musa harnessing research to improve livelihood 6-9 July, 2004, Penang, Malaysia

Cherian A.K., Praveen, S. and Ahlawat, Y.S. 2004. Cloning the coat protein gene of the Cucumovirus causing mosaic disease in India. International congress on Musa: Harnessing research to improve livelihoods-6-9 July, Penang, Malaysia

Childs, J.F.L. 1950. The cachexia disease of Orlando tangelo. *Plant Dis. Reptr*. 34: 295–298.

Childs, J.F.L. 1952. Cachexia disease: Its bud transmission and relation to xyloporasis and tristeza. *Phytopathology,* 42: 265–268.

Childs, J.F.L. 1978. Rio Grande gummosis of citrus trees. Part I. A brief review of the history and occurrence of Rio Grande gummosis. *Plant Dis. Reptr.,* 62: 390–394.

Childs, J.F.L. 1978. Rio Grande gummosis of citrus trees. Part II. The causal relations of Rio Grande gummosis. *Plant Dis. Reptr.* 62: 395–399.

Childs, J.F.L. 1979. Florida citrus blight, Part I. Some causal aspects of citrus blight, Part II. occurrence of citrus blight outside florida. *Plant Dis. Reptr.,* 63: 565–569.

Chu, P.W.G. and Francki, R.I.B. 1979. The chemical subunit of tobacco ring spot virus coat protein. *Virology,* 93: 398–412.

Chung, K.R. and Brlansky, R.H. 2009. Citrus diseases exotic to florida: Citrus variegated chlorosis. Fact Sheet, one of a series of the Plant Pathology Department, Florida Cooperative Extension Service, Institute of Food and Agricultural Sciences, University of Florida. p. 223.

Cinar A., Korkmaz, S. and Kersting, U. 1994. Presence of new whitefly-born citrus disease of possible viral aetiology in Turkey. *FAO Plant Prot. Bull.*, 42: 73–75.

Cinar, A., Kersting, U., Onelge, N., Korkmaz, S. and Sas, G. 1993. Citrus virus and virus-like diseases in eastern Mediterranean region of Turkey. *In*: Proc. 12th Conf. IOCV, IOCV, Riverside. pp. 397–400.

Clark, C.C. and da Graca, J.V. 2000. Detection of citrus vein enation virus using cereal yellow dwarf virus ELISA kits. *In*: Proc. 14th Conf. IOCV, IOCV, Riverside, CA. pp. 357–359.

Clark, M.F., Davies, D.L., Buss, S.L., and Morton, A. 1989. Serological discrimination among Mycoplasma-like organisms using polyclonal and monoclonal antibodies. *Acta Hort.*, 235: 107–113.

Cohen, M. 1968. Exocortis virus as a possible factor in producing dwarf citrus trees. *Proc. Interl. Soc. Citriculture,* 81: 115–119.

Cohen, M. and Wutscher, H.K. 1979. Diagnosis of trees with citrus blight (YTD). *Proc. Internatl. Soc. Citriculture,* 3: 884–886.

Coit, J.E. 1928. Sunblotch of the avocado.*California Avocado Asson. Yearbook*. p. 28.

Colariccio, A., Lovisolo, O., Chagas, C.M., Galleti, S.R., Rossetti, V. and Kitajima, E.W. 1995. Mechanical transmission and ultrastructural aspects of citrus leprosis disease. *Fitopatol. Bras.,* 20: 208–213.

Coletta-Filho, H.D., Targon, M.L.P.N., Takita, M.A., De Negri Jr, J.D., Pompeu, J., Machado, M.A. 2004. First report of the causal agent of huanglongbing ("*Candidatus* Liberibacter asiaticus") in Brazil. *Plant Dis.,* 88: 1382.

Corbett, M.K. and Grant, T.J. 1967. Purification of citrus variegation virus. *Phytopathology,* 57: 137.

Cordero, C., Luis, M., Llauger, R., Teixeira, D.C., Eveillard, S., Batista, L., Pena, I., lopez, D., Perez, D., Cueto, J.R. and Bove, J.M. 2010. Citrus huanglongbing in Cuba: Diagnostics, characterization and management. *In*: 18th Conf. IOCV, IOCV, Riverside, CA. (Abs) p. 30.

Cronje, P., Dabek, A.J., Jones, P. and Tymon, A.M. 2000. First report of a phytoplasma associated with a disease of date palms in North Africa. *Plant Pathology,* 49: 6–801.

Costa, A.S., Grant, T.J. and Moreira, 1949. Investigacoes sobre a tristeza dos citros. *Bragantia,* 9: 59–80.

Cui, P.F., Gu, C.F. and Roistacher, C.N. 1991. Occurrence of satsuma dwarf virus in Zhejiang Province, China. *Plant Dis.,* 75: 242–244.

da Clara Henriques, M.I.E. 1994. Virus diseases of olive: An overlook. *Acta Hort.* (ISHS) 356: 379–385.

da Graça, J.V. and van Vuuren, S.P. 1980. The use of high glasshouse temperature to increase the rate of avocado sunblotch symptom development in indicator seedlings. *Plant Dis.,* 65: 46–47.

da Graca, J.V. and S.B. Maharaj 1991. Citrus vein enation virus, a possible luteovirus, *In:* Proc. 11th Conf. IOCV., IOCV, Riverside, CA. pp. 391–394.

da Graca, J.V., Lee, R.F., Moreno, P., Civerolo, E.L. and Derrick, K.S. 1991. Comparison of isolates of citrus ringspot, psorosis and other virus-like agents of citrus. *Plant Dis.,* 75: 613–616.

Dahal, G., Hughes J.D.A., Thottappilly, G. and Lockhart B.E.L. 1998. Effect of temperature on symptom expression and reliability of banana streak badnavirus detection in naturally infected plantain and banana (*Musa* spp.). *Plant Dis.*, 82: 16–21.

Dahal, G., Dahal, F., Pasberg-Gauhl, C., Hughes, J.D.A., Thottapilly, G. and Lockhart, B.E.L. 1999. Evaluation of micropropagated plantain and banana (*Musa* spp.) for banana streak badnavirus incidence under field and screenhouse conditions in Nigeria. *Ann Appl Biol.*, 134:181–191.

Dahal, G., Ortiz, R., Tenkouano, A., Hughes, J.D.A., Thottappilly, G., Vuylsteke, D. and Lockhart, B.E.L. 2000. Relationship between natural occurrence of banana streak badnavirus and symptom expression, relative concentration of viral antigen, and yield characteristic of some micropropagated Musa spp. *Plant Pathology*, 49: 68.

Dakshinamurti, V. and Reddy, G.S. 1975. Mosaic - a transmissible disorder of sweet oranges. *Indian Phytopath*, 28: 398–399.

Dakshinamurty, V. and Subbayya, J. 1981. Studies on transmission of citrus mosaic – a new disease of *Citrus sinensis* in Andhra Pradesh. (Abs) 3rd International Symp. Plant Pathology, New Delhi.

Dale, J.L. 1988. Rapid compression technique for detecting Mycoplasma-like organisms in leaf midrib sieve tubes by fluorescence microscopy. *Phytopathology*, 78: 118–120.

Dale, J.L. and Allen, R.N. 1979. Avocado affected by sunblotch disease contains low molecular weight ribonucleic acid. *Australian Plant Pathol.*, 3: 3–4.

Dale, J.L., Allen, R.N. and Symons, R.H. 1982. Avocado sunblotch viroid. *CMI/AAB Descriptions of Plant Viruses*, No. 254.

Dale, J.L., Phillips, D.A. and Parry, J.N. 1986. Double stranded RNA in Banana plants with bunchy top disease. *J. Gen. Virology*, 67: 371–375.

Dale, J.L. 1987. Banana bunchy top: An economically important tropical plant virus disease. *Adv. Virus Res.* 33: 301–325.

Dallot, S., Acuña, P., Rivera, C., Ramírez, P., Côte, F., Lockhart, B.E.L. and Caruana, M.L. 2001. Evidence that the proliferation stage of micropropagation procedure is determinant in the expression of Banana streak virus integrated into the genome of the FHIA 21 hybrid (M *hybrid (Musa AAAB). Archives of Virology*, 146: 2179–2190.

Damsteegt, V.D., Postnikova, E.N., Stone, A.L., Kuhlmann, M., Wilson, C., Sechler, A., Schaad, N.W., Brlansky, R.H. and Schneider, W.L. 2010. *Murraya paniculata* and related species as potential hosts and inoculum reservoirs of *Candidatus* Liberibacter asiaticus', causal agent of Huanglongbing. *Plant Dis.*, 94: 528–533.

Daniells, J.W., Geering, A.D., Bryde, N.J. and Thomas, J.E. 2001. The effect of banana Streak virus on the growth and yield of desert bananas in tropical Australia. *Ann. Appl. Biol.*, 139: 51–60.

Daniells, J., Thomas, J.E. and Smith, M. 1995. Seed transmission of banana streak virus confirmed. *Infomusa*, 4: 1.

Das, A.K. 2004. Rapid detection of *Candidatus* Liberibacter asiaticus, the bacterium associated with citrus huanglongbing (greening) disease using PCR. *Curr. Sci.*, 87: 1183–1185.

Das, A.K. 2009. Molecular identification and characterization of citrus greening bacterium, *Candidatus* Liberibacter asiaticus associated with decline of Nagpur mandarin orange in Vidarbha region, Maharashtra. *Curr. Sci.*, 96: 890–892.

Davino, M. and Garnsey, S.M. 1984. Purification, characterization and serology of a Mild strain of citrus infectious variegation virus from Florida. *In*: Proc. 9th Conf. IOCV, IOCV, Riverside, CA. pp. 196–203.

de Coll, O.R., Remes Lenicov, A.M.M., Agostini, J.P. and Paradell, S. 2000. Detection of *Xylella fastidiosa* in weeds and sharpshooters in orange groves affected with citrus variegated chlorosis in Misiones, Argentina. *In*: Proc. 12th Conf. IOCV, IOCV, Riverside, CA.

de Jesus Jr., W.C., Teixeira, D.C., Tanaka, F.A.O., Martins, E.C., Barros, C.C.P., Nagata, T. and rezende, J.A.M. 2005. Purification of Tymo-Like Virus Particles Isolated from Orange Trees with Sudden Death. Abs. 16th Conf. IOCV, IOCV. Riversides, CA.

Del Serrone, P. and Barba, M. 1997. Olive witches' broom: A new olive disorder associated with phytoplasmas. *In*: Abs. ISHS XVII Internatl Symp. virus dis. fruit trees, Bethesda, US, 1997-06-23/27. p. 119.

Deng, X., Zhou, G., Li, H., Chen, J. and Civerrolo, E.L. 2007. Detection of *Candidatus* Liberibacter asiaticus from Wampee (*Clausena lansium* Skees) by nested PCR. Online plant Healthy Progress doi:10.1094/PHP-2007-0419-01-BR.

Derrick, K.S., Brlansky, R.H., Graca, J.V. da, Lee, R.F., Timmer, L.W., Nguyen, T.K. 1988. Partial characterization of a virus associated with citrus ringspot. *Phytopathology*, 78: 1298–1301.

Derrick, K.S., Lee, R.F., Brlansky, R.H., Timmer, L.W., Hewitt, B.G. and Barthe, G.A. 1991. Proteins and filamentous structures associated with citrus blight. *In*: Proc. 11th. Conf. IOCV. IOCV, Riverside, CA pp. 265–268.

Derrick, K.S., Barthe,G.A., Hewitt, B.G., Lee, R.F. and Albrigo, L.G. 1992. Detection of citrus blight by serological assays. *Proc. Fla. State Hortic. Soc.* 105: 26–28.

Derrick, K.S., Beretta, M.J. and Barthe, G.A. 2006. Detection of an Idaeovirus in Citrus with implications as to the cause of citrus blight. *Proc. Fla. State Hort. Soc.*, 119: 69–72.

DeSimone, E.R., Laranjeira, F.F., Neri, F.M., Cunniffe, N.J. and Gilligan, C.A. 2010. Modelling spread and control of Bahia bark scaling of citrus. *In*: Abs. 18th Conf. IOCV held in Campinas/SP, Brazil. p. 49.

Desjardins, R. 1969. Purification, electron microscopy and serology of citrus viruses—A review? *In*: Proc. 1st Intern. *Citrus Symp.*, 3: 1481–1488.

Desjardins, P.R. and French, J.V. 1970. Purification of the citrus infectious variegation virus by density gradient electrophoresis. *Virology*, 40: 746–151.

Desjardins, P.R. and Wallace, J.M. 1966. Host to citrus infectious variegation including wide diversity of virus strains. *Phytopathology* 56: 347–348.

Desjardins, P.R., Drake, R.J. and French, J.V. 1969. Transmission of citrus ringspot virus to citrus and noncitrus hosts by dodder (*Cuscuta subinclusa* Dur. and Hilg.). *Plant Dis. Reptr.* 53: 947–948.

Desjardins, P.R., Drake, R.J., Atkins, E.L. and Bergh, O. 1979. Pollen transmission of avocado sunblotch virus experimentally demonstrated. *California Agric.*, 33: 14–15.

Desjardins, P.R., Drake, R.J. and Swiecki. 1980. Infectivity studies of Avocado sunblotch disease causal gent, possibly a viroid rather than a virus. *Plant Dis.*, 64: 313–315.

Dhanya, M.K., Rajagopalan, B., Umamaheshwaran, K. and Ayisha, R. 2007. Comparison of detection methods for banana bract mosaic virus in banana (*Musa* sp.). *World J. Agric. Sci.*, 3: 659–662.

Dickson, R.C., Flock, R.A. and Johanson, M.M. 1951. Insect transmission of citrus quick decline. *J. Econo. Ent.*, 44: 172–176.

Diekmann, M. and Putter, C.A.J. 1996. *FAO/IPGRI* Technical Guidelines for the Safe Movement of Germplasm, No. 15, Musa *(2nd edition)*, FAO of the United Nations/ International Plant Genetic Resources Institute, Rome.

Di Franco, A., Gallitelli, D., Volvos, C. and Martelli, G.P. 1989. Partial characterization of artichoke virus. *Phytopath.* 35: 45–46.

Ding, F., Wang, G., Yi, G., Zhong, Y., Zeng, J., Zhou, B. 2006. Infection of Wampee and lemon by the citrus huanglongbing pathogen (*Candidatrus* Liberibacter asiaticus) in China. *J. Plant Pathol.*, 87: 207–212.

Djelouah, K., Potere, O., D'Onghia, A.M. and Savino, V. 2000. Production of monoclonal antibodies to citrus psorosis virus. *In*: Proc. 14th Conf. IOCV, IOCV, Riverside, CA. pp. 152–158.

Dodds, J.A., Morris, T.J., and Jordan, R.L. 1984. Plant viral double stranded RNA. *Ann. Rev. Phytopathol.*, 22: 151–168.

Dollet, M. 1980. Report on research on the etiology of blast in oil and coconut palms. *Oleagineux* 35: 304.

Dominguez, A., Guerri, J., Cambra, M., Navarro, L., Moreno, P., and Pena, L. 2000. Efficient production of transgenic citrus plants expressing the coat protein gene of citrus tristeza virus. *Plant Cell Reports,* 19: 427–433.

D'onghia, A.M., Djelouah, K., Frasheri, D., and Potere, O. 2001. Detection of citrus psorosis virus by direct tissue blot immunoassay. *J. Plant Pathol.,* 83: 139–142.

Duan, Y.P., Gottwald, T., Zhou, L.J., Gabriel, D.W. 2008. First report of dodder transmission of *Candidatus* Liberibacter asiaticus to tomato (*Lycopersicon esculentum*). *Plant Dis.*, 92: 831.

Duan, Y. Lijuan Zhou, David G. Hall, Wenbin Li, Harshavardhan Doddapaneni, Hong Lin, Li Liu, Chery M. Vahling, Dean W. Gabriel, Kelly P. Williams, Allan Dickerman, Yijun Sun, and Tim Gottwald. 2009. Complete Genome Sequenceof Citrus Huanglongbing Bacterium, '*Candidatus* Liberibacter asiaticus' Obtained Through Metagenomics. *Molecular plant- microbe interactions* 22: 1011–1020.

Duan, Y.P., Vahling, C.M. and Zhou, L.J. 2010. Genomics and functional genomics of citrus huanglongbing bacterium '*Candidatus* Liberibacter asiaticus. In: Abs. 18th Conf. IOCV held in Campinas/S.P., brazil. p. 18.

Du Charme, E.P. and Knorr, L.C. 1954. Vascular pits and pegs associated with diseases in citrus. *Plant Dis. Reptr.* 38: 127–142.

Duran-Vila, N., Pina, J.A., Balester, J.F., Juarez, J., Roistacher, C.N., Rivera-Bustamante, R. and Semancik, J.S. 1988. The citrus exocortis disease: A complex of viroid RNAs. *In*: Proc. 10th Conf. IOCV, IOCV Riverside. pp. 152–164.

Dura-Vila, N., Pina, J.A., Molins, M.I. and Navarro, L. 1991. A new indexing method for cachexia. PP. 224-229. In Proc. 11th Conf., IOCV, IOCV, Riverside, CA.

Duran Vila, N., Pina, J.A. and Navarro, L. 1993. Improved indexing of citrus viroids.. *In*: Proc. 12th Conf. IOCV, IOCV, Riverside, CA. pp. 202–211.

Elarosi, H., Nagieb, H., El-Said, M., A. and Jaheen, N. 1983. Al-Wijam, a decline disease of date palm. *In*: Proc. First Symposium on Date Palm in Saudi Arabia, King Faisal University, Al-Hassa, Saudi Arabia. pp. 388–402.

El-Dougdoug, K.A., Osman, M.E., Hayam, S.A., Rehab, A.D. and Reham, M.E. 2010. Biological and molecular detection of HSVd - infecting peach and pear trees in Egypt. *Australian J. Basic and Appl. Sci.*, 4: 19–26.

Elleuch, A., Khouaja, F.D., Hamdi, I., Bsais, N., Perreault, J.P., Marrakchi, M. and Fakhfakh, H. 2006. Sequence analysis of three citrus viroids infecting a single Tunisian citrus tree (*Citrus, reticulata* - Clementine). Genet. Mol. Biol 29, No. 4- Sao Poulo.

Eloja, A.L., Velasco, L.G. and Agati, J.A. 1962. Studies on the abaca mosaic disease. I. Sap transmission of abaca mosaic virus. *Phil. J. Agric.*, 27: 75–84.

Eloja, A.L. and Tinsley, T.W. 1963. Abaca mosaic virus and its relationship to sugarcane mosaic virus. *Ann. Appl. Biol.*, 51: 253–258.

Espino, T.M., Exconde, S.B., Zipagan, F.B., and Espino, R.R.C. 1990. Banana bract mosaic, a new disease of banana. II. Isolation and purification for monoclonal antibody production. *Philipp. Agric.* 73: 61–68.

Estelitta, S., Radhakrishnan, T.C. and Sheela Paul, T. 1996. Infectious chlorosis disease of banana in Kerala. *InfoMusa,* 5: 25–26.

Etienne, J. and Aubert, B. 1980. Biological control of psyllid vectors of greening disease on Reunion Island. *In*: Proc. 8th Conf. IOCV, IOCV, Riverside, CA. pp. 118–121.

FAOSTAT-FAO statistical Database 2005

Fauquet, C.M., Mayo, M.A., Maniloff, J., Desselberger, U. and Ball, L.A. *eds.* 2005. 8th Report of the International Committee on Taxonomy of viruses. Academic press, 1162, pp, Elsevier publication 27th May 2005.

Fawcett, H.S. 1911. Scaly bark or nail-head rust of citrus. *Florida Agricultura Experimental Station, Bulletin,* 106: 41.

Fawcett, H.S. 1932. New angles on treatment of bark lesions of citrus. *California Citrograph,* 17: 106–108.

Fawcett, H.S. 1933. New symptoms of psorosis, indicating a virus disease of citrus. *Phytopathology,* 23: 930.

Fawcett, H.S. 1936. *In*: *citrus Diseases and their Control.* McGraw Hill, New York, London, p. 656.

Fawcett, H.S. 1946. Stubborn disease of citrus, a virosis. *Phytopathology,* 36: 675–677.

Fawcett, H.S. and Bitancourt, A.A. 1943. Comparative symptomatology of psorosis varieties on citrus in California. *Phytopath.,* 33: 837–864.

Fawcett, H.S. and Wallace, J.M. 1946. Evidence of virus nature of citrus quick decline. *Calif. Citrog.,* 32: 88–89.

Fawcett, H.S. and Klotz, L.J. 1939. Infectious variegation of citrus. *Phytopathology,* 29: 911–912.

Fawcett, H.S., Perry, J.C. and Johnston, J.C. 1944. The stubborn disease of citrus. *Calif. Citrograph,* 29: 146–147.

Febres, V.J., Niblett, C.l., Lee, R.F. and Moore, G.A. 2002. Characterization of grapefruit plants (*Citrus paradisi* Macf.) transformed with citrus tristeza closterovirus genes. Springer Verlag.

Febres, V.J., Moore, G.A. and Lee, R.F. 2005. Development of pathogen-drived transgenic Citrus for resistance against Citrus tristeza virus. *In:* Proc. 16th. Conf. IOCV. IOCV, Riverside, CA. p. 490.

Fernandez Valiela, M.V., Fortugno, C. and Corizzi, F. 1965. Incidence of bud-union crease in citrus trees grafted on trifoliate rootstock in the Delta del Parana and San Pedro areas of Argentina. *In*: Proc. 3rd Conf. IOCV, IOCV, Riverside, CA. pp. 182–186.

Flores, R. 1988. Detection of citrus exocortis viroid in natural and experimental citrus hosts by biochemical methods. *In*: Proc. 10th Conf. IOCV, IOCV, Riverside, CA pp. 192–196.

Flores, R. and Rodriguez, J.L. 1981. Altered pattern of root-formation on cuttings of *Gynura aurantiaca* infected by citrus exocortis viroid. *Phytopathology,* 71: 964–966.

Foguet, J.L. and Oste, C.A. 1968. Disorder of trifoliate orange rootstock in Tacuman, Argentina. *In*: 4th Conf. IOCV, IOCV, Riverside, CA. pp. 183–189.

Font, C., Curbelo, I., Fernandez, J., Valdes, S., Pereira, D. 1998. Detection de particulas bacilliforme en hybridos de platano (*Musa* spp.) FHIA 21 en Cuba. Fitosanidad, 2: 3–5.

Font, I., Abad, P., Dally, E.L., Davis, R.E. and Jordá, C. 1998. Nueva enfermedad en el olivar español. *Phytoma España,* 102: 211–212.

Francki, R.I.B., Mossop, E.W. and Hatta, T. 1979. Cucumber mosaic virus. CMI/AAB Description of plant viruses No. 213. Association of Applied Biologists, Wellesbourne, U.K.

Fraser, L.R. 1958. Virus diseases in Australia. *Proc. Linn. Soc. NSW,* 73: 9–19.

Fraser, L.R. 1961. Lemon crinkly leaf virus. *In*: Proc. 2nd Conf. IOCV, IOCV, Riverside, CA. pp. 205–210.

Fraser, L.R. and Daljit Singh. 1966. Greening virus – new threat to citrus industry. *Indian Hort.,* 10: 21–22.

Fraser, L.R. and Singh, D. 1968. Citrus dieback in India - the contribution of greening virus. *In:* Proc. 4th Conf: IOCV, IOCV, Riverside, CA. pp. 141–144.

Fraser, L.R., Levitt, E.C. and Cox, J.E. 1961. Relationship between exocortis and stunting of citrus varieties on *Poncirus trifoliata* rootstock. *In*: Proc. 2nd Conf. IOCV, IOCV, Riverside, CA. pp. 34–39.

Fraser, L.R., Daljit Singh, Capoor, S.P. and Nariani, T.K. 1966. Greening virus, the likely cause of citrus dieback in India. *FAO Plant Prot. Bull.,* 14: 127–130.

Fraser, L.R., Broadbent, P. and Cox, J.E. 1976. Gummy pitting of *Poncirus trifoliata*: its association with dwarfing of citrus in New south Wales. *In*: Proc. 7th Conf. IOCV, IOCV, Riverside, CA. pp. 147–151.

Fraser, L.R., singh, D., Capoor, S.P. and Nariani, T.K. 1966. Greening virus, the likely cause of citrus dieback in India. *FAO Plant Prot. Bull.,* 14: 127–130.

Frison, E.A. and Putter, C.A.J. (*eds.*) 1989. FAO/IBPGR technical guidelines for the safe movement of Musa germplasm. FAO/IBPGR, Rome, Italy.

Frison, E.A. and Taher, M.M. 1991. FAO/IBPGR Technical guidelines for the safe movement of citrus germplasm, p. 45. FAO/IBPGR, Rome, Italy.

Fudl-Allah, A.E.C., Calavan, E.C., and Igwegbe, E.C.K. 1971. Culture of a mycoplasma-like organism associated with stubborn disease of citrus. *Phytopathology,* 61: 1321.

Galipienso, L., Navarro, L., Ballester-Olmos, J.F., Pina, J., Moreno, P. and Guerri J., 2000. Host range and symptomatology of graft-transmissible pathogen causing bud Union crease of citrus on trifoliate rootstocks. *Plant Pathology,* 49: 308–314.

Galipienso, L., Vives, M.C., Moreno, P., Milne R.G., Navarro, L., Guerri J., 2001. Partial characterization of citrus leaf blotch virus, a new virus from Nagami kumquat. *Archives of Virology,* 146: 357–368.

Galipienso, L., Vives, M.C., Navarro, L., Moreno, P., and Guerri, J. 2004. Detection of citrus leaf blotch virus using Digoxigenin-labeled cDNA probes and PCR. *European J. Plant Path.,* 110: 175–181.

Gallitelli, D., Martelli, G.P. and Rana, G.L. 1984. Artichoke vein banding virus. *CMI/ AAB Descr. Pl.Viruses.*

Gallitelli, D. and savino, V. 1985. Olive latent virus-1, an isometric virus with a single RNA species isolated from olive in Apulia, Southern Itali. *Ann. Appl. Biol.,* 106: 295–303.

Gambley, C.F. and Thomas, J.E. 2001. Molecular characterisation of Banana mild mosaic virus, a new filamentous virus in *Musa* spp. *Arch. Virol.* 146: 1369–1379

Gambley, C.F., Thomas, J.E., Magnaye, L.V. and Herradura, L. 2004. Abaca mosaic virus: A distinct strain of sugarcane mosaic virus. *Australian Pl. Path.,* 33: 475–484

Garcia, M.L., Bo, E.D., Grau, O. and Milne, R.G. 1994. The closely related citrus ringspot and citrus psorosis viruses have particles of novel filamentous morphology. *J. Gen. Virol.,* 75: 3585–3590.

García, M. L., De La Torre, M. E., Dal Bo, E., Djelouah, K., Rouag, N., Luisoni,E., Milne, R.G. and Grau, O. 1997. Detection of citrus psorosis-ringspot virus using RT-PCR and DAS-ELISA. *Plant Pathol.*, 46: 830–836.

Garnier, M. and Bové, J.M. 1983. Transmission of the organism associated with citrus greening disease from sweet orange to periwinkle by dodder. *Phytopathology*, 73: 1358–1363.

Garnier, M. and Bove, J.M. 1993. Citrus greening disease and the greening bacterium. *In*: Proc. 12th Conf. IOCV, IOCV, Riverside, CA. pp. 212–219.

Garnier, M., Bové, J.M. 1996. Distribution of the huanglongbing (greening) liberobacter species in fifteen African and Asian countries. *In*: *Proc.* 13th Conf. IOCV, IOCV, Riverside, CA. pp. 388–391.

Garnier, M., Danel, N. and Bové, J.M. 1984. The greening organism is a Gram-negative bacterium. *In*: Proc. 9th Conf IOCV, IOCV, Riverside, CA. pp. 115–125..

Garnier, M., Zreik, L. and Bové, J.M. 1991. Witches' broom disease of lime trees in Oman: Transmission of a mycoplasma-like organism (MLO) to periwinkle and citrus and the production of monoclonal antibodies against the MLO. *In*: Proc. 11th Conf. IOCV,, IOCV Riverside, CA. pp. 448–453.

Garnier, M. Villechanoux, S., Jagoueix, S. and Bove, J.M. 1993. Nature of greening bacterium – like organism (BLO): Taxonomic characterization by use of cloned DNA fragments. *In*: Proc. 12th Conf. IOCV, IOCV, Riversised, CA. pp. 250–257.

Garnier, M., Chang, C.J., Zreik, L., Rossetti, V. and Bove, J.M.1993. Citrus variegated chlorosis: Serological detection of *Xylella fastidiosa*, the bacterium associated with the disease. *In*: Proc. 12th Conf. IOCV, IOCV, Riverside, CA. pp. 301–305.

Garnier, M., Jagoueix-Eveillard, S., Cronje, P., Le Roux, H., Bové, J.M. 2000. Genomic characterisation of a liberibacter present in an ornamental Rutaceous tree, *Calodendrum capense,* in the Western Cape province of South Africa. Proposal for a "*Candidatus* Liberibacter africanus subsp. capensis". *Int. J. Sys. Evol. Microbiol.* 50: 2119–2125.

Garnsey, S.M. 1968. A citrus crinkly-leaf-type virus recently discovered in Florida. *Proc. Fla. State Hort. Soc.*, 81: 79–84.

Garnsey, S.M. 1968. Additional non-citrus hosts for the Florida isolate of citrus variegation virus. *Phytopathology*, 58: 1433–1434.

Garnsey, S.M. 1968. Exocortis virus in citrus can be spread by contaminated tools. *Citrus Indus.* 49: 13–16

Garnsey, S.M.1974. Purification and serology of a Florida isolate of citrus variegation Virus. *In* Proc. 6th Conf. IOCV, IOCV, Riverside, pp. 169–75.

Garnsey, S.M. 1974. Mechanical transmission of a virus that produces tatter leaf symptoms in *Citrus excelsa*. *In*: Proceedings 6th Conference IOCV, IOCV, Riverside, CA. pp. 137–140.

Garnsey, S.M. 1975. Purification and properties of citrus leaf rugose virus. *Phytopathology*, 65: 50–57.

Garnsey, S.M. 1988. Vein enation (woody gall). *In*: *Compendium of citrus diseases* (*Eds*). Whiteside, J.O.; Garnsey, S.M.; Timmer, L.W.), p. 50.

Garnsey, S.M. 1993. Indexing procedure for 15 viruses. *In*: Agric. Handbook No. 333. pp. 5–15

Garnsey, S.M. and Jones, J.W. 1967. Mechanical transmission of exocortis virus with contaminated budding tools. *Plant Dis. Reptr.*, 51: 410–413.

Garnsey, S.M. and Drake, R.J. 1968. Indexing of tatter leaf with cow pea. *In*: Agric. Handbook 333 USDA Washington D.C. pp. 80–82.

Garnsey, S.M. and Purcifull, D.E. 1969. Serological detection of a citrus virus in leaf extracts from field trees. *Proc. Fl. State Hort. Soc.*, 82: 56–60

Garnsey, S.M. and Muller, G.W. 1988. Efficiency of mechanical transmission of citrus tristeza virus. *In*: Proc. 10th Conf. IOCV, IOCV, Riverside, CA. pp. 46–54.

Garnsey, S.M. and Timmer, L.W. 1988. Local lesion isolate of citrus ringspot virus induces psorosis bark scaling. *In*: Proc. 10th Conf. IOCV, IOCV, Riverside. pp. 334-339.

Garnsey, S.M., Permar, T.A., Cambra, M and Henderson, C.T. 1993. Direct tissue Blot immunoassay (DTBIA) for detection of citrus tristeza virus (CTV). *In*: Proc. 12th Conf. IOCV, IOCV Riverside, CA. pp. 39–50.

Garnsey, S.M., , Civerolo, E.L., Gumpf, D.J, Paul, C., Hilf, M.E., Lee, R.F., Brlansky, R.H. , Yokomi, R.K., and Hartung, J.S. 2005. Biological characterization of an international collection of *Citrus tristeza virus* (CTV) isolates. *In*: Proc. 16th Conf. IOCV, IOCV, Riverside, CA. pp. 75–93.

Ge, X. and Scott, S.W. 1994. The nucleotide sequence of citrus leaf rugose ilarvirus RNA-2. *J. Gen. Virol.*, 75: 2841–2846.

Geering, A.D.W., McMichael, L.A., Dietzegen, R.G. and Thomas, J.E. 2000. Genetic diversity among banana streak virus isolates from Australia. *Phytopathology*, 90: 921–927.

Geering, A.D.W., Olszewski, N.E., Dahal, G., Thomas, J.E., Lockhart, B.E.L. 2001. Analysis of the distribution and structure of integrated Banana streak virus DNA in a range of Musa cultivars.. *Mol Plant Pathol.*, 2: 207–213

Gallitelli, D. and Savino, V. 1985. A single RNA virus isolated from olive in Apulia (Southern Itali) *Ann. Appl. Biol.* 106: 295–303.

Gerola, F.M., Lombardo, G. and Catara, A. 2005. Histological localization of citrus infectious variegation virus(CVV) in *Phaseolus vulgaris*. *Protoplasma*, 67: 319–326.

Ghosh, D.K., Das, A.K., Singh, S., Singh, S.J. and Ahlawat, Y.S. 1999a. Occurrence of witches' broom -a new phytoplasmal disease of acid lime (*Citrus aurantifolia*) in India. *Plant Dis.*, 83: 302.

Ghosh, D.K., Das, A.K., Singh, S., Singh, S.J. and Ahlawat, Y.S. 1999b. Association of a phytoplasma with witches' broom, a new disease of acid lime (*Citrus aurantifolia*). *Curr. Sci.*, 77: 174–177.

Ghosh, D.K., Mathur, S., Gupta, K., Ahlawat, Y.S. and Ramachandran, P. 2002. Viroid infection of citrus in central India. *Indian Phytopath*, 55: 290–293.

Ghosh, D.K., Das, A.K. and Singh, S. 2003. Individual and mixed infection of citrus tristeza virus and greening librobacter and their association with sweet orange decline in the state of Maharashtra. *J. Mycol. Pl. Pathol.*, 33: 69–72.

Ghosh, D.K., Aglave, B., Bhanare, K., and Baranwal, V.K. 2007. PCR based detection of citrus yellow mosaic disease from Vidarbha region of Maharashtra. *Indian Phytopath.*, 60: 520–526.

Ghosh, D.K., Aglave, K., and Baranwal, V.K. 2008. Simultaneous detection of one RNA and one DNA virus from naturally infected citrus plants using duplex PCR technique. *Curr. Sci.*, 94: 1314–1318.

Ghosh, D.K., Balaji Aglave, Roy, A. and Ahlawat, Y.S. 2009. Molecular cloning, sequencing, and phylogenetic analysis of coat protein gene of a biological distinct citrus tristeza virus isolate occurring in central India. *J. Plant Biochemistry and Biotechnology*, 18: 105–108.

Ghosh, D.K., Bhose, S., Manimekalai, R. and Gowda, S. 2013. Molecular detection of *Candidatus* Phytoplasma spp. Causing witches' broom disease in acid lime (*Citrus aurantifolia*) in India. *J. Plant Biochem. Biotechnol.*, 22: 343–347.

Ghosh, D.K., Bhose, S., Mukherjee, K., and Baranwal, V.K. 2013. Sequence and evolutionary analysis of ribosomal DNA from Huanglongbing (HLB) isolates of Western India. *Phytoparasitica DOI 10.1007/s12600-013-0290-4.*

Gillings, M.R., Broadbent, P. and Gollnow, B.I. 1988. Biochemical indexing for citrus exocortis viroid. *In:* Proc. 10th Conf. IOCV, IOCV, Riverside, CA. pp. 178–187.

Gillings, M.R., Broadbent, P., Gollnow, B.I. and Lakeland, C. 1991. Viroids in Australian citrus. *Australian J. Plant Physiology,* 18: 559–570.

Girmaldi, V. and Catara, A. 1989. Detection of citrus tristeza and greening in Pakistan through electron microscopy. *J. Phytopathol.,* 162: 17–21.

Girmaldi, V. and Catara, A. 1996. Association of a filamentous virus with yellow vein clearing of lemon. *In*: Proc. 13th Conf. IOCV, IOCV, Riverside, CA. pp. 343–345.

Gold, A.H. 1972. Seed transmission of banana viruses. *Phytopathology* 62: 760.

Golmohammadi, M., Aliabadi, A.A., Bertolini, E. Llop, P. and Lopez, M.M. 2010. Current situation and impact of huaglongbing (HLB) caused by *Candidatus* Liberibacter asiaticus in the South of Iran. *In*: Abs. 18th Conf. IOCV held in Campinas/SP, Brazil, Nov. 7-12, 2010. p. 75.

Gonsalves, D. 1998. Control of papaya ringspot virus in papaya: A case study. *Ann. Appl. Phytopathology,* 36: 415–437

Gonsalves, D. and Garnsey, S.M. 1974. Infectivity of the multiple nucleoprotein and RNA components of citrus leaf rugose virus. *Virology,* 61: 343–53

Gonsalves, D. and Garnsey, S.M. 1975. Functional equivalence of an RNA component and coat protein infectivity of citrus leaf rugose virus. *Virology,* 64: 23–31.

Gonsalves, D. and Garnsey, S.M. 1975a. Nucleic acid components of citrus variegation virus and their activation by coat protein. *Virology,* 67: 311–318.

Gonsalves, D. and Garnsey, S.M. 1975b. Infectivity of heterologous RNA-Protein mixture from alfalfa mosaic, citrus leaf rugose, citrus variegation and tobacco streak viruses. *Virology,* 67: 319–326.

Gottwald, T.R., Hall, D.G., Beattie, G.A.C., Ichinose, K., Nguyen, M.C., Le, Q.D., Bar-Joseph, M., Lapointe, S., Stover, E., Parker, P.E., McCollum, G. and Hilf, M.E. 2010. Investigations of the effect of guava as a possible tool in the control/ management of huanglongbing. *In*: Proc. 17th Conf. IOCV, IOCV, Riverside, CA. pp. 98–109.

Gouda, K.A., Baranwal, V.K. and Ahlawat, Y.S. 2006. Simplified DNA extraction and improved PCR based detection of Citrus greening bacterium. *J. Plant Biochem. and Biotech.,* 15: 117–121.

Grant, T.J. and Corbett, M.K. 1960. Mechanical transmission of the infectious variegation virus of citrus. *Nature,* 188: 519–520.

Grant, T. J. and Smith, P. F.1960. Infectious variegation of citrus found in Florida, *Plant Dis., Reptr.,* 44: 426–429.

Grevena, S.J., Lopes, R.S., Paiva, P.E.B., Yamamoto, P.T. and Roberto, S.R. 1998. The *Xylella fastidiosa* vectors. *In*: Citrus variegated chlorosis, Bebedouro-SP. pp. 36-53.

Grimaldi, V. and Catara, A. 1989. Detection of citrus tristeza and greening in Pakistan through electron microscopy. *J. Phytopathol.,* 162: 17–21.

Grimaldi, V. and Catara, A. 1996. Association of filamentous virus with yellow vein clearing of lemon. *In*: Proc. 13th Conf. IOCV, IOCV, Riverside, CA. pp. 343–345.

Groot, T.V.M., Freitas-Astua, J. and Kitajima, E.W. 2006. *Brevipalpus phoenicis* transmits Citrus leprosis virus, cytoplasmic type (CiLV-C) to common bean (*Phaseolus vulgaris*) under experimental conditions. *Virus Reviews and Research,* 11: 67.

Gross, H.J., Domdey, H., Lossow, C., Jank, P. Raba, M., Alberty, H and Sanger, H.L. 1978. Nucleotide sequence and secondary structure of potato spindle viroid. *Nature, Lond.,* 273: 203–208.

Gross, H.J., Krupp, G., Domday, H., Raba, M., Alberty, H., Lossow, C.H., Ramm, K. and Sanger, H.L. 1982. Nucleotide sequence and secondary structure of citrus exocortis and chrysanthemum stunt viroid. *Eup. J. Biochem.,* 121: 249–257

Guardo M., Sorrentino G., Marletta T. and Caruso A. 2007a. First report of Citrus leaf blotch virus on Kumquat in Italy. *Plant Dis.,* 91: 1054.

Guardo, M., Castellano, M.A., Savino, V. and Caruso, A. 2007b. First detection of Citrus leaf blotch virus in Italy. *J. Plant Pathol.* 89: 813.

Guardo, M., Potere, O., Castellano, M.A., Savino, V. and Caruso, A. 2009. A new herbaceous host of citrus leaf blotch virus. *J. plant Pathology,* 91: 485–488

Guerra-Moreno, A.S., Manjunath, K.L., Brlansky, R.H. and Lee, R.F. 2005. Citrus Leprosis Symptoms can be Associated with the Presence of Two Different viruses: Cytoplasmic and Nuclear, the Former having a Multipartite RNA Genome. *In*: Proc. 16th Conf. IOCV, IOCV, Riverside, CA pp. 230–239.

Guerra-Moreno, A. S., Keremane, M.L., Brlansky, R.H. and Lee, R.F. 2007. The Complete sequence of the cytoplasmic Citrus leprosis virus (CCLV) and its genome organization. *Phytopathology,* 97: S174.

Gumpf, D.J. 1988. Stubborn diseases of citrus caused by *Spiroplasma citri*. *In*: Mycoplasma diseases of crops. Basic and applied aspects (*Eds.* by Maramorosch, K. and Raychaudhuri, S.P.), Springer-Verlag, New York, USA. pp. 327-342.

Gupta, P. 1986. Eradication of mosaic disease and rapid clonal multiplication of bananas and plantain through meristem tip culture. *Plant Cell Tissue Organ Culture,* 6: 33–39.

Gupta, O.P., Nauriyal, J.P. and Knorr, L.C. 1972. Citrus greening in the Indian Punjab. *In;* Proc. 5th Conf: IOCV, IOCV, Riverside, CA. pp. 14–18.

Gutierrez, M.A., Luth, D. and Murre, G.A. 1997. Factors affecting Agrobacterium-mediated transformation in Citrus and production of sour orange (*Citrus aurantium* L.) plants expressing the coat protein gene of citrus tristeza virus. *Plant Cell Reports,* 16: 745–753

Haas, A.R.C., Klotz, L.J. and Johnston, J.C. 1944. Acorn disease in oranges. *Calif. Citrograph* 29: 168–169.

Hailstones, D. L., Bryant, K.L., Broadbent, P. and Zhou, C. 2000. Detection of Citrus tatter leaf virus with reverse transcription-polymerase chain reaction. *Aust. Plant Pathol.,* 29: 240–248.

Hajeri, S., Ramadugu, C., Manjunath, K., Vidalakis, G. and Lee, R. 2010. Nucleotide sequence and genome organization of Dweet mottle virus and its relationship to members of the family betaflexiviridae. *Arch. Virol.,* 155: 1523–1527.

Halbert, S.E. 2005. The discovery of huanglongbing in Florida. Proceedings of 2nd International Citrus Canker and Huanglongbing Research Workshop, Florida Citrus Mutual, Orlando 2005, H-3.

Hanold, D. and Randle, J.W. 1990. Detection of nucleotide sequences related to coconut cadang- cadang viroid in coconut and oil palm and other monocotyledons in the Pacific region. *In*: 8th Intl. conf. virol. Berlin (Abs). p. 93.

Hanold, D. and Randles, J.W. 1991. Coconut cadang-cadang disease and its viroid agent. *Plant Dis.,* 75: 330–335

Hanold, D. and Randles, J.W. 1991. Detection of coconut cadang cadang viroid like sequences in oil and coconut palm and other monocotyledons in the southwest pacific. *Ann. Appl. Biol.,* 118: 139–151.

Hanold, D., Langridge, P. and randles, J. W. 1988. The use of cloned sequences for the identification of coconut foliar decay disease-associated DNA. *J. Gen. Virol.,* 69: 1323–1329

Harding, R.M., Thomas M., Burns, G.F. and Dale, J.L. 1991. Virus-like particles associated with banana bunchy top disease contain small single- stranded DNA. *J. Gen. Virol.,* 72: 225–230.

Harding, R.M. and Teakle, D.S. 1988. Autofluorescence of necrotic phloem cells and laticifers, and a reduced latex flow: new symptoms for papaw dieback disease in Australia. *Aust. J. Agric. Res.,* 39: 857–862

Harding, R.M., Thomas M., Burns, G.F., Dietzgen, G.R. and Dale, J.L. 1993. Nucleotide sequence of one component of the banana bunchy top virus genome contains a putative replicase gene. *J. Gen. Virol.,* 74: 323–328.

Harper, G. and Hull, R. 1998. Cloning and sequence analysis of banana streak virus DNA. *Virus Genes,* 17: 271–278.

Harper, G, Darren Hart , Sarah Moult and Roger Hull. 2004. Banana streak virus is very diverse in Uganda. *Virus research,* 100: 51–56

Harper, G, Dahal, G, Thottappilly, G. and Hull, R. 1999a. Detection of episomal banana streak badnavirus by IC-PCR. *J. Virol. Methods,* 79: 1–8

Harper, G, Osuji, J.O., Heslop-Harrison, J.S. and Hull, R. 1999b. Integration of banana streak badnavirus into the Musa genome: Molecular and cytogenetic evidence. *Virology* 255: 207–213.

Harper, S.J., Chooi, K.M. and Pearson, M.N. 2008. First report of Citrus leaf blotch virus in New Zealand. *Plant Dis.,* 92:1470.

Harper, S.J., Dawson, T.E. and Pearson, M.N. 2010. Isolates of citrus tristeza virus that overcome *Poncirus trifoliata* resistance comprise a novel strain. *Arch. Virol.,* 155: 471–480.

Harris, H.C. 1977. *In*: Plant health and quarantine in international transfer of genetic resources (*Eds.* by Hewitt, W.B. and Chiarappa, L.). CRC Press, Cleveland, USA.

Harrison, B. D. and Murant, A.F. 1977. Nepovirus group. CMI AAB Descr. Plant Viruses, No. 185.

Hartung, J.S., Halbert, S.E., Pelz-Stelinski, K., Brlansky, R.H., Chen, C. and Gmitter, F.G. 2010. Lack of evidence for transmission of '*Candidatus* Liberibacter asiaticus' through citrus seed taken from affected fruit. *Plant Dis.,* 94: 1200-1205.

Haseloff, J. and Symons, R. H. 1981. Chrysanthemum stunt viroid: Primary sequence and secondary structure. *Nucleic Acid Res.,* 9: 2741–2752.

Haseloff, N. M. and Symons, R.H. 1982. Viroid RNAs of cadang cadang disease of coconut. *Nature,* 299: 316–321

He, C.X., Li, W.B., Ayres, A.J., hartung, J.S., Miranda, V.S. and Teixeira, D.C. 2000. Distribution of *Xylella fastidiosa* in citrus rootstocks and transmission of citrus variegated chlorosis between sweet orange plants through natural root grafts. *Plant Dis.,* 84: 622–626.

Hermoso de Mendoza, A., Pina, J.A., Ballester-Olmos, J.F. and Navarro, L. 1993. Persistent transmission of citrus vein enation virus by *Aphis gossypii* and *Myzus persicae*. In Proc. 12th Conf. IOCV, IOCV, Riverside, CA. pp. 361–362.

Hibino, H., Tsuchizaki, T., Usugi, T. and Saito, Y. 1977. Fine structures and developmental process of tubules produced by mulberry ringspot virus and satsuma dwarf virus infections. *Ann. Phytopathol. Soc. Japan,* 43: 255–264.

Hoa, N.V. and Ahlawat, Y.S. 2004. Characterisation of four isolates of Indian citrus ringspot virus. *Indian Phytopath,* 57: 296–302.

Hoa, N.V., Ahlawat, Y.S. and Pant, R.P. 2004. Production of virus-free Kinnow mandarin and Mosambi sweet orange nucleus planting material through shoot tip grafting. *Indian Phytopath,* 57: 482–487

Hodgson, R.A.J.; Wall, G.C. and Randles, J.W. 1998. Specific identification of coconut tinangaja viroid for differential field diagnosis of viroids in coconut palm. *Phytopathology,* 88: 774–781.

Horne, W.T. and Parker, E.R. 1931. The avocado disease called sunblotch. *Phytopathology,* 21: 235–238

Howard, F.W.; Thomas, D.L.; Donselman, H.M.; Collins, M.E. 1979. Susceptibilities of palm species to mycoplasma-like organism-associated diseases in Florida. *FAO Plant Prot. Bull,* 27: 109–117.

Howard, F.W. 1983. World distribution and possible geographic origin of palm lethal yellowing and its vectors. *FAO Plant Prot. Bull.,* 31: 101–113.

Howard, F.W. 1990. Evaluation of grasses for cultural control of *Myndus crudus,* a vector of lethal yellowing of palms. *Entomologia Experimentalis et Applicata,* 56: 131–137.

Howd, D. S., Hartung, J.S. and Brlansky, R.H. 2002. Ultrastructure of citrus cholorotic dwarf-infected leaves and bark. *In*: Proc. 14th Conf. IOCV, IOCV, Riverside, CA. pp. 373–377.

Hu J.S., Li, H.P., Barry, K. and Wang. M. 1995. Comparison of dot blot, ELISA and RT-PCR for detection of two Cucumber Mosaic Virus isolates infecting banana in Hawaii, *Plant Dis.,* 79: 902–906.

Huang, Q. and Hartung, J.S.2001. Cloning and sequence analysis of an infectious clone of citrus yellow mosaic virus that can infect sweet orange via Agrobacterium-mediated inoculation. *J. Gen. Virol.,* 82: 2549–2558.

Hull, R., Harper, G. and Lockhart, B.E.L. 2000. Viral sequences integrated into plant genomes. *Trends Plant Sci.,* 5: 362–365.

Hull, R. and Matthews, R.E.F. 2002. Plant Virology. Academic Press, San Diego, CA, USA, p. 1001.

Hume, H.H. 1901. Some fungus diseases of citrus and others fruits. *Proc. Florida State Hort. Soc.,* 14: 64–70.

Hung, T.H., Wu, M.L. and Su, H.J. 2000. Identification of alternative hosts of the fastidious bacterium causing citrus greening disease. *J. Phytopathol.,* 148: 231.

Hutton, R., Broadbent, P. and Bevington, K. 2000. Viroid dwarfing for high density plantings. *Hort. Rev.,* 24: 277–317.

Igwegbe, E.C.K. and Calavan, E.C.1970. Occurrence of mycoplasma-like bodies in Phloem of stubborn infected citrus seedlings. *Phytopathology,* 60: 1525–1526.

Index of Viruses – Pospiviroidae. 2006. *In*: ICTVdB - The Universal Virus Database, version 4. Büchen-Osmond, C (Ed), Columbia University, New York, USA. http://www.ncbi.nlm.nih.gov/ICTVdb/Ictv/fs_index.htm

Ishigai, T. and Jinno, M. 1958. On citrus mosaic. *Ann.Phytopathol.Soc. Japan* 23: 29.

Isoda, T., Gyoutoku, Y. and Yamada, K. 1991. [Studies on citrus virus diseases. VIII. post-measure of orchard attacked by satsuma dwarf virus]. *Proc. Assoc. Plant Prot. Kyushu* , 37: 55-58.

Ito, T., Ieki, H. and Ozaki, K. 2002. Simultaneous detection of six citrus viroids and Apple stem grooving virus from citrus plants by multiplex reverse transcription polymerase chain reaction. *J. Virol. Methods* 106: 235–239.

Iwanami,T., Kondo,Y.,Kobayashi, M., Han,S.S. and Karasev, A.V. 2001. Sequence diversity and interrelationships among isolates of satsuma dwarf-related viruses. *Arch. Virol.* 146: 807-813.

Iwanami, T. and Koizumi, M. 2000. Satsuma dwarf virus group. *In*: Compendium of citrus disease, 2nd edn. (*Eds.*) Timmer, LW., Garnsey, S.M. and Graham, J.H. APS Press.

Iwanami, T. and Ieki, H.1996. Nucleotide sequence of the coat protein genes of citrus mosaic virus. *Virus Res.*, 42: 181–186.

Iwanami, T., Kano, T. and Koizumi, M. 1992. Spherical virus-like particles associated with vein enation on Yuzu (*Citrus junos*). *Bull. Fruit Tree Res. Stat.*, 23:137–143.

Iwanami, T., Koizumi,M. and Ieki, H. 1993. Diversity of properties among satsuma dwarf virus and related viruses. *Ann. Phytopathol.Soc. Japan*, 59: 642–650.

Iwanami, T., Yamao, F., Seno, T. and Ieki, H.1996. Nucleotide sequence of the 3'terminal region of RNA1 of satsuma dwarf virus. *Ann. Phytopathol. Soc. Japan*, 62: 4–10.

Iwanami, T., Kondo, Y. and Karasev, A.V.1999. Nucleotide sequences and taxonomy of satsuma dwarf virus. *J. Gen. Virol.*, 80: 793–797

Izawa, H. 1966. Investigation on withering disease of *Citrus unshiu* Marc. (in Gamagori district, Aichi Prefecture). *Bull. Aichi Hort. Expt. Stn.*, 5: 1–9.

Jagouiex, S., Bove, J.M. and Garnier, M. 1996. PCR detection of two '*Candidatus*' Liberobacter species associated with greening disease of citrus. *Mol. and Cell. Probes*, 10: 43-50.

Jain, R.K., Pappu, H.R., pappu, S.S., Varma, A. and Ram, R.D. 1998. Molecular characterization of papaya ringspot potyvirus isolates from India. *Annl. Appl. Biol.*, 132: 413-425.

Jain, R.K., Sharma, J., Sivakumar, A.S., Sharma, P.K., Byadgi, A.S., Verma, A.K. and Varma, A. 2004. Variability in coat protein gene of papaya ringspot virus isolates from multiple locations in India. *Arch. Virol.*, 149: 2435–2442.

Jaufeerally-Fakim, Y., Ashwin, K. and Harper, G. 2006. Genetic variants of banana streak virus in Mauritius. *Virus Research*, 115: 91–98.

Javer, E., Acina-Mambole, I., Font,C.,Quiala,I.,González,G.,Echemendía, A.L. and Teycheney, P.Y. 2009. Identification of *Banana streak virus* species Goldfinger, Imové, Mysore and Obino l'Ewaï in *Musa* spp. in Cuba . *Plant Pathology*, 58: 787.

Jones D.R. and Lockhart B.E.L. 1993. Banana streak disease. *Musa* disease fact sheet No. 1. International Network for the Improvement of Banana and Plantain, Montpellier, France.

Joshi, D.M. and Joshi, H. V. 1974. The control of Banana mosaic. *Indian J. Hort. Sci.*, 31: 382-385.

Julia, J.F. 1979. Isolation and identification of insects carrying juvenile diseases of the coconut and the oil palm in the Ivory cost. *Oleagineux* 34: 385–393.

Kamat, M.N. and Patel, M.K. 1957. Notes on two important plant diseases in Bombay State. *Bull. Pl. Prot.* New Delhi 3:16

Kanrar, S., Afunian, M., Loannis, G.G., Tzanetakis, E. and Vidalakis, G. 2010. Molecular characterization of viruses associated with yellow vein and vein enation disease of citrus. *In*: (Abs) 18th Conf. IOCV held at Campinas/SP, Brazil, Nov.7-12, 2010. p. 47

Kapur, S.P., Dhillon, R.S., Cheema, H.S. and Deol, I.S. 1974. Citrus exocortis in the Punjab. *Punjab Agric. Univ. J. Res.*, 11: 378–379.

Karan, M., Harding, R.M. and Dale, J.L. 1997a. Association of banana bunchy top virus DNA components 2-6 with bunchy top diseases. *Molecular Plant Pathology* On-line [http://www.bspp.org.uk/mppol/] 1997/0624karan

Karan, M., Harding, R.M. and Dale, J.L. 1997b. Evidence of two groups of banana bunchy top virus isolates. *J. Gen. Virol.*, 75: 3541–3546.

Karasev, A.N., Boyko, V.P., Gowda, S., Nikolaev, O.N., Hilf, M.E., Koonin, E.N., Niblett, C.L., Cline, K., Gumpf, D., lee, R.F., Garnsey, S.M., Lewandowski, D.J. and Dawson, W.O. 1995. Complete sequence of the citrus tristeza virus RNA genome. *Virology*, 208: 511–520.

Kawai, A. and Nishio, T. 1990. Detection of citrus tatter leaf virus by enzyme-linked immunosorbent assay (ELISA). *Ann. Phytopathol. Soc. Japan*, 56: 342–345.

Kawai, A., Tsukamoto, T., Namba, S. and Nishio, T. 1996. Citrus tatter leaf virus: A review of its properties and the development of a serological detection system. *In*: Proc. 13th Conf. IOCV, IOCV, Riverside, CA. pp. 339–342

Ke, C., Ke, S., Wu, R.J., Yang, H. and Hsu, P.T. 1993. Purification and serology of the organism associated with citus huanglungbin. *In*: Proc. 12th Conf. IOCV, IOCV, Riversised, CA. pp. 220–223

Keese, P. and Symons, R.H. 1985. Domains in viroids: evidence of intermolecular RNA rearrangement and their contribution to viroid evolution. *Proc. Natl. Acad. Sci.* USA, 82: 4582–4586.

Keese, P., Osorio-Keese, M.E. and Simons, R.H. 1988. Coconut tinangaja viroid: Sequence homology with coconut cadang-cadang viroid and other potato spindle tuber viroid-related RNAs. *Virology*, 162: 508–510.

Kersting, U., Baspinar, H., Cinar, A., Sengonca, C. and Uygun, N.1993. New findings on the epidemiology of *Spiroplasma citri* in the eastern Mediterranean region of Turkey. *In*: Proc. 12th Conf. IOCV, IOCV, Riverside, CA. pp. 336–341.

Kersting, U. Korkmaz, S., Cinar, A., Ertugrul, B., Onelge, N., and Garnsey, S.M. 1996. Citrus chlorotic dwarf, a new whitefly transmitted disease of eastern Mediterranean region of Turkey. *In*: Proc. 13th Conf. IOCV, IOCV, Riverside. pp. 220–225.

Khan, I.A. and Grosser, J.W. 2004. Regeneration and characterization of somatic hybrid plants of *Citrus sinensis* (sweet orange) and *Citrus micrantha*, a progenitor species of lime. *Euphytica*, 137: 271–278

Khurana, S.M.P. 1974. Studies on three virus diseases of papaya in Gorakhpur, India. Abs. 9th Interl. Hort. Cong. Warsa (Poland) 14: 260.

Kiranmai, G., Lavakumar, P., Hema, M., Venkatramana, M., Kishnaprasadji, J., Rao, M. and Sreenivasalu, P. 2005. Partial characterization of a potyvirus causing bract mosaic of banana in Andhra Pradesh. *Indian J. Virol.*, 16: 7–11.

Kiranmai, G., Sreenivasulu, P. and Nayudu, M.V. 1996. Comparison of three different tests for detection of cucumber mosaic cucumovirus in banana (*Musa paradisiacal* L.) *Curr. Sci.*, 71: 764–767.

Kiranmai, G., satyanarayna, T. and Sreenivasulu, P. 1998. Molecular cloning and detection of cucumber mosaic cucumovirus (CMV) causing infectious chlorosis disease of banana using DNA probe. *Curr. Sci.*, 74: 356–359.

Kishi, K. and Tanaka, S.1964. Studies on the indicator plants for citrus viruses. 2. Mechanical transmission of the virus causing satsuma dwarf to sesame (*Sesamum indicum* L.). *Ann. Phytopathol. Soc. Jap.*, 29: 142–148.

Kitajima, E.W. 2010. Citrus viruses and bud transmitted bacteria in Brazil. History and Prospectives. *In*: 18th Conf. IOCV held at Campinas/SP, Brazil, 7-12 Nov. 2010

Kitajima, E.W., Silva, D.M., Oliviera, A.R., Muller, G.W. and Costa, A.S. 1964. Thread-like particles associated with tristeza disease of citrus. *Nature*, 201: 1011–1012.

Kitajima, E.W., Chagas, C.M. and Rodrigues, J.C.V. 2003. *Brevipalpus*-Transmitted plant virus and virus-like diseases: Cytopathology and some recent cases. *Experim. and Appl. Acarology*, 30: 135–160.

Kitajima, E.W., Muller, G.W., Costa, A.S. and Yuki, W. 1972. Short rod-like particles associated with citrus leprosis. *Virology,* 50: 254–258.

Klotz, L.J. and Fawcett, H.S. 1941. Color handbook of citrus diseases.Univ. Calif. Press, Berkeley and Los Angeles.

Knorr, L.C. 1968. Studies on the etiology of leprosis in citrus. *In*: Proc. 4th Conf. IOCV, IOCV, Riverside, CA. pp. 332–341.

Koizumi, M. 1984. Elimination of tatter leaf-citrange stunt virus from satsuma mandarin by shoot-tip grafting following pre-heat-treatment. *In*: Proc. 9th Conference IOCV, IOCV, Riverside, USA. pp. 229–233.

Koizumi, M., Kano, T., Leki, H. and Mae, H. 1988. China laurestine: A symptomless carrier of satsuma dwarf virus which accelerates natural transmission in fields. *In*: Proc. 10th Conf. IOCV., IOCV, Riverside, CA. pp. 348–352.

Korkmaz, S., Cýnar, A., Bozan, O. and Kersting, U. 1994. Distribution and natural transmission of a new whitefly-borne virus disease of citrus in the eastern Mediterranean region of Turkey. *In*: Proc. 9th Congr. Mediterr. pp. 437–439.

Korkmaz, S. Cinar, A., Kersting, U. and Garnsey, S.M. 1995. Citrus chlorotic dwarf : A whitefly transmitted virus-like disease of citrus in Turkey. *Plant Dis.*, 79: 1074.

Krueger, R., Lee, R.F. and Bash, J.A. 2005. Dweet mottle virus and citrus leaf blotch virus. *Subtropics Newsletter V3,* 1: 7–9.

Kubaa,R.A.,Djelouah, K. and D'Onghia, A.M. 2010. Study on the genetic variabilityof citrus infectious variegation virus isolates from different Mediterranean areas. *In*: Abs. 18th Conf. IOCV, IOCV, Riverside, CA. p. 84.

Kubiriba, J., Legg, J.P., Tushemereirwe, W. and Adipala, E. 2001. Vector transmission of Banana streak virus in the screenhouse in Uganda. *Ann Appl Biol.,* 139: 37–43.

Kubo, K.S., Arena, G.D., Kitajima, E.W., Machado, M.A. and Freitas-Astua, J. 2010. Transmission of citrus leprosis virus C (CiLV) to *Arabidopsis thaliana* by *Brevipalpus phoenicis* mite. *In*: Abs. 18th Conf. IOCV held in Campinas/SP, Brazil, November 7-12, 2010 p. 85.

Labroussaa, F., arricau-Bouvery, N., Dubrana, M.B.M, Duret, S., Bove, J.M., Renaudin, J. and Saillard, C. 2010. *Spiroplasma citri*-insect vector interactions. *In*: Abs. 18th Conf. IOCV held in Campinas/SP, Nov.7-12, 2010. p. 8.

Lafleche, D. and Bove, J.M. 1970. Mycoplasmes dans les agrumes attients de "greening" de "stubborn" ou de maladiessimilaires. *Fruits,* 25: 455–465.

Laird, E.F. and Weathers, L.G.1961. *Aphis gossypii,* a vector of citrus vein enation virus. *Plant Dis. Reptr.,* 45: 877.

Lassoudie're, A. 1974. La mosaÃ¯que dite "a' tirets" du bananier Poyo en CÃ´te d'Ivoire. *Fruits,* 29: 349–357.

Lebsky, Vladimir, arevik, P. and Laura S.R. 2009. Application of scanning electron microscopy for diagnosis of phytoplasmas in single and mixed infections in papaya. *In*: 21st internl. Conf. on virus and other graft transmissible diseases of fruit crops. *Berichte aus dem JKI* 148. p. 29.

Lee, I.M. and Davis R.E. 1984. New media for rapid growth of *Spiroplasma citri* and corn stunt spiroplasma. *Phytopathology,* 74: 84–89.

Lee, R.F., Marais, L.J., Timmer, L.W. and Graham, J.H. 1984. Syringe injection of water into the trunk: A rapid diagnostic test for citrus blight. *Plant Dis.,* 68: 511–513.

Lee, R.F., Derrick, K.S., Beretta, M.J.G., Chagas, C.M. and Rossetti, V. 1991. Citrus variegated chlorosis: A new destructive disease of citrus in Brazil. *Citrus Ind.,* 72: 12–13.

Lee, R.F., Derrick, K.S., Futch, S.H. and Tucker, D.P.H. 1993. Graft transmission of citrus measles. *In*: Proc. 12th Conf. IOCV, IOCV, Riverside, CA. pp. 371–374.

Lee, R. F., Beretta, M.J.G., Hartung, J.S., Hooker, M. E. and Derrick, K.S. 1993. Citrus variegated chlorosis: Confirmation of a *Xylella fastidiosa* as the causal agent. *Summa Phytopathol.* 19: 123–125.

Lee, R.F, Pappu, H.R., Pappu, S.S., Roch-Pena, M.A., Febris, E.J., Manjunath, K.L., Nicolaeva, O.V., Karasev, A.N., Cevik, B., Akbulut, M., Benscher, D., Anderson, E.J. Price, M., Ochaoa, F. and Niblett, C.L. 1997. Progress on strain differentiation of citrus tristeza virus. *Rev. Mex. Fitopatol.,* 79: 79–87.

Lee, R.F. and Bar-Joseph, M. 2000. Tristeza. *In*: Compendium of citrus diseases, 2nd ed. (*Eds.* Timmers, L.W., Garnsey, S.M. and Grahameds, J.H.). APS, St.Paul, MN. pp. 61–63.

Lee, R. F., Febres V. J. and Moore G. A. 2007. Resistance to Citrus tristeza virus *via* RNA interference. *Plant Cell Reports* (published on line 20 Sep 07).

Lee, Y.S., Su, H.J. and Hung, T.H. 2003. First report of papaya leaf curl virus infecting papaya plant in Taiwan. *Plant Dis.* 87: 204.

Leki, H. and Yamada, S. 1984. Inactivation of tristeza virus, satsuma dwarf virus and citrus tatter leaf virus by heat treatment. *Bull. Fruit Tree Res. Stn., Japan B,* 11: 71–87.

Leki, H. and Ito, T. 1996. Occurrence of concave gum on Hyuganatsu *(Citrus tamitrana)* in Japan. *In*: Proc. 13th Conf. IOCV, IOCV, Riverside, CA. pp. 346–348.

Le Provost, G., Iskra-Caruana, M.L., Acina, I. and Teycheney, P.Y. 2006. Improved detection of episomal banana streak viruses by multiplex immunocapture PCR. *J. Virol. Methods,* 13:77

Levy, L. and Hadidi, A.1993. Direct nucleotide sequencing of PCR-amplified DNAs of the closely related citrus viroids IIa and IIb (Cachexia). *In*: Proc. 12th Conf. IOCV, IOCV, Riverside, CA. pp. 180–186

Li, W. B., Pria,W.D. Jr., Teixera, D.C., Miranda, V.S., Ayres, A.J., Franco, C.F., Costa, M.G., He, C.X., Costa, P.I. and Hartung, J.S. 2001. Coffee leaf scorch caused by a strain of *Xylella fastidiosa* from citrus. *Plant Dis.,* 85: 501–505.

Lima, J.E.O. and Borducchi, A.S. 1982. Observation on citrus blight in Sao Paulo, Brazil. *Proc. Fla, State Hort. Soc.* 95: 72–75

Lin, K.H. 1956. Observation on yellow shoot of citrus. Etiological study of yellow shoot of citrus. *Acta Phytopathol Sinica,* 2: 1–42.

Liu,K.H., Song, Z., Zhou, Y., Li, Z.A. and Zhou, C.Y. 2011. Detection of Citrus tatter leaf virus by Real-Time RT-PCR. *In*: Abs. 18th Conf. IOCV, IOCV, Riverside, CA

Locali, E.C., Freitas-Astua, J., Souza, A.A., Takita, M.A., Astua-Monge, G., Antonioli, R., Kitajima, E.W. and Machado, M.A. 2003. Development of a molecular tool for the diagnosis of leprosis, a major threat to citrus production in the Americas. *Plant Dis.,* 87: 1317–1321.

Locali-Fabris, E.C., Freitas-Astua, J., Souza, A.A., Takita, M.A., Astua-Monge, G., Antonioni-Luizon,R., Rodrigues, V., Targon, M.L.P.N. and Machado, M.A. 2006. Complete nucleotide sequence, genomic organization and phylogenetic analysis of citrus leprosis virus, cytoplasmic type. *J. Gen. Virol.,* 87: 2721–2729.

Lockhart, B.E.L. 1986. Purification and serology of a bacilliform virus associated With banana streak disease. *Phytopathology,* 76: 995–999.

Lockhart, B.E.L. 1995. Technical Bulletin 143, Food & Fertilizer Technology Center, Taiwan, 1995.

Lockhart, B.E.L.1998. Occurrence of a bacilliform virus related serologically to banana streak virus. *Plant Dis.,* 72: 230–233.

Lockhart, B.E.L. 1990. Evidence for a circular double-stranded DNA genome in a second group of plant viruses. *Phytopathology,* 80: 12–131.

Lockhart, B.E.L. and Autrey, J.C. 1988. Occurrence in sugarcane of a bacilliform virus related serologically to banana streak virus. *Plant Dis.,* 72: 230–233.

Lockhart, B.E.L. and Jones, D.R. 1999. Banana streak. *In*: Diseases of Banana, Abaca and enset., edited by Jones D.R. (CABI Publishing, Wellingford, U.K. pp. 263–274

Loconsole, G., Saldarelli, P., Doddapaneni, H., Savino, V., Martelli, G.P. and Saponari M. 2012. Identification of a single-stranded DNA virus associated with citrus chlorotic dwarf disease, a new member in the family Geminiviridae. *Virology,* 432: 162–172.

Long, J.K., Fraser, L.R. and Cox, J.E. 1972. Possible value of close planted, virus dwarfed orange trees. *In*: Proc. 5th Conf. IOCV, IOCv, Riverside, CA. pp. 262–267.

Lopes, S.A., Martins, E.C. and Frare, G.F. 2005. Deteção de *Candidatus* Liberibacter americanus em *Murraya paniculata. Summa Phytopathologica,* 31: 48–49.

Lopes, S.A., Martins, E.C. and Frare, G.F. 2006. Deteção de *Candidatus* Liberibacter asiaticus em *Murraya paniculata. Fitopatologia Brasileira,* 31: 303.

Lopes, A.S., frare, G.F., Sausa, M.C., fernandes, N.G. and Ayres, A.J. 2010. A new strategy to control citrus variegated chlorosis disease. *In*: Abs. 18th Conf. IOCV held in Campinas/S.P., brazil, Nov. 7-12, 2010. p. 13.

Lovisolo, O., Colariccio, A., Chagas, C.M., Rossetti, V., Kitajima, E.W. and Harakava, R. 1996. Partial characterization of citrus leprosis virus. *In*: Proc. 13th Conf. IOCV, IOCV, Riverside, CA. pp. 178–188.

Maccheroni, W., Alegria, M.C., greggio, C.C., Piazza, J.P., Kamla, R.F., Zacharias, P.R.A., Bar-Joseph, M., ferro, J.A. and da silva, A.C.R. 2005. A new *Tymoviridae* virus associated with citrus sudden death disease in Brazil. Proc. 16th Conf.IOCV, IOCV, Riverside, CA (Abs)

Magee, C.J.P. 1927. Investigations on the bunchy top disease of bananas. *Bull. Coun. Sci. Industry. Res. Aug.* Cairo, Egypt. 30: 64.

Magee, C.J.P. 1940. Transmission of the banana bunchy top virus. *J. Aust. Inst. Agric. Sci.,* 6: 109–110.

Magee, C.J.P. 1940. Transmission of infectious chlorosis or hear-rot of the banana and its relationship to cucumber mosaic virus *J. Aust. Inst. Agric. Sci.,* 6: 44–47.

Magee, C.J.P. 1953. Some aspects of the bunchy top disease of banana and other *Musa* spp. *J. Proc. R. Soc. N.S.W.,* 87: 3–18.

Magnaye, L.V. and Espino, R.R.C. 1990. Banana bract mosaic, a new disease of banana.I. Symptomatology. *Philipp. Agric.,* 73: 55–59.

Maharaj, S.B. and da Graca, J.V. 1989. Transmission of citrus vein enation virus by *Toxoptera citricidus. Phytophylactica,* 21: 81–82.

Maheshwari, Y., Verma, H.N. and Mandal, B. 2012. Molecular cloning and Prokaryotic expression of the genes encoding antibody to the recombinant coat protein of papaya ringspot virus. *In*: XXI National Conf. on Immunobiology and Management of viral diseases in 21st Century. Held at IVRI Mukteshwar 8-10 Nov., 2012. p. 36.

Majorana, G and Schwarz, R.E. 1972. Studies on two mechanically transmissible citrus viruses. *In*: Proc. 5th Conf. IOCV, IOCV, Riverside, CA. pp. 188–192.

Mali, V.R. and Deshpande, G.O. 1976. Heart rot a virus disease of banana in Marathwada. *Indian J. Mycol. and Pl. Pathol.* 6: 23–26.

Mali, V.R. and Rajegore, S.B. 1980. A cucumber mosaic virus of banana in India. *Phytopath. Z.,* 98: 127–136.

Mali, V.R. and Chaudhuri, K.G. 1993. Occurrence of citrus virus and virus-like diseases in Maharashtra, India. *In*: Abs. 12th Conf. IOCV, IOCV, Riverside, CA. p. 466.

Mali, V.R., Choudhury, K.G. and. Rane, S.D. 1975. Blastomania - a new bud transmissible disease of *Citrus*. *Curr. Sci.*, 44: 627–628.

Mali, V.R., Chaudhuri, K.G. and Rane, S.D. 1975. Vein enation and woody gall diseases of citrus. *FAO Plant Prot. Bull.*, 23: 190–191.

Mali, V.R., Chaudhuri, K.G. and Rane, S.D. 1976. The vein-enation virus disease of citrus in India. *Indian Phytopatho,* 29: 43–45.

Mali V.R., Chaudhury, K.G. and Rane, S.D. 1976. Leaf curl virus disease of Citrus in India. *Sci. and Cult.*, 42: 525–527.

Mandal, B. 2010. Elucidation of etiology of foorkey disease of large cardamom. *In*: XIX Nat. Conf. On Recent trends in viral disease problems and management. March 18-20, 2010 at S.V. University Tirupati p. 34.

Manimekalai, R., Soumya, V.P., Satish Kumar, R., Selvarajan, R., Reddy, M.K., Sasikala, M.,Rajeev, K., Baranwal, V.K. and Thomas, G.V. 2009. First report of molecular evidence for the phytoplasma association in root (wilt) disease of coconut in India. *In*: 5th Intl. Conf. Pl. Path. Globalized era, Nov. 10-13, 2009, New Delhi, India (Abst). p. 266.

Manimekalai, R., Soumya, V.P., Satish Kumar, R., Selvarajan, R., Reddy, M.K., Thomas, G.V. Sasikala, M., Rajeev, K., and Baranwal, V.K. 2010. Molecular detection of 16SrX1group phytoplasma associated with root (wilt) disease of coconut (*Cocos nucifera*) in India. *Plant Dis.* 94: 636 (Abs).

Manjunath, K.L. 1987. Studies on Vein enation virus disease of *Citrus* in South India. *Indian J. Plant Path.*, 5: 121–125.

Manjunath, K.L., Sawant, S.D. and sawant, I.S. 1989. Stem pitting in coorg mandarin caused by citrus tristeza virus. *In*: Abstr. Symp. Indian Phytopath. Soc. (Southern Chapter). p. 17.

Manjunath, K.L., Pappu, H.R., Lee, R.F., Niblett, C.L. and Civerolo, E. 1993. Studies on the coat protein genes of four isolates of citrus tristeza closterovirus from India: Cloning, sequencing and expression. *In*: Proc. 12th Conf., IOCV, IOCV, Riverside, CA. pp. 20–27.

Manjunath, K. L., Chandrika, R., Warren, C., Niblett, C.L. and Lee, R.F. 2005. Complete nucleotide sequence of a new genotype of Citrus tristeza virus from an isolate having a mixed infection. *In:* Proc. 16th Conf. IOCV. IOCV, Riverside. p. 487

Manjunath, K.L., Rangel, E., Guerra-Moreno, A.S., Brlansky, R.H. and Lee, R.F. 2006. Serological detection of the cytoplasmic Citrus leprosis virus (CiLV) from Infected citrus plants. *Phytopathology,* 96: 163.

Marais, L.J. and Lee, R.F. 1986. Citrange stunt virus associated with decline of shamouti on Swingle citrumelo rootstock in South Africa. *Plant Dis.* 70: 892.

Marais, L.J. and Lee, R.F. 1991. Experimental transmission of citrus blight in South Africa.. In: Proc. 11th Conf. IOCV. IOCV, Riverside, CA. pp. 261–264.

Marathe, T.S., Summanwar, A.S. and Ram, R.D. 1982. New additional hosts of *Pentalonia nigronervosa*. *Curr. Sci.,* 51: 435–436.

Maria, C. Z., Peña, E., Reyes, C.A., Figueroa, J., Stein, B., Grau, O. and Garcia, M.L. 2006. Detection of Citrus psorosis virus in the northwestern citrus production area of Argentina by using an improved TAS-ELISA. *J. Virol. Methods,* 137: 245–251.

Martelli, G.P. 2008. Infectious diseases and certification of olive: An overview. *EPPO Bull.* 29: 127–133.

Martin, S., Alioto, D.,Milne R.G., Garnsey, S. M., Garcia M.L., José, G. and Moreno, P. 2004. Detection of Citrus psorosis virus by ELISA, molecular Hybridization,

RT-PCR and immunosorbent electron microscopy and its association with citrus psorosis disease. *European J. Plant Pathology* 110: 747–757.

Martin, S., Lopez, C., Garcia, M.L., Naum-Ongania, G., Grau, O., Flores, R., Moreno, P. and Guerri, J. 2005. The complete nucleotide sequence of a Spanish isolate of citrus psorosis virus: Comparative analysis with other ophioviruses. *Archives of Virology,* 150: 167–176.

Martin, S., Garcia, M.L.,Troisi, A., Rubio, L., Legarreta, G., Grou, O., Alioto, D., Moreno, P. and Guerri, J. 2006. Genetic variation of populations of citrus psorosis virus. *J. Gen. Virol.,* 87: 3097–3102.

Martinez, A.L. and Wallace, J.M. 1967. Citrus leaf-mottle yellows disease in the Philippines and transmission of the causal virus by a psyllid: *Diaphorina citri. Plant Dis. Rep.,* 51: 692–695.

Mathen, K., Rajan, P., Radhakrishnan Nair, C.P., Sasikala, M., Guna sekharan, M., Govindankutty, M.P. and Solomon, J.J. 1990. Transmission of root (wilt) disease to coconut seedlings through *Stephanitis typical* (Distant) (Heteroptera: Tingidae). *Trop. Agric.,* 67: 69–73.

Maureen M., Fitch, M., Richard, M., Manshardt, Gonsalves, D., Slightom, J.L. and Sanford, J.C. 1992. Virus resistant papaya plants derived from tissues bombarded with the coat protein gene of papaya ringspot virus. *Bio/Technology,* 10:1466–1472.

Mayo M.A. and Haenni A.L. 2006. Report from the 36th and 37th Meetings of the Executive Committee of the International Committee on Taxonomy of Viruses. *Archives of Virology,* 151: 1031–1037.

McLean, A.P.D. 1954. Citrus vein enation virus. *S. Afr. J. Sci.,* 50: 147–151

McClean, A.P.D. 1974. Abnormal bud union between some sweet oranges and rough lemon rootstock: Evidence that it is caused by a transmissible pathogen. *In*: Proc. 6th Conf. IOCV, IOCV, Riverside CA. pp. 203–210

McCoy, R.E. 1972. Remission of lethal yellowing in coconut palm treated with tetracycline antibiotics. *Plant Dis. Reptr.,* 56: 1019–1021.

McLean, A.P.D. and Oberholzer, P.C.J.1965. Citrus psylla, a vector of the greening disease of sweet orange. *S. African J. Agric. Sci.,* 8: 297–298.

Medberry, S.L., Lockhart, B.E.L. and Olszewski, N.E. 1990. Properties of Commelina yellow mottle virus' complete DNA sequence: Genomic discontinuities and transcript suggests that it is a pararetrovirus. *Nucleic Acids Research* 18: 5055-5513.

Mello, A.F., Wayadande, A., Yokomi, R.K., Fletcher, J. 2009. Transmission of different strains of *Spiroplasma citri* to carrot and citrus by *Circulifer tenellus* Baker (Hemiptera:Cicadellidae). *J. Economic Entomol.,* 102: 1417–1422.

Meneghini, M. 1946. Sobre a natureza a transmissibilidale do doencia"tristeza" dos citrus. *O. Biologico,* 12: 285–287.

Menon, K. P. A. and Pandalai, K. M .1958. The coconut palm, A monograph. Indian Central Coconut Committee, Emakulum. p. 384.

Milne, R.G., Djelouah, K., Garcia, M.L., Dal Bo, E. and Grau. O. 1996. Structure of Citrus Ringspot-Psorosis-Associated Virus Particles: Implications for Diagnosis and Taxonomy *In*: Proc. 13th Conf. IOCV, IOCV, Riverside, CA. pp. 189–197.

Milne, R.G., Garcia, M.L. and Grau, O. 2000. Genus ophiovirus. Type species Citrus psorosis virus (CPsV). *In*: Van Regenmortel, M.H.V., Fauquet, C.M., Bishop, D.H.L., Carstens, E.B., Estes, M.K., Lemon, S.M., Maniloff, J., Mayo, M.A., McGeoch, D.J., Pringle, C.R., Wickner, R.B. *Eds. Virus Taxonomy*. 7th Report of the International Committee on Taxonomy of Viruses. San Diego, CA: Academic Press, pp. 627–31

Mishra, M.D., Hammond, R.W., Owens, R.A., Smith, D.R. and Diener, T.O. 1991. Indian bunchy top disease of tomato plant is caused by a distinct strain of citrus excocortis Viroid. *J. Gen. Virol.* 72: 1–5.

Miyakawa, T. 1968. Susceptibility of trifoliate orange and some other citrus varieties to satsuma dwarf virus. *In*: Proc. 4th Conf. IOCV., IOCV, Riverside, CA. p. 153–159.

Miyakawa, T. 1980a. Thermo-therapy for some citrus cultivars infected by tatter leaf virus. *Bull.Tokushima Hort.Exp. Station,* 9: 7–11.

Miyakawa, T. 1980b. Occurrence and varietal distribution of tatter leaf- citrange stunt virus and its effects on Japanese citrus. *In*: Proc. 8th Conf. IOCV, IOCV, Riverside, CA. pp. 220–224.

Miyakawa, T. and Matsui, C. 1976. A bud union abnormality of Satsuma mandarin on *Poncirus trifoliata* rootstock in Japan. *In*: Proc. 7th Conf. IOCV, IOCV, Riverside, CA. pp. 125–131

Mohan, S. and Lakshmanan, P. 1988. Banana Chlorosis Disease. *Phytoparasitica,* 16: 281–290.

Mokashi, A.N., Byadgi, A.S., Kunkalikar, S., Hegde, R., and Ambika, D.S. 2005. Survey, purification and detection of banana bract mosaic virus disease. *Karnataka J. Agric. Sci.,* 18: 177–180.

Moore, G. A., Luth, D., Febres,V.J., Garnsey, S.M.and Niblett, C.L. 2000. Agrobacterium-Mediated transformation of grapefruit (*Citrus paradisi* Macf.) with genes from citrus tristeza virus. *Acta Hort.,* 535: 237–243.

Moreira, S. 1942. Observacoes sobre a 'tristeza' dos citrus ou podridao das radicelas. *O. Biologico.,* 8: 269–272.

Munez, A.R. 1992. Symptomatology, transmission and purification of banana bract mosaic virus (BBMV) in 'Giant Cavendish' banana. M.S. thesis, University of Philippines, Los Baffos.

Murcia, N., Serra, P., Olmos, A. and Duran-Vila, N. 2008. A novel hybridization approach for detection of citrus viroids. *Molecular and Cellular Probes,* 23: 95–102.

Murray, R.G.E. and Stackebrandt, E. 1995. Taxonomic note: Implementation of the provisional status Candidatus for incompletely described Prokaryotes. *Intl. J. Syst. Bacteriol.,* 47: 240–241.

Murray, S, Thomas, J.E. and Dietzgen, R.G. 2001. Development of a multiplex immunocapture PCR with colourimetric detection for viruses of banana. *J. Virol. Methods,* 89: 75–88.

Myrie, W.A., Douglas, L., Harrison, N.A., McLaughlin, W. and James M, 2012. First report of lethal yellowing disease associated with subgroup 16SrIV, a phytoplasma on St. Kitts in the Lesser Antilles. *New Disease Reports,* 26: 25. [http://dx.doi.org/10.5197/j.2044-0588.2012.026.025]

Nageswara-Rao, M., Miyata, S., Ghosh, D., Irey, M., Garnsey, S.M. and Gowda, S. 2013. Development of rapid, sensitive and non-radioactive tissue-blot diagnostic method for the detection of citrus greening. *Molecular and Cellular Probes* XXX:1-8. (htpp://dx.doi.org./10.1016/j.mcp.2013.04.003)

Nagpal, R.L. 1959. Tristeza found in Bombay state. *Calif: Citrog.,* 44: 392, 402–405.

Naidu, P.H. and Reddy, M.R.S. 1985. Movement and distribution of citrus mosaic in host tissue. *Indian Phytopath.* 38:782

Nariani, T.K. 1956. Leaf curl of papaya. *Indian Phytopath.* 9: 151–157.

Nariani, T.K., Sahambi, H.S. and Chona, B.L. 1965. Occurrence of tristeza virus in citrus in northern India. *Indian Phytopath.* 18: 220.

Nariani, T.K., Raychaudhuri, S.P. and Bhalla, R.B. 1966. Citrus tristeza virus in northern and Central India. *Indian Phytopath.,* 19: 397–399.

Nariani, T.K., Raychaudhuri, S.P. and Bhalla, R.P. 1967. Greening virus of citrus in India. *Indian Phytopath.* 20: 146–150

Nariani, T.K. and Raychaudhuri, S.P. 1968. Occurrence of tristeza and greening viruses in Bihar, West Bengal and Sikkim. *Indian Phytopath.* 21: 343–344.

Nariani, T.K., Raychaudhuri, S.P. and Sharma, B.C. 1968. Exocortis in citrus in India. *Pl. Dis.Reptr.* 52: 834.

Nariani, T.K., Viswanath, S.M. and Menon, M.R. 1970. Occurrence of tristeza and greening Diseases of citrus in Kerala. *Agric. Res. J. Kerala,* 8: 123–124

Nario Kusano. 2006. Simple and rapid diagnosis of Satsuma dwarf by Immune-chromatographic assay using monoclonal antibodies. *Plant Protection,* 60: 491-495

Navarro, L., Jfiarez, J., Ballester, J.F. and Pina, J.A. 1980. Elimination of some citrus pathogens producing psorosis like leaf symptoms by shoot-tip gtrafting *in vitro. In*: Proc. 8th Conf. IOCV, IOCV, Riverside, CA. pp. 162–166.

Navarro, L., Civerolo, E.L., Juarez, J. and Garnsey, S.M. 1989. Improvement therapy methods for citrus germplasm exchange. *In*: Abs. 11th Conf. IOCV, IOCV, Riverside, CA.

Navarro L., Pina J.A., Ballester-Olmos J.F., Moreno P. and Cambra M. 1984. A new graft transmissible disease found in Nagami kumquat. *In*: Proc. 9th Conf. IOCV, IOCV, Riverside, CA. pp. 234–240

Navarro, L., Pina, J.A., Juárez, J. and Ballester, J.F. 1993. Elimination of a bud union abnormality of sweet orange grafted on rough lemon by shot-tip grafting *in vitro. In*: Proc. 12th Conf. IOCV, IOCV, Riverside, CA. pp. 375–378.

Navas-Castillo, J., Moreno, P., Cambra, M. and Derrick, K. 1993. Partial purification of a virus associated with a Spanish isolate of citrus ringspot. *Plant Pathology,* 42: 339–346.

Ndowora, T. Dahal, G., LaFleur, D., Harper, G., Hull, R. Olszewski, N.E. and Lockhart, B.E.L. 1999. Evidence that badnavirus infection in *Musa* can originate from integrated pararetroviral sequences. *Virology,* 255: 214–220.

Nishio, T., Kawai, A., Kato, M. and Kobayashi, T. 1982. A sap-transmissible closterovirus in citrus imported from China and Formosa. *Res. Bull. Plant Prot.Service Japan,* 18: 11–18.

Nishio, T., Kawai, A., Takahashi, T., Namba, S. and Yamashita, S. 1989. Purification and properties of citrus tatter leaf virus. *Annl.Phytopathol. Soc.,* 55: 254–258.

Noe-Becerra, E., Cardenas, E., Lozoya, H. and Mosquedo, R. 1999. Rhabdovirus in Papaya (*Carica papaya* L.) in the Southeast of Mexico. *Agronomia–Meroamericana* 10 : 85–90.

Norman, P.A, and Grant, T.J. 1956. Transmission of tristeza virus by aphids in Florida. *Proc. Fla. State hort.Soc.* 69: 38–42.

Noronha Jr, N.C., Bento, J. M. S. and Parra, J. R. P. 2008. Guava and citrus plant volatiles. Fundecitrus workshop on HLB-resistant citrus. *Araraquara,* February 2008.

Nour-Eldin, F. 1956. Phloem discoloration of sweet orange. *Phytopathology,* 46: 238-239.

Nour-Eldin, F. 1959. Citrus virus research in Egypt. *In*: Citrus virus diseases. Berkeley, Div. Agric. Sci. Univ. Calif. Berkeley, CA. pp. 219–227.

Nour-Eldin, F. 1968. Gummy bark of sweet orange. Indexing procedures for 15 virus diseases of citrus trees. *In*: Agricultural handbook No. 33. Washington, D.C., Agr. Res. Service.USDA. pp. 50–53

Nozu, Y.; Usugi, T. and Nishimori, K. 1986. Production of monoclonal antibodies to satsuma dwarf virus. *Annl. Phytopathol. Soc. Japan,* 52: 86–89.

Nunes, M.A., Kitajima, E.W., Freitas-Astua, J., Hill, M.E. and Gottwald, T.R. 2006. Infecção de hibisco e malvavisco pelo vírus da leprose dos citros através de *Brevipalpus phoenicis* (Acari: Tenuipalpidae). *In: Resumos Simpósio Brasileiro de Acarologia I. Viçosa, Universidade Federal de Viçosa.* p. 218.

Nutman, F.J. and Robert, P.M. 1955. Lethal yellowing the "unknown disease"of Cococut palms in Jamaica. *Emp. J. Exp. Agric.,* 23: 257–267.

OEPP/EPPO. 1995. Certification schemes No. 12. Pathogen-tested citrus trees and rootstocks. *OEPP/EPPO Bulletin* 25:4.

Ohira, K., Ito, T., Kawai, A., Namba, S., Kusumi, T. and Tsuchizaki, T. 1994. Nucleotide sequence of the 3'-terminal region of citrus tatter leaf virus RNA *Virus Genes,* 8: 169–172.

Ohira, K., Nimba, S., Rosanov, M., Kusumi, T. and Tsuchizaki, T. 1995. Complete sequence of an infectious full length cDNA clone of citrus tatter leaf Capillovirus:comparative sequence analysis of Capillovirus genomes. *J. Gen.Virol.,* 76: 2305–2309.

Oldfield, G.N. 1988. Ecological associations of *Spiroplasma citri* with insects, plants and other plant mycoplasmas in the western United States. *In*: *Mycoplasma diseases of crops. Basic and applied aspects* (*Eds.* by Maramorosch, K. and Raychaudhuri, S.P.), Springer-Verlag, New York, USA. pp. 175–191.

Oldfield, G.N., Kaloostian, G.H., Pierce, H.D., Calavan, E.C., Granett, A.L. and Blue, R.L. 1976. Beet leafhopper transmits Citrus stubborn disease. *California Agriculture,* 30 :6–15.

Oldfield, G.N., Sullivan, D.A. and Calavan, E.C. 1984. Aspects of transmission of the citrus stubborn pathogen by *Scaphytopius nitrides* (DeLong). *In*: Proc. 9th Conf. IOCV, IOCV, Riverside CA. pp. 131–136.

Omoto, C. 1998. Acaricide resistance management of leprosis mite (*Brevipalpus phoenicis*) in Brazilian citrus. *Pesticide Sci.,* 52: 189–198.

Önelge, N. 2003. First report of yellow vein clearing of lemons in Turkey. *J. Turk. Phytopathol.,* 32: 53–55.

Onelge, N., Bozan, O., Gok, M. and Satar, S. 2010. Yellow vein clearing of lemons in Turkey.. Proc. 17th Conf. IOCV, IOCV, Riverside, CA. pp. 227–228.

Onelge, N. and Çýnar, A. 2010. Virus and virus-like diseases in turkey citriculture. Proc. 17th Conf. IOCV, IOCV, Riverside, CA pp. 233–236.

Palacio, A. and Duran-Vila, N. 2000. Citrus cachexia disease: Molecular characterization of its viroid agents. *In*: Proc. 14th Conf. IOCV, IOCV, Riverside, CA. pp. 273–281.

Palacio-Bielsa, A., Foissac, X. and Duran-Vila, N. 2000. Indexing of citrus viroids by imprint hybridization. *Europ. J. Plant Pathol.,* 105: 897–903.

Palukaitis, P., Hatta, T., Alexander, D-McE., and Symons, R.H. 1979. Characteisation of a viroid associated with avocado sunblotch disease. *Virology,* 99: 145–151.

Palukaitis, P., Rakowski, A.G., Alexander, D. McE. and Symons, R.H. 1981. Rapid indexing of the sunblotch disease of avocados using a complementary DNA probe to avocado sunblotch viroid. *Ann.Appl. Biol.* 98, 439–449.

Palukaitis, P. and Symons, R.H. 1980. Purification and characterization of the circular and linear forms of chrysanthemum stunt viroid. *J. Gen. Virol.* 46: 477–489.

Palukaitis, P., Roossinck, M. J., Dietzgen, R.G., and Francki, R.I.B. 1992. Cucumber mosaic virus. *Adv. Virus Res.,* 48: 349–377.

Pandey, P. K., Chavan, V.M. and Ahlawat, Y.S.1987. Nasnaran –an indicator host of severe strain of citrus tristiza virus. *In*: Proc. 5th Nat. Citrus Seminar, Jorhat. p. 23.

Pandey, P.K., Nimbalkar, M.R. and Dhale, M.G. 1986. Observations on a mosaic disease of citrus in Maharashtra State. *J. Maharashtra Agric. Univ.*, 11: 254 .

Pant, R.P. and Ahlawat, Y.S. 1997. Partial characterization of citrus mosaic virus. *Indian Phytopath.*, 50: 557–564.

Pant, R. P. and Ahlawat, Y.S. 1998. Partial characterisation of a filamentous virus associated with ringspot disease of citrus. *Indian. Phytopath.*, 51: 225–232.

Pant, R.P., Ahlawat, Y.S. and Milne, R.G. 1997. Studies on Citrus ringspot virus in India. *In*: National Symposium on Citriculture, Nov. 17-19, 1997 at NRCC,Nagpur (abstract). pp. 82–83.

Pant, R.P., Baranwal, V.K., Singh, S.P. and Ahlawat, Y.S. 2000. Mechanical transmission of Indian citrus ringspot virus (ICRSV) *In*: *Proc.* Internatl. Symp. Citriculture, Nagpur. pp. 1021–1023.

Papasolomontos, A. 1965. The present status of impietratura, a citrus disease in Cyprus. *Plant Dis. Reptr.*, 49: 111–113.

Pappu, S.S., Pappu, H.R., Niblett, C.L. and Bird, J. 1991. Molecular cloning and nucleotide sequence of isolates of cucumber mosaic virus from bananas and plantains in Puerto Rico. *Phytopathology*, 81: 1154.

Pappu, H.R., Pappu, S.S., Niblett, C.L., Lee, R.F., and Civerolo, E. 1993. Comparative sequence analysis of the coat protein of biologically distinct citrus tristeza closterovirus isolates. *Virus Genes*, 7: 255–264.

Parmeshwari, B., Mangranthia, S.K., Parveen, S. and Jain, R.K. 2009. Complete Genome sequence and organization of papaya ringspot virus pathotypes P and W from India. *In*: 5th Intl. Conf. Pl. Path. Globalized era, Nov. 10-13, 2009, NewDelhi, India (Abst). p. 124.

Parthasarathy, M.V. 1974. Mycoplasma-like organisms associated with lethal yellowing disease of palms. *Phytopathology*, 64: 667–674

Pascon, R.C., Kitajima, J.P., Breton, M.C., Assumpção, L., Greggio, C., Zanca, A.S., Okura, V.K., Alegria, M.C., Camargo, M.E., Silva, G.G.C., Cardozo, J.C., Vallim, M.A.I., Franco, S.F., Silva, V.H., Jordão Jr., H., Oliveira, F., Giachetto, P.F., Ferrari, F., Aguilar-Vildoso, C.L., Franchiscini, F.J.B., Silva, J.M.F., Arruda, P., Ferro, J.A., Reinach, F. and Silva, A.C.R. 2006. The complete nucleotide sequence and genomic organization of Citrus leprosis associated virus, cytoplasmic type (CiLV-C). *Virus Genes*, 32: 289-298.

Patil, B.P. and Warke, D.C. 1968. A note on existence of exocortis virus in India. *Curr. Sci.*, 37: 469–470.

Permar, T., Garnsey, S.M., Gumpf, D.J. and Lee, R. 1990. A monoclonal antibody that discriminates strains of citrus tristeza virus. *Phytopathology*, 80: 224–228.

Peña, L. and Navarro, L. 1999. Transgenic citrus. *In*: *Biotech. Agric. and Forestry* Y.P.S. Bajaj, Springer, Berlin. pp. 39–54.

Peng de W., Zheng, G.H., Zheng, Z.Z., Tong, Q.S. and Ming, Y.L.2013. Orchid fleck virus: An unclassified bipartite, negative-sense RNA plant virus. *Arch. Virol.*, 158: 313–323.

Pérez-M. B. E. and Ochoa, A. 1998. Regeneration of transgenic plants of Mexican lime from *Agrobacterium rhizogenes* –transformed tissues. *Plant Cell Rep.* 17: 591–596.

Petri, L. 1931. Variegatura infecttiva delle foglie di "citrus vulgaris". Risso. *Boll. R. Stazione Pat. Veg.*, N.S., 11: 105–114.

Philippe, G., Noa-Carrazana, J.C., Lescot, M., Lheureux, F., Lockhart, B.E.L., Pietro, T.M.P. and Iskra-Caruana, M.L. 2008. A single Banana streak virus integration event in the banana genome as the origin of infectious endogenous pararetrovirus (EPRV). *J. Virol.*, 82: 6697–6710.

Plavsic-Banjac, B., Hunt, P. and Maramorosch, K. 1972. Mycoplasma-like bodies associated with lethal yellowing disease of coconut palms. *Phytopathology,* 62: 298–299.

Pooler, M. R. and Hartung, J.S. 1995. Specific PCR detection and identification of *Xylella fastidiosa* strains causing citrus variegated chlorosis. *Curr. Microbiol.,* 31: 377–381.

Powell, C.A., Pelosi, R.R., Sonoda, R.M. and Lee, R.F. 1998. A psorosis-like agent prevalent in Florida's grapefruit groves and budwood sources. *Plant Dis.,* 82: 208–209.

Price, W.C. 1963. Transmission of psorosis virus by dodder. *In*: Proc. 3rd Conf. IOCV, IOCV, Riverside, CA. pp. 162–166.

Price, W.C. 1970. En. C.M.I./A.A.B. Description of plant viruses no. 33.

Pringle, C.R. 1998. Virus taxonomy San Diego 1998. *Arch. Virol.* 143:1449-1459

Radha, K. and Lal, S.B. 1972. Diagnostic symptoms of root (wilt) disease of coconut. *Indian J. agric. Sci.,* 42: 410–413.

Raj, S.K., Saxena, H., Hallan, V. and Singh, B.P. 1998. Reverse transcription-polymerase chain reaction (RT-PCR) for direct detection of Cucumber mosaic virus (CMV) in gladiolus. *Biochem. Mol. Biol. Int.* 44: 89–95.

Rajan, P. and Mathen, K. 1984. *Sophonia greeni* (distant-Nirvanidae-Jassoides) on leaves of coconut palm (*Cocos nucifera*). *J. Plant Crops,* 12: 178–179.

Rajan, P. and Mathen, K.1985. *Proutista moesta* (westwood) and other additions to insect fauna on coconut palm. *J. Plant Crops,* 13: 135–136.

Ramapandu, S. and Rajamannar, M. 1983. Symptomatology and indexing of tatipaka disease of coconut. *Indian Phytopath.* 36: 608–612

Ram, R.D. and summanwar, A.S. 1984. *Colocasia esculenta* (L.) Schott. A reservoir of bunchy top of banana. *Curr. Sci.,* 53: 145–146.

Ramachandran, P., Pandey, P.K., Ahlawat, Y.S., Varma, A. and Kapur, S.P. 1993. Viroid diseases of citrus in India. *In*: Proc. 12th Conf. IOCV, IOCV, Riverside. pp. 438–440.

Ramachandran, P. Mathur, S., Agarwal, J. and Ahlawat, Y.S. 2003. Diagnostics for Citrus exocortis and Hop stunt viroids associated with corky vein disease in citrus. *Indian Phytopath.* 56: 428-433.

Ramachandran, P., Agarwal, J., Roy, A., Ghosh, D.K. , Das, D.R. and Ahlawat, Y.S. 2005. First record of a hop stunt viroid variant of Nagpur mandarin and Mosambi sweet orange trees or rough lemon and Rangpur lime rootstocks. *New disease reports, British Society for Plant Pathology. Vol 10, August 04-January 05*

Ramjos, C.S. and Zamora, A.B. 1990. Elimination of banana bunchy top infection from banana (*Musa* sp. Cv. Lacatan) by heat pretreatment and meristem culture. *Phil. J. Crop Sci.,* 15: 119–123.

Rana, G.L., Quacquarelli, A and Martelli, G.P. 1974. Some properties of citrus variegation virus. *In*: Proc. 6th Conf. IOCV, IOCV, Riverside, CA. pp. 165–168.

Randles, J.W. 1975. Association of two nucleic acid species with cadang-cadang disease of coconut palm. *Phytopathology,* 65: 163–167.

Randles, J.W. and Hanold, D. 1989. Coconut foliar decay virus particles are 20 nm icosahedra. *Intervirology,* 30: 177–180.

Randles, J.W., Julia, J.F., Calvez, C. and Dollet, M. 1986. Association of a single stranded DNA with the foliar decay disease of coconut palm in Vanuatu. *Phytopathology,* 76: 889–894.

Randles, J.W., Miller,D. C., Morin, J. P., Rohde, W. and Hanold, D. 1992. Localisation of coconut foliar decay virus in coconut palm. *Ann. Appl. Biol.,* 121: 601–617.

Rangel, B., Krueger, R.R. and Lee, R.F. 2005. Current research on *Spiroplasma citri* in California. *In*: Proc.16th Conf. IOCV, IOCV, Riverside, CA. pp. 439–442.

Rao, D.G. 1980. A short note on bunchy top disease of banana. *In*: Nat. Sem. Banana Prod. Tech., Faculty of Hort. TNAU, Coimbatore, India. pp. 142–143

Rao, D.G. 2006. Studies on a new strain of banana mosaic virus in south India. National Sem. Banana Produc. Tech. CABI Abstract.

Raychaudhuri, S.P. and Nariani, T.K. 1989. (2nd Ed.). Virus and Mycoplasma diseases of plants in India. Malhotra Pub. House, New Delhi.

Raychaudhuri, S.P., Nariani, T.K. and Lele, V.C. 1967. Citrus dieback complex – a serious threat to citrus industry in India. Internl. Symp. on sub tropical and Tropical Horticulture, New Delhi.

Raychudhuri, S.P., Nariani, T.K. and Lele, V.C. 1969. Citrus dieback problem in India. *Proc. First Internl. Cirus Symp.*, 3: 1433–1437.

Raychaudburi, S.P., Nariani, T.K., Lele, V.C. and Singh, G.R. 1972. Greening and citrus decline in India. *In*: Proc. 5th Conf: IOCV, IOCV, Riverside, CA. pp. 35-37.

Raychaudhuri, S.P., Nariani, T.K. and Ahlawat, Y.S.1977. Dieback of citrus in India. *Proc. Interl. Soc. Citriculture,* 3: 914–918.

Reddy, G.S. 1965. Citrus decline in south India. *In Proc. 3rd Conf: IOCV,* IOCV, Riverside, CA. p. 225.

Reddy, G.S. and Govinda Rao, P. 1961. Is there tristeza in Andhra Pradesh, India? *In*: Proc. 2nd Conf: IOCV, IOCV, Riverside, CA. pp. 132–135.

Reddy, B.V.B. 1997. Characterization of citrus mosaic virus and to develop methods for its quick detection. Ph.D thesis, Division of Plant Pathology, IARI, New Delhi, India.

Reddy, M. R. S., Naidu, P.H. and Raju, D.G. 1985. Dodder laurel transmission of citrus mosaic from sweet orange to acid lime. *Indian Phytopath.*, 38: 172.

Reddy, G.S., Dakshinamurti, V., and Reddy, V.R.K. 1974. Yellow corky vein: First report on a graft transmissible disorder of "Sathgudi" in Andhra Pradesh. *Indian Phytopath,* 27: 82–84.

Reddy, G.S. and Naidu, P.H. 1989. Transmission of citrus yellow vein. *Indian Phytopath.* 42: 455–456.

Reichel, H, Martínez, A.K. Arroyave, J.A., Sedano, R., Morales, F.J., Duterme, O., Kummert, J. and Lepoivre, P. 2003. First Report of *Banana mild mosaic virus* isolated from Plantains (*Musa* AAB) in Colombia. *Plant Dis.,* 87: 1150.3.

Reichert, I. and Perlberger, P. 1934. Xyloporosis, the new citrus disease. Jewish Agency for Palestine Agr. Expt. Sta. *Rehovot, Bull.* 12: 1–50.

Reinking, O.A. 1919. Diseases of economic plants in south China. *Philippine Agr.* 8: 109–135

Rethinam, P., Rajamannar, M., and Narasimhachari, C.L. 1989. Tatipaka disease of coconut in Andhra Prqadesh. *Indian Coconut J.,* 20: 1–4

Reuther, W., Calavan, E.C. and Carman, G.E. 1978. The citrus Industry Vol. IV. (revised ed). Division of Agricultural sciences, University of California, Riverside, CA. p. 362.

Reuther, W., Calavan, E.C. and Carman, G.E. 1978. (Edtors). *In*: The citrus Industry. Dept. of Agric. Sci., Univ. California, Riverside. pp. 170-171

Rivera, R., Ramirez, P. and Pereira, R. 1993. Preliminary characterization of viruses infecting banana in Costa Rica. *In*: Proc. Workshop Biotechnol. Appl. Banana plantain improve. International network for the improvement of banana andplantain, Montpellier, France, pp. 63–68.

Robert, K., Bash, J., Benjamin, R. and Lee, R.F. 2005. Dweet mottle virus and citrus leaf blotch virus in California. IOCV Newsletter.

Rocha-Pena, M.A. and Lee, R.F. 1991. Serological techniques for detection of citrus tristeza virus. *J. Virol.Methods,* 34: 311–331.

Rodoni, B.C., Ahlawat, Y.S., Varma, A., dale, J.L. and Harding, R.M. 1997. Identification and characterization of banana bract mosaic virus in India. *Plant Dis.,* 81: 669-672.

Rodrigues, J. C. V., Kitajima, E.W., Childers, C.C. and Chagas, C.M. 2003. Citrus leprosis virus vectored by *Brevipalpus phoenicis* (Acari: Tenuipalpidae) on citrus in Brazil. *Exp. Appl. Acar.,* 30: 161–179.

Rodrigues, J.C.V. 2000. Relações patogeno-vector-planta no sitema leprose dos citrus. Ph.D. Thesis. 168 pp. da Universidade de São Paulo, Piracicaba, Brasil.

Rodrigues, J.C.V., Locali, E., Freitas-Astúa, J. and Kitajima, E.W. 2005. Transmissibility of citrus leprosis virus by *Brevipalpus phoenicis* in *Solanum violaefolium. Plant Dis.,* 89: 911

Rodrigues Neto, J., Berian, L.O.S., Baptista, C.R. and Oliveira, A.R. 1992. Seed transmission of an organism associated to CVC-V. RAIB-23a 27/11/92 Instituto Biologico, Arq., do Instituro Biologico (Suplemento).

Rohde, W., Randles, J.W., Langridge, P. and Hanold, D. 1990. Nucleotide sequence of a circular single stranded DNA associated with coconut foliar decay virus. *Virology,* 176: 648–651.

Roistacher, C.N. 1977. Elimination of citrus pathogens in propagative budwood. I. Budwood selection, indexing and thermotherapy. Proc. *Int. Soc. Citriculture,* 3: 965–972.

Roistacher, C.N. 1988. Citrus tatter leaf virus: further evidence for single virus complex. *In*: Proceedings 10th Conf. IOCV, IOCV, Riverside,CA. pp. 353-369.

Roistacher, C.N. 1988. The cachexia and xyloporosis diseases of citrus - a review. *In*: Proc. 10th Conf: IOCV, IOCV, Riverside, CA. pp. 116–124.

Roistacher, C.N. 1991. Graft transmissible diseases of citrus. Handbook of detection and diagnosis. FAO Rome, p. 286.

Roistacher, C.N., 1993. Psorosis – A review. *In*: Proc. 12th Conf. IOCV, IOCV, Riverside,CA pp. 139-154.

Roistacher C.N. and Calavan, E.C. 1965. Cross protection studies with strains of concave gum and psorosis viruses. Proc. 3rd Conf. IOCV, IOCV, Riverside, CA. pp. 154-161.

Roistacher, C.N.and Kitto, S.L. 1977. Elimination of additional citrus viruses by shoot-tip grafting *in vitro. Plant Dis. Reptr.,* 61: 594–596.

Roistacher, C.N. and Dodds, J.A. 1993. Failure of 100 mild citrus tristeza virus isolates from California to cross protect against a challenge by severe sweet orange stem pitting isolates. *In*: Proc. 12th Conf. IOCV, IOCV, Riverside, CA. pp. 100–107.

Roistacher C.N., Nauer E.M. and Wagner R.C. 1980. Transmissibility of cachexia, Dweet mottle, psorosis and infectious variegation viruses on knife blades and its prevention. PP. 225-229 In; Proc. 8th Conf. IOCV, IOCV, Riverside,CA

Roistacher, C.N., Gumpf, D.J., Nauer, E.M. and Gonzalea, R.1983. Cachexia disease: virus or viroid. *Citrograph,* 68: 111–113.

Roistacher, C.N., D'Onghia, A.M. and Djelouah, K. 2000. Defining psorosis by biological indexing and ELISA. *In*: Proc. 14th Conf. IOCV, IOCV, Riverside,CA. pp. 144–151.

Roistacher, C.N., Bash, J. and Gumpf, D.J. 2000. Continued attempts over a 22-year period to separate components of the citrus tatter leaf-citrange stunt virus complex. *In*: Proc. 13th Conf. IOCV, IOCV, Riverside, CA. pp. 179–184.

Román R.C.F., Gutiérrez E.M.A., Pérez-M., E., Gutiérrez C.R., Gómez L.M.A., Palacios, T, E.C. and Robles G.M.M. 2005. Expression of different genes to confer resistance to pathogens in transgenic citrus plants. *In*: Proc. 16th Conf. IOCV, IOCV, Riverside, CA pp. 392–398.

Roossinck, M.J. 2002. Evolutionary history of Cucumber mosaic virus deduced by phylogenetic analayses. *J. Virol*, 76: 3382–3387.

Roossinck, M.J. Zhang, L. and Hellward, K. 1999. Rearrangements in the 5' nontranslated region and phylogenetic analyses of Cucumber mosaic virus RNA3 ndicate radial evolution of three subgroups. *J. Virol.*, 6: 6752–6758.

Rossetti, V. 1980. Differenciacao entre cancro citrico e outras molestias. *Citrus*, 1: 23–26.

Rossetti, V. 1993. Citrus variegated chlorosis, a new severe disease in Brazil : A reviw. *In*: Proc. 12th Conf. IOCV, IOCV, Riverside, CA. pp. 449–452

Rossetti, V. 1996. Transmission of citrus leprosis disease (CL) – A review. *In*: Proc. 13th Conf. IOCV, IOCV, Riverside, CA. pp. 331–335.

Rossetti, V. 2000. Citrus variegated chlorosis (CVC) in Brazil – an overview. *In*: Proc. 14th Conf. IOCV, IOCV, Riverside, CA. pp. 213–215

Rossetti, V., Beretta, M.J.G., Lefevre, A.F.V. and Trindade, M.L.B. 1988. Evaluation of possible recovery of declinio affected citrus trees. *In*: Proc. 10th Conf. IOCV, IOCV, Riverside, CA. pp. 396–399.

Rossetti, V., Garnier, M., Bove, J.M., Beretta, M.J.G., Teixeira, A.R.R., Quaggio, J.M. and De negri, J.D. 1990. Presence de bacteries dan le xyleme d'orangers attaints de chlorose variegee, une nouvelle maladie des agrumes au Brazil. *C.R. Acad. Sci. Paris* 310, serie III: 345–349

Rosner, A., Bar-Joseph, M., Moskovitz, M. and Mevarech, M. 1983. Diagnosis of specific viral RNA sequences in plant extract by hybridization with polynucleotide kinase mediated P32 labeled double stranded RNA probe. *Phytopathology*, 73: 609–702.

Roy, A. and Ramchandran, P. 2003. Occurrence of a Hop stunt viroid (HSVd)variant in Yellowcorky vein disease of citrus in India. *Curr. Sci.*, 85: 1608–1612

Roy, A. and Ramachandran, P. 2006. Characterization of citrus exocortis viroid variant In yellow corky vein disease of citrus in India. *Curr. Sci.*, 91: 798–803.

Roy, A., Ramachandran, P., and Brlansky, R.H. 2003. Grouping and comparison of Indian citrus tristeza virus isolates based on coat protein gene sequences and restriction analysis patterns. *Arch. Virol.*, 148: 707–722.

Roy, A., Fyad, A, Barthe, G. And Brlansky, R.H. 2005. A multiplex polymerase chain reaction method for reliable, sensitive and simultaneous detection of multiple viruses in citrus trees. *J. Virol. Methods* 129: 47–55.

Roy, A. and Brlansky, R.H. 2010. Genome analysis of an orange stem pitting *Citrus tristeza virus* isolate reveals a novel recombinant genotype. *Virus Res.*, 151: 118–130.

Roy, A., Ananthakrishnan, G., Hartung J.S. and Brlansky, R.H. 2010. Development and application of a multiplex reverse transcription polymerase chain reaction assay for screening a global collection of *Citrus tristeza virus* isolates. *Phytopathology*, 100: 1077–1088.

Roy, A., Choudhari, N., Guillermo, L.M., Shao, J., Govindarajalu, A., Achor, D., Wei, G., Picton, D.D., Levi, L., Nakhla, M.K., Hartung, J.S and Brlansky, R.H. 2013. A novel virus of the genus *Cilevirus* causing symptoms similar to citrus leprosis. *Phytopathology*, 103: 488–500.

Roy, A., Stone, A., Otero-Colina, G., Wei, G., Choudhary, N., Achor, D., Shao, J., Levy, L., Nakhla, M.K., Hollingsworth, C.R., Hartung, J.S., Schneider, W.L., and

Brlansky, R.H. 2013. Genome assembly of citrus leprosies virus nuclear type reveals a close association with orchid fleck virus. *Genome Announc.*, 1(4): e00519–13. doi:10.1128/genomeA00519-13.

Roy, Gourgopal, Jain, R.K., Bhat, A.I. and Varma, A. 1999. Comparative host range and serological studies of papaya ringspot potyvirus isolates. *Indian Phytopath.* 52:14-17.

Ruggieri, G. 1955. Le arance impietrate. *Riv. Agrumic.*, 1: 65–69.

Rustem, F.A. and Ahlawat, Y.S. 1999. Yellow corky vein – a viroid disease of citrus in India. *Indian J. agric. Sci.*, 69: 789–794.

Rustici, G., Accotto, G.P. Noris, E., Mesenga, V., Luisoni, E. and Milne, R.G. 2000a. Indian citrus ringspot virus: A proposed new species with some affinities to potex, Carla, fovea and allexiviruses. *Arch. Virol.*, 145: 1895–1908.

Rustici, G., Milne, R.G. and Accotto, G.P. 2000b. Nucleotide sequence, genome organization and phylogenetic analysis of Indian citrus ringspot virus. *Arch. Virol.*, 147: 2215–2224

Rustici, G., Noris, E., Accotto, G.P., Luisoni, E., Milne, R.G., Pant, R.P., and Ahlawat, Y.S. 2000c. Further characterization and detection of Indian citrus ringspot virus. *In*: Proc. 14th conf. IOCV, IOCV, Riverside, California. pp. 360–362.

Saadi, I., Namsi, A., Ben Mahamoud, O., Takrouni, M.L., Zouba, A., Bové, J.M. and Duran-Vila, N. 2006. First report of 'Maladie des feuilles cassantes' (brittle leaf disease) of date palm in Algeria. *New Disease Reports*, Volume 13: February 2006 - July 2006. http://www.bspp.org.uk/ndr/july2006/2006-14.asp

Saglio, P.D., Lafleche, D., Bonissol, C. and Bove, J.M. 1971. Isolement et culture *in vitro* des mycoplasmes associes au 'stubborn' des agrumes et leur observation au microscope electronique. Compy. *Rend. Acad. Sci. Paris, series D*, 272: 1387-1390.

Saglio, P., hospital, M.L., Lafleche, D., Dupont, G., Bove, J.M., Tully, J.G. and Freundt, E.A. 1973. *Spiroplasma citri* gen. and sp. N.:A mycoplasma-like organism associated with "stubborn" disease of citrus. *Intern. J. syst. Bacteriology*, 23: 191–204.

Saillard, C. and Bové, J.M. 1983. Application of ELISA to spiroplasma detection and classification. *In*: *Methods in mycoplasmology* (*Ed*. by Razin, S. and Tully, J.G.), Vol. 1, pp. 471-476. Academic Press, New York, USA.

Saillard, C., Nhami, A., Moreno, P., garnier, M. and Bove, J.M. 1996. *Spiroplasma citri* detection by immunocapture PCR. *In*: Proc. 13th Conf. IOCV, IOCV, Riverside, CA. p. 413.

Saito, Y. and Hibino, H. 1962. Electron microscopy of satsuma dwarf virus in host cells. In: Proc. 5th Conf. IOCV., IOCV, Riverside, CA. pp. 217–222.

Salehi, M., Izadpanah, K., and Taghizadeh, M. 2001. Witches broom disease on lime in Iran: New distribution areas, herbaceous hosts and transmission trial. *In*: Proc. 15th Conf. IOCV, IOCV, Riverside, CA. pp. 293–296.

Salibe, A.A. 1959. Leaf curl – a transmissible virus disease of citrus. *Plant Dis. Reptr.*, 43: 1081–1083.

Salibe, A.A. 1965. Susceptibility of citrus varieties to leaf curl virus. *In*: Proc. 3rd IOCV, IOCV, Riverside, CA. pp. 175-178.

Salibe, A.A. and Cortez, R.E. 1966. Studies on the leaf mottling disease of citrus in the Philippines. *FAO Plant Prot. Bull.*, 14: 141–144.

Samraj, K.S., Menon, M.R., and Christudas, S.P. 1996. Kokkan a new disease of Banana. *Agric. Res. J. Kerala*, 4: 116.

Sankaralingam, A., Baranwal, V.K., Ahlawat, Y.S., Renuka Devi; and Ramiah, M. 2006. RT-PCR detection and molecular characterization of Banana bract mosaic virus from the pseudostem and bract of banana. *Arch. Phytopath. and Plant Prot.*, 39: 273–281.

Sánchez de la Torre, M.E., García,M.L., Riva, O., Dal Bó,E., Jones,L.,Zandomeni, R. and Grau. O. 2000. Genome organization of the top component of citrus psorosis virus and identification of the coat protein gene. *In*: Proc. 14th Conf. IOCV, IOCV, Riverside pp. 345–346

Santos, M.A., Wulff, N.A., Toloy, R.S., Ayres, A.J., and Rezende, J.A.M. 2010. Experimental transmission of citrus sudden death associated virus CSDaV to herbaceous plants. *In*: Abs. 18th Conf. IOCV held in Campinas/SP, Brazil, Nov. 7-12, 2010. p. 104.

Saponari, M., Manjunath, K.L. and Yokomi, R.K. 2008. Quantitative detection of Citrus tristeza virus (CTV) in citrus and aphids by real-time reverse transcription-PCR (TaqMan®). *J. Virological Methods,* 147: 43–53.

Sasikala, M., Mathen, K., Govindankutti, M.P., Solomon, J.J. and Geetha, L. 1989. Transmission of a Mycoplasma-like organism from *Cocos nucifera* with root (wilt) disease to *Catharanthus roseus* by *Cassytha filiformis*. *Eur. J. Plant Pathol.*, 94: 191–194.

Sasikala, M., Prakash, V.R., Sapna, V.P., Mayilvagnan, M. and Leena, S.N. 2005. Refinement of ELISA and its use in early detection of coconut root (wilt) disease. *Cord,* 212: 37–44.

Satyanarayana, T., Mohammad, A.R., Hilf, M.E., Gowda, S., Dawson, W and Garnsey, S.M. 2009. Molecular characterization of citrus tatter leaf virus historically associated with Meyer lemon trees.: Complete genome sequence and development of biologically active *in vitro* transcripts. *Phytopathology,* 99: 423–431.

Savino,V., Gallitelli, D. and Barba, M. 1983. Olive latent ringspot virus, a newly recognised virus infecting olive in Italy. *Ann. appl. Biol.*, 103: 243–249.

Savino,V., Piazzola, P., Di Franco, A. and Martelli, G.P. 1984. Olive latent virus-2, a newly recognized virus with differently shaped particles. *In*: Proc. 6th Cong. Mediter. Phytopathological Union, Cairo. pp. 24–26

Saxena, S. and Singh, K.N. 2005. Papaya leaf curl resistant transgenic papaya for improved papaya production. Presented at 74 th SBC (I) Annual meeting held at CDRI, lucknow during Novemver 07-10, 2005.

Saxena, S., Srivastava, K.M., Raizada, R.K., Santha Devi and Singh. B.P. 1994." E.M. Based on phytopathological studies: An aid for a Papaya virus identification". Presented at the XIX Conference of the Electron microscope Society of India, held at NPL, New Delhi, Dec., 21–23, 1994.

Saxena, S., Srivastava, K.M., Raizada, R.K. Zaidi, A.A. and Singh, B.P. 1995. "Biochemical and immunological characterization of Papaya ring spot virus occuring in India". Presented at Indian Phytopathological Society held at NDUAT Kumarganj, Faizabad during January 17–18, 1995.

Saxena, S., Hallan, V., Singh, B.P. and Sane, P.V. 1998a. Leaf curl disease of *Carica Papaya* from India may be caused by a bipartite geminivirus. *Plant Dis.*, 82: 126.

Saxena, S., Hallan, V., Singh, B.P. and Sane, P.V. 1998b. Nucleotide sequence and intergeminiviral homologies of the DNA-A of papaya leaf curl geminivirus from India. *Biochem. Mol. Biol. Int.*, 45: 101–113.

Saxena ,S., Srivastava, N. and Chandra, R. 2004 "Polymerase Chain Reaction (PCR) and sequence analysis based diagnostics to detect papaya leaf curl geminivirus. Presented In Zonal Meeting of IPS (Indian Phytopathological Society. & National Symposium on "Detection and Management of Plant Diseases Using Conventional and Modern Tools" held on31st December 04, at Lucknow.

Scaramuzzi, G. and Catara, A.A. 1968. Studies on sour orange stem pitting in Sicily. *In*: Proc. 4th Conf. IOCV, IOCV, Riverside, CA. pp. 201–205.

Schneider, H. 1959. The anatomy of tristeza virus infected citrus. *In*: Wallace, J.M. (*Ed*) Citrus virus diseases. Univ. Cal. Div.Agr. Sci. Berkeley, CA. pp. 73–84.

Schwartz, R.E. 1970. Multiple sprouting disease of citrus. *Plant Dis. Reptr.* 54: 1003-1007.

Schwinghamer, M. W. and Broadbent, P. 1986. Detection of viroids in dwarfed orange trees by transmission to Chrysanthemum. *Phytopathology,* 77: 210–215.

Schwinghamer, M.W. and Broadbent, P. 1987. Association of viroids with graft transmissible dwarfing symptoms in Australian orange trees. *Phytopathology,* 77: 205–207.

Scott, S.W. and Ge, X. 1995a. The complete nucleotide sequence of the RNA-3 of citrus leaf rugose and citrus variegation ilarviruses. *J. Gen. Virol.,* 76: 957–963

Scott, S.W. and Ge, X. 1995b. The nucleotide sequence of citrus leaf rugose virus RNA-1. *J. Gen. Virol.* 76: 3233–3238.

Searle, C.M. 1969. A preliminary report on off- type trees. Mazoe Citrus Estates, (Rhodesia), p. 15 .

Selvarajan, R., Sathiamoorthy, S., Dayakar, S., Balasubramanian, V., Viswanathan, R. and Ahlawat, Y.S. 2002. Purification and detection of Indian isolates of banana bunchy top virus (BBTV). *In*: Proc. 54th crop protection and WTO-an Indian perspective. p. 30.

Selvarajan, R.,Balasubramanian, V., Dayakar, S., Uma, S., Ahlawat, Y.S. and Sathiamoorthy, S. 2002. Molecular diagnosis of banana bunchy top nanavirus (Indian isolates) coat protein gene by polymerase chain reaction. *In*: Global Conf. on Banana and Plantain. p. 142

Selvarajan, R.,Balasubramanian, V., Dayakar, S., Kavitha, K.S., Uma, S., Sathiamoorthy, S. and Ahlawat, Y.S. 2004. Detection of geographical isolates of banana bunchy top virus in India. *In*: 1st Internl. Cong. On Musa: Harnessing research to improve livelihood 6-9 July, 2004, Penand, Malaysia

Selvarajan, R.,Balasubramanian, V., Kavitha, K., Kavitha, K.S., Sathiamoorthy, S. and Ahlawat, Y.S. 2008. Detection of banana bunchy top virus and banana streak Mysore virus by PCR. Impact of storing virus infected banana samples. *Indian J.Virol.* 19: 155–159.

Selvarajan, R., Balasubramanian, V., Jryabaskaran, K.J., Pandey, S.D. and Mustaffa, M.M. 2009. Management of bract mosaic disease through higher dose of fertilizers in banana cv. Ney Poovan (AB). *Indian J. Hort.,* 66: 301–305.

Selvarajan, R., Sheeba, M.M., Balasubramanian, V., Rajmohan, R., Lakshmi, D.N. and Sasireka, T. 2010. Molecular characterization of geographically different Banana bunchy top virus isolates in India. *Indian J. Virol.,* 21: 110–116.

Selvarajan, R., Balasubramanian,, V., Dayakar, S., Sathiamoorthy, S. and Ahlawat, Y.S. 2010. Evaluation of immunological and molecular techniques for the detection of different isolates of Banana bunchy top virus in India. *Indian Phytopath,* 63: 333–336.

Selvarajan, R.,Mary Sheeba, M. and Balasubramanian, V. 2011. Simultaneous detection of episomal banana streak Mysore virus and banana bunchy top virus using multiplex RT PCR. *Curr. Sci.,* 100: 31–34.

Semancik, J.S. and Weathers, L.G. 1965. Partial purification of a mechanically transmissible virus associated with tatter leaf of citrus. *Phytopathology,* 55: 1354–1358.

Semancik, J.S., Roistacher, C.N., Rivera-Bustamante, R. and Duran-Vila, N. 1988. Citrus cachexia viroid, a new viroid of citrus: relationship to viroids of the exocortis disease complex. *J. Gen. Virol.,* 69: 3059–3068.

Shanta, P., Joseph, T. and Lal, S.B. 1964. Transmission of root (wilt) disease of coconut. *Indian coconut J.* 18: 25–28.

Sharman, M., Thomas, J.E. and Dietzgen, R.G. 2001. Development of a multiplex immunocapture PCR with colourimetric detection for viruses of banana. *J. Virol. Methods,* 89: 75–88.

Sharma, S.R. and Singh, S.J. 1988. Association of mycoplasma-like organisms with Citrus blastomania disease. *Indian J. Pl. Pathol.,* 6: 86–92.

Sharma, D.C. and Pandey, P.K. 1983. Leaf yellow mid vein a previously undescribed virus disease of citrus. *Interl. J. Trop. Plant Dis.* 1:149-152

Sharma, Jyoti, Jain, R.K and Varma, A. 2004a. Detection of papaya ringspot virus on naturally infected plants by reverse transcriptase polymerase chain reaction. *Indian Phytopath,* 57: 237–239.

Sharma, Jyoti, Mandal, B.,Jain, R.K., Sudhakar, D., Balasubramanyam, P., Kumar, N. and Varma, A. 2004b. Agrobacterium mediated transformation protocol to develop transgenic papaya resistant to papaya ringspot virus. *Indian Phytopath,* 57 : 347.

Shimizu, S. and Miyoshi, T. 2003. Association of satsuma dwarf virus with dwarf-Like symptoms on novel citrus varieties in Ehime Prefecture. *Bull. Ehime Prefec.Fruit Tree Exp. Stn.,* 16: 41–46.

Silva, A.M.R., Kitajima,E.W., Sousa, M.V. and Resende, R.O. 1997. Papaya lethal yellowing virus: a possible member of the Tombusvirus genus. *Fitopatologia–Brasileira,* 22: 529–534.

Singh, S.J., Selvarajan, R. and Singh, H.P. 2000. Detection of bract mosaic virus (Kokkan disease) by electron microscopy and serology. *In*: Banana improvement, production and utilization. Proc. Conf. AIPUB NRCB (H.P.Singh and K.L.Chadha *Eds*) Trichy, India. pp. 381–383

Singh, Z.; Jones, R.A.C. and Jones, M.G.K. 1995. Identification of cucumber mosaic virus subgroup- I isolates from banana plants affected by infectious chlorosis disease using RT-PCR. *Plant Dis.,* 79: 713–716.

Singh, J., Baranwal, V.K. and Ahlawat, Y.S. 2004. Detection of citrus yellow mosaic virus in its vector *planococcus citri* (mealy bug) by polymerase chain reaction. *Indian Phytopath.,* 57: 335

Singh, J., Baranwal, V.K. and Ahlawat, Y.S. 2009. Molecular characterization and full length genome sequencing of citrus yellow mosaic virus associated with Rangpur lime cultivar. *In*: 5th Intl. Conf. Pl. Path. In globalized era. Nov. 10-13, New Delhi, India (Abst.) p. 296.

Singh, R.P. and Clark, M.C. 1973. Similarity of host response to both potato spindle tuber and citrus exocortis viruses. *FAO Plant Protection Bulletin.,* 21: 121–125.

Singh, R.P., Avila, A.C., Dusi, A.N., Boucher, A., Trindade, D.R., Van Slobbe, W.G., Ribeiro, S.G. and Fonseca, M.E.N. 1988. Association of viroid like nucleic acids with the fatal yellowing disease of oil palm. *Fitopatol. Brazil,* 13: 392–394.

Singh, R.P., Dilworth, A.D., Ao, X., Singh, M. and Baranwal, V.K. 2009. *Citrus exocortis viroid* transmission through commercially-distributed seeds of *Impatiens* and *Verbena* plants. *European J. Plant Pathol.* 124: 691–694.

Sofy, A.R., Soliman, A.M., Mousa, A.A., Ghazal, S.A. and El-Dougdoug, K.A. 2010. "First record of *Citrus viroid II* (CVd-II) associated with gummy bark disease in sweet orange (*Citrus sinensis*) in Egypt". *New Dis. Reports,* 21: 24.

Solomon, J.J., Govindankutty, M.P. and Mathen, K. 1989. Detection of mycoplasma-like organisms in root (wilt) disease affected coconut and its putative vector. In: Plant diseases caused by fastidious prokaryotes (3rd regional workshop on Plant mycoplasma) Raychaudhuri, S.P. and Varma Anupam (*Eds.*). Today and Tomorrow publishers, New delhi. pp. 89–87.

Solomon, J.J., Govindankutty, M.P and Nienhaus, F. 1983. Association of mycoplasma-like organisms with the coconut root (wilt) disease in India. *Z. Pftkrankh. PftSchutz* 90: 295–297.

Storm, l.W. and Streets, R.B. 1963. Transmission of stubborn and psorosis-A viruses of citrus to an herbaceous host. *Phytopath.*, 53: 1142.

Su, H.J., Wu, R.Y. and Tsao, L.Y. 1993. Ecology of banana bunchy top virus disease. *In*: Proc. Imterl. Symp. on recent devt. in banana cultivation technology. 14-18 Dec., 1992, Taiwan banana Res. Instt, Chiuju Pingtung, Taiwan (Valmayer, R.V., Huang, S.C., Ploetz, R., Lee, S.W. and Rao, V.N. *Eds*). INIBAP/ASPNET< Montpellier, France. pp. 308–312.

Su, H.J., Hung, T.H. and Wu, M.L. 1997. First report of banana streak virus infecting banana cultivars (*Musa* spp.) in Taiwan. *Plant Dis.*, 81: 550–551.

Sullivan, D.A., Oldfield, G.N., Eastman, C.E., Fletcher, J. and Gumpf, D.J. 1987. Transmission of a citrus-infecting strain of *Spiroplasma citri* to horseradish. *Plant Dis.* 71: 469.

Sumiya, K.V., Baranwal, V.K. and Ahlawat, Y.S. 2004. isolation and characterization of antiviral principles from seed of *celosia cristata* inhibiting tobacco streak ilarvirus. *Indian Phytopath.* 57: 528.

Summanwar, A.S. and Marathe, T.S. 1982. Diagnostic technique for the detection of bunchy top and infectious chlorosis in banana suckers. *Curr. Sci.*, 52: 47–49.

Swingle, W.T. and Reece, P.C. 1967. The botany of citrus and its wild Relatives. *In*: Reuther, W., Webber, H. and Batchelor, L. (*Eds*.). The citrus Industry Vol.1. Revised ed. Univ. California press, California, pp. 190–430.

Swingle, W.T. and webber, H.J. 1896. The principal diseases of citrus fruits in Florida. *USDA Div. Veg. Physiol. Pathol. Bull.* 8: 42.

Symons, R.H. 1981. Avocado sunblotch viroid: Primary sequence and proposed Secondary structure. *Nucleic acid Research,* 9: 6527–6537.

Takahara, T., Kawase, K., Ono, S., Iwagaki, I., Hirose, K. and Yoshinaga, K. 1988. Rootstocks for Ponkan (*Citrus reticulata* Blanco) in relation to tatter leaf virus. *Bull. Fruit Tree Res. Station* D, 10: 35–45.

Tarafdar, A., Kumari, S. and Biswas, K.K. 2009. Agrobacterium mediated transformation of *Citrus aurantifolia* (Kagzi lime) using coat protein gene of citrus tristeza virus. *In*: 5th Intl. Conf. Pl. Path. Globalized era. November 10-13, 2009, new Delhi, India (Abst.) p. 163.

Tanaka, H. 1972. Mechanical transmission of viruses of satsuma dwarf and natsudaidai dwarf from citrus to citrus. *Ann. Phytopathol. Soc. Japan,* 38: 156–160.

Tanaka, S. and Kishi, K. 1963. Studies on indicator plants for citrus viruses. I. mechanical inoculation on leguminous plants with sap from Satsuma dwarf tree. *Annl. Phytopath. Soc. Japan,* 28: 262–269.

Tanaka, S. and Nakanishi, K. 1972. *Ann. Phytopath. Soc. Japan* 38: 212.

Tanaka, H. and Imada, J. 1974. Mechanical transmission of viruses of satsuma dwarf, Citrus mosaic, navel infectious mottling and natsudaidai dwarf to herbaceous plants. *In*: Pro.6th Conf. IOCV, IOCV, Riverside, CA, pp. 141–145.

Tanaka, H. and Imada, J. 1974. *Physalis floridana,* a good production host for Satsuma dwaf virus. *Plant Dis. Reptr.*, 58: 603–605.

Tanaka, H. and Imada, J. 1976. Purification of viruses of citrus mosaic and navel orange infectious mottling. *In*: Proc. 7th Conf. IOCV, IOCV, Riverside, CA pp. 116–118.

Tanaka, H. and Yamada, S. 1972. Evidence for a relationship among the viruses of Satsuma dwarf, Citrus mosaic, Navel-infectious-mottling, Natsudaidai dwarf,

Citrus variegation, and Citrus crinkly leaf. *In:* Proc. 5th Conf. IOCV. IOCV, Riverside, CA. pp. 71–76.

Tanaka, S., Shikata, E. and Sasaki, A. 1969. Studies on Hassaku dwarf virus. *In*: Proc. First Internl. *Citrus Symp.*, 3: 1445–1448.

Tanaka, H., Yamada, S. and Kishi, K.1971. Symptoms and occurrence of Navel orange infectious mottling and Natsudai-dai dwarf. *Bull. Hort. Res. Stn. B*, 11: 141–147..

Tanner, J.D., Kunta,M., da Graca, J.V., Skaria, M., and Nelson, S.D. 2011. Evidence of A low rate of seed transmission of citrus tatter leaf virus in citrus. *In*: Proc. 18th Conf. IOCV, IOCV, Riverside, CA.

Terranova, G., Caruso, A. and Reforgiato Recupero, G.1991. Susceptibility and Symptomatology of Mapo Tangelo to Xyloporosis. *In*: Proc. 11th Conf. IOCV, IOCV, Riverside, CA. pp. 209–213.

Texeira, D.C., Ayres, A.J., Kitajima, E.W., Tanaka, F.A.O., Danet, J.L.,. Jagoueix-Eveillard, S,, Saillard, C. and Bové, J.M. 2005a. First report of a huanglongbing-like disease of citrus in Sao Paulo State, Brazil, and Association of a New liberibacter Species, "*Candidatus* Liberibacter americanus", with the Disease. *Plant Dis.*, 89: 107.

Teixeira, D.C., Saillard, C., Eveillard S., , Danet J.L., Ayres, A.J. and Bové, J.M. 2005b. A new liberibacter species, *Candidatus* Liberibacter americanus sp. nov., is associated with citrus huanglongbing (greening disease) in São Paulo state, Brazil. *In*: Proc. 16th conf. IOCV, IOCV, Riverside, CA. pp. 325–340.

Teycheney, P., Acina, I., Lockhart, B.E.L. and Candresse, T. 2007. Detection of *Banana mild mosaic virus* and Banana virus X by polyvalent degenerate oligonucleotide RT-PCR (PDO-RT-PCR). *J. Virol. Method*, 142: 41–49.

Thind, S. K. and Arora, P.K. 1997. Incidence and behaviour of ring spot virus in citrus. *Indian J. Virol.*, 13: 159–160.

Thind, S. K., Sharma, J.N., Arora, P.K. and Cheema, S.S. 1998. Effect of citrus ringspot virus on quality of Kinnow mandarin fruits. *Indian J. Virol.*, 14: 91–92.

Thiribhuvanamala, G., Doraiswamy, S. and Ganapathy, T. 2005. Detection of Banana bunchy top virus in aphid vector using double antibody sandwich (DAS) – ELISA.. *Indian J. Virol.*, 16: 12–14.

Timmer, L.W. and Powell, C.A. 2006. Rio Grande gummosis. www.ivia.es/iocv

Timmer, L.W., Garnsey, S.M. and Graham, J.H. 2000. 2nd Ed. Cristacortis. *In*: Compendium of citrus diseases. APS Press, St. Paul, Minnesota, USA.

Timmer, L.W., Garnsey, S.M. and McKitchie, J.J. 1978. Comparative symptomatology of Florida and Texas isolates of citrus ringspot virus on citrus and herbaceous plants. *Plant Dis. Reptr.*, 62: 1054–1058.

Timmer, L., Brlansky, R.H., and Lee, R.F., Graham, J.H., Agostini, J.P., Fischer, H.U. and Casafus, C. 1984. Characteristics of citrus trees affected with blight in Florida, declinamiento in Aregentina and by declinio in Brazil. *Proc. Internl. Soc. Citriculture*, 1: 371–374.

Timmer, L.W., Brlansky, R.H., Derrick, K.S. and Lee, R.F. 1991. Transmission of citrus blight by root graft inoculation. *In*: Proc. 11th Conf. IOCV, IOCV, Riverside, CA. pp. 277–282.

Timmer, L.W., Lee, R.F., Brlansky, R.H., Graham, J.H., Albrigo, L.G., Derrick, K.S. and Tucker, D.P.H. 1992. The infectious nature of citrus blight. *Proc. Fla. State Hortic. Soc.* 105: 21–26.

Timmer, L.W., Garnsey, S.M. and Graham, J.H. (*eds.*) 2000. Compendium of citrus diseases, Second Edition. APS Press, st. Paul, MN. pp. 65–66.

Thomas, K.M. and Krishnaswamy, C.S. 1939. Leaf crinkle-A transmissible disease of Papaya. *Curr. Sci.*, 8: 316.

Thomas, J.E., Iskra-Caruana, M.L. and Jones, D.R. 1994. Banana bunchy top disease. Musa Disease Fact Sheet No. 4. Montpellier, France.

Thomas, J.E., Geering, A.E.W., Kessling, A.T. and White, M. 1997. Puriification, properties, diagnosis of banana bract mosaic potyvirus and its distinction from abaca mosaic potyvirus. *Phytopathology,* 87: 698–705.

Thottappilli, G., Dahal, G. and Lockhart, B.E.L.1998. Studies on a Nigerian isolate of a banana streak badnavirus.I. Purification and enzyme-linked immunosorbent assay. *Ann. Appl. Biol.,* 132: 253–261.

Tomar, S.P.S., Chavan, V.M., Raj Verma, Zote, K.K. and Jasbir Singh. 2009. Weeds as alternate host plants for papaya ringspot virus (PRSV-P). *In*: 5th Intl. Conf. Pl. Path. Globalized era, Nov. 10-13, 2009, New Delhi, India (Abst). p. 307.

Tomlinson, J.A., Carter, A.L., Faithful, E.M and Webb, M.J.W. 1973. Purification and serology of the cucumber mosaic virus. *Ann. appl. Biol.,* 74: 181–182.

Tripathi, S., Jon, Y., Suzuki, S., Ferreira, A. and Gonsalves, D. 2008. *Papaya ringspot virus*-P: characteristics, pathogenicity, sequence variability and control. *Molecular Plant Pathology* 9: 269–280.

Tsai, J.H. and Maramorosch, K. 1987. Lethal yellowing in palms. *In*: mycoplasma diseases of woody plants, Raychaudhuri, S.P. and Narain rishi (*Eds.*) Malhotra Publishing house, New Delhi, pp. 29–39.

Tucker, D.P.H., Lee, R.F., Timmer, L.W., Albrigo, L.G. and Brlansky, R.H. 1984. Experimental transmission of citrus blight. *Plant Dis.,* 68: 979–980.

Tushmereirwe, W. K., Karamura, E.B. and Karyyeija, R. 1996. banana streak virus (BSV) and an associated filamentous virus (unidentified) disease complex of highland bananas in Uganda. *Infomusa,* 5: 9–12.

Ushiyama, K. 1981. Studies on satsuma dwarf disease. II. Spread of the disease by top working satsuma trees with infected scions]. *Bull. Kanagawa Hort. Exp. Stn.,* 28: 24–30.

Ushiyama, K. and Ohgaki, T. 1970. Studies on the satsuma dwarf virus disease. 1. Survey in Kanagawa Prefecture. *Bull. Kanagawa Hort. Exp. Stn.,* 18: 57–65.

Usugi, T. and Saito, Y. 1977. [Some properties of satsuma dwarf virus]. *Ann. Phytopathol. Soc. Japan,* 43: 137–144.

Usugi, T. and Saito, Y. 1979. Citrus mosaic virus. *CMI/AAB Descriptions of Plant Viruses* No. 208. Association of Applied Biologists, Wellesbourne, UK.

Usugi, T. and Tsuchizaki, T. 1982. [Detection of citrus mosaic virus by enzyme-linked immunosorbent assay]. *Ann. Phytopathol. Soc. Japan,* 48: 330–332.

Usugi, T.,Yamamoto, S. and Tsuchizaki, T. *1986.* Morphology, host range and serological properties of citrus mosaic virus causing mosaic diseases in Satsuma mandarins. *Ann. Phytopathol. Soc. Japan.,* 52: 349–354.

van Regenmortel, M. H. V., Fauquet, C.M., Bishop, D.H.L., Carstens, E.B., stes, M.K., Lemon, S.M., Maniloff, J., Mayo, M.A., McGeoch, D.J., Pringle, C.R. and Wickner, R.B. 2000. Virus taxonomy: The classification and nomenclature of viruses. The Seventh Report of the International Committee on Taxonomy of Viruses. Academic Press, San Diego, CA, USA. p. 1167.

Varghese, M.K.1934. Diseases of coconut palm. Dept. Agric & Fisheries, Travancore p. 105.

Varma, P.M., Rao, D.G. and Capoor, S.P. 1960. Additional vectors of tristeza disease of citrus in India. *Curr. Sci.,* 29: 359.

Varma, P.M., Rao, D.G and Capoor, S.P.1965.Transmission of tristeza virus by *Aphis craccivora* Koch. and *Dactynotus Jaceae* L. *Indian. J.Agric. Sci.,* 35: 85–89.

Varma, A., Ahlawat, Y.S., Chakarborty, N.K., Garnier, M. and Bove, J.M. 1993. Detection of greening BLO by electron microscopy, DNA hybridization in citrus leaves with and without mottle from various regions in India. *In*: Proc. 12th Conf. IOCV, IOCV, Riverside, CA. pp. 280–285

Vasudeva, R.S. and Capoor, S.P. 1958. Citrus decline in Bombay state. *FAO Plant Prot. Bull.*, 6: 91

Vasudeva, R.S., Varma, P.M., Capoor, S.P. and Rao, D.G. 1959. Transmission of citrus decline virus by *Toxoptera Citricidus* Kirk in India. *Curr. Sci.*, 28: 418–419.

Vidya, T., Sivaprasad, Y., Naresh Kumar, C.V.M., Mahathi, N. and Sai Gopal, D.V.R. 2010. Biotin labeled probe for sensitive and rapid detection of papaya ringspot virus (PRSV-P) by dot blot hybridization. *In*: Nat. Conf. on Recent Trends in viral Disease Management. Sri Venkateswara Univ. Tirupati, March 18-20, 2010. (Abs) p. 111.

Vijayakumari, N., Ghosh, D.K., Das, A.K. and Singh, S. 2006a. Elimination of CTV and greening pathogens through *in vitro* shoot grafting from citrus exotic germplasm. *Indian j. Agric. Sci.* 76: 209-210

Vijayakumari, N., Ghosh, D.K., Das, A.K. and Singh, S. 2006b. Production of disease-free planting material of Nagpur mandarin (*Citrus reticulata* Blanco) by shoot tip grafting. *Indian Phytopath,* 59: 48–51

Villalobos, W., Moreira, L., Derrick, K.S., Beretta, M.J.G., Lee, R.F. and Rivera, C. 2005. Occurrence of citrus blight in Costa Rica. P 509 *In:* Proc. 16th. Conf. IOCV. IOCV, Riverside

Visvader, J. E., Gould, A. R., Bruening, G. E. and Symons, R. H. 1982. Citrus exocortis viroid: Sequence and secondary structure of an Australian isolate. *FEBS Letters,* 137: 288–292.

Vives M.C., Galipienso L., Navarro L., Moreno P. and Guerri J. 2001. The nucleotide sequence and genomic organization of citrus leaf blotch virus: Candidate type species for a new virus genus. *Virology,* 287: 225–233.

Vives, M. C., Rubio, L., Galipienso, L., Navarro, L., Moreno, P. and Guerri, J. 2002. Low genetic variation between isolates of Citrus leaf blotch virus from different host species and different geographical origins. *J. Gen. Virol.*, 83: 2587–2591

Vives M.C., Pina J.A., Juarez J., Navarro L., Moreno P., Guerri J., 2005. Dweet mottle disease is probably caused by Citrus leaf blotch virus. *In*: Proc. 16th Conf. IOCV, IOCV, Riverside, CA. pp. 251–256.

Vogel, R. and Bové , J.M. 1964. Stem pitting sur bigaradier et sur oranger Tarocco en Corse: Une maladie à virus. *Fruits,* 19: 264–274.

Vogel, R. and Bove, J.M. 1968. Cristacortis, a new virus disease inducing stem pitting on sour orange and other citrus species. *In*: Proc. 4th Conf. IOCV, IOCV, Riverside, CA. pp. 221–228.

Vogel, R. and Bove, J.M. 1972. Reaction of cristacortis virus to other citrus viruses. *In*: Proc. 5th Conf. IOCV, IOCV, Riverside, CA. pp. 78–184.

Vogel, R. and Bove, J.M. 1974. Studies on the cause of leaf symptoms associated with cristacortis disease of citrus. *In* Proc. 6th Conf. IOCV, IOCV, Riverside, CA. pp. 131–134.

Vogel, R. and Bove, J.M. 1980. Pollen transmission to citrus of the agent inducing cristacortis stem pitting and psorosis young leaf symptoms. *In*: Proc. 8th conf. IOCV, IOCV, Riverside, CA. pp. 188–190.

Vogel, R. and Bové, J.M. 1976. Transmission de maladies infectieuses d'agrumes à agrumes par le pollen d'arbres appliqué sous l'écorce de plantes saines. *C.R. Acd. Sci. Paris, Série D,* 283: 1409–1412.

Wadlaw, C.W. 1972. Banana diseases including plantations and abaca. 2nd Ed. *Longman*, 69: 70–93

Wallace, J.M. 1957. Virus strain interference in relation of symptoms of psorosis disease of citrus. *Hilgardia,* 27: 223–246.

Wallace, J.M. 1958. The sunblotch disease of avocados. *J. Rio Grande Valley Hort. Soc.,* 12: 67–74.

Wallace, J.M. 1959. (ed) Citrus Virus Diseases. Univ. Calif. Div. Agric. Sci., Berkeley, California.

Wallace, J. M. 1959. A half century of research on psorosis. *In*: Wallace, J.M. (*Ed.*) Citrus virus diseases. Univ. Calif. Div. Agr. Sci., Berkeley, Calif. pp. 5–21.

Wallace, J.M.1968. Recent developments in Citrus psorosis diseases. *In*: Proc. 4th Conf. IOCV, IOCV, Riverside, CA. pp. 1–9.

Wallace, J.M. and Drake, R.J. 1951. Recent developments in studies of quick decline and related diseases. *Phytopathology,* 41: 785–793.

Wallace, J.M. and Drake, R.J. 1953. Seed transmission of the avocado sunblotch virus. *Citrus Leaves,* 33: 18–20.

Wallace, J. M. and Drake, R.J. 1953. A virus-induced vein enation in citrus. *Citrus Leaves,* 33: 22–24.

Wallace, J.M. and Drake, R.J. 1959. Citrus vein enation. *In*: Citrus Virus Diseases. Wallace, J.M. (*Ed.*). pp. 163–165.

Wallace, J.M. and Drake, R.J. 1960. Woody galls on citrus associated with vein-enation virus infection. *Plant Dis. Reptr.,* 44: 580–584.

Wallace, J.M. and Drake, R.J. 1962. Further studies on the woody gall disease of citrus. *Phytopathology,* 52: 756.

Wallace, J. M. and Drake. R.J. 1962. Tatter leaf, a previously undescribed virus effect on citrus. *Plant Dis. Reptr.,* 46: 211–212.

Wallace, J. M. and Drake, R.J. 1963. New information on symptom effects and host range of the citrus tatter leaf virus. *Plant Dis. Reptr.,* 47: 352–53.

Wallace, J.M. and Drake,R.J. 1968. Citrange stunt and ringspot –two previously undescribed virus diseases of citrus.. *In*: Proc. 4th Conf. IOCV, IOCV, Riverside, CA. pp. 177–183.

Wallace, J.M. and Drake, R.J. 1972. Use of seedling yellows recovery and protection phenomena in producing tristeza-tolerant, susceptible scion-root stock combinations. *In*: Proc. 5th Conf. IOCV, IOCV, Riverside, CA. pp. 137–143.

Wang, H.L., Yeh, S.D., Chiu, R.J. and Gonsalves, D. 1987. Effectiveness of cross-protection by mild mutants of papaya ringspot virus for control of ringspot disease of papaya in Taiwan. *Plant Dis.,* 71: 491–497.

Wang, X.Y., Xie, Y. and Zhou, X.P. 2004. Molecular characterization of two distinct begomoviruses from papaya in China. *Virus Genes,* 29: 303–309.

Weathers, L.G.1957. A vein yellowing disease of citrus caused by a graft transmissible virus. *Plant Dis. Reptr.,* 41: 741–742.

Weathers, L.G. 1960. Yellow vein disease of citrus and studies of interactions between yellow vein and other viruses of citrus. *Virology,* 11: 753–764.

Weathers, L.G. 1965a. Transmission of exocortis virus of citrus by *Cuscuta subinclusa. Plant Dis. Reptr.,* 49: 189–190.

Weathers, L.G. 1965b. Petunia, an herbaceous host of exocortis virus of citrus. *Phytopath,.* 55: 1081.

Weathers, L.G., Greer Jr., F.C. and Harjung, M.K. 1967. Transmission of exocortis virus of citrus to herbaceous plants. *Plant Dis. Reptr.,* 51: 868–871.

Webber, H.J. 1925. A comparative study of the citrus industry of South Africa. *S. Africa Dep. Agr. Bull.,* 6: 106.

Weng, Z., Liu, X., Gowda, S., Barthelson, R.A., Galbraith, D.W., Dawson, W.O., and Xiong, Z. 2010. Extreme genome stability of citrus tristeza virus. *In*: Abstract 18th Conf. IOCV, Campinas/SP, Nov.7-12, 2010. p. 25.

Whiteside, J.O., Timmer, L.W. and Graham, J.H. 1988. *In*: Compendium of citrus diseases (J.O. Whiteside, S.M. Garnsey and L.W. Timmer, *Eds*) APS Press st. Paul, MN. pp. 62–63.

Wu, R.Y. and Su, H.J. 1990. Purification and characterization of banana bunchy top virus. *Phytopathology*, 128: 153–160.

Wu, R.Y. and Su, H.J. 1991. Regeneration of healthy banana plantlets from banana bunchy top virus-infected tissues cultured at high temperature. *Plant Pathol.*, 40: 4–7.

Wutscher, H.K., Cohen, M. and Young, R.H. 1977. Zinc and water-soluble phenolic levels in the wood for the diagnosis of citrus blight. *Plant Dis. Reptr.*, 6: 572–576.

Yamada, S. and Sawamura, K. 1952. Studies on the dwarf disease of Satsuma orange, *Citrus unshiu* Marcovitch (Preliminary report). *Hort. Div., Tokai-Kinki Agric. Exp. Stn., Bull.*, 1: 61–71.

Yamaguchi, A. 1984. Time requirement for graft transmission of Satsuma dwarf virus. *Bull. Fruit Tree Res. Stn*. Japan B, 11: 63–69.

Yamamoto, S. and Yamaguchi, A. 1980. Spread of citrus mosaic through distribution of a new clone of satsuma mandarin, *In*: Proc. 8th Conf. IOCV., IOCV, Riverside. pp. 230–231.

Yasir Iftikhar, Zafar Iqbal, Salman Ahmed, Awan, A.R., Usman Saleem, Sarwar, G. and Moeen-ud-Din. 2010. Effect of environmental factors on yellow vein clearing virus incidence in lemon. *J. Agric. Res.* 48: 87–92.

Yedidyah, S. 1937. Citrus growing (Hebrew) 2nd ed., Tel Aviv. Yeh, C.D., Gonslaves, D., and Providenti, R. 1984. Comparative studies on host Range and serology of papaya ringspot virus and watermelon mosaic virus-1. *Phytopathology* 74: 1081–1085. Copyright © 2008 Blackwell publishing Ltd.

Yeh, S.D.; Gonsalves, D.; Wang, H.L.; Namba, R. and Chiu, R.J. 1988. Control of papaya ringspot virus by cross protection. *Plant Dis.*, 72: 375–380.

Yemewar, S.I. and Mali, V.R. 1980. On the identity of a sap transmissible virus of papaya in Marathwada. *Indian J. Mycol. and Pl. Pathol.*, 10: 155–160.

Yokomi, R.K., Garnsey, S.M., Young, R.H. and Grimm, G.R.1984. Spatial and temporal analysis of citrus blight incidence in 'Valensia' orange groves in central Florida. In: Proc. 9th Conf. IOCV, IOCV, Riverside, CA. pp. 260-269.

Yokomi, R. K., Saponari, M., and Sieburth, P.J. 2010. Rapid differentiation and identification of potential severe strains of Citrus tristeza virus by real-time reverse transcription-polymerase chain reaction assays. *Phytopathology*, 100: 319–327.

Yokomi, R.K., Mello,, A.F., Fletcher,, J., Saponari,, M. 2010. Estimation of citrus stubborn disease incidence in citrus groves by real-time PCR. 17th Conf. IOCV, IOCV, Riverside (Available: http://www.ivia.es/iocv/)

Yora, K., Raychaudhuri, S.P. and Ahlawat, Y.S. 1977. Infectious variegation of citrus in India. *California Citrograph.*, 62: 342.

Yoshikawa, N., IImaizumi, M., Takahashi, T. and Inouye, N. 1993. Striking similarities Between the nucleotide sequence and genome organization of citrus tatter leaf virus and apple stem grooving capilloviruses. *J. Gen Virol.* 74: 2743–2747.

Yot Dauthy, D. and Bove, J.M. 1968. Purification of citrus crinkly leaf virus. *In*: Proc. 4th Conf. IOCV, IOCV, Riverside, CA. pp. 255–263

Young, R.H.; Wutscher, H.E. and Albrigo, L.G. 1980. Relationship between water translocation and zinc accumulation in citrus trees with and without blight. *J. American Soc. Hort. Sci.*, 105: 444–447.

Zentmyer, G.A. 1955. Diseases of the avocado. California Avocado Society Yearbook, pp. 44–58

Zhou, L.J., Gabriel, D.W., Duan, Y.P., Halbert, S.E. and Dixon, W.N. 2007. First report of dodder transmission of huanglongbing from naturally infected *Murraya paniculata* to citrus. *Plant Dis.*, 91: 227.

Zreik, L., Carle, P., Bove, J.M. and Garnier, M. 1995. Characterization of mycoplasma-like organism associated with witches' broom disease of lime and proposition of a *Candidatus* taxon for the organism "*Candidatus* Phytopasma aurantifolia". *Intern. J. Syst. Bacteriol.*, 45: 449–453.

5

VIRUS DISEASES OF ARID FRUITS

Several fruits are known to be cultivated in arid climate. Among them important ones are: Ber (*Ziziphus mauritiana* Lamk.), Pomegranate (*Punica granatum* L.), Anola (*Emblica officinalis* Gaertn.), Custard apple (*Annona squamosa* L.), Date palm (*Phoenix dactylifera* L.), fig (*Ficus carica* L.), Phalsa (*Grewia subinaequalis* L.), Lasoda/Lehsua/Gonda (*Cordia myxa* Roxb.), Jamun (*Syzygium cumunii* Skeels), Karonda (*Carisa carandus* L.), Bael (*Aegle marmelos* Corr.), Wood apple (*Feronia limonia* L.), Mulberry (*Morus* spp.), Kair (*Capparis apphylla* Roth.) and Pilu (*Salvadora oleoides* Decae.). Most of these fruit species are poorly studied for virus and virus-like pathogens. The information available on these fruits will be discussed.

A. VIRUS DISEASES OF FIG TREE (*FICUS CARICA*)

Fig belongs to family Moraceae. The fruit is soft and sweet, full of small seeds and often eaten dried. Fresh figs are delicious and often jam and chutney is made from them. As figs have to be picked when ripe you only can eat them fresh in the country of origin. Today there are more than 600 different fig types. They are often dried for preservation, since the fresh fruits are highly perishable. The fig flowers develop inside the fruit and cannot be seen.

Mosaic is the most wide spread viral disease of fig in India and USA. Recently four new viruses have been isolated from fig tree showing mosaic symptoms from USA. The first one is European Mountain ash ringspot associated virus.The second is a member of the genus badnavirus and is related most closely to citrus yellow mosaic virus reported from India (Ahlawat *et al.*, 1996). the other two viruses belong to the family

Closteroviridae (Tzanetakis, and Martin, 2009). Detection protocols for all the four viruses are available (Tzanetakis and Martin, 2009).

FIG MOSAIC

A severe mosaic disease of *Ficus palmata* Forsk., is known to occur in Punjab and Himachal Pradesh of India (Nagaich and Vashisth, 1962). The virus was also graft transmitted to *F. carica* Linn., *F. nemosalis* wall. and *Morus indica* Linn. (Vashisth and Nagaich, 1965). The leaves of inoculated *M. indica* showed faint chlorotic spots gradually changing to mosaic mottling of the leaves (Fig. 49). These symptoms can be seen better against light. Mottling of the leaves are common symptoms with most fig cultivars. Leaves and fruits may be dwarfed and some leaves may be malformed. On leaves, mosaic spots will appear yellow and may cover large areas of the leaf. Mosaic spots on the fruit may be more subtle in appearance in comparison with the leaf symptoms. In some cultivars, premature defoliation and fruit drop can occur. The virus was transmitted to periwinkle plants by mite vector (Credi *et al.*, 2012).

Fig. 49: Fig leaf with mosaic symptoms

Fig mosaic virus is transmitted by grafting and by vegetative cuttings. The virus is also transmitted by an eriophyid mite *Aceria ficus (Cotte)* both in India and U.S.A. (Vashisth and Nagaich, 1968; Flock and Wallace, 1955). After inoculation plants will show symptoms in about 6 months (Vashisth and Nagaich, 1968). Graft and mite vector transmission clearly indicated that fig mosaic is caused by a virus. A four segmented negative strand RNA virus in the newly formed genus *Emaravirus* is associated with fig mosaic disease. The members of *Emaravirus* consists of six RNA segments (Ishikawa *et al.*, 2012). The Clean propagation stock should always be used since this virus is spread by vegetative cuttings. Controlling or managing the mite vector may help reduce incidence of the disease.

FIG LATENT VIRUS-1

Fig latent virus-1 is wide spread in fig orchards of south Italy. The virus is transmitted mechanically to some herbaceous hosts without showing any visible symptoms. It is seed transmitted in fig seeds upto 80-100%. The causal virus has filamentous particles ca 700 nm long which are present in cytoplasm of parenchyma cells of infected fig plants. The viral genome is a single stranded positive sense RNA about 8000 nt in size, 6500 of which have been sequenced. Genomic RNA has four open reading frames. A 42 kDa putative movement protein (ORF 2), the 46 kDa coat protein (ORF 3), and a 12 kDa protein with nucleic acid binding properties have been observed. The genome structure and organization resembles to members of *Trichovirus*, family *flexiviridae*. The virus can be detected by immunosorbent electron microscopy using homologous antibodies (Giuliano *et al.*, 2009).

ROSETTE OF *FICUS AMPLESSIMA*

A rosette disease in *F. amplessima* is found in Karnataka and Maharashtra with an incidence from 11.11 to 30.76%. The symptoms are cluster of apical rosette, short internodes, excessive shoot proliferation, mild chlorosis of leaves, pendulus twigs and upright growth habit. The disease is graft transmitted. In electron microscopy, pleomorphic bodies measuring 80-800 nm were seen in sieve tubes of phloem tissues. PCR test with oligonucleotide primers showed an amplification of 572bp. The disease showed remission of symptoms with tetracycline therapy suggesting involvement of phytoplasma. This phytoplasma belongs to elm yellows group phytoplasmas (Suryanarayan *et al.*, 2009).

B. VIRUS/PHYTOPLASMA DISEASES OF BER

The ber or jujube belongs to genus *Ziziphus* of the family *Rhamnaceae* having more than 600 species. The major cultivated species are Indian jujube (*Ziziphus mauritiana*) and Chinese jujube (*Ziziphus jujuba*). The species under cultivation in India are Umran and Gola but new varieties like CHES Godhra, Sel-1 also developed. There are numerous wild species of *Ziziphus* in Asia specially in China and the Indian subcontinent. No virus disease has been recorded in jujube but a phytoplasma and a viroid disease are known to occur on jujube.

WITCHES' BROOM DISEASE

Jujube witches' broom (JWB) is a very serious disease in jujube tree cultivation in Asian countries such as Korea, Japan, India and China. Witches' broom has been reported both from Indian and Chinese jujubes (Kim, 1965; Zhu *et al.*, 1998; Pandey *et al.*, 1976; Khan *et al.*, 2008). Affected trees become weak and produce small, yellow leaves. The leaves developed in clusters on twigs giving witches broom appearance (Plate 64). Witches' broom disease of Chinese jujube (*Ziziphus jujuba* Mill.) is found in China and Korea.

The disease is transmitted to healthy seedlings by budding and grafting with diseased scions. The phytoplasma is transmitted by a leaf hopper, *Hishimonoides chinensis* Anufrive. The phytoplasma is transmitted to periwinkle (*Catharanthus roseus*) through dodder and also by grafting (Zhang *et al.*, 1991; Lee *et al.*, 2012). The causal agent has been confirmed to be a mycoplasma-like organism now known as phytoplasma (Zhang *et al.*, 1991; Zhu *et al.*, 1998).

JWB belongs to 16S rV-B group (Elm yellows group) based on RFLP and sequence analyses of 16S rRNA gene (Han and Cha., 2002; Lee *et al.*, 2000). The genetic relationships based on 16S rDNA sequences between Korean and Chinese JWB phytoplasma were very similar. The phytoplasma has been classified as *Candidatus* Phytoplasma ziziphi (Jung *et al.*, 2003).The witches broom phytoplasma can be detected by serological tests and also by nested PCR and by fluorescent microscopy (Lee *et al.*, 2012).

Jujube witches' broom disease can be contained by treatment with 500 and 1000 ppm oxytetracycline (Wang *et al.*, 1981). Winter jujube, a species that originated in China, is the most prominent elite variety of jujube (*Zizyphus jujuba* Mill.). Witches' broom resistant transgenic plants have been developed to winter jujube (Gu *et al.*, 2008).

HOP STUNT VIROID ON JUJUBE

Hop stunt viroid (HSVd) has been detected in many woody species including grapevine, citrus, peach, plum, apricot, and almond. This viroid has also been reported in jujube trees (*Ziziphus jujuba*) from Hebei and Shanxi provinces of China. Low molecular weight RNAs were extracted for two-dimensional polyacrylamide gel electrophoresis (2D-PAGE), hybridisation, and reverse transcription polymerase chain reaction (RT-PCR) assays. The sequences of the HSVd isolates from jujube trees shared 92.6–92.8% homology with the first reported HSVd sequence (GenBank: X00009). Using alignment and phylogenetic analyses, the HSVd sequence variants from jujube were classified in the plum subgroup and were found to be more conserved than other HSVd subgroups (Zhang *et al.*, 2009).

C. VIRUS-LIKE DISEASES OF POMEGRANATE (*Punica granatum* L.)

In India, pomegranate had covered an area of 122,000 hectares with a production of 858,000 tonnes. The fruit is fetching a lot of foreign exchange for the country as sizeable quantity of fruits is being exported. India exported 35.2 thousand tonnes of fruits valued at Rs 911 million. In India, the commercial varieties of pomegranate are Jodhpur Red, Muskat, Ganesh, Jyoti, Bassein Seedless and new varieties developed are Ganesh, GKVK-1 (Jyoti) and G-27. Fortunately no virus disease is known to infect pomegranate. However, viroid like RNAs were detected in four varieties of pomegranate by Gomez and Pallas (2001) and hope stunt viroid was reported by Gazel and Caglayan (2009).

D. VIRUS DISEASES OF BAEL (*Aegle marmelos* Corr.)

Indian Bael (*Aegle marmelos*), a native plant of India, has spread over wide areas of southeast Asia. The ripe fruit and unripe fruit, as well as the roots, leaves and branches have all been used in traditional medicine. In Ayurveda, the ripe fruit has been used for chronic diarrhea and dysentery, as a tonic for the heart and brain, and as adjuvant treatment of dysentery. A decoction of the roots has been used to treat melancholia, intermittent fevers, and palpitations; the roots have mainly been used as an ingredient of the Ayurvedic medicine, *dashmool*. The leaves have been given as a febrifuge, and as a poultice for the treatment of eye disorders and ulcers; and administration of fresh leaves has been used for weakness of the heart, dropsy, and beriberi. In India, Mirzapuri, NB 5, and NB 9 are the main variety under cultivation. As such no virus or virus-like disease has been reported on bael but it is host of citrus tristeza *Closterovirus* (Lee and Bar-Joseph, 2000).

E. VIRUS DISEASES OF MULBERRY

Mulberry (*Morus* spp.) belongs to family *Moraceae*.The white mulberry (*M. alba* L.) is native to eastern and central China. The red or American mulberry (*M. rubra* L.) is native to Eastern United States. The black mulberry (*M. nigra* L.) is native to western Asia and has been grown for its fruits in Europe since before Roman times. Other species are: Korean Mulberry (*Morus australis*), Himalayan Mulberry (*M. laevigata*) and there are hybrid between *Morus alba* and *M. rubra*. Mulberry foliage is the sole food for silk worm (*Bombyx mori* L.). A few viruses or virus-like diseases are reported in mulberry causing an adverse effect on silkworm development.

MULBERRY MOSAIC

In India, a disease with typical mosaic symptoms, often associated with curling or puckering of leaves, was first observed on mulberry variety 'China black" in Kalimpong hills of west Bengal (Raychaudhuri *et al.*, 1962). Later studies revealed that mosaic disease is more commonly distributed in North Bengal. The causal virus infects large number of mulberry varieties and clones (Raychaudhuri *et al.*, 1965). The virus associated with mosaic disease is an unassigned virus. Mulberry mosaic spread in the field by the use of infected cuttings. The disease is transmissible by mechanical inoculation from mulberry to mulberry. It is also transmitted by aphid vectors, *Rhopalosiphum maydis, Myzus persicae* and *Aphis gossypii* in non-persistent manner (Chatterjee and Raychaudhuri, 1963, 1965).

In crude extract the virus can withstand an exposure upto 50°C for 10 min but not at 60°C. Dilution end-point is between 1:2000 and 1:4000 and longevity *in vitro* is 10 and 15 days at room temperature (28-30°C). No other information is available on mulberry mosaic virus.

YELLOW-NET VEIN

This disease was first observed on mulberry variety 'Koksu' in Kalimpong. Later the disease was recorded on mulberry plantations at Siliguri, Berhampur and Malda districts of West Bengal in India. The characteristic symptoms of the disease appear in March-April as yellowing of leaf veins but interveinal portion remains green. The disease is transmissible by pen and inarch grafting (Raychaudhuri *et al.*, 1961). Several *Morus* species/varieties and clones are susceptible to yellow net-vein infection. The vector of the disease is a whitefly, *Bemisia* sp. (Raychaudhuri *et al.*, 1966). It was interesting to note that when symptomatic plants carrying yellow net vein symptoms maintained in glasshouse at Kalimpong, the symptoms disappeared and the plants showed only mosaic symptoms (Ahlawat-unpublished). Therefore, further studies are necessary to establish yellow net vein disease as a separate disease from mosaic.

MULBERRY RINGSPOT

A ring spot disease on mulberry has been reported from Japan and perhaps remain restricted in Japan as it has not been reported from other countries. The main symptoms of the disease are mosaic and ringspot on leaves of affected plants. Some isolates may produce enations. The leaves also become filiform (Tsuchizaki *et al.*, 1971). Soybean, cowpea, pea, bean, *Chenopodium quinoa, Nicotiana clevelandii* and sesame were infected with this virus by sap inoculation The virus is transmitted by a nematode vector, *Longidorus martini* (Yogita and Komuro, 1972). The virus is seed transmitted in *Glycine max* cv. Mikawashima (about 10%). *Vigna sinensis* cv. Black Eye is a suitable host for virus assay. Symptoms of the disease on various hosts are given below:

- *Morus alba* (mulberry). Systemically infected leaves develop mosaic, ringspots or enations.
- *Vigna sinensis* cv. Black Eye (cowpea). Necrotic spots or ringspots in inoculated leaves followed by mosaic or necrosis of leaves and necrosis of stems.
- *Glycine max* (soybean). Systemically infected leaves develop mosaic or chlorotic spots.
- *Pisum sativum* (pea). Chlorotic spotting or mosaic of systemically infected leaves.
- *Chenopodium quinoa*. Indistinct chlorotic local lesions may form; systemically infected leaves show mosaic.

Virus particles are isometric about 22 nm in diameter. The genome of the virus is RNA. Purified preparations sedimented as three components, apparently empty protein shells (T) and two kinds of nucleoprotein (M and B) (Tsuchizaki *et al.*, 1971). The virus belongs to NEPO viruses (Comoviridae). Serological diagnosis in gel diffusion test has been done with polyclonal antisera. The virus is not serologically related to nine nematode borne viruses.

MULBERRY LATENT VIRUS

This virus disease has also been reported on *M. alba* from Japan. No visible symptoms are known from infected mulberry trees. The virus is transmitted by mechanical inoculation from mulberry to herbaceous hosts. It is also transmitted by grafting but not transmitted in seeds of *Chenopodium quinoa*. Following species were susceptible to virus infection: *Beta vulgaris, Chenopodium amaranticolor, Chenopodium quinoa, Gomphrena globosa, Morus alba, Nicotiana clevelandii, Sesamum indicum, Spinacia oleracea. Morus alba* and *Chenopodium quinoa* are used as assay hosts (Tsuchizaki (1976). Symptoms on diagnostic hosts are given below:

- *Chenopodium quinoa* — Local and systemic chlorotic spotting, mosaic and malformation.
- *C. amaranticolor* — Green ringspots followed by symptomless systemic infection.

The virus particles associated with the disease are filamentous and straight with a length of 700 nm and a width of 12 nm. Axial canal is indistinct. Basic helix is obscure (Tsuchizaki, 1976). The virus is serologically related to *Carnation latent virus*. The virus belongs to *Carlavirus* in family *Flexiviridae*. The genome of the virus is linear, positive-sense, single-stranded RNA (monopartite). The virus codes for 6 ORF(s).

MULBERRY DWARF

The dwarf disease of mulberry tree is an important and of common occurrence in Japan and China. The disease was considered to be caused by a virus as the disease was transmitted by graft and also by a leafhopper vector, *Hishimonus discigulus* (Walker).

SYMPTOMS

The symptoms produced by mulberry dwarf are mainly of two types. In the first category accessory and lateral buds sprout early in the season and on the top of the tree. There is reduction in the leaf size and leaves turn to golden yellow. Proliferated branches become slender and internodes become shorter. Affected trees die within 2-3 year after getting infection. The second category known as common dwarf is characterized with reduced leaf size without serration and lobing. Leaf surface become crinkled. Infected branches grow slowly making the tree stunted. Axillary and secondary buds appear early in the season forming group of small branches showing witches broom appearance. With the advancement of the disease, the leaves become small and yellow. Floral parts are converted into vegetative shoot. No flower develops on severely affected trees. Root system is poor and infected tree may decline over a period of several years.

TRANSMISSION

Mulberry dwarf disease is transmitted by crown-grafting and by leafhopper vectors, *Hishimonas sellatus* and *Hishimonoides sellatiformis* (Ishiie and Matsuno, 1971a,b). *H. sellatus* can acquire the causal pathogen in 3h. There was no distinct relationship between the acquiring rate of causal agent and the sexual difference of the leafhopper (Ishiie and Matsuno, 1971a). Transovarial transmission of mulberry dwarf (MD) agent occurs in *Hishimonoides sellatiformis* (Kawakita *et al.*, 2000).

THE PATHOGEN

The pathogen causing the dwarf disease of mulberry is a phytoplasma which can be detected in ultrathin sections by electron microscopy (Doi *et al.*, 1967). The phytoplasmas are found in the sieve tube, companion, and parenchyma cells of the phloem tissues of infected plants. The organism is pleomorphic in shape, ranging from 50 nm to 160 nm in diameter, and with bilamina unit membrane of 8-10 nm in thickness which is filled with nuclear strands or fibrous material.

DIAGNOSIS AND MANAGEMENT

The MD phytoplasma was found in all species by nested polymerase chain reaction using specific primers for the 16Sr I-group phytoplasma and by electron microscopy (Hua Jiang *et al.*, 2005; Wei *et al.*, 2000). Use of phytoplasma-free planting material is important to manage MD phytoplasma. However, the tetracyclines and N-triacontanol (TRIA) when applied over a period of 1 year to dwarf mulberry plants, symptoms of the disease did not reappear for at least 3 years (Daiqun Sunzhi-huan, 2007).

F. VIRUS DISEASES OF PINEAPPLE (*Ananas comosus*)

Pineapple (*Ananas comosus* Linn. Merrill) is native to tropical America and it is cultivated in many parts of the world including India for its delicious fruit having pronounced flavour and also for fibre. Pineapple cultivars are generally grouped under three categories: Cayenne, Queen and Spanish (Singh, 2004). In India, following varieties are more popular: Giant Kew, Queen, Mauritius, Hy. 7, Selection 1/82 and 2/82. Yellow spot and wilt are common diseases of pineapple but pineapple wilt is economically important.

PINEAPPLE WILT

The association of pineapple wilt and mealybug, *Pseudococcus brevipes* was first described by Carter, 1933, 1951 from Hawaii.

SYMPTOMS

Symptoms expression of wilt disease depends on environmental conditions, mealybug populations, and pineapple genotype. The symptoms of the disease first appear at the tip of leaves as yellowish-red to bright red colour followed by yellowing and flaccidity of young leaves and necrosis of leaf tips. The discolouration moves down to leaves and finally the leaves become completely yellow or red. Affected plants show wilting at this stage. The root system show rotting. Infected plants become stunted and rarely develop any mature fruit. The outbreak of the diseases appears at warm temperature as it helps the mealybug vector to build up its population. The host range of the virus is limited to pineapple cultivars only.

TRANSMISSION

The primary method of transmission is through contaminated planting material. Wilt of pineapple is transmitted by the pink pineapple mealybug, *Dysmicoccus brevipes*, and the gray pineapple mealybug, *D. neobrevipes* (Sether and Hu, 1997). Mealybugs transmitted the causal virus in a semi-persistent manner (Sether *et al.*, 1998). The mealybugs moved from plant to plant by attendant ants (Sanford, 1988).

THE PATHOGEN

The Closterovirus-like particles with pineapple wilt have been reported in Hawaii, South Africa, Australia, Brazil, France, Malaysia, Taiwan and India. However, alongwith Clostero-like virus particles occasionally some bacilliform particles (133 × 33nm) were seen in partially purified preparations from both asymptomatic and wilted pineapple leaves (Wakman *et al.*, 1995). The Closterovirus particles are filamentous, flexuous with a length of 1200-1500 nm and a width of 12 nm. Axial canal is indistinct. The virus belongs to Closterovirus group. The relationship of the Clostero-like and bacilliform viruses to yield loss and mealybug wilt in pineapples is unknown (Wakman *et al.*, 1995). The genome of Closterovirus is monopartite with one particle size of linear, positive-sense, single-stranded RNA.The complete genome is 7100 nucleotides long.

DIAGNOSIS AND MANAGEMENT

Both poly and monoclonal antibodies have been prepared against Closterovirus and used for disease diagnosis in serological tests like tissue blot immunoassay and immunosorbent electron microscopy (Gunasinghe and German, 1989; Hu *et al.*, 1996, 1997). The management is achieved by the use of planting material from wilt-free areas and control of mealybug vectors. The virus can be eliminated from planting stock by hot water treatment of "crowns" for up to 120 minutes at 50°C. Plant resistant varieties if available should be used. 'Singapore Spanish' shows some resistance. The disease is reduced in thickly populated crop and with higher doses of NPK fertilizer (Singh and Sastry, 1975).

PINEAPPLE YELLOW SPOT

Pineapple yellow spot disease has been shown to be identical to the tomato spotted wilt Tospovirus. It infects over 100 species of plants. One of the symptoms in the fruit is a blackened, dry cavity in the side of the fruit due to one or more "eyes" having died. This is known as "dead eye". Infection of very young fruit results in an irregular arrangement of fruitlets as some fail to develop. Attacked fruits may also fail to develop a crown. Thrips are vectors of this virus. The concentric rings are developed on crown of leaves due to feeding by thrips. These spots enlarge and the infection spreads into the fruit itself, by which time the crown will often have dried out. The flesh of fruits thus infected will be discoloured and necrotic below the butt of the crown. The whole fruit may eventually become affected. When other host plants including weed hosts of the virus are grown near to pineapple plants, the incidence of the yellow spot virus is considerable enhanced due to thrips migrating into the pineapple field (Linford, 1932).

To manage yellow spot disease control of weeds in and around pineapple fields is important. The fruits showing early symptoms of the disease should be removed to avoid the spread of infections into other fruits. Control thrips by spraying mild insecticide is helpful.

G. VIRUS DISEASES OF GRAPEVINE

Cultivated grapes are believed to have been introduced into India by the Persian invaders in 1300 AD. Grape cultivation is one of the most remunerative farming enterprises in India and several other countries of the world. The famous wines in the world are made from grapes in Bordeaux, France. The famous fungicide, Bordeaux mixture was also evolved from this place based on grape cultivation and their infection by powdery mildew. The area and production of grapes are given in the following table.

Area and production of different varieties of grapes in India

Variety	*Area (ha)*	*Production (t)*
Anab-e-Shahi (white, seeded)	3,000	135,000
Bangalore Blue Syn. Isabella (black, seeded)	4,500	180,000
Bhokri (white, seeded)	500	15,000
Flame Seedless (red, seedless)	500	10,000
Gulabi Syn. Muscat Hamburg (purple, seeded)	1,000	30,000
Perlette (white, seedless)	1,500	60,000
Sharad Seedless - A mutant of Kishmish Chorni (black, seedless)	1,000	20,000
Thomson Seedless and its mutants (white, seedless)	22,000	550,000
Total	**34,000**	**1,000,000**

Source: NRC for Grapes

Approximately 85 percent of the total production, irrespective of the variety, is consumed fresh. About 120,000 tonnes of Thompson Seedless and its mutants, namely, Tas-A-Ganesh, Sonaka and Manik Chaman are dried for raisins. Some 20,000 tonnes of Bangalore Blue are crushed to make juice, and 10,000 tonnes of Bangalore Blue, Cabernet Sauvignon, Chenin Blanc, Chardonnay, Merlot, Pinot Noir and Uni Blanc are crushed to process into wine. The variety under cultivation in India are: Thompson Seedless, Anab-e-shahi, Bangalore Blue, Black Champa, Kishmish Charni, (Sharad Seedless), Arka Kanchan, Arka Shyam, Arkavati, Arka Hans, Arka Nilmani, Tas-e-Ganesh Manak Chaman, Sonaka, and Dilkush.

Virus and virus-like diseases are one of the major constraints in grapevine cultivation worldwide. These diseases not only affect the quality and quantity of yield but also reduce the productive life of vineyards. Grapevine is reported as a natural host of several viruses affecting other crops. These viruses are: arabis mosaic, tomato black ring, peach rosette mosaic, tobacco mosaic, apple mosaic, tomato bushy stunt and tomato ringspot. In addition, several other viruses/phytoplasmas were recorded from grapevine such as asteroid mosaic (Hewitt and Goheen, 1959), grapevine corky bark (Hewitt, 1968), enation disease (Refatti, 1966), Fan leaf (Cadman *et al.*, 1960), leaf roll (Hewitt, 1968) and vein clearing (Goheen and Hewitt, 1962).

Viruses infecting grapevines can be broadly divided into 2 groups based on their economic significance and geographical distribution. The first group consists of major diseases (also called traditional diseases) that are of great economic concern to viticulture and the global wine industry. These include grapevine leafroll disease (GLD), rugose wood (RW) complex, and nematode-borne viruses. The second group consists of minor diseases that are of less economic significance and/or limited geographical distribution. The majority of grapevine-infecting viruses fall into this latter category. Nearly 60 different documented viruses and virus-like pathogens infect grapes worldwide (Martin *et al.*, 2005). Some viruses are spread by insect vectors and others by soil-inhabiting nematode vectors. Since grapes are propagated by vegetative cuttings to maintain varietal integrity and clonal identity (trueness-to-type), all these viruses can be spread from place to place through the distribution of infected vegetative cuttings. The important virus and virus-like diseases will be discussed in this chapter.

GRAPEVINE FANLEAF

Fanleaf disease of grapevine first reported in *Vitis vinifera* from Austria by Rathay (1883) is now worldwide in distribution. This disease is also called as infectious degeneration.

SYMPTOMS

The symptoms of the disease vary from variety to variety. The affected vines get stunted with short internodes which develop in a zigzag fashion. Affected nodes sometimes develop 2-3 buds. The leaves on affected vines are smaller than normal with widened petiolar sinuses. Because of this and reduction of leaf area between the main veins, the leaf lamina appears like a half closed fan. Fan shaped leaves may or may not show mosaic and vein-banding symptoms. New growth on infected vines often is dwarfed on weak, flattened, crooked canes. Young leaves are mottled and deformed, showing excessive serration and a broad sinus at the petiole. Some strains cause bright yellow patterns on leaves, often without the leaf deformation. The symptoms of yellow mosaic or vein banding of major veins are developed on leaves of some varieties. Together, these symptoms resemble much like 2,4-D injury.

TRANSMISSION

Fan leaf virus is transmitted by nematodes; *Xiphinema index* (Hewitt *et al.*, 1958) and *X. italiae* (Cohn *et al.*, 1970). Among the nematode-transmitted viruses, grapevine fan leaf virus (GFLV) is the most widespread around the world. The virus is also transmitted by mechanical inoculation (Hewitt *et al.*, 1962). It is seed transmitted in *Chenopodium amaranticolor,* C. *quinoa* and *Glycine max* (Dias, 1963, Cory and Hewitt, 1968).

HOST RANGE

Fan leaf virus has a wide host range among perennial plants and trees. The symptoms on *Vitis vinifera, V. rupestris* and many other *Vitis* spp. and interspecific hybrids appear as green or yellow mosaic, ring and line patterns and flecks, leaf and nodal malformation. The symptoms remain systemic. Fan leaf virus infects More than 30 herbaceous plant species in families, *Amaranthaceae, Chenopodiaceae, Cucurbitaceae, Leguminoceae, Solanaceae* and *viridaceae* (Martelli and Hewitt, 1963; Taylor and Hewitt, 1964). The important herbaceous host plants are given below with symptoms on them.

- *Chenopodium amaranticolor* and *C. quinoa:* Systemic mottling and leaf malformation.
- *Cucumis sativus, Phaseolus vulgaris* cv: Bountiful: Systemic chlorotic or necrotic mosaic, mottling, flecking or ringspots. Immune to some strains.
- *Gomphrena globosa:* Systemic chlorotic spots, leaves twisted.
- *Nicotiana benthamiana:* Systemic mottling, leaf malformation, stunting.
- *Nicotiana clevelandii:* Systemic mottling and stunting.

The virus can be assayed on *Chenopodium amaranticolor* and *Gomphrena globosa*.

THE PATHOGEN

The causal agent of fanleaf disease is a *Nepovirus*. Grapevine fan leaf virus (GFLV) is a +tive sense single stranded RNA virus and belongs to family *Comoviridae*, genus *Nepovirus* and species *grapevine fan leaf virus*. Virions are isometric; not enveloped; 30 nm in diameter; angular in profile; without a conspicuous capsomere arrangement (Dias, 1963). The genome size of fanleaf virus is 11.116 kb and consists of two functional single stranded RNAs of 7.342 kb and 3.774 kb (Quacquarelli *et al.*, 1976). The multipartite genome is divided among different particles, each segment encapsidated separately and the segments are distributed between 2 different types of particles. The largest particles contain one molecule each of RNA-1 and RNA-2. The medium sized particles contain one molecule of RNA-2, or two molecules of RNA-2. The smallest particles contain no nucleic acid and are found in the top (sedimenting component T) band after sedimentation (Quacquarelli *et al., 1976*). Some strains of GFLV have an additional satellite RNA (Fuchs *et al.*, 1991).

DIAGNOSIS AND MANAGEMENT

Fan leaf virus can be detected by grafting on indicators (*Vitis rupestris* St. George), or by mechanical inoculation to herbaceous plants (*Chenopodium amaranticolor* and *Gomphrena globosa*). The virus can also be detected by serology in ELISA and immunosorbent electron microscopy (ISEM) (Bovey *et al.*, 1980). RT-PCR and molecular hybridization have also been used for detection of fan leaf virus (Fuchs *et al.*, 1991). For management of fan leaf virus and other Nepoviruses, it is very important to use Nepovirus-free propagating materials for planting new vineyards. This can be obtained through selection and production of virus free stocks by heat therapy and/or meristem culture (Barlass *et al.*, 1982; Goheen and Luhn,

1973). Since nematodes and Nepoviruses have a broad host range, it is also important to test soil for nematodes before new plantings or replacing an existing crop with new grapevines. Rootstocks and/or cultivars resistant to the virus must be used if available. Development of transgenic plants are important for economically manage this disease (Golles *et al.*, 1997, 2000).

GRAPEVINE LEAFROLL-ASSOCIATED VIRUSES

Grapevine leaf roll disease (GLD) is probably the most wide-spread and destructive virus disease of grapevines world-wide. It occurs widely in almost all grape-growing regions and accounts for about 62% of the global grape production losses due to virus diseases. There are currently 11 different viruses associated with leafroll, but Grapevine leafroll-associated viruses 1 and 3 are most commonly found. In India, leaf roll virus disease has been recently reported (Kumar *et al.*, 2012). The Indian isolate has been identified as an *Ampelovirus* antigenically related to Grapevine leafroll –associated virus 3 (GLRaV-3). Leafroll virus has impacts on both vine health and grape quality. Quality impacts included delayed maturity of grapes, a 25-50% reduction in sugar content, and poorly coloured fruit (Golino *et al.*, 1992). Five viruses, designated grapevine leafroll-associated 1,2,3,4 and 5 Closteroviruses have been associated with grapevine leafroll disease (Candresse and Martelli, 1995).

SYMPTOMS

Typical symptoms on leaves develop as reddening of the leaves between major veins in red varieties, and yellowing of the leaves between major veins in white varieties. Leaves become thick, brittle, and roll downwards (Plate 65). Other symptoms include delayed foliation, production of fewer, smaller clusters of berries and poor growth of vines. American *Vitis* species and their hybrids used as rootstock can be symptomless carriers (Goheen, 1970). Leafroll will cause graft union necrosis in grafted plants. Grapevine leaf roll associated viruses (GLRaVs) may infect a grapevine individually or as mixed infections of different combinations. Mixed infections often result in synergistic effects leading to more severe damage to grapevines than single infections.

TRANSMISSION

The disease is transmitted by grafting. Mealybugs; *Pseudococcus longispinus* and *Planococcus ficus* (Rosciglione and Gugerli, 1989, Tanne *et al.*, 1989) and soft scales, *Pulvinaria vitis, Parthenolecanium corni* and *Neopulvinaria innumberabilis* (Martelli, 2000) have been reported to transmit leafroll associated viruses. The common grape mealybug (*Pseudococcus maritimus*) is also a vector of GLRaV-3 (Charles *et al.*, 2009).

THE PATHOGEN

Several phloem limited closteroviruses with particle length of 1800-2200 nm are associated with leafroll disease (Hu *et al.*, 1990). ICTV approved GLRaV-3 as the type of the genus *Ampelovirus* in the family *Closteroviridae*. The genome is monopartite. Only one particle size of linear, positive-sense, single-stranded RNA is found. The viral genome encodes structural proteins and non-structural proteins. The genome is partially sequenced (Habili *et al.*, 1995; Kumar *et al.*, 2012).

DIAGNOSIS AND MANAGEMENT

The virus can be detected by enzyme-linked immunosorbent assay (ELISA) using antibody to GLRaV-3 (Zimmermann *et al.*, 1990) and by slot blot hybridization analysis using double-stranded RNA as target and labeled GLRaV-3-specific cDNA as probe. A cDNA clone from an Italian isolate of GLRaV-3 hybridized in Northern blots with three major dsRNAs of 19.5, 1.9, and 0.9 kbp extracted from leafroll-infected vines (Habili *et al.*, 1995). The virus was also detected in an immunocapture reverse transcriptase polymerase chain reaction (Kumar *et al.*, 2013). Micrografting is an effective method for viral indexing of grapevines. The method can be used in conjunction with wood indexing for post-entry quarantine to identify infected material and reject it much earlier than is currently possible (Ranjith *et al.*, 2005). The diseases can be managed by using healthy planting material and control of mealy bugs.

ARABIS MOSAIC VIRUS INFECTING GRAPES

Arabis mosaic virus can also infect grapes in addition to many other herbaceous and woody hosts such as raspberry, strawberry, rhubarb, cherry, peach, and plums. It is occurring worldwide wherever *Vitis vinifera* and hybrid rootstocks of grapevine are grown.

Symptoms on infected trees appear as leaf mottling and flecking, stunting and leaf deformation including enations. Symptoms intensity depends on variety, rootstock and environmental conditions. There is heavy dropping of fruits resulting in yield loss up to 50%. Virus infection is characterized by loose clusters and abundant small berries. The infections may be latent and plants do not show symptoms. The disease primarily spread by the movement of propagation material from infected plants. Arabis Mosaic virus is transmitted by the nematode *Xiphinema diversicaudatum* (Jha and Posnette, 1961).

Arabis mosaic virus is polyhedral in shape and is a NEPO virus. The RNA 2 of the Arabis mosaic virus (ArMV) encodes a polyprotein from which protein 2A is released by proteolytic cleavage at the N-terminus. The 2A gene of 19 ArMV isolates were subsequently cloned and sequenced. These 19 isolates and those from data banks were classified into four groups based on the size of the protein 2A which ranged from 233 to 280 amino acids, and sequence identities. Sequence variability was mainly located in the N-terminus of the proteins, whereas the core region and the C-terminus were conserved (Wetzel, 2002).

The ELISA technique is an efficient method for the detection of this virus in grapevine. Nine Agrobacterium-mediated transformation were achieved with different constructs containing chimeric coat protein (CP) genes of grapevine fanleaf virus (GFLV). Putative transformed tissues were selected by continued proliferation on kanamycin containing medium. Selected embryos were germinated. Integration of the different CP genes was confirmed by PCR. Parallel experiments with *Nicotiana benthamiana*, expressing the full-length CP gene of GFLV, showed protection against virus inoculation (Gölles *et al.*, 2000).

RUGOSE WOOD COMPLEX

Rugose wood complex was reported as ealy as 1966 (Graniti and Martelli, 1966). Rugose wood (RW) affects woody cylinder causing pitting, grooving when grafted onto specific *Vitis* indicator hosts.

Following effects have been observed in infected grapevines: delayed bud opening, swelling of trunk above the bud union, reduced vigour and yield. Affected vines may decline and die. Some cultvars and rootstocks are symptomless carriers (Granity and Martelli, 1966, Goheen, 1988). Diseases can be differentiated into 4 different types of disorders based on their effects on woody grapevine cylinders when grafted onto specific *Vitis* indicator hosts:

1. Rupestris stem pitting disorder
2. Kober stem grooving disorder
3. LN 33 stem grooving disorder
4. Corky bark disorder

Rugose wood disease is transmitted by grafting and by mechanical inoculation but some viruses of the comlex are transmitted by mealybugs, *Pseudococcus longispinus*, *P. ficus* and *Planococcus citri* (Rosciglione and Castellano, 1985). The typical symptoms of corky bark are grooving and pitting on the woody cylinder of *Vitis rupestris* and LN 33 (Martelly, 1993). The aetiology of RW complex is multifaceted and not yet completely understood. At least 4 viruses have been found associated with different types of RW diseases. They are grapevine rupestris stem pitting-associated virus (GRSPaV), grapevine virus A (GVA), grapevine virus B (GVB), and grapevine virus D (GVD) (Rosa and Rowhani, 1966). A closterovirus with particles of 800 nm long is associated with GVA and a closterovirus with particle size of 1400-2200 is associated with corky bark (Conti *et al.*, 1980; Namba *et al.*, 1991). GRSPaV, GVA, GVB, and GVD may also occur as mixed infections among themselves and with different GLRaVs. RW has variable growth effect on different rootstocks (Rosa *et al.*, 1911). Grafting to *V. rupestris,* LN 33 and Kober 5BB can differentiate the diseases of this complex. Rupestris stem pitting and corky bark will be described in following pages.

RUPESTRIS STEM PITTING VIRUS

Rupestris stem pitting associated virus-1 is a common and widespread virus of grapevines. The incidence of this disease in Washington State was observed upto 4.6% (Goheen, 1988). Rupestris stem pitting causes a slow decline in growth resulting in reduced yield. No leaf discoloration is observed. The disease affects only grafted vines causing small pits in the wood on rootstock. Ungrafted vines may be infected, but usually do not show symptoms. It can occur as distinct variants that show different symptoms in diverse *Vitis* species. The severity of this disease is more pronounced in vines that are infected with other viruses of the rugose wood complex, including Grapevine virus A (GVA) and Grapevine virus B (GVB). GVA affected vines may show swelling at the graft union and GVB is associated with corky bark disease, which may cause an incompatibility to develop at the graft union. Rootstocks of *Vitis rupestris* and its hybrids develop pitting in the xylem which can be observed on woody tissues after bark is removed and can be confused with corky bark symptoms. On self-rooted grapevines, Grapevine Rupestris stem pitting-associated virus (GRSPaV) intensifies the symptoms similar to be associated with leafroll disease. Spread is mainly through propagation. There is no known insect vector for Rupestris stem pitting virus. GVB can be transmitted by some species of mealybugs.

GRSPaV is a member of the genus *Foveavirus* within the family *Flexiviridae.* Flexuous virus particles of 723 nm length were detected in infected grapevine (Petrovic *et al.*, 2003). GRSPaV is closely associated with the disease. Rupestris stem pitting is frequently detected in grapevines worldwide (Meng *et al.*, 2006). The genome of Rupestris stem pitting virus-1(RSPaV-1) consists of 8726 nt excluding a poly(A) tail at the 3' terminus. Comparison of the genome structure and nucleotide and amino acid sequences indicated similarities of RSPaV-1 to apple stem pitting virus, and to a lesser extent, to potato virus M carlavirus (Meng *et al.*, 1998). ELISA, RT-PCR assays have been used to detect the virus (Meng *et al.*, 2003; Zhang *et al.*, 1998). Disease management depends upon use of certified healthy, virus-free stock. Vines showing suspicious symptoms should be removed.

CORKY BARK

Corky bark occurs worldwide. It remains in a latent state in many grapevine cultivars but can show symptoms only after infected buds are grafted onto virus-susceptible *Phylloxera*-resistant rootstocks. Grafted plant show union disorders including pitting, grooving, lesions or necrosis of rootstock and/or scion. Affected plants may die. In commercial vineyards, corky bark symptoms vary in severity from very mild effects to the death of vines depending on the rootstock used. Shoot growth is delayed and foliage becomes discolored, turning red or bronze/yellow. The wood at the base of canes may swell slightly, causing the bark to split. Depending on the cultivar, various wood grooving may also occur. On self-rooted grapevines, the symptoms of corky bark are very mild. Fruit maturity is delayed and yields are much reduced. Stem symptoms may develop in both scion and rootstock or in only one of the two. The biological indicator for the disease is the grapevine rootstock hybrid LN33, a cross of Couderc 1613 x *V. vinifera* cv Thompson Seedless.

Closterovirus-like particles have been reported to be associated with grapevine corky bark-associated virus (GCBaV). Electron microscopy of purified preparations revealed the presence of flexuous rod-shaped virus-like particles that were about 13 nm in diameter and between 1,400 and 2,000 nm long, with a helical pitch of 3.4 nm. Polyclonal antisera were produced which reacted in ELISA with corky bark-affected grapevines (Boscia *et al.*, 1993; Azzam *et al.*, 1991; Namba *et al.*, 1991). The disease can be prevented by selecting plants with healthy scions and rootstock (Tanne *et al.*, 1997). Fragmented shoot apex culture of grapevine has been shown to regenerate plants free from stem pitting and corky bark (Barless, 1987).

PEACH ROSETTE MOSAIC VIRUS INFECTING GRAPES

Peach rosette mosaic virus (PRMV) was reported on grapevine cultivars 'Concord', 'Catawba' and 'Niagara' from vineyards in Michigan (Ramsdell and Myers, 1974). The virus causes degeneration of grapes (*Vitis lebrusca* cv Concord). Infected vines are usually umbrella-shaped because the virus causes the canes to grow somewhat crookedly. Internodes are shorter than normal and leaves are distorted. Berry cluster shelling occurs on vines that have been infected for several years. 'Delaware' grapevines are less susceptible to PRMV than are 'Concord', 'Catawba' and 'Niagara' varieties. PRMV is primarily a problem in *Vitis labrusca,* but some cultivars of *V. vinifera,* and French-American hybrids are also susceptible. The virus is seed-borne in grapes (Ramsdell and Myers, 1978). This virus is also soil-borne and is spread from plant to plant by the root-feeding dagger nematodes (*Xiphinema americanum*) and *Longidorus diadecturus* (Allen *et al.*, 1982). The virus is mechanically transmitted to herbaceous hosts in Chenopodiaceae, Cucurbitaceae, Leguminosae and Solanaceae. Dandelion, curly dock and carolina horsenettle are weed hosts for the virus. The virus belongs to Nepovirus group. Virus particles are isometric and range from 24 to 31 nm, averaging 28 nm. Control of this disease includes removal of infected vines, soil fumigation of vineyard sites to kill the dagger nematodes vector and planting with certified virus-tested clean stock.

TOMATO RINGSPOT VIRUS IN GRAPES

French hybrid grapevines (especially blue-fruited varieties) are susceptible to infection by Tomato ringspot virus (TmRSV) and tobacco ringspot virus (TRSV). In infected vines initially a few leaves may show mottling and ringspot. The leaves are reduced in size and rosette due to shortened internodes (Young *et al.*, 1986). Finally, growth is very stunted and limited to basal suckers, and the vine eventually dies. Fruit clusters are reduced in size with aborted berries. Removal of bark from trunks and stems of diseased vines may reveal thickened spongy phloem tissue with several necrotic pits (Uyemoto, 1975). The viruses are introduced into vineyards with infected planting stock or by dispersal of seed from infected weeds (Uyemoto and Gilmer, 1972). The dagger nematode, *Xiphinema americanum* is the vector of this disease. The virus is also seed borne in some herbaceous hosts like petunia. Control is similar to peach rosette mosaic virus. The nematodes can retain the virus for long periods.

OTHER VIRUS DISEASES OF GRAPES

AJINASHIKA DISEASE

This virus disease was reported from Japan (Namba *et al.*, 1979). There are no recognizable symptoms on affected plants but sugar contents of berries are reduced making the fruits unmarketable. The disease is transmitted by grafting and restricted to *Vitis vinifera* cv. Koshu. Virions are isometric; not enveloped; 25 nm in diameter; angular in profile; without a conspicuous capsomere arrangement (Namba *et al.*, 1986, 1991). Virions found in phloem and companion cells in sieve tubes; in cytoplasm, in nuclei, and in cell vacuoles. Possibly a luteovirus but there is no serological relationships to barley yellow dwarf and potato leafroll luteoviruses. The genome is not segmented and contains a single molecule of linear positive-sense, single-stranded RNA. The genome has been sequenced and is about 6800 nucleotides long. The virus can be detected by ELISA and ISEM and also by grafting on cv. Koshu. No information is available on its management. Prolong heat treatment can eliminate the disease.

FLECK (MARBRURE) DISEASE

Fleck disease was first reported from Europe as 'Marbrure' in 1966 (Vuittenez *et al.*, 1966). Fleck disease is worldwide in its distribution. In *Vitis rupestris*, symptoms appear as clearing of leaf veins and malformation of leaves. *Vitis* species and their hybrids in Europe and America are symptomless (Hewitt *et al.*, 1972). The associated virus is restricted in its host range only to genus *Vitis* and is transmitted by grafting. Natural spread of the disease was reported from South Africa (Engelbrecht and Kasdorf, 1990). Grapevine fleck virus is isometric, phloem-limited RNA virus of 30 nm in diameter (Boscia *et al.*, 1991; Boulila *et al.*, 1990). Grapevine fleck virus is a positive-sense, single-stranded RNA virus with a genome size of 7,564 nucleotides (Sabanadzovic *et al.*, 2001). ICTV approved virus is the type of the genus *Maculavirus* in the family *Tymoviridae*. The virus can be detected by grafting on *V. rupestris* and also by ELISA and ISEM (Boscia *et al.*, 1991) and also in RT-PCR (Naidu and Mekuria, 2010). The disease may be eradicated by thermotherapy and meristem tip culture.

STUNT DISEASE

Stunt disease of grapevine occurs in Japan. Main symptoms of the disease are short internodes, smalling of leaves, leaf rolling, shelled branches and stunting. Fruit setting is impaired and few. Susceptible host species are found in the Family *Vitidaceae*. The following species were susceptible to experimental virus infection: *Vitis thunbergii, Vitis vinifera* cv Campbell early. The virus is transmitted by grafting and by leahopper, *Arboridia apicalis* in a persistent manner. An isometric, phloem limited RNA virus of 25 nm is associated with the disease (Namba *et al.*, 1986). The virus can be detected by grafting on *V. vinifera* cv. Campbell Early and *V. thunbergii* which will develop leaf rolling and stunting symptoms. The disease can be managed by the control of insect vector and use of virus-free stock for grafting.

LINE PATTERN

Line pattern disease of grapes has been reported from Hungary on *Vitis vinifera*. Affected vines show bright yellow spots or blotches, marginal rings or maple-leaf line pattern on leaves. Affected vines are less vigourous and yield is reduced. The virus is transmitted by grafting and also by mechanical inoculation to *Cucumis sativus* and *Nicotiana glutinosa*. These hosts show chlorotic lesions and systemic mottle upon inoculation. Grapevine line pattern virus is a possible member of ilarvirus group. Virus particles are quasispherical to bacilliform with multipartite RNA genome (Lehoczky *et al.*, 1989). The virus can be detected by graft transmission to cv. Jubileum 75 or by mechanical inoculation to herbaceous hosts.

YELLOW DWARF

Yellow dwarf disease on grapes was reported from Taiwan. Affected grape plants remain stunted and leaves show chlorotic to yellow mottling with necrotic spots. An enveloped virus of 80 nm in diameter resembling to tomato spotted wilt virus was observed in thin sections of affected plants (Chen *et al.*, 1981). No other information is available on this disease.

YELLOW MOTTLE

Yellow mottle disease in grapevines has been reported from several countries of Europe. The leaves of affected vines show various types of discoloration including speckling, rings and lines (Bercks *et al.*, 1973). The virus has a very wide host range in wild and cultivated plant species. The disease is transmitted by grafting and sap inoculation. Alfalfa mosaic virus has been shown to be associated with yellow mottle. The genome of the virus is tripartite RNA with four types of bacilliform particles with 18 nm wide and 30, 35, 43 and 57 nm long. The virus can be detected by grafting to cv. Chardonnay or Veltliner rouge precoce (Beczner and Lehoczky, 1981). Mechanical inoculation to diagnostic herbaceous hosts, *Phaseolus vulgaris* (necrotic lesions), *Chenopodium amaranticolor* and *C. quinoa* (chlorotic lesions, systemic mosaic and top necrosis), *Ocimum basilicum* (Systemic bright yellow mottling) can also detect the virus. The virus can also be detected in ELISA systems.

OTHER VIRUS-LIKE DISEASES OF UNKNOWN AETIOLOGY IN GRAPE

ASTEROID MOSAIC

Asteroid mosaic disease has been reported from USA, Italy, South Africa and Greece (Kyriakopoulou, 1991, Refatti, 1970). Symptoms on grapes develop as translucent spots with a star-like shape between veins. The yield and vigour of affected plants reduced (Hewitt and Goheen, 1959). The disease is transmitted by graft and can be detected by grafting to *Vitis rupestris*. No other information is available on this disease.

BUSHY STUNT

In bushy stunt disease, proliferation of apical and axillary buds of affected plants results in stunted and bushy growth of European grapevines. Yield is reduced. American rootstocks are symptomless carriers (Savino *et al.*, 1991). The hostrange is restricted to *Vitis* only. The disease is transmitted by grafting. Disease diagnosis is done by grafting to cv. Italia and Sangiovese. Thermotherapy has been suggested for management of the disease.

ENATION DISEASE

Enation disease has been reported from Australia, Europe, Israel, New Zealand, South Africa, Turkey, USA and Venezuela. Symptoms on basal leaves of affected vines appear as malformation with laminar or cup shaped enations on the underside of leaves (Hewitt, 1954). There is delay in bud breaking and bushy growth of shoots. The disease is transmitted by grafting (Garau *et al.*, 1989). Disease diagnosis is done by grafting to LN 33 (Martelli *et al.*, 1966).

INFECTIOUS GRAFT INCOMPATABILITY

This disorder was reported from France and was suspected to be caused by a virus. The symptoms appear on *Vitis vinifera* scions grafted onto some rootstock varieties (Durquety *et al*, 1977). The disorder is graft transmitted (Legin and Walter, 1986). The disorder can be detected by grafting to Kober 5BB and managed by thermotherapy and meristem culture.

RODITIS LEAF DISCOLORATION

This disease has been reported from Greece. Symptoms of the disease are: Yellowish and/or reddish discolorations along the veins or interveinal areas of leaves, malformation, reduced leaf size and number of bunches (Rumbos and Avgelis, 1989). The disease is transmitted by grafting and mechanical inoculation to a wide range of wild and cultivated hosts. Although the actual cause of the disease is not known but grapevine fan leaf and carnation mottle viruses have been found associated with the disease (Avgelis and Rumbos, 1991). The disease can be detected by grafting on cv. Mission.

SUMMER MOTTLE (VEIN MOTTLE)

Summer mottle-a graft transmissible disease was reported from Australia. Symptoms of the disease were pale green vein banding similar to that caused by vein mosaic (Krake and Woodham, 1978). The disease is restricted to *Vitis* only and can be detected by grafting on cv. Sideritis, Cabernet franc and Mission.

VEIN MOSAIC

Vein mosaic disease on grapes has been reported from Europe and Mediterranean countries. The disease is graft transmitted and suspected to be caused by a virus. Symptoms on infected plants develop as pale green discoloration along the veins, producing a vein banding effect. Several American rootstocks and European grape cultivars are symptomless carriers (Legin and Vuittenez, 1973). Grafting on *Vitis riparia* Gloire de Montpellier can detect the disease. Thermotherapy has been recommended for disease management.

VEIN NECROSIS

This disease is known to occur in Europe, Mediterranean countries and USA.The disease is symptomless in most *Vitis* spp. but hybrid *Vitis rupestris* X *Vitis berlandieri* 110R develop necrosis of the veinlets on the underside of leaves and occasionally necrosis of the shoot tips (Legin and Vuittenez, 1973). The disease is graft transmitted and can be detected to 110R. Thermotherapy has been recommended for disease management (Savino *et al.*, 1985).

PHYTOPLASMA DISEASES OF GRAPEVINE

Phytoplasmas are microscopic plant pathogens, similar to bacteria but much smaller and lacking cell walls. They live in the vascular system of plants and are spread by leafhoppers and planthoppers. There are several diseases of grapevine caused by phytoplasmas like Flavescence dorée, Bois noir (black wood, vergilbungskrankheit) and Grapevine yellows (American grapevine yellows and Australian grapevine yellows).

Grapevine phytoplasma belonged to the aster yellows (AY) phytoplasma group and for identification primers are normally designed for specific amplification of a part of the 16S rRNA gene from all known phytoplasmas. Management of phytoplasma-infected plants has mainly focussed on controlling the insect vectors and on roguing infected plants and weeds. The production of genetically engineered plants by introducing disease-resistance genes into cultivated crops togheter with the use of resistance inducer microorganisms to reduce the disease symptoms, represent potential tools to control phytoplasma diseases. The taxonomic position is: Bacteria, Tenericutes, Mollicutes, Phytoplasma.

FLAVESCENCE DORÉE

Flavescence dorée (FD) is the most important and destructive phytoplasma disease of grapevines (Caudwell, 1957). It is known to occur in France, Italy and Spain and possibly in USA (Angelini *et al.*, 2001; Martini *et al.*, 1999; Batlle *et al.*, 1997; Maixner and Pearson, 1991). Flavescence dorée is a disease of high economic importance in France. The disease can reduce the vitality of vines. It is also responsible for yield reductions, and reduced wine quality due to high acid and low sugar contents of fruit from infected plants. The disease is spreading fast affecting up to 80-100% of vines within a few years (Belli, 1985). Flavescence dorée has affected the viticulture in France and Corsica.

SYMPTOMS

The symptoms of the FD disease appear both on shoots and leaves. On shoots symptoms develop as reduced growth, absence of shooting, brittle shoots, necrosis of apical and lateral buds, numerous black pustules on shoots of susceptible cultivars and absence of lignification in autumn. In leaves of affected plants, symptoms include leaf roll, creamy yellow ring spots along the veins, vein chlorosis and necrosis. There is withering of flowers and inflourescence which dry out and fall off. A few rootstock varieties are symptomless carriers that have been responsible for spread of flavescence dorée in Europe. Popular cultivars such as Chardonnay, Cabernet Sauvignon, Pinot noir, Riesling, Sauvignon blanc, and Sémillon are highly susceptible to flavescence dorée and may be killed due to FD infection.

TRANSMISSION

Flavescence dorée is transmitted by grafting and by the leafhopper vector, *Scaphoideus titanus* (Mori *et al.*, 2002; Bressan *et al.*, 2006). *S. titanus* spends its whole life cycle on grapevines. Both nymphs and adults are able to acquire the phytoplasma while feeding.

THE PATHOGEN

Flavescence dorée (FD) is caused by a phytoplasma which belongs to the elm yellows group (EY or 16SrV) (Daire *et al.*, 1997; Angelini *et al.*, 2001).

Pathogen associated with FD has been classified as *Candidatus* Phytoplasma vitis (Lee *et al.*, 2004). Grapevine yellows disease in Virginia closely resembles flavescence dorée and other grapevine yellows diseases, but the phytoplasmas infecting grapevines in Virginia are distinct from other grapevine yellows pathogens and designated as VGYIII (Caudwell, 1990). It was most closely related to spirea stunt (SPI), walnut witches' broom (WWB), and poinsettia branch-inducing (PoiB1) phytoplasmas in subgroups E, G, and H, respectively (Davis *et al.*, 1998).

DIAGNOSIS AND MANAGEMENT

FD can be detected by grafting to Baco 22A or transmission by vector to herbaceous hosts (Caudwell, 1977). It can also be detected by ELISA and dot blot hybridization (Boudon-Padieu *et al.*, 1989) and by PCR and other molecular methods (Marzachi *et al.*, 2001; Margaria *et al.*, 2007; Daire *et al.*, 1997). The phytoplasma can be detected in viruliferous vector, *Scaphoideus titanus* by dot blot hybridization using DNA probes (Bertaccini *et al.*, 1993). In France, flavescence dorée is managed by effective quarantine measures, destroying all infected plants and control of the vector with insecticide sprays in the affected zone. Use only planting material from reliable sources and produced in areas where phytoplasma diseases are not present have also been recommended. Hot water treatment at 50^0 Celsius for a minimum period of 35 minutes is an effective measure to manage this disease (Caudwell *et al.*, 1992).

GRAPEVINE YELLOWS (GY)

Grapevine yellows (GY) is a term that is used to refer to any of several grapevine diseases that are currently attributed to infection of grapevine plants by pathogens known as phytoplasmas. Grapevine yellows is serious disease of grapevines all over the world (Boudon-Padieu, 2003). Two types of grapevine yellows (GY) are known as American grapevine yellows that occurs in New York and Virginia, and an Australian grapevine yellows. Mixed phytoplasma infections can also occur.

Typical symptoms of yellows phytoplasma include discoloration and necrosis of leaf veins and leaf blades, downward curling of leaves, lack or incomplete lignification of shoots, stunting and necrosis of shoots, abortion of inflorescences and shrivelling of berries. There is callose deposition at the sieve plates and subsequent degeneration of the phloem. All grapevine yellows diseases have similar symptoms but symptoms are not uniform, and may appear on some or all shoots of infected vines. Leaf symptoms in white cultivars appear as small, yellow spots along the main veins. These spots enlarge to form yellow bands along the veins, which gradually extend over large parts of the leaf. Red cultivars develop a similar pattern on the leaves, but the discolorations are reddish. Numerous small black pustules are formed along the diseased branches of susceptible cultivars. Fruit set is reduced on grapevines infected early in the season. Premature berry drop occurs in some cultivars. The GY disease is transmitted through contaminated planting material. *Oncopsis alni* (Schrank) has been reported the vector of yellows phytoplasma (Maixner *et al.*, 2000). Like other phytoplasmas GY is also transmitted to *Catharanthus roseus* by dodder.

Grapevine yellows (GY) disease is caused by phytoplasma. Presence of phytoplasma bodies in mature sieve elements of infected plants are seen in electron microscope. The PCR and restriction fragment length polymorphism tests suggest that only one type of phytoplasma is associated with grapevine yellows virus in Australia (Lia *et al.*, 2004). Restriction profiles of grapevine phytoplasmas using Mse I suggest that Australian grapevine yellows (AGY) is unique but most closely resembles those phytoplasmas associated with grapevine diseases in the stolbur group (Padovan *et al.*, 1996; Daire *et al.*, 1997). Sequence comparisons and phylogenetic analysis of the 16S rRNA genes and the 16S/23S spacer regions of the phytoplasmas associated with Australian grapevine yellows, papaya dieback and Phormium yellow leaf diseases revealed minimal nucleotide differences between them resulting in the formation of a monophyletic group and these phytoplasmas have been put under '*Candidatus* Phytoplasma australiense' (Davis *et al.*, 1997; Lia *et al.*, 2004). Phytoplasmas associated with grapevine yellows in Virginia belong to group 16SrI, subgroup A (tomato big bud phytoplasma subgroup), and group 16SrIII, new subgroup I (Davis *et al.*, 1998). Phytoplasmas associated with Australian grapevine yellows (AGY) can be detected using the polymerase chain reaction (PCR) (Prince *et al.*, 1992; Gibbs *et al.*, 1999). For control, Use only planting material from reliable, approved sources and produced in areas where phytoplasma diseases are not present.

BOIS NOIR (BLACK WOOD, VERGILBUNGSKRANKHEIT)

Bois noir (BN) disease is widespread in many parts of Europe especially in France and Germany, but is less serious due to a slower rate of spread.

Diseases similar to bois noir have also been reported in Switzerland, Italy, Greece, Hungary, Israel, and are suspected to occur in Chile and South Africa. The Bois noir disease is also reported from grapevine in Spain (Lavina *et al.*, 1995). The disease is cultivar and climate-dependent (Uyemoto *et al.*, 2009). Exceptionally sensitive varieties include Chardonnay, Sauvignon Blanc, Pinot Noir, and Cabernet Sauvignon. Bois noir is transmitted by a few planthopper vectors, the main one being *Hyalesthes obsoletus* (Maixner, 1994). This insect prefers herbaceous weeds over grapevines and transmits the phytoplasma from these weeds to grapevines (Mori *et al.*, 2007). Bindweed (*Convulvulus arvensis*), nettle (*Urtica dioica*) and hedge bindweed (*Calystegia sepium*) are the main reservoirs in Europe. Since the vector only feeds incidentally on grape, bois noir is a much slower spreading disease than flavescence dorée (FD).

A phytoplasma associated with Bois noir disease in France is similar to FD phytoplasma. Both causes similar symptoms in grapevine but the vector of FD, *Scaphoideus titanus* can not transmit the BN phytoplasma. Bois noir has been attributed to '*Candidatus* Phytoplasma solani'-related strains. These strains have been classified in group 16SrXII (stolbur phytoplasmas group) subgroup A (formerly called subgroup 16SrI-G). The recommended procedure to prevent the spread of Bois noir is a hot water treatment of all plant propagation material.

LITTLE LEAF

In India, little leaf disease was observed on "Merlot Noir" variety of grapevine in Bangalore (Singh and Sastry, 1975). The leaves on affected vines become extremely small and develop light yellow colour. Internodes are reduced and stem develops in zigzag fashion. Axillary buds proliferates. No flower or fruit develop on affected vines. Roots become stunted with a few rootlets. Temporary remission of disease symptoms was obtained by heat and tetracycline therapy. The disease was presumed to be caused by phytoplasma (Singh and Sastry, 1975). No other information is available on this disease.

VIROID DISEASES OF GRAPEVINE

GRAPEVINE YELLOW SPECKLE VIROID

Grapevine yellow speckle viroid (GYSVd) was first detected in Japan from commercial grapevines. The viroid seems to be spreading in Japanese commercial vineyards but without causing noticeable disease symptoms. The viroid is now worldwide in distribution (Semancik *et al.*, 1987). Two distinct viroid, grapevine yellow speckle viroid-1(GYSVd-1) and 2 (GYSVd-2) are known to cause this disease (Sano *et al.*, 2000). Symptoms on the leaves of infected plants ranged from a few chrome yellow spots to yellowish-green vein banding (Taylor and Woodham, 1972). Plants infected with GYSVd-1 may remain symptomless, while GYSVd-2 show the above mentioned symptoms (Koltunow *et al.*, 1989).

Both the viroids (GYSVd-1 and GYSVd-2) have single stranded circular RNA and are associate with grapevines yellow speckle disease. The RNA of GYSVd-1 contains 367 nucleotide residues and GYSVd-2 has 363 nucleotide residues, the potential to form the rod-like secondary structure characteristic of viroids. GYSVd has 37% sequence homology with the recently described apple scar skin viroid with 330 residues and has some sequence homology with the viroids in the potato spindle tuber viroid (PSTVd) group (Koltunow and Rezaian, 1988).

Two closely related circular RNA molecules previously named grapevine yellow speckle viroid (GYSVd) and grapevine viroid 1B (GVd1B) have a rod-like structure typical of viroids as observed in electron microscopy after spreading under non-denaturing conditions. Purified GYSVd and GVd1B replicated independently in inoculated grapevine seedlings and some of the infected plants developed yellow speckle symptoms indicating that both viroids can cause grapevine yellow speckle disease. Plus-sense RNA transcripts derived from a dimeric GYSVd cDNA clone induced yellow speckle symptoms in a grapevine seedling confirming the role of GYSVd in the yellow speckle disease (Koltunow *et al.*, 1989). Two oligonucleotide probes were synthesized for the detection of the two related viroids. The probes which could detect each viroid individually were used to assess correlations between the occurrence of these viroids and the incidence of the disease (Koltunow *et al.*, 1989). These viroids are transmitted by grafting, micro-injection and slash inoculation and by contaminated operation tools. Both the viroids can be detected in two dimentional gel electrophoresis. Nucleic acid hybridization and polymerase chain reaction (PCR). These viroids can be eliminated by meristem tip culture at 10°C (Duran-Vila *et al.*, 1988). Plants grown by seed are free from these viroids.

OTHER VIROIDS INFECTING GRAPEVINES

Citrus exocortis and hop stunt viroids are known to infect grapevine worldwide (Garcia-Arena *et al.*, 1987; Sano *et al.*, 1985). Another viroid infecting grapes is Australian grapevine viroid (AGVd) is mechanically transmissible infecting cucumber, tomato and grapevine. AGVd has 369 nucleotide residues. All the three viroids do not induce visible symptoms in grapevine (Rezaian, 1990).

REFERENCES

Ahlawat, Y.S., Pant, R.P., Lockhart, B.E.L., Srivastava, M., Chakraborty, N.K. and Varma, A. 1996. Association of a badnavirus with citrus mosaic disease in India. *Plant Disease,* 80: 590–592.

Allen, W.R., Van Schagen, J.G. and Eveleigh, E.S. 1982. Transmission of peach rosette mosaic virus to peach, grape and cucumber by *Longidorus diadecturus* obtained from diseased orchards in Ontario. *Canadian J. Plant Pathol.,* 4: 16–18.

Angelini, E., Clair, D., Borgo, M., Bertaccini, A. and Boudon-Padieu, E. 2001. Flavescence dorée in France and Italy: Occurrence of closely related phytoplasma isolates and their near relationships to palatinate grapevine yellows and an alder yellows phytoplasma. *Vitis,* 40: 79–86.

Avgelis, A.D. and Rumbos, I.C. 1991. Carnation mottle virus isolated from vines affected with "Roditis leaf discoloration". *In*: Proc. 10th Meeting ICVG, 1990. Volos, Greece.

Azzam, O.I., Namba, S., Boscia, D., Maixner, M., Hu, J.S., Golino, D., and Gonsalves, D. 1991. purification and properties of closterovirus-like particles associated with grapevine corky bark disease. *Phytopathology,* 81: 964–970.

Barlass, M. 1987. Elimination of stem pitting and corky bark diseases from grapevine by fragmented shoot apex culture. *Annl. Appl. Biol.,* 110: 653–56.

Barlass, M., Skene, K.G.M., Woodham, R.C. and Krake, L.R. 1982. Regeneration of virus-free grapevines using *in vitro* apical culture. *Ann. Appl. Biol.,* 101: 291–295.

Batlle, A., Laviña, A., Kuszala, C., Clair, D., Larrue, J. and Boudon-Padieu, E. 1997. Detection of flavescence doree phytoplasma in grapevine in northern Spain. *Vitis,* 36: 211–212.

Beczner, L. and Lehoczky, J. 1981. Grapevine disease in Hungary caused by alfalfa mosaic virus infection. *Acta Phytopathol. Acad. Sci. Hung.*, 16: 119–128.

Belli, G., Fortusini, A. and Rui, D. 1985. Recent spread of flavescence dorée and its vector in vineyards of northern Italy. *Phytopathol. Mediterr.*, 24: 189–191.

Bercks, R., Lesemann, D. and Querfurth, G. 1973. Uber den Nachweis des alfalfa mosaic virus in einer weinrebe. *Phytopathol. Z.*, 76: 166–171.

Bertaccini, A., Arzone, A., Alma, A., Bosco, D. and Vibio, M. 1993. Detection of Mycoplasma-like organisms in *Scaphoideus titanus* Ball reared on Flavescence doree infected grapevine by dot hybridizations using DNA probes. *Phytopathologia mediterranea,* 32: 20–24.

Boscia, D., Martelli, G.P., Savino, V. and Castellano, M.A. 1991. Identification of the agent of grapevine fleck disease. *Vitis,* 30: 97–105.

Boscia, D., Savino, V., Minafra, A., Namba, S., Elicio, V., Castellano, M., Gonsalves, D. and Martelli, G. 1993. Properties of a filamentous virus isolated from grapevines affected by corky bark. *Archives of Virology,* 130: 109–120.

Boudon-Padieu, E. 2003. The situation of grapevine yellows and a current research directions: Distribution, diversity, vector, diffusion and control. *In*: XIV meeting of ICGV, Locorotondo (BA), Italy. pp. 47–53.

Boudon-Padieu, E., Larrue, J. and Caudwell, A. 1989. ELISA and dot blot detection of flavescence doree – MLO in individual leafhopper vector during latency and inoculative state. *Curr. Microbiol.*, 19: 357–364.

Boulila, M., Boscia, D., Di Terlizzi, B.,Castellano,M.A., Minafra, A., Savino, V. and Martelli, G.P. 1990. Some properties of a phloem limited non-mechanically transmissible grapevine virus. *J. Phytopathol.*, 129: 151–158.

Bovey, R., Brugger, J.J. and Gugerli, P. 1980. Detection of grapevine fan leaf virus in grapevine tissue extracts by enzyme-linked immunosorbent assay (ELISA) and immune electron microscopy (IEM). *In*: Proc. 7th meeting ICVG, 1980, Niagara falls. pp. 259–275.

Bressan, A., Larrue, J. and Boudon Padieu, E. 2006. Patterns of phytoplasma-infected and -infective *Scaphoideus titanus* leafhoppers in vineyards with high incidence of Flavescence dorée. *Entomol. Exp. Appl.*, 119: 61–69.

Cadman, C.H., Dias, H.F. and Harrison B.D. 1960. Sap transmissible viruses associated with diseases of grapewine in Europe and North America. *Nature* (Lond), 187: 577.

Candresse T. and Martelli, G.P. 1995. "Genus Closterovirus". *In*: Murphy, F.A., Fauquet, C.M., Bishop, D.H.L., Ghabrial, S.A., Jarwis, A.W., Martelli, G.P., Mayo, M.A. and Summers, M.D. (*Eds.*) Virus Taxonomy. 6th Report of the International Committee on the Taxonomy of Viruses, Vienna, Austria: Springer. pp. 461–464.

Carter, W. 1933. The pineapple mealy bug, *Pseudococcus brevipes,* and wilt of pineapples. *Phytopathology,* 23: 207–242.

Carter, W. 1951. The feeding sequence of *Pseudococcus brevipes* (Ckl) in relation to mealybug wilt of pineapples in Hawaii. *Phytopathology,* 41: 769–780.

Caudwell, A. 1957. Deux années d'étude sur la flavescence dorée, nouvelle maladie grave de la vigne. *Ann. Amélior. Plantes,* 4: 359–363.

Caudwell, A. 1977. Aspects statistiques des epreuves d'infectivite chez les jaunisses (Yellows) des plantes et chez les viroses transmises selon le mode persistent interet de la Feve (Vicea faba) comme plante-test pour les jaunisses. *Ann. Phytopathol.*, 9: 141–159.

Caudwell, A. 1990. Epidemiology and characterization of flavescence doree and other yellows diseases. *Agronomie,* 10: 655–663.

Caudwell, A., Larrue, J., Valat, C. and Grenan, S. 1992. Hot water treatments against flavescence doree on dormant wood. Proc. 10th Meeting of ICVG, Volos, Greece.

Charles, J.G., Froud, K.J., Van den Brink, R. and Allan, D.J. 2009 Mealybugs and the spread of grapevine leafroll-associated virus 3 (GLRaV-3) in a New Zealand vineyard. *Australasian Plant Pathol.*, 38: 576–583.

Chatterjee, S.N. and Raychaudhuri, S.P. 1963. Additional vectors of mulberry mosaic. *Indian Phytopath.*, 16: 243–244.

Chatterjee, S.N. and Raychaudhuri, S.P. 1965.A note on aphid transmission of mosaic disease of mulberry. *Indian Phytopath.*, 18: 319–320.

Chen, H.L., Tzeng, D.S. and Chen, M.J. 1981. Preliminary studies on the grapevine yellow dwarf, a new virus disease in Taiwan. *Nat. Sci. Counc. Monthly ROC*, 9: 584–591.

Cohn, E., Tanne, E. and Nitzany, F.E. 1970. *Xiphinema italae*, a new vector of grapevine fanleaf virus. *Phytopathology*, 60: 181–182.

Conti, M., Milne, R.G., Luisoni, E. and Boccardo, G. 1980. A closterovirus from a stem pitting –diseased grapevine. *Phytopathology*, 70: 394–399.

Cory, L. and Hewitt, W.B. 1968. Some grapevine viruses in pollen and seed. *Phytopathology*, 58: 1316–1320.

Credi,R., Terlizzi, F., Beber, R., Poggi Pollini, C. and Ratti, C. 2012. Transmission trials of the fig mosaic disease agent. *In*: Abs. 22nd Interl. Conf. Virus and other graft transmissible diseases of fruit crops. Held in Rome June 3-8, 2012. p. 122.

Daire, X., Clair, D., Reinert, W. and Boudon-Padieu, E. 1997. Detection and differentiation of grapevine yellows phytoplasmas belonging to the elm yellows group and to the stolbur subgroup by PCR amplification of non-ribosomal DNA. *European J. Plant Path.*, 103 :507–514

Daiqun Sun Zhi-Huan. 2007. Suppressive effects of N-triacontanol on symptoms of mulberry dwarf disease and on the causal phytoplasma. *Plant Pathology*, 44: 979–981.

Davis, R.E., Dally, E., Gundersen, D.E., Lee, I.M. and Habili, N. 1997. "*Candidatus* phytoplasma australiense," a new phytoplasma taxon associated with australian grapevine yellows. *Intl. J. Syst. and Evol. Microbiology*, 47: 262–269.

Davis, R. E., Jomantiener, R., Dally, E. L. and Wolf, T.K. 1998. Phytoplasmas associated with grapevine yellows in Virginia belong to group 16SrI, subgroup A (tomato big bud phytoplasma subgroup), and group 16SrIII, new subgroup I. *Vitis*, 37: 131–137.

Dias, H.F. 1963. Host range and properties of grapevine fanleaf and grapevine yellow mosaic viruses. *Ann. appl. Biol.*, 51: 85–95.

Doi, Y., Teranaka, M., Yora, K., and Asuyama, H. 1967. Mycoplasma or PLT group-like microorganisms found in the phloem element of plants infected with mulberry dwarf, potato witches broom, aster yellows or paulwinia witches broom. *Ann. Phytopath. Soc. Japan*, 33: 259–266.

Duran-Vila, N., Juarez, J. and Arregui, J.M. 1988. Production of viroid free grapevines by shoot tip culture. *Am. J. Enol. Vitic.*, 39: 217–220.

Durquety, P.M., Ruchaud, C., Gazeau, J.P. and Fallot, J. 1977. Le clone et sesreactions au greffage. II. Nouvelles recherches sur incompatibilite clonale d'Abouriou greffe sur 5BB. Autres cas chez la vigne. *Progr. Agric. Vitic.*, 94: 420–427.

Engelbrecht, D.J. and Kasdorf, G.G.F. 1990. Field spread of corky bark, fleck, leaf roll and shiraj decline diseases and associated viruses in South African grapevines. *Phytophylactica*, 22: 347–354.

Flock, R.A. and Wallace, J.M. 1955. Transmission of fig mosaic by the eriophyid mite. *Aceria ficus*. *Phytopathology*, 45: 52–54.

Fuchs, M., Pinck, M., Etienne, L., Pinck, L. and walter, B. 1991. Characterization and detection of grapevine fan leaf virus using cDNA probe. *Phytopathology*, 81: 559–565.

Garau, R., Prota, U. and Cugusi, M. 1989. Studies on reproduction of enation symptoms by grafting in Sardinia. *In*: Proc. 9th meeting ICVG. Kiryat Anavim. pp. 203–206.

Garcia-Arena, F., Pallas, V. and Flores, R. 1987. The sequence of a viroid ftom grapevine closely related to the severe isolate of citrus exocortis viroid. *Nucl. Acid Res.*, 15: 4203–4210.

Gazel, M., Caglayan, K., Ulubas Serce, C. and Durgac, C. 2009. Detection of hop stunt viroid in pomegranate (*Punica granatum* L.) trees in the east Mediterranean region of Turkey. *Acta Hort.*, 818: 273–275.

Gibbs, K.S., Constable, F.E., Moran, J.R. and Padovan, A.C. 1999. Phytoplasmas in Australian grapevines - detection, differentiation and associated diseases. *Vitis*, 38: 107–114

Giuliano, G., Angelantonio, M., antonietta, C.M., angelo, D., donato, B., Toufic, E., Michele, D., and Paolo, M.G. 2009. Fig latent virus-1, a new putative member of the family flexiviridae. *In*: 21st internl. Conf. on virus and other graft transmissible diseases of fruit crops. *Berichte aus dem JKI*, p. 30–148.

Goheen, A.C. 1970. Grape leafroll. In:Virus diseases of small fruits and grapevine. *Ed*. Frazier, N.W. Univ. Calif., Div. Agric. Sci. Berkeley. pp. 209–212.

Goheen, A.C. 1988. Rupestris stem pitting. *In*: Compendium of grape diseases. *Eds*. Pearson, R.G. and Goheen, A.C. APS, St. Paul, MN. p. 53.

Goheen, A.C. and Hewitt, W.B. 1962. Grapevine vein banding virus. *Am. J. Ent. Vitic.*, 13: 73–77.

Goheen, A.C. and Luhn, C.F. 1973. Heat inactivation of viruses in grapevine. *Rev. Pathol.Veg. Ser.* IV, 9: 287–289.

Golles, R., Moser, R., Puhringer, H., Katinger, H., Machado,A., Tsolova, V. and Bouquet, A. 1997. Transformation of somatic embryos of *Vitis* Sp. With different constructs containing nucleotide sequences from nepovirus coat protein gene. *Acta Hort.*, 447: 265–272.

Gölles, R., Moser, R., Pühringer, H., Katinger, H., Laimer da Câmara Machado, M., Minafra, A., Savino, V., Saldarelli, P. and da Câmara Machado, G.P.M. 2000. Transgenic mosaic virus, grapevine virus A and grapevine virus B. *Acta Hort.* (ISHS), 528: 307–314.

Golles, R., Moser, R., Puhringer, H., Katinger, H., Machado, A., Minafra, A., Savino, V., Saldarelli, P. and Martelli, G.P. 2000. Transgenic grapevines expressing coat protein gene sequences of grapevine fanleaf virus, Arabis mosaic virus, grapevine virus A and grapevine virus B. VII Internl. Symp. On Grapevine Genetics and Breeding. *Acta Hort.*, 528: 305.

Golino, G., Uyemoto, J.K., and Goheen, A.C. 1992. Grape virus diseases, *In*: Grape Pest Management, Second Edition. D.L. and Flaherty *et al*., *eds*. University of California: Oakland, CA. pp. 101–109.

Gomez, G. and Pallas, V. 2001. Detection of viroid-like RNAs in pomegranate (*Punica granatum* L.). 18th Int. Symp. on Fruit Tree *Virus Diseases*. *Ed*. Clark, M.F., *Acta Hort.* (ISHS) 550: 321–326.

Graniti, A. and Martelli, G.P. 1966. Further observation on Legno riccio rugose wood, a graft transmissible stem pitting of grapewine. *In*: Proc. Intl. Conf. Virus vectors, perennial hosts and Vitis. Div. Agric. Sci., Univ. Calif. Davis pp. 168–179.

Gu, X.F., Meng, H., Qi, G. and Zhang, J.R. 2008. Agrobacterium mediated transformation of the winter jujube (*Ziziphus jujuba* Mill.). *Plant Cell Tissue Organ Cult.*, 94: 23–32.

Gunasinghe, U.B. and German, T.L. 1989. Purification and partial characterization of a virus from pineapple. *Phytopathology*, 79: 1337–1341.

Habili, N., Fazeli, C.F., Ewart, A., Hamilton, R., Cirami, R., Saldarelli, P., Minafra, A. and Rezaian, M.A. 1995. Natural spread and molecular analysis of grapevine leafroll-associated virus 3 in Australia. *Phytopathology*, 85: 1418–1422.

Han, S. and Cha B. 2002. Genetic similarity between Jujube witches broom and Mulberry dwarf phytoplasmas transmitted by *Hishimonas sellatus* Uhler. *Plant Pathol. J.*, 18: 98–101.

Hewitt, W.B. 1954. Some virus and virus-like diseases of the grapevine. *Calif. Dept. Agric. Bull.*, 43: 47–64.

Hewitt, W.B. 1968. Viruses and virus diseases of the grapevine. *Rev. Appl. Mycol.*, 47: 433–455.

Hewitt, W.B. and Goheen, A.C. 1959. Asteroid mosaic of grapevine in California. *Phytopathology* 49: 541.

Hewitt, W.B., Raski, D.J. and Goheen, A.C. 1958. Nematode vector of soil-borne fan leaf virus of grapevine. *Phytopathology* 48: 586–595.

Hewitt, W.B., Goheen, A.C., Raski, D.J. and Gooding, G.V. Jr. 1962. Studies on virus diseases of grapevine in California. *Vitis*, 3: 57–83.

Hewitt, W.B., Goheen, A.C., Cory, L. and Luhn, C. 1972. Grapevine fleck disease, latent in many varieties, is transmitted by graft inoculation. *Ann Phytopathol. Nohors. Serie.*, pp. 43–47.

Hu, J.S., Gonsalves, D. and Teliz, D. 1990. Characterization of closterovirus-like particles associated with grapevine fanleaf disease. *J. Phytopathol.*, 128: 1–14.

Hu, J.S., Sether, D.M., and Ullman, D.E. 1996. Detection of pineapple closterovirus in pineapple plants and mealybugs using monoclonal antibodies. *Plant Pathol.*, 45: 829–836.

Hu, J.S., Sether, D.M., Liu, X.P., Wang, M., Zee, F., and Ullman, D.E. 1997. Use of a tissue blotting immunoassay to examine the distribution of pineapple closterovirus in Hawaii. *Plant Dis.*, 81: 1150–1154.

Hua Jiang, Tomoaki Saiki, Kenji Watanabe, Hiromu Kawakita and Mamoru Sato. 2005. Possible vector insect of mulberry dwarf phytoplasma,*Tautoneura mori* Matsumura. *J. Gen. Plant Pathol.*, 71: 370–372.

Ishiie, T. and Matsuno, M. 1971a. Studies on the transmission of mulberry dwarf disease by a vector leafhopper, *Hishimonas sellatus* Uhler: 1. on the acquisition of the causal agent in non-viruliferous vector. *Ann. Phytopathol. Soc. Japan*, 37: 136–140.

Ishiie, T. and Matsuno, M. 1971b. Studies on the transmission of mulberry dwarf disease by a vector leafhopper, *Hishimonas sellatus* Uhler: II. on the inoculation of the causal agent by viruliferous vector. *Ann. Phytopathol. Soc. Japan*, 37: 141–146.

Ishikawa, K., Maejima, K., Komatsu, K., Kitazawa, Y., Hashimoto, M., Tanaka, D., Yamaji, Y. and Namba, S. 2012. Novel genomic RNA segments of fig mosaic virus, a newly identified *Emaravirus* infecting the common fig (*Ficus carica*). *In*: Abs. 22nd Interl. Conf. Virus and other graft transmissible diseases of fruit crops. p. 128.

Jha, A. and Posnette, A.F. 1961. Transmission of Arabis mosaic virus by the nematode *Xiphinema diversicaudatum (Micol.)*. *Virology*, 13: 119–123.

Jung, H.Y., Sawayanagi, T., Kakizava, S., Nishigava, H., Wei, W., Oshima, K., Miyata, S., Ugaki, M., Hibi, T. And Namba, S. 2003. *Candidatus* Phytoplasma ziziphi, a novel phytoplasma taxon associated with zuzube witches broom disease. *Int. J. Syst. Evol. Microbiol.*, 53: 1037–1041.

Kawakita, H., Saiki, T., Wei, W., Mitsuhashi, W., Watanabe, K. and Sato, M. 2000. Identification of mulberry dwarf phytoplasmas in the genital organs and eggs of leafhopper. *Hishimonoides sellatiformis. Phytopathology* 90 : 909–914.

Kazuya Ishikawa, Kensaku Maejima, Ken Komatsu, Osamu Netsu, Takuya Keima, Takuya Shiraishi, Yukari Okano, Masayoshi Hashimoto, Yasuyuki Yamaji and Shigetou Namba. 2013. Experimental and bioinformatic evidence that raspberry leaf blotch emaravirus P4 is a movement protein of the 30K superfamily *J. Gen. Virol.*, 94: 2117–2128.

Khan, M.S., Raj, S.K. and Snehi, S.K. 2008. Natural occurrence of '*Candidatus* Phytoplasma ziziphi' isolates in two species of jujube trees (*Ziziphus* spp.) in India. New Disease Reports 17: 14.

Kim, C.J. 1965. Witches' broom of zuzube tree (*Ziziphus jujube* Mill.var. inermis Rehd.). Transmission by grafting. *Korean J. Microbiol.*, 3: 1–6.

Koltunow, A.M. and Rezaian, M.A. 1988. Grapevine yellow speckle viroid: Structural features of a new viroid group . *Nucleic acid Research*, 16: 849–864.

Koltunow, A.M., Krake, L.R., Johanson, S.D. and Rezaian, M.A. 1989. Two related viroids cause grapevine yellow speckle disease independently. *J. Gen. Virol.*, 70 : 3411–3419.

Krake, L.R. and Woodham, R.C. 1978. Grapevine vein summer mottle: A new graft transmissible disease. *Vitis*, 17: 266–270.

Kumar, S., Baranwal, V.K., Singh, P., Jain, R.K., Sawant, S.D. and Singh, S.K. 2012. Characterization of a Grapevine leafroll-associated virus 3 from India showing incongruence in its phylogeny. *Virus Genes*, 45: 195–200.

Kumar, S., Rai, R., Kumar, P.V. and Baranwal, V.K. 2013. Current scenario of grapevine viruses in India. *In*: Abs. Asia-Pacific Congress of Virology held at Amity Univ. India p. 50.

Kyriakopoulou, P.E. 1991. Symptoms of grapevine asteroid mosaic in Greece. *In*: Proc. 10th Meeting ICVG, 1990. Volos, Greece.

Lavina, A., Batlle, A., Larrue, J., Daire, X., Clair, D. and Boudon-Padieu, E. 1995. First report of grapevine bois noir phytoplasma in Spain. *Plant Disease*, 79: 1075.

Legin, R. and Vuittenez, A. 1973. Comparaison des symptoms et transmission par greffage d'une mosaique nervaire de *Vitis vinifera* de la marbrure de *V. rupestris* et d'une affection necrotique des nervures de I'hybride *Rup-Berl.* 110R. *Riv. Patol. Veg. Ser.* IV 9: 57–63.

Legin, R. and Walter, B. 1986. Erude de phenomenes d'incompatabilite au greffage chez la vigne. *Progr. Agric. Vitic.*, 103: 279–283.

Lee, R.F. and Bar-Joseph, M. 2000. Tristeza. P.61 In: Compendium of citrus diseases 2nd edition. (*Eds.*) Timmer, L.W., Garnsey, S.M. and Graham, J.H. APS Press, St.Paul, MN.

Lee, I.M., Davis, R.E., and Gundersen-Rindal, D.E. 2000. Phytoplasma-Phytopathogenic mollicutes. *Ann. Rev. Microbiology*, 54: 221–255.

Lee, I.M., Martini, M, Marcone, C., Zhu, S.F. 2004. Classification of phytoplasma strains in the elm yellows group (16SrV) and proposal of *Candidatus* Phytoplasma ulmi for the phytoplasma associated with elm yellows.Intl. *J. Systemic Evol. Microbiol.*, 54: 337–347.

Lee, S.,Chul-eung, Kim, Byeongjin, Cha. 2012. Migration and distribution of graft Inoculated Jujube witches' broom phytoplasma within a *Catharanthus roseus* Plant. *Plant Pathol. J.*, 28: 191–196.

Lehoczky, J., Boscia, D., Burgyan, J., Castellano, M.A., Beczner, L. and Farkas, G. 1989. Line pattern, a novel virus disease of grapevine in Hungary. *In*: Proc. 9th meeting ICVG. 1987, Kiryat Anavim. pp. 23–30.

Lia W. Liefting, Anna C. Padovan, Karen S. Gibb, Ross E. Beever, Mark T. Andersen, Richard D. Newcomb, David L. Beck and Richard L.S. Forster. 2004. *Candidatus Phytoplasma australiense'* is the phytoplasma associated with Australian grapevine yellows, papaya dieback and Phormium yellow leaf diseases. *European J. Plant Path.*, 104: 619–623.

Linford, M.B. 1932. Transmission of the pineapple yellow spot virus by *thrips tabaci* lind. *Phytopath*, 22: 301–324.

Maixner, M. 1994. Transmission of German grapevine yellows (Vergilbungskrenkhiet) Bythe planthopper *Hyalesthes obsoletus(Auchenorrhyncha cixiidae). Vitis*, 33: 103-104.

Maixner, M. and Pearson, R.C. 1991. Studies on *Scaphytopius titanus*, a possible vector of grapevine yellows in New York. Proc. 10th Meeting ICVG, Volos, Greece.

Maixner, M., Reinert, W. and Darimont, H. 2000. Transmission of grapevine yellows by *Oncopsis alni* (Schrank) (*Auchenorrhyncha*: *Macropsinae*). *Vitis*, 39: 83–84.

Margaria, P., Rosa, C., Marzachi, C., Turina, M. and Palmano, S. 2007. Detection of flavescence dorée phytoplasma in grapes by reverse-transcription PCR. *Plant Dis.*, 91: 1496–1501

Martelli, G.P. 1993. Rugose wood complex. *In*: Graft transmissible diseases of grapevine. Handbook for detection and diagnosis pp. 45-54 edited by G.P. Martelli Rome, FAO of the United Nations.

Martelli, G.P. 2000. Major graft-transmissible diseases of grapevines: Nature, diagnosis and sanitation. Proc 50th Annu Am Soc Enol Viticulture meeting, Seattle, WA. *Am. J. Enol. Vitic.*, 51: 231–236.

Martelli, G.P. and Hewitt, W.B. 1963. Comparative studies on some Italian and Californian virus diseases of grapevine. *Phytopathol. Medit.*, 2: 275–284.

Martelli, G.P., Graniti, A., Lamberti, F. and Quacquarelli, A. 1966. Transmissione per innesto della "malattia delle enazioni" della vite. *Phytopathol. Medit.*, 5: 122-124

Martini, M. Murari E. Mori N. and Bertaccini A. 1999. Identification and epidemic distribution of two flavescence dorée-related phytoplasmas in Veneto (Italy). *Plant Dis.*, 83: 925–930.

Martin, R.R., Eastwell, K.C., Wagner, A. Lamprecht, S., and Tzanetakis, I.E. 2005. Survey for viruses of grapevine in Oregon and Washington. *Plant Disease*, 89: 763–766.

Marzachiet, C., Palermo, S., Boarino, A., Veratti, F., D'Aquilio, M., A. Loria, A. and Boccardo, G.. 2001. Optimization of a one-step PCR assay for the diagnosis of Flavescence dorée-related phytoplasmas in field-grown grapevines and vector populations. *Vitis*, 40: 213–217.

Meng, B., Pang, S.Z., Forsline,, P.L., McFerson, J.R. and Gonsalves, D. 1998. Nucleotide sequence and genome structure of grapevine rupestris stem pitting associated virus-1 reveal similarities to apple stem pitting virus. *J. Gen. Virol.*, 79: 2059–2069.

Meng, B., Credi, R., Petrovic, N., Tomazic, I. and Gonsalves, D. 2003. Antiserum to recombinant virus coat protein detects Rupestris stem pitting-associated virus in grapevine. *Plant Dis.*, 87: 515–522.

Meng, B., Rebelo, A.R. and Fisher, H. 2006. Genetic diversity analyses of grapevine Rupestris stem pitting-associated virus reveal distinct population structures in scion versus rootstock varieties. *J. Gen. Virol.*, 87: 1725–33.

Mori, N., A. Bressan, M. Martini, M. Guadagnini, V. Girolami, and A. Bertaccini. 2002. Experimental transmission by *Scaphoideus titanus* Ball of two Flavescence dorée-type phytoplasmas. *Vitis,* 41: 99–102.

Mori, N., Pavan, F., Bondawalli, R., Reggiani, N., Paltrinieri, S. and Bertaccini,A. 2007. Factors affecting the spread of "Bois Noir" disease in grapevine cv. Lambrusco in Northern Itali. *Bull. Insectology,* 60: 295–296.

Nagaich, B.B. and Vashisth, K.S. 1962. Mosaic of *Ficus* spp. in India. *Curr. Sci.,* 31: 166-167.

Nagaich, B.B. and Vashisth, K.S. 1965. Additional virus diseases of temperate fruits in Shimla hill. *Indian Phytopath.,* 18: 288–290

Naidu, R.A. and Mekuria, T.A. 2010. First report of grapevine fleck virus from Washington Vineyards. *Plant Dis.,* 94: 784.

Namba, S., Yamashita, S., Doi, Y. and Yora, K. 1979. A small spherical virus associated with the ajinashika disease of koshu grapevine. *Ann. Phytopath. Soc. Japan,* 45: 70–73

Namba, S., Iwanami, T., Yamashita, S., Doi, Y. and Hatamoto, M. 1986. Three phloem limited viruses of grapevine: Direct fluorescence detection. *Food* and *fertilizer technology centre Taiwan Tech. Bull.,* 92: 1–17.

Namba, S., Yamashita, S., Tsuchizaki, D., Boscia, D. and Gonsalves, D. 1991. Purification and properties of a small spherical virus associated with grapevine Ajinashika disease. *Plant Dis.,* 75: 1249–1253.

Namba S, Boscia D, Azzam, O., Maixner, M., Hu, J., Golino, D. and Gonsalves D. 1991. Purification and properties of closterovirus-like particles associated with grapevine corky bark disease. *Phytopathology,* 81: 964–970.

Padovan A. C., Gibbs, K. S., Daire, X. and Boudon-padieu, E. 1996. A comparison of the phytoplasma associated with Australian grapevine yellows to other phytoplasmas in grapevine. *Vitis,* 35: 189–194

Pandey, P.K., Singh, A.B., Nimbalkar, M.R. and Marathe, T.S.. 1976. A witches broom disease of Jujube from India. *Plant Dis. Reptr.,* 60: 301–303.

Petrovic, N., Meng, B., Rawnikar, M., Mavric, I. and Gonsalves, D. 2003. First detection of Rupestris stem pitting-associated virus particles by antibody to a recombinant coat protein. *Plant Dis.,* 87: 510–514.

Prince J.P., Davis R.E., Mogen B.D., Dally E.L., Lee I.-M., Caudwell A., Boudon-Padieu, E., di Terlizzi, B., Savino, V., Barba, M., Bertaccini, A., Osler, R., and Carraro, L. 1992. Detection and differentiation of European grapevine yellows MLOs using the polymerase chain reaction. *Phytopathology,* 82: 1170.

Quacquarelli, A., Gallitelli, D., Savino, V. and Martelli, G.P. 1976. Properties of grapevine fanleaf virus. *J. gen. Virol.,* 32: 349–360.

Ramsdell, D.C. and Myers, R.L. 1974. Peach rosette mosaic virus, symptomatology, and nematodes associated with grapevine degeneration in Michigan. *Phytopathology,* 64: 1174–1178.

Ramsdell, D.C. and Myers, R.L., 1978. Epidemiology of peach rosette mosaic virus in a Concord grape vineyard. *Phytopathology,* 68: 447–450.

Ranjith Pathirana and Marian J. McKenzie. 2005. Early detection of grapevine leafroll virus in *Vitis vinifera* using *in vitro* micrografting. *Plant Cell, Tissue and Organ Culture,* 81: 11–18.

Rathay, E. 1883. Önol. Lehranstalt Klosterneuburg, 3.

Raychaudhuri, S.P., Ganguli, B. and Basu, A.N. 1965. Further studies on the mosaic disease of mulberry. *Plant Dis. Reptr.,* 49: 981.

Raychaudhuri, S.P.,Chatterjee, S.N. and Dhar, H.K. 1962. A mosaic disease of mulberry. *Indian Phytopath,* 15: 187–189.

Raychaudhuri, S.P., Chatterjee, S.N. and Dhar, H.K. 1961. Preliminary note on the recurrence of yellow net vein disease of mulberry. *Indian Phytopath.*,14: 94–95.

Raychaudhuri, S.P., Ganguli, B. and Basu, A.N. 1966. Virus diseases of mulberry in India. *In* : Proceedings of 1st Symposium on Plant Pathology, Indian Pathological Society, IARI, New Delhi, India.

Refatti, E. 1966. Su Una possible correlazione frail virus del complesso del aniciamento a la malattia delle enazioni nella vite. *Riv. Veg. Pakia. Ser.*, 2: 207-217.

Refatti, E. 1970. Asteroid mosaic of grapevine. *In*: Virus diseases of small fruits and grapevine. *Ed.* N.W. Frazier. Univ. Calif. Div. Agric. Sci. Berkeley. pp. 212–214.

Rezaian, M.A. 1990. Australian grapevine viroid-evidence for extensive recombination between viroids. *Nucl. Acid Res.*, 18: 1813–1818.

Rosa, C. and Rowhani, A. 1966. Research Project on "Etiology of rugose wood complex"Department of Plant Pathology, Univ. Calif., Davis. CA, USA

Rosa, C., Jimenez, J.F., Margaria, P. and Rowhani, A. 2011. Symptomatology and effects of viruses associated with rugose wood complex on the growth of four different rootstocks. *Am. J. Enol. Vitis,* 62: 207–213.

Rosciglione, B. and Castellano, M.A. 1985. Further evidence that mealybugs can transmit grapevine virus A (CVA) to herbaceous hosts. *Phytopathol. Medit.*, 24: 186–188.

Rosciglione, B. and Gugerli, P. 1989. Transmission of grapevine leaf roll disease and an associated Closterovirus to healthy grapevine by the mealybug *Planococcus ficus* Signoret. *In*: Proc. 9thmeeting ICVG 1987, Kiryat Anavim. pp. 67–69.

Rumbos, I.C. and Avgelis, A.D. 1989. Roditis leaf discoloration-a new virus disease of grapevine; symptomatology and transmission to indicator plants. *J. Phytopathol.,* 125: 274–278

Sabanadzovic, S., Abou Ghanem-Sabanadzovic, N., Saldarelli, P. and Martelli, G.P. 2001. Complete nucleotide sequence and genome organization of Grapevine fleck virus. *J. Gen. Virol.,* 82: 2009–2015.

Sanford, W.G. 1988. Mealybug wilt, mealybugs, and ants on pineapple. *Plant Dis.*, 72: 558–565.

Sano, T., Ohshima, K., Uyeda, I., Shikata, E., Meshi, T. and Okada, Y. 1985. Nucleotide sequence of grapevine viroid: A grapevine isolate hop stunt viroid. *Proc. Japan Acad. Ser.* B, 61: 265–268

Sano, T., Kobayashi, T., Ishiguro, A. and Motomura, Y. 2000. Two types of grapevine yellow speckle viroid 1 isolated from commercial grapevine had the nucleotide sequence of yellow speckle symptom-inducing type. *J. Gen. Pl. Pathology,* 66: 68–70.

Savino, V., Boscia, D., and Martelli, G.P. 1985. Incidence of some graft transmissible virus-like diseases of grapevine in visually selected and heat treated stocks from southern Italy. *Phytopathol. Medit.,* 24: 204–207

Savino, V., Di Terlizzi, B., Rivieccio, S. and Di Silvio, F. 1991. Presence in conal rootstocks of a graft transmissible factor that induces stunting and bushy growth in European grapevines. *In*: Proc. 10th Meeting ICVG, 1990. Volos, Greece.

Semancik, J.S., Rivera-Bustamante, R. and Goheen, A.C. 1987. Widespread occurrence of viroid-like RNAs in grapevine. *Am. J. Enol. Vitic,* 38: 35–40.

Sether, D. M. and Hu, J. S. 1997. Transmission of pineapple closterovirus (PCV) by two species of mealybug. (Abstr.) *Phytopathology* (Suppl.), 87: S88.

Sether, D.M., Ullman, D.E. and Hu, J.S. 1998. Transmission of pineapple mealybug wilt–associated virus by two species of mealybug (*Dysmicoccus* spp.). *Phytopathology,* 88: 1224–1230.

Singh, S.J. 2004. Virus diseases of pineapple. *In*: Naqvi, S.A.M.H. (*Ed.*) Diseases of fruits and vegetables Volume I. Diagnosis and Management. Springer Netherlands, pp. 639–651.

Singh, S.J.and Sastry, K.S.M. 1975. Effect of different fertilizers and spacings on the incidence of pineapple wilt virus. *Indian J. Mycol. and Plant Pathol.*, 5: 156–160

Singh, S.J., Sastry, K.S.M. and Sastry, K.S., 1975. Little leaf disease of grapevine. *Curr. Sci.*, 44: 26–28.

Suryanarayana, V., Pradeep kumar, V.H. and Reddy, M.K. 2009. Rosette of *Ficus amplicissima*, a new phytoplasma disease – biological and molecular confirmatory. *In*: 5th Intl. Conf. Pl. Path. Globalized era, Nov. 10- 13, 2009, New Delhi, India (Abst). p. 281.

Tanne, E., Ben-Dov, Y. and Raccah, B. 1989. Transmission of closterovirus-like particles by mealybug ((Pseudococcidae) in Israel. *In*: Proc. 9th meeting ICVG 1987, Kiryat Anavim. pp. 71–73.

Tanne, E., Bazak, H. and Dubitzky, E. 1997. Epidemiology, spread, rootstock sensitivity and economical impact of corky bark disease in grapevines. *In*: Proceedings of the 12th Meeting of the International Council for the study of viruses and virus like diseases of grapevines". Lisbon, Portugal. p. 127.

Taylor, R.H. and Hewitt, W.B. 1964. Properties and serological relationship of Australian and Californean soil borne viruses of grapevine and arabis mosaic virus. *Australian J. Agric. Res.*, 15: 571–585.

Taylor, R.N. and Woodham, R. C. 1972. Grapevine yellow speckle - a newly recognized graft-transmissible disease of *Vitis*. *Australian Journal of Agricultural Research*, 23: 447–452.

Tsuchizaki, T. 1976. Mulberry latent virus isolated from mulberry (*Morus alba* L.). *Ann. Phytopath. Soc. Japan*, 42: 304–309

Tsuchizaki, T., Hibino, H. and Saito, Y. 1971. Mulberry ringspot virus isolated from mulberry showing ringspot symptoms. *Ann. Phytopath. Soc. Japan*, 37: 266–271.

Tzanetakis, I. E. and Martin, R.R. 2009. New viruses identified in fig trees exihibiting fig mosaic disease. *In*: 21st internl. Conf. on virus and other graft transmissible diseases of fruit crops. *Berichte aus dem JKI* 148. pp. 29–30.

Uyemoto, J.K. and Gilmer, R.M. 1972. Spread of tomato ringspot virus in Baconoir grapevines in Newyork. *Plant Dis. Reptr.*, 56: 1062–64.

Uyemoto, J.K. 1975. A severe outbreak of virus-induced grapevine in cascade grapes in New York. *Plant Dis. Reptr.*, 59: 98–101.

Uyemoto, J.K., Martelli, G.P. and Rowhani 2009. Grapevine viruses, virus-like diseases and other disorders. *In*: Virus diseases of plants: Grape, potato and wheat. Image collection and image resource CD-Rom. APS Press St. Paul, MN 55121

Vashisth, K.S. and Nagaich, B.B. 1965. *Morus indica* – An additional host of fig mosaic. *Indian Phytopath*. 18: 315.

Vashisth, K.S. and Nagaich, B.B. 1968. *Aceria ficus* (Cotte) as vector of fig mosaic in India. *Indian J. Entom.*, 30: 322.

Vuittenez, A., Legin, R. and Kuszala, J. 1966. Observations surune mosaique de la vigne, probablement independente du virus du court-noue. *Ann. Epiphyt.* hors serie, 17: 67–73.

Wakman, W., Teakle, D.S., Thomas, J.E. and Dietzgen, R.G. 1995. Presence of a clostero-like virus and a bacilliform virus in pineapple plants in Australia. *Australian J. Agric. Research*, 46: 947–958

Wang, Q.X.S., Chen, Z. and Zhang, J. 1981. On the witches' broom disease of *Ziziphus jujuba* Mill [J]. *Acta Phytopathologica Sinica*, 1981–01.

Wei, W., Kawakita, H. and Sato, M. 2000. Detection of a small population of mulberry dwarf (MD)-phytoplasmas in symptomless-mulberry trees by nested PCR. *Journal of Sericultural Science of Japan,* 69: 261–269.

Wetzel,T., Fuch, M.,Bobko, S.M. and Krczal, G. 2002. Size and sequence variability of the arabis mosaic virus Protein 2A. *Arch. Virol.,* 147: 1643–1653.

Yogita, H. and Komuro, Y. 1972. Transmission of Mulberry ringspot virus by *Longidorus martini* Memy. *Ann. Phytopath. Soc. Japan,* 38: 275–283.

Young, I.L., Deng, T.C. and Chen, M.J. 1986. Sap transmissible viruses associated with grapewine yellow mottle disease in Taiwan. *J. Agric. Res. China,* 35: 504–510.

Zhang, X.G., Chen, Z.W., Chen, Z.A., Sun, S.M., Zhang, F.G., and Tiang, X.D. 1991. Study on the infection of MLO of Jujube witches' broom disease to periwinkle (*Vinca rosea*). *Intl. J. Trop. Plant Dis.,* 9: 251–256.

Zhang, Y.P., Uyemoto, J.K., Golino, D.A. and Rowhani, A. 1998. Nucleotide sequence and RT-PCR detection of a virus associated with grapevine rupestris stem-pitting disease. *Phytopathology,* 88: 1231–1237.

Zhang, B., Liu, G.Y., Liu, C., Wu, Z., Jiang, D. and Li, S. 2009. Characterization of hop stunt viroid (HSVd) isolated from jujube trees (*Ziziphus jujuba*). *European J. Plant Path.,* 125-665-669.

Zhu, S.F., Hadidi, A., Gundersen, D.E., Lee, I.M. and Zhang, C.L. 1998. Characterization of the phytoplasmas associated with cherry lethal yellows and jujube witches' broom diseases in China. *Acta Hort.* (ISHS) 472: 701–714.

Zimmermann, D., Bass, P., Legin, R. and Walter, B. 1990. Characterization and serological detection of four closterovirus-like particles associated with leaf roll disease of grapevine. *J. Phytopathol.,* 130: 205–218.

6

VIRUS DISEASES OF SMALL FRUITS

Small fruits are grown world over. Among small fruits important ones are: strawberry, raspberry, blackberry, Himalyan raspberry, boysenberry, thimbleberry, wineberry, blue berry, cranberry, gooseberry, black currant and red current. The virus and phytoplasma diseases of these crops will be discussed in this chapter.

A. VIRUS DISEASES OF STRAWBERRY

Strawberry (*Fragaria virginiana, F. vesca, F. moschata*) belongs to family Rosaceae and is cultivated throughout the world for delicacy of its fruits and fruit products. Strawberry is basically a temperate crop but due to availability of some varieties suitable for tropical region, now it is grown successfully in plains of some Indian states. The use of plastic tunnels during night (5 pm to 9 am) and bed planting coupled with micro irrigation system proved to be optimum for harvesting fruit yield of 18-20 t/ha in North Indian Plains.

Strawberry is highly susceptible to virus and virus-like diseases. Virus diseases that affected cultivated strawberry *Fragaria x ananassa* Duch. were first noted in the 1920's. Currently, more than thirty viruses and phytoplasmas have been reported in *Fragaria* (Spiegel and Martin, 1998). Although single virus infections can result in little to no observable symptoms in strawberry plantings, multiple virus infections in strawberry plants are responsible for reduction in fruit yield, and diminished fruit quality.

The viral/phytoplasma diseases of strawberry are transmitted by aphids, leaf hopper or by nematodes. Diseases like Necrotic shock,

Strawberry leafroll, and Strawberry feather-leaf are graft transmitted and do not have insect vector.

STRAWBERRY CHLOROTIC FLECK

Chlorotic fleck disease of strawberry was reported from Louisiana in USA. This disease remains symptom-less in most of the commercial cultivars but in *Fragaria vesca* and *F. virginiana* indicator plants, symptoms appear as downward curling and distortion and sometimes chlorotic lesions on leaves of *F. vesca.* The disease agent is transmissible by leaflet grafting and the vector is cotton aphid, *Aphis gossypii,* which transmits it in non-persistent manner.

Three viruses have been cloned and sequenced from dsRNA isolated from the chlorotic fleck (CF) infected plants. Two of them being the *Criniviruses* associated with pallidosis disease (Strawberry pallidosis associated virus and Beet pseudo yellows virus) and the third, a previously unidentified member of the family *Closteroviridae* (Tzanetakis and Martin, 2008). The genome organization is similar to Beet yellows virus but its sequence is most closely related to Citrus tristeza virus (Tzanetakis and Martin, 2007). The associated viruses can be managed by combination of thermotherapy (8 weeks at 37°C) and meristem tip culture (McGrew, 1980).

STRAWBERRY MOTTLE

Strawberry mottle virus disease (SMoV) is widely distributed in strawberry all over the world. Severe strains of SMoV may reduce yield up to 30% and losses can be up to 80% in mixed infections with other viruses (Thompson and Jelkmann, 2003; Martin and Tzanetakis, 2006).

SYMPTOMS

All *Fragaria* species are susceptible and the virus is symptomless in most commercial cultivars. Severe strains may reduce vigour and fruit yield and cause mottle to severe degeneration and necrosis of one or two young leaves in *F. vesca.* Leaves become short and distorted and petioles are irregular in growth. Intermediate strains can cause mottle of one or more leaflets of the youngest leaf and clorosis or net veins. In some cases, leaves curl downward. Mild strains will cause general diffuse mottle (Mellor and Krezel, 1987). Under experimental conditions following hosts were susceptible to mottle virus infection: *Potentilla adscherica, P. sundaica, P. canadensis, P. sterilis, P. taurica, Chenopodium quinoa, C. amaranticolor; C. album, C. tetragonoloba, Duchesnea chrysantha, D. indica* and *Gomphrena globosa*. These hosts developed variable symptoms as Veinal chlorosis, distortion of veins of youngest leaves. Chlorotic and necrotic lesions, diffuse veinal chlorosis and mottling of young leaves, crinkling and curling of leaves. Leaf mottling is observed on indicator plants of *Fragaria vesca* 'EMC' and 'UC4', and *F. virginiana* 'UC12'.

TRANSMISSION

The causal virus is transmitted by several species of aphid and by grafting. The most efficient aphid vectors are *Chaetosiphon frafaefolii, C. thomasi, C. minor, C. jakobi* and *Aphis gossypii*. The virus is transmitted in semi-persistent manner (Ward *et al.*, 2008). The virus is also transmitted by dodder, *Cuscuta subinclusa* and *C. gorovoni* and also by mechanical inoculation to herbaceous hosts.

THE PATHOGEN

The virus particles are isometric with a diameter of 32-37 nm and with icosahedral symmetry (Kitajima *et al.*, 1971, Yoshikawa and Converse, 1991). The genome of the virus is bipartite, linear, positive-sense, single-stranded RNA. Sequencing revealed that the genome of strawberry mottle virus consists of two genomic components, RNA1 and RNA2. Strawberry mottle virus was reported to be a tentative member of a new family of bipartite picorna-like viruses (Thompson Jr. *et al.*, 2002). Strawberry mottle virus has been classified as: Order: *Picornavirales,* Family: *Secoviridae,* Genus: unassigned, species: *Strawberry mottle virus*. Synonyms is Strawberry mild crinkle virus.

DIAGNOSIS AND MANAGEMENT

The virus can be detected by inoculation to indicator hosts, by aphid transmission and using conventional and real-time RT-PCR (Thompson

Jr *et al.*, 2003a, Ratti *et al.*, 2006, Ward *et al.*, 2008).The virus can also be detected by multiplex PCR from mixed infections (Thompson *et al.*, 2003). The virus can be eliminated by raising plants in tissue culture using Gamborg's medium with the addition of 2,4-dioxohexahydro-1,3,5-triazine (Kondakova and Schuster, 2008). Strawberry mottle virus can be eliminated by thermotherapy at 37°C for 10-14 days (Mellor and Krezal, 1987).

STRAWBERRY DECLINE

A general and rapid strawberry plant decline became a major problem for strawberry growers from 2001 to 2004 along the west coast of North America (Martin and Tzanetakis, 2006). Symptoms on declining plants appear as red or chlorotic flecking on petioles and runners with a general reddening of the leaves, leaf necrosis and cessation of plant growth. Yield losses up to 100% was noted in severely affected plantations. Affected plants may collapse and die. Older leaves of affected plants often develop a red colour, and newer leaves show a distortion and some cultivars develop red coloured lesions on the major veins and petioles and produce small leaves. Roots become stunted resulting failure of the development of new leaves and fruits (Martin and Tzanetakis, 2006).

The disease was attributed to be caused by viruses (Tzanetakis and Martin, 2004). It is now known that Strawberry decline is caused by a combination of whitefly and aphid transmitted viruses. The whitefly transmitted viruses present are: Beet pseudo yellows virus, Strawberry pallidosis associated virus, and a *Crinivirus* (Martin and Tzanetakis, 2006). The aphid-borne viruses are Strawberry mottle, Strawberry crinkle, Strawberry mild yellow edge and Strawberry vein-banding. These viruses will be discussed in following pages.The aphid-borne viruses and their vector(s) have very limited host range. Except Strawberry mottle virus, aphid borne viruses are transmitted efficiently by the strawberry aphid. Strawberry chlorotic fleck associated virus and Strawberry mottle virus are transmitted by the cotton aphid (*Aphis gossypii*).

Best management will include an integrated approach focussed on maintenance and use of virus-indexed nursery material, testing of plant material exhibiting suspicious symptoms, elimination of weed and crop sources of virus to the extent possible near strawberry fields, and reducing vector populations through biological or insecticidal methods. Knowing that the decline is caused by a combination of whitefly and aphid transmitted viruses, the first and most important strategy is to start with certified planting material (Martin and Tzanetakis, 2006).

STRAWBERRY CRINKLE

The distribution of strawberry crinkle disease is world-wide and has become the limited factor in many countries for production of strawberry.

SYMPTOMS

The symptoms appear as translucent spots which later become opaque, yellow, reddish, or necrotic. In *F. vesca*, these spots may be sunken or swollen and cause angular bending. Leaflets are distorted and crinkled, petioles and leaves are reduced in size (Converse, 1987). The symptoms of severe strain appear as small, scattered opaque, chlorotic to necrotic spots along the veins of affected leaves. Symptoms of mild strains develop as leaf epinasty which is evident during early growth stage. Angular epinasty of leaflets is reliable and early symptom of the disease. Streaks on petals is a highly diagnostic symptoms of crinkle disease. These streaks later become necrotic and affected petals may be deformed. The 'MI', and 'UC-11' clones are the best indicator of the disease (Frazier *et al.*, 1987).

TRANSMISSION

Crinkle disease is transmitted by grafting and through aphid vectors, *Chaetosiphon fragaefolii, C. jakobi* and also by strawberry root aphid, *Aphis forbesi* in a persistent manner (Frazier *et al.*, 1987). There is a 14 days minimum latent period from acquisition to transmission. The virus multiply inside the aphid body. Infectious virus can even be recovered from aphids frozen at -54°C for two years (Frazier *et al.*, 1987).

HOST RANGE

Natural hosts of the virus include *Fragaria vesca, F. vesca* var. semperflorens, *F. virginiana, F. chiloensis, F. x ananassa* (cultivated strawberry). The symptoms on these hosts are: chlorotic and necrotic spotting of veins, epinasty, deformation, local lesions in petioles and stolons, streaking and deformation of petals and malformation of leaflets. In cultivated strawberry usually no symptoms develop (Frazier and Mellor, 1970).

THE PATHOGEN

The virus particles are bacilliform typical of rhabdoviruses. The particle size is 42 X 190-380 nm and are enveloped. The virus belongs to group V(–) ssRNA, order *Mononegavirales,* family *Rhabdoviridae,* genus *Cytorhabdovirus* and species *strawberry crinkle virus* (Sylvester *et al.,* 1976; Richardson *et al.,* 1972). The genome is not segmented and contains a single molecule of linear single-stranded RNA. The viral genome encodes structural proteins and non-structural proteins. Lipids are present but they are of host origin. Virions accumulate in the cell cytoplasm (Hunter *et al.,* 1990).

DIAGNOSIS AND MANAGEMENT

Diagnosis of the disease is done either by graft or aphid transmission to sensitive indicator plants. *F. vesca* cvs UC-4, UC-5, UC-6, and *F. vesca* var. semperflorens cv. Alpine are all very sensitive and show the diagnostic petal-streak and petiole lesion symptoms. Other symptoms are chlorotic to necrotic irregular spots on veins, epinasty, crinkling, distortion and uneven expansion of leaflets. Indicators of *F. virginiana* are cvs M1, UC-10, UC-11 and UC-12 (Frazier *et al.,* 1987). The virus can be managed by: isolation of planting from inoculation sources, use of virus-free certified planting material, thermotherapy and meristem tip culture (Posnette and Cropley, 1958; Miller and Belkengren, 1962).

STRAWBERRY MILD YELLOW EDGE DISEASE

Strawberry mild yellow edge is one of the widespread and common virus disease in cultivated strawberries. It has been reported from several countries in Western Europe, Asia, North and South America, Oceania and South Africa. Strawberry mild yellow edge virus (SMYEV) is also known as strawberry virus-2.

SYMPTOMS

Infected *Fragaria vesca* plants develop chlorotic flecks in young leaves, gradual loss of vigour, chlorotic mottling and interveinal necrosis of older leaves. The wild species *F. virginiana, F. vesca* and some clones of *F. chiloensis* show symptoms but *F. ovalis* is a symptomless carrier. Most strawberry cultivars are symptomless carriers of the disease. The disease has been found in nature on *F. chiloensis* in Chile and in several *Duchesnea* in Japan.

TRANSMISSION

Mild yellow edge disease is transmitted by graft. Virus is transmitted by the aphid, *Chaetosiphon fragaraefoli* in persistent manner (Krezal, 1979). Other vectos of the disease are : *C. thomasi, C. jacobi, C. minor* (Frazier, 1975), *Macrosiphum rosae* (Mellor and Forbes, 1960) and *Myzus persicae* (Converse, 1987). The virus is also transmitted by mechanical inoculations.

HOST RANGE

Mild yellow edge disease has been experimentally transmitted to *Chenopodium quinoa* and *C. murale* but without any symptoms (Jelkmann *et al.*, 1990). The virus is also transmitted by grafting to *Rubus rosaefolius*. The assay or indicator hosts are clones of *Fragaria vesca* UC4 (W), UC5 (W), EMC (W), *F. vesca* var. semperfloreus 'Alpine' (W). Symptoms in these hosts appear as mottling of young leaves, epinasty, chlorotic flecking, vein necrosis of matured leaves and premature senescence of older leaves. *F. virgin*iana UC-10 and UC-11 are also useful indicators.

THE PATHOGEN

Recent investigations have shown that strawberry mild yellow edge disease is probably caused by a virus complex consisting of a potexvirus with rod shaped particle of 14 × 500 nm which is constantly associated with the disease (Jelkmann *et al.*, 1990, Hepp and Martin, 1991) as well as a virus originally designated strawberry mild yellow edge luteovirus with isometric particles of 25 nm in diameter, but this is now recognized as a strain or synonym of soybean dwarf luteovirus (Randles and Rattigen, 1995; Yoshikawa *et al.*, 1984; Martin and Converse, 1985; Spiegel *et al.*, 1986). The virus particles are filamentous, flexuous with a length of 482 nm and a width of 13 nm with monopartite,linear, positive-sense, single-stranded RNA genome. The virus is a tentative member of the genus *Potexvirus* in the family *Flexiviridae*. The complete genome is 5966 nucleotides long. The genome has an intergenic poly (A) region (Jelkmann *et al.*, 1991). The virus codes for 5 ORF(s). The genome Encodes proteins involved in cell to cell movement.

DIAGNOSIS AND MANAGEMENT

Transmission by grafting or by its aphid vector *Chaetosiphon fragaefolii* to sensitive clones of *F. vesca* is used for detection and identification of the disease. The potexvirus can be detected by ELISA and by immunoelectron microscopy (Jelkmann *et al.*, 1990; Hepp and Martin, 1991). It can also be detected in RT-PCR (Hadidi *et al.*, 1993). Control of the virus in the field can be achieved by minimizing aphid population using insecticide (Converse and Alniazee, 1988). Thermotherapy (9 weeks at 38°C) or meristem culture, combined with planting of certified virus-free material are useful to eliminate the disease. The technique stimulated the development of side crowns, which could then be excised and rooted in sand at normal greenhouse temperatures (Converse *et al.*, 1987).

STRAWBERRY LATENT C DISEASE

Strawberry latent C virus disease occurs in East Asia, Canada, Japan, and the United States of America. Normally no obvious symptoms are observed on cultivated strawberries except when the disease is in combination with other virus diseases, such as crinkle, mottle, veinbanding or yellows. Mixed infections can cause moderate to severe degeneration. Most of the cultivars of strawberry are symptomless to strawberry latent C virus but infection may cause severe epinasty in young leaves followed by dwarfing, mottling and distortion of mature leaves in susceptible cultivars such as clone UC5. *Fragaria vesca* clone UC4 and UC6 are symptomless. Strawberry latent C virus is transmitted by grafting, dodder and also by an aphid vector, *Chaetosiphon fragaefolii*. The disease is limited to cultivated strawberries and related species (*Fragaria vesca, F. virginiana, F. nilgerrensis* and *F. chiloensis*). Some weed species like *Potentilla canadensis, P. anserina* were infected experimentally. Assay or indicator host for the virus is *Fragaria vesca* clone UC5 (W).

Virus particles are bullet-shaped and measure 68 nm × 190-380 nm with helical symmetry. The virus belongs to the genus *Rhabdovirus*, family *Rhabdoviridae* and order *Mononegavirales*. This rhabdovirus, has apparently no close relationship with the similarly named strawberry latent virus strains A and B, which are strains of strawberry crinkle cytorhabdovirus (EPPO/CABI, 1996a). The rhabdovirus particles associated with this disease accumulate in the perinuclear space and in the nuclei, whereas strawberry crinkle rhabdovirus particles are located in the cytoplasm (Yoshikawa *et al.*, 1986). The genome is not segmented and contains a single molecule of linear single-stranded RNA. The viral genome encodes structural and non-structural proteins. Infection of this virus can be detected by graft-transmission to indicator clones of *F. vesca* or *F. virginiana* (McGrew, 1987). The disease can be managed by the use of certified virus-free planting material. An internationally agreed certification scheme has been developed by EPPO (OEPP/EPPO, 1994). The virus is also managed by thermotherapy and meristem tip culture.

STRAWBERRY VEIN BANDING

Strawberry vein banding disease was first reported by Frazier (1955). It is now found in Africa, Australia, Brazil, Europe, Japan and North America (Schöniger, 1958).

SYMPTOMS

The vein banding virus causes vein banding symptoms in strawberry. The virus naturally occurs in *Fragaria vesca, F. virginiana, F. chiloensis* and *F. x ananassa* showing yellow vein banding, discontinuous banding, streaking, spotting of older leaves and twisting of leaflets. In association with other strawberry diseases like crinkle, yellows and leaf curl, severe symptoms appear (Miller and Frazier, 1970). In *Fragaria vesca* clone UC-6 and *F. virginiana* clone UC-12, the main symptoms are mild leaf curl, epinasty of midribs and petioles, and twisting of leaflets. Discontinuous chlorotic banding of the main veins are used as diagnostic symptoms of the disease followed by necrosis and stunting (Frazier, 1974b). The following species were susceptible to experimental virus infection: *Fragaria chiloensis, Fragaria vesca, Fragaria virginiana, Fragaria x ananassa, Sanguisorba minor* (Morris *et al.*, 1980). *Fragaria vesca* clone UC-6 and *F. virginiana* clone UC-12 are the indicator hosts of the virus (Frazier, 1974b).

TRANSMISSION

The virus is transmitted by grafting (Frazier, 1974a) and dodder (*Cuscuta subinclusa*) (Frazier, 1955), but not by inoculation with sap. It is readily transmitted by aphid vectors in semi-persistent manner (Frazier, 1955; Mellor and Forbes, 1960). Following Aphid species are the vectors of this virus: *Amphorophora agathonica, Aphis rubifolii, Aulacorthum solani, Chaetosiphon fragaefolii, C. jacobi, C. tetrahodus, C. thomasi, Macrosiphum pelargonii* and *M. rosae,* and *Myzus ascalonicus, M. ornatus* and *M. persicae*. *Chaetosiphon fragaefolii, C. jacobi* and *C. thomasi* are the most efficient vectors.

THE PATHOGEN

The virus particles are isometric 40 to 50 nm in diameter commonly found in cytoplasmic inclusions. Virus is a tentative member of the genus *Caulimovirus* in the family *Caulimoviridae* (Kitajima *et al.*, 1973; Morris *et al.*, 1980). The virus genome is double-stranded DNA. The complete DNA sequence (7876 nucleotides) of strawberry vein banding virus (SVBV) has been determined. Seven open reading frames were detected that potentially code for proteins of calculated weight 37.8; 18.3; 16.6; 56.0; 81.1; 59.0 and 12.6 kDa, respectively. The virus is closely related to Cauliflower mosaic virus, figwort mosaic virus and carnation etched ring virus than with other caulimoviruses (Petrzik *et al.*, 1998). The virus is serologically related to cauliflower mosaic virus.

DIAGNOSIS AND MANAGEMENT

The virus can be detected in ELISA and ISEM in addition to inoculation to susceptible indicator hosts (Morris *et al.*, 1980). PCR and dot blot hybridization assays have been used for virus diagnosis (Stenger *et al.*, 1988; Mahmoudpour, 2004). The virus can be eliminated from mother plants by tissue culture (Miller and Belkengren, 1962) or by heat treatment for 10 days at 42°C (Bolton, 1967). Use of virus-free certified stock is important for management of the disease.

STRAWBERRY PALLIDOSIS

Pallidosis disease of strawberry has been reported from united states, Australia and Canada (Frazier and Stubbs, 1969). In field, symptoms of the disease were observed in *Fragaria virginiana* UC-10, but not on *F. vesca* UC-4 and UC-5 indicators suggesting strawberry pallidosis disease is a potentially significant strawberry disease (Hokanson *et al.*, 2000). Pallidosis is generally remains latent in most commercial cultivars of strawberry. However, mild symptoms are developed in cvs. Marshall, Lassen, Hood and Northwest but a severe isolate significantly reduced root and runner production in otherwise symptomless 'Northwest' plants (Converse and Volk, 1990). Graft inoculated *F. virginiana* plants show leaf epinasty, distortion, chlorosis and dwarfing and can be used as indicator host. Strawberry pallidosis is known to be graft transmissible and may be spread by the leafhopper *Coelidia olitoria* (Frazier, 1975). It may also be pollen-borne. The following species were experimentally susceptible to virus infection: *Fragaria chiloensis, F.vesca, F. virginiana,* and *Fragaria x ananassa*.

Flexuous virus particles of 10-11 nm in diameter were observed in phloem cells of infected plants (Henriques and Schlegel, 1975). The dsRNA pattern suggested the virus might be a member of the family *Closteroviridae* and genus *Crinivirus* (Yoshikawa and Converse, 1990; Tzanetakis *et al.*, 2003). Pallidosis can be detected by grafting on indicator host (*F. virginiana* clones). It can also be detected in PCR system. The plants can be made free from pallidosis by shoot apex culture using excised shoot apices of 0.23 to 1.0 mm in length (Mullin *et al.*, 1974, 1975).

STRAWBERRY VEIN YELLOWING

Vein yellowing disease in strawberry has been reported from Japan (Yoshikawa and Inouye, 1988). On cultivated strawberry and *Fragaria vesca,* the symptoms appear as bright yellow veins. The disease is transmitted by leaf grafting. No other mode of transmission is known. The host range is restricted to *Fragaria* spp. only. *F. vesca* clones UC 4, UC 5 and UC 6 are used as indicator hosts. Probably a virus is associated with this disease.

STRAWBERRY LEAF ROLL

Leafroll disease of strawberry is of minor importance as the distribution is limited to Canada, United states of America and Kazakhstan, USSR (Bolton, 1964, Eliseeva *et al.*, 1974). The incidence of the disease was observed upto 60% (Bolton, 1964). Most of the commercial cultivars and *Fragaria* X *ananassa* Duch. are natural hosts of the virus. Affected plants do not bear any fruit and produce only a few distorted runners. The characteristic symptoms of the disease are downward rolling of leaflet margins. Sometimes the margins are completely rolled and appear like a tube. The leaves are chlorotic rugose with net veins and the petioles become shortened. Serrated margins, leaf reduction in size, diffuse chlorotic areas, mainvein banding are other symptoms of the disease. The symptom appearance is like a peacock pattern and normally known as peacock symptoms. The symptoms are recovered from infected plants in due course of time (Frazier, 1987).

The causal virus is transmitted by grafting (Frazier, 1974b). The incubation period for the development of symptoms of the disease is from 42-180 days (Frazier, 1974a). The causal agent of the leaf roll disease is not known. The disease can be detected on indicator (*F. vesca* and *F. virginiana* clones) plants by grafting. Roguing out of infected plants is the only control measure of the disease (Bolton, 1970).

STRAWBERRY FEATHER-LEAF

Feathery leaf disease was reported in strawberry from USA (McGrew, 1970). It is also known as 'Sparkle virus disease' and is of minor importance. *Fragaria vesca* clones show the typical feathery leaf symptoms. The symptoms include dwarfing, strap-like leaves due to reduction in leaf size, leaves show rugosity and serrated margins. The disease is graft transmitted and produces better symptoms on 'Alpine' and UC-1 indicator hosts. More than one disease causing entity is involved in this disease (Frazier, 1987). Feather leaf can be detected by grafting on indicator hosts and can be controlled by heat therapy (38°C for 5-9 weeks) followed by culture of 28-300 mg excised bud (McGrew, 1970).

STRAWBERRY PSEUDO MILD YELLOW-EDGE VIRUS

Strawberry pseudo mild yellow- edge virus (SPMYEV) has been reported from Japan and USA. Infected strawberry cultivars do not show any symptoms. Susceptible hosts are *Fragaria* spp., *Duchesnea chrysanth, D. indica* and *Rubus parvifolius*. The virus is transmitted by aphids, *Chaetosiphon fragaefolii* and *Aphis gossypii* in a semipersistent manner (Frazier, 1987). A carlavirus with rod shaped particles is associated with the disease (Yoshikawa and Inouye, 1986). The virus can be detected in ELISA and immunoblot assay (Yoshikawa *et al.*, 1986) and also by leaflet grafting on *F. virginiana* clone UC-10 and *F.vesca* clones UC-4 and UC-5. Inoculated clones may show mottled leaves, vein yellowing and necrosis. No control is known.

TOBACCO NECROSIS VIRUS INFECTING STRAWBERRY

Tobacco necrosis virus (TNV) is present world wide (Kassanis, 1967; Komuro *et al.*, 1973). The symptoms are necrosis of small roots, dwarfing, leaf malformation, necrosis and mottling (Ficcioli, 1969, 1970). The virus is transmitted by mechanical inoculation. Virus is also transmitted by fungus, *Olpidium brassicae* of the order *Chytridiales*. The indicator or diagnostic hosts are: *Chenopodium amaranticolor, C. quinoa, Lactuca sativa, Cucumis sativus, Phaseolus vulgaris, Vigna radiata, V. unguiculata, Nicotiana clevelandii, N. tabacum*. The symptoms in these hosts are necrotic local lesions.

TNV particles are round with icosahedral symmetry and has a diameter of 28.8-32 nm. Virus preparations contain more than one particle component (*i.e.*, satellite virus TNV of 17 nm is associated with virus and depends on TNV for its replication) (Finlay and Teakle, 1969). The Mr of the genome constitutes 19% of the virion by weight. The genome is monopartite. Only one particle size of linear, positive-sense, single-stranded RNA was recovered. The complete genome of the virus is 3762 nucleotides long. The virus can be detected by indexing on indicator/ diagnostic hosts which will develop characterstic symptoms upon inoculation. Use of sterilized soil for raising nursery through seed is important to avoid TNV infection.

TOBACCO STREAK VIRUS INFECTING STRAWBERRY

Tobacco streak virus (TSV) has been reported worldwide on different hosts but on strawberry TSV has been reported from United states and Australia (Fulton, 1971). In India, tobacco streak virus was reported on Sunflower and some other hosts (Ramiah *et al.*, 2001). Tobacco streak virus induces a necrotic reaction when grafted on *Fragaria vesca* indicator clones and hence also called as strawberry necrotic shock disease (Frazier *et al.*, 1962; Stace-Smith and Frazier, 1971). In strawberry cultivars, infected plants remain symptomless, although a reduction in vigour and fruit yield was noted (Spiegel *et al.*, 1995). The virus is transmitted mechanically to several hosts. It is also transmitted through pollen, seed (Johnson *et al.* 1983), and by thrips, *Thrips tabaci* and *Frankniella occidentalis.* (Costa and Lima Neto, 1976; Kaiser *et al.*, 1982). It is also transmitted by dodder, *Cuscuta campestris* (Fulton, 1948).

Tobacco streak virus (TSV) infects over 30 monocotyledonous and dicotyledonous plant species (Edwardson and Christie, 1991; Brunt *et al.*, 1996). In experimentally inoculated plants symptoms appear as leaflet epinasty, petiole stunting, necrosis of tips of leaves, or etched rings and lines on the leaf surface (Greber, 1979). *F. vesca* clones are used as indicators. *Vigna unguiculata* sp. cylindrica, *Dolichos biflorus, Beta patellaris, Phaseolus vulgaris* cv. Black Turtle and *Gomphrena globosa* are used as local lesion hosts and for assay of TSV.

TSV is the type member of the genus *Ilarvirus*, within the family *Bromoviridae* (Roossinck *et al.*, 2000). The virus has quasi-isometric particles of three major sizes: 27, 30, and 35 nm in diameter (Lister *et al.*, 1972) and contains at least four ssRNA species of 3.49, 2.96, 2.2 and approx 0.9 kb and a major protein of M_r 28,000. The complete nucleotide sequences of the 3 genomic species of the WC isolate have been determined; they contain 3,491 nt (RNA-1), 2,926 nt (RNA-2) (Scott *et al.*, 1998) and 2,205 nt (RNA-

3) (Cornelissen *et al.*, 1984). The subgenomic RNA-4 is a copy of the coat protein gene located at the 3' end of RNA-3. (Scott *et al.*, 1998).

The virus can be detected by inoculated indicator clones of *F. vesca* by leaf graft or the herbaceous indicator hosts by sap inoculation. The virus can also be detected by ELISA and by nucleic acid hybridization tests using cDNA probes (Stenger *et al.*, 1987) or by PCR. Management of the virus is difficult because the native hosts act as reservoir of the virus. The virus can be eradicated by heat therapy and shoot-tip culture.

VIRUS DISEASES TRANSMITTED BY NEMATODES IN STRAWBERRY

Nematode transmitted virus diseases are important in strawberry. These are: Arabis mosaic virus, Raspberry ringspot virus, Strawberry latent ringspot virus, Tomato black ring virus, Tomato ringspot virus.

STRAWBERRY LATENT RINGSPOT VIRUS

Strawberry latent ringspot virus (SLRSV) is mostly found in combination with Arabis mosaic virus. Both viruses are transmitted by the same nematode, *Xiphinema diversicaudatum* (Jha and Posnette, 1959). Both these viruses have wide host range and are widely distributed in Europe, Australia, New Zealand, Israel, Turkey and North America (Murant, 1981; Fry and Wood, 1973; Allen *et al.*, 1970; Hanson and Campbell, 1979) .

SYMPTOMS

Infected plants of susceptible cultivars show stunting with short internodes with a rosette appearance and yellow blotches on leaves. The disease is usually latent in strawberries but some strawberry cultivars show varying degrees of mottling, small and narrow leaves often folded upward and decline (Rüdel, 1985; Pelet, 1989). Leaves become patchy and reddish in color. There is decrease in yield. The virulent strain of the virus in Italy induces a disease known as 'willow leaf rosette' (Belli *et al.*, 1986). In mixed infection with Arabis mosaic virus, affected plants also show symptoms similar to Arabis mosaic (Murant, 1981). In mixed infections with Prune dwarf virus, variable symptoms appear in peach and cherry trees (Scotto La Massese *et al.*, 1973; Ragozzino and Alioto, 1992). Mixed infection with Cucumber green mottle virus, symptoms of short branches, leaf rolling and upward branching appear in apricot (Blattny and Janeckova, 1980).

TRANSMISSION

The Virus is transmitted by mechanical inoculation and by grafting. It is also transmitted by seeds upto 70% in several plant species such as *Mentha arvensis* and *Chenopodium quinoa* etc. Virus is transmitted by nematodes vector, *Xiphinema diversicaudatum* (Lister, 1964; Lamberti *et al.*, 1986). Both adults and larvae transmit the virus and the virus may be retained for up to 84 days in the vector kept without plants (Brown, 1985).

HOST RANGE

Strawberry latent ring spot virus has a very wide host range (Schmelzer, 1969; Murant, 1976). There are large number of natural hosts such as: *Fragaria vesca* (strawberry), *Apium graveolens, Robinia pseudoacacia, Rosa* spp. *Euonymus europaeus, Aesculus carnea, Rubus idaeus, R. fruticosus, Ribes nigrum, R. rubrum, Asparagus officinalis, Vitis vinifera, Capsella bursa-pastoris, Delphinium* spp., *Lamium amplexicaule, Narcissus* spp., *Prunus domestica, P. persica, Rheum rhaponticum, Sambucus nigra, Senecio vulgaris, Stellaria media, Trifolium repens* and *Urtica dioica.* Symptoms on these hosts appear as mottling and death of plant, strap leaf, mosaic, chlorotic ringspots and stunting and line pattern. Under experimental conditions, susceptible host species are found in about 30 Families. *Chenopodium amaranticolor, C. murale, C. quinoa* and *Cucumis sativus* are used as diagnostic hosts. Symptoms on inoculated hosts appear as chlorotic or necrotic local lesions, systemic chlorosis and deformation, necrosis or faint chlorotic mottle. *Chenopodium murale and C. amaranticolor* are used as indicator hosts.

THE PATHOGEN

Virus particles are isometric with a diameter of 30 nm, round and exhibits icosahedral symmetry (Lister, 1964; Richter and Proll, 1970; Allen *et al.*, 1970). The virus belongs to Order: *Picornavirales*, Family: *Secoviridae*, Genus: *Sadwavirus*. However, current literature indicates that it may be a *Cheravirus*. The genome of the virus is bipartite, linear, positive-sense, single-stranded RNA. Segements are distributed among 2 particle types. The complete genome is 11338 nucleotides long (Mayo *et al.*, 1974). Virions contain 38% nucleic acid (B); 62% protein (T).

DIAGNOSIS AND MANAGEMENT

The virus can be detected by sap inoculation to *Chenopodium quinoa, C. amaranticolor* and *Cucumis sativus* and by graft inoculation to *Prunus persica* 'GF305'. It can also be detected by serological methods like ELISA and also by PCR (TingWei *et al.*, 2008). All isolates studied seem to be serologically identical (Murant, 1976), but not all populations of *X.*

diversicaudatum are equally efficient in transmitting the virus. Strawberry planting material should be virus-free. Diseased plants should be eradicated and where possible soil should be fumigated.

ARABIS MOSAIC VIRUS INFECTING STRAWBERRY

This disease was first reported by Smith and Markham (1944). The disease is now distributed in many countries of Europe, Asia, South Africa, Canada, Australia and New Zealand. Arabis mosaic virus (AMV) is damaging if occurs in combination with strawberry latent ringspot virus (SLRSV). The virus causes yellow dwarf of raspberry, mosaic and yellow crinkle of strawberry, stunt mottle of cucumber, chlorotic stunt of lettuce, stunting and necrosis of celery, and yellow net of *Forsythia intermedia*, and is one of the causes of mosaic of rhubarb.

SYMPTOMS

The conspicuous symptoms of the disease are leaf mottling, flecking, stunting and several forms of deformation including enations. The virus infects about 93 wild and cultivated plant species (Schmelzer, 1963). Several host of the virus are symptomless carier. It is also reported that AMV infects the roots of gymnosperms (*Chamaecyparis lawsoniana*).

TRANSMISSION

The virus is transmitted by grafting, through dodder, *Cuscuta subinclusa* and *C. californica* and occasionally by *C. campestris* (Schmelzer, 1963). The virus is seed-borne in at least 15 plant spp. in 12 plant families. Many plants infected through seed show no symptoms (Lister and Murant, 1967; Murant and Lister, 1967). The vectors of the virus are nematodes, *Xiphinema diversicaudatum* (Jha and Posnette, 1959; Harrison and Cadman, 1969) and *X. coxi* (Fritzsche and Schmidt, 1963). The nematodes retain virus for at least 31 days when kept in fallow soil (Jha, 1961).

HOST RANGE

The virus occurs naturally in many species of wild and cultivated monocotyledonous and dicotyledonous plants. It was mechanically transmissible to 93 species in 28 dicotyledonous families (Schmelzer, 1963). It occurs in many other plants including sugarbeet, hop, horseradish, narcissus, rose, *Sambucus nigra, Ligustrum vulgare,* white clover and grapevine. Grapevine plants more commonly contain grapevine fanleaf virus, which is distantly serologically related to Arabis mosaic virus. Following herbaceous hosts were susceptible by mechanical inoculation: *Chenopodium amaranticolor* and *C. quinoa* (Murrant, 1970), *Cucumis sativus, Nicotiana tabacum* cv. White Burley, *Phaseolus vulgaris cv.* the Prince and *Petunia hybrida.* These hosts developed variable symptoms from local lesions, to systemic chlorotic or necrotic rings, vein banding, line pattern and mottling. *Chenopodium amaranticolor* can be used as a assay host.

THE PATHOGEN

The virus particles are isometric and about 30 nm in diameter. The virus belongs to family *Secoviridae* (*Comoviridae*), genus *Nepovirus* and species *Arabis mosaic virus* (Murant, 1970). Arabis mosaic virus has a bipartite genome with two RNA-species of molecular weights 2.4 and 1.4 x 106. There is a single coat protein with a molecular weight of 54 x 103 (Murant, 1981; Francki *et al.*, 1985).

DIAGNOSIS AND MANAGEMENT

The virus can be detected by inoculation to herbaceous indicator plants, *Chenopodium quinoa* and *C. amaranticolor* which develop chlorotic local lesions followed by a systemic mottle (Murant, 1970). ELISA is widely used for serodiagnosis. It can also be detected by electron microscopy and in immunosorbent electron microscopy (Roberts and Brown, 1980). Strawberry latent ringspot virus (SLRSV) and Arabis mosaic virus (ArMV) were detected by ELISA and RT-PCR using virus-specific antibodies and primers respectively (Kulshretha *et al.*, 2005). Another means of detecting AMV is the use of cDNA-clones in dot-hybridization tests (Jelkmann *et al.*, 1988; Steinkellner *et al.*, 1989). The virus can be managed by developing virus-free planting material through a strict certification scheme. Soil fumigation and/or fallow for at least one year or nematodes control with nematcides seem to be good for management of Arabis mosaic nepovirus (Dale and Hendy, 1967).

TOMATO BLACK RING VIRUS INFECTING STRAWBERRY

Tomato black ring disease was first described from Europe (Smith, 1946) and now the disease is known to be present in several countries in Europe. (France, Finland, Germany, Greece, Hungary, Ireland, Netherlands,

Moldova, Norway, Poland, Romania, Russia (European), Sweden, Turkey, UK (England, Scotland) and Yugoslavia). In Asia, the disease was reported from India, Japan and Turkey. The disease was intercepted in Kenya, USA, Brazil and Canada.

SYMPTOMS

The symptoms of tomato blackring virus (TBRV) disease in strawberry cv. Talisman appear as chlorotic streaks, chlorotic spots and ringspots on leaves of infected plants. In strawberry cv. Auchincruive Climax, chlorotic spots without any definite shape appear on leaves (Converse and Stace-Smith, 1987). Plants infected with tomato black ring virus (TBRV) are patchily distributed in crops because of the slow migration of the soil-inhabiting vectors, *Longidorus* spp.

TRANSMISSION

Black ring virus (TBRV) in strawberry is transmitted by grafting and through seeds in at least 24 plant species. Virus is transmitted in seed of raspberry and strawberry from either male or female parent and dissemination of the virus infection by the seed (Murant and Lister, 1967). Different TBRV isolates are transmitted by different vector species. For example, in the UK, isolates of TBRV from Scotland are transmitted most efficiently by *Longidorus elongatus*, whilst those in England (and also on the mainland of Europe) are transmitted largely by *L. attenuatus* (Harrison *et al.*, 1961; Harrison, 1964; Taylor and Murant, 1969). Both larvae and adults of *L. elongatus* can transmit the disease but the adult does not pass the virus to its progeny, nor is the virus retained when the nematode moults (Murant, 1970). *Longidorus elongatus* survives in fallow soil and retains infectivity up to about 9 weeks.

HOST RANGE

Tomato black ring virus causes ringspot diseases of bean, sugar beet, lettuce, raspberry and strawberry, yellow vein of celery, 'bouquet' and pseudo-aucuba of potato, shoot-stunting of peach and also infects many other plants, including cabbage, grapevine, lucerne and *Sambucus nigra* etc. TBRV infects large number of plant species. For example, an isolate from *Robinia pseudoacacia* infected 76 species in 29 dicotyledonous families by mechanical inoculation (Schmelzer, 1963). In naturally infected plants through nematode, symptoms appear as patches of poor growth which slowly extend in size each year. Depending on the cultivar, natural infection in *Rubus* and *Fragaria* may induce chlorotic mottling and/or ringspots in leaves (Murant, 1987; Murant and Lister, 1987). *Chenopodium amaranticolor, C. quinoa, Nicotiana rustica, N. tabacum* cvs. White Burley or Xanthi-nc, *Phaeolus vulgaris* cv. Prince, and *Petunia hybrid* are susceptible by mechanical inoculation. Inoculated leaves of these hosts develop local lesions followed by systemic infection. *C. amaranticolor and C. quinoa* are the suitable assay host.

THE PATHOGEN

The virus particles are isometric, about 28 nm in diameter with a bipartite genome (Harrison and Nixon, 1960, Stace-Smith, 1984). The two types of RNA present in TBRV are single stranded, positive sense. Both genomic RNA species are necessary for infectivity (Randles *et al.*, 1977). The particles have RNA about 28% (M) and 38% (B) of the particle weight. The virus belongs to order *Picornavirales,* family *Comoviridae* and genus *Nepovirus* with species *tomato black ring virus*. Different satellites have been described for different TBRV isolates (Fritsch *et al.*, 1984).

DIAGNOSIS AND MANAGEMENT

Tomato black ring nepovirus is readily detected by sap inoculation to many herbaceous test plants. In test species such as *Chenopodium quinoa, C. amaranticolor* and *Nicotiana clevelandii,* TBRV induces chlorotic or necrotic local lesions and systemic necrosis, depending on the virus isolate. Serological tests are most reliable to detect TBRV which can be detected in precipitin tube and in gel-diffusion tests (Hollings, 1965). Tomato black ring virus shares many properties with other nepoviruses (arabis mosaic, grapevine fanleaf, raspberry ringspot, strawberry latent ringspot, tobacco ringspot and tomato ringspot) but is unrelated to them serologically. To manage the virus, use of healthy planting material (seed or vegetative material) at sites free from infestation with vector nematode, *Longidorus sp.* is important. At sites contaminated with such infective nematodes, soil treatment with nematicides prior to planting can be used as a control measure (Trudgill and Alphey, 1976; Taylor and Murant, 1968). Continued weed control is an additional and recommended treatment as they are the natural hosts for both virus and the vector.

TOMATO RINGSPOT VIRUS IN STRAWBERRY

Tomato ringspot virus infects strawberry in most of the areas where strawberry is grown. Tomato ringspot disease is found in several countries of North and South America, central America and Caribbean, Asia, Oceania and Europe. Tomato ringspot virus remains symptomless in *Fragaria chiloensis* along the coast of northern California (Frazier *et al.*,

1961). The virus infects both wild and cultivated strawberry. There are several synonyms of this virus such as Tobacco ringspot No. 2, Nicotiana virus 13, Peach yellow bud mosaic virus (strain), Blackberry (Himalaya) mosaic virus, Winter peach mosaic virus, and Grape yellow vein virus (strain).

SYMPTOMS

The commercial cultivars of strawberry normally did not show symptoms but sometime may show dwarfing of plants and mottling of leaves. Runner production is reduced. Symptoms of crinkling, mottling, mild yellow edge and death of infected plants often occur in case of severe infection (Converse and Stace-Smith, 1987).

TRANSMISSION

Tomato ringspot virus of strawberry is readily transmissible by grafting and by sap inoculation to herbaceous hosts. It is also pollen transmitted in *Pelargonium* (Scarborough and Smith, 1977). The virus is transmitted occasionally through seed of tomatoes, tobacco and grapes, and frequently through seed of *Gomphrena globosa*, strawberries, *Pelargonium* and soyabean cv. Lincoln. The virus may spread by natural root grafts. The vector of the virus is *Xiphinema americanum sensu lato*, the American dagger nematode. The following *Xiphinema* species have been suggested as being possible vectors: *X. americanum, X. californicum, X. incognitum, X. occidium, X. rivesi, X. thornei, X. utahense* (Brown, 1989).

HOST RANGE

Natural infection of the virus has been observed in wild and cultivated strawberries, raspberries, *Rubus laciniatus*, grapes, peaches, cherries and other *Prunus* spp., black currants, gooseberries, *Pelargonium, Hydrangea, Gladiolus* and *Fraxinus americana.* The experimental host range is very wide in plant species of more than 35 dicotyledonous and monocotyledonous families (Stace-Smith, 1962). *Fragaria vesca* clones 'EMC', 'EMK', 'EM-1', 'UC-1', 'UC-3', 'UC-4', 'UC-5' and 'UC-6' develop epinasty, blotchy mottling and necrosis of leaf blade, thickening of petioles upon leaf graft inoculation (Frazier and Mellor, 1970b). Tomato ringspot virus is lethal to *F. virginiana* cvs 'UC-10' and 'UC-11'. Weeds like *Taraxacum officinale* and *Stellaria media* has been shown to be reservoir of the virus (Converse, 1981). Symptoms on some of the hosts are described below:

- *Raspberries:* Canes become stunted, fruit size is reduced, Chlorotic ringspots on leaves, epinasty and death of new canes, smaller canes and fruits (Smith, 1972).

- *Grapes:* Ringspot, mottling and reduction in size of leaves, short internodes resulting in rosetting (Yang *et al.*, 1986). Abortion and reduction in fruit size. Bark scaling of trunk and shoots with thick and spongy phloem tissue with numerous necrotic pits (Uyemoto, 1975).
- *Tomato:* Curling, brown necrotic rings and sinuous lines on leaves. Necrotic streaks and rings on stem and petiole, fruits show grey to brown, corky partial or concentric rings (Smith, 1972).
- *Peach:* Young leaves show pale green to pale yellow, oblong, feather-edged blotches along the veins, rosette formation of leaves which may show mottling symptoms. Stem pitting symptoms may develop in some cultivar (Teliz *et al.*, 1966).
- *Pelargonium:* Ringspot, chlorotic flecks and mottling on young leaves. Oak-leaf pattern on older leaves, flowers may be distorted (Ryden, 1972)

Chenopodium quinoa and *C. amaranticolor, Cucumis sativus, Nicotiana tabacum* cv. Xanthi-NC, *Phaseolus vulgaris, Vigna unguiculata, Lycopersicon esculentum, Nicotiana clevelandii* and *Petunia x hybrida* are used as indicator hosts. In these hosts, symptoms develop as local lesions (necrotic or chlorotic) followed by systemic infection in some cases. The virus can be assayed on *Chenopodium amaranticolor, C. quinoa, Vigna unguiculata, Nicotiana clevelandii* and *N. tabacum.*

THE PATHOGEN

Tomato ringspot virus is a nepovirus measuring 28 nm in diameter. It is a single stranded RNA virus and belons to family *Comoviridae* (Stace-Smith, 1984). The genome is segmented; bipartite, segements are distributed among 2 particle types of linear, positive-sense, single-stranded RNA. The complete genome is 15800 nucleotides long. Both RNAs have been sequenced. Complete sequence of RNA 1 is 8500 nucleotides long and RNA-2 sequence is 7300 nucleotides long. The viral genome encodes structural proteins and non-structural proteins (Allen and Dias, 1977).

DIAGNOSIS AND MANAGEMENT

The virus can be detected by mechanical transmission to indicator herbaceous hosts. Serological methods, such as direct and indirect ELISA (Converse, 1981) and ISEM can be used to detect the virus. Following management practices have been recommended:

- Control nematode vector by fumigation or avoid infested soil for planting (Murant and Taylor, 1965; Thompson and McKenry, 1975).

- Use of virus-free nursery stock
- Nurseries should be raised in fumigated soil.
- Use immune cultivars like 'British Sovereign and 'Sparta' (Mellor and Stace-Smith, 1963).

OTHER MINOR DISEASES OF STRAWBERRY

There are some diseases found on strawberry but have not been studied in detail. Such diseases are: strawberry necrosis, strawberry stunt, strawberry vein necrosis, NEPO virus-2 and tobacco mosaic. There is another group of virus and virus-like pathogens which infects strawberry experimentally such as apple mosaic virus, cherry leaf roll virus, cherry rasp leaf virus, plum line pattern, peach X, Rubus Chinese seed born virus and some others. The disorders like strawberry june yellows is a nongraft transmissible disorder. Readers may consult Agriculture Handbook No.631'Virus diseases of small fruits' edited by R.H. Converse and published by USDA, 1987 and CMI/AAB description of plant viruses, Commonwealth Agricultural Bureaux, Slough.

PHYTOPLASMA DISEASES OF STRAWBERRY

Phytoplasmas are wall-less, plant-pathogenic organisms that are classified in the class Mollicutes. In diseased plants, phytoplasmas are found in the sieve elements of phloem tissue. Phloem-feeding insects, mainly leafhoppers, transmit phytoplasmas from plant to plant. Fifteen phytoplasma groups (16Sr groups) and more than 40 subgroups have been delineated on the basis of results from RFLP analysis of 16SrRNA gene sequences. Diseases like Aster yellows, Maladie du bord jaune, Strawberry green petal, Strawberry lethal decline, Strawberry yellows, Strawberry witches'-broom are caused by phytoplasmas. A new disease of strawberry (*Fragaria × ananassa*), marginal chlorosis, was observed in Spain in 1984 and in France in 1988 was demonstrated to be associated with two phloem restricted phytoplasmas: *Candidatus* Phytoplasma fragariae and stolbur phytoplasma (Nourrisseau *et al.*, 1993; Danet *et al.*, 2012). Another disease known as strawberry yellows is caused by rickettsia-like organism will also be described in this chapter.

STRAWBERRY GREEN PETAL

Strawberry green petal disease is found in Europe, North America, Canada, Australia, New Zealand and Russia. This disease is similar to clover phyllody. Based on symptoms it was named as 'green petal' (Posnette, 1953). The agent causing phyllody (virescence) in clover flowers was transmitted by *Cuscuta subinclusa* to *Fragaria vesca* and *Duchesnia indica* plants which then produced symptoms of strawberry green-petal disease suggesting that both the diseases are same (Frazier and Posnette, 1957). The incidence of green petal disease varies from year to year depending on climatic conditions and the cultivars grown. The disease is economically important as either affected plants do not produce fruits at all or if a few fruits develop they are not marketable.

SYMPTOMS

The most striking and characteristic symptoms of the disease are green petals of the flower in infected plants. Petals of the flower are reduced in size, become dark green and leaf-like (phylloid). The flowers are mostly sterile. The leaves become smaller in size and cupped and petioles are extremely short at later stages of infection. Affected plants develop only a few stunted runners and develop one or two plants. The natural hosts of the green petal pathogen are *Trifolium* species (*T. repens, T. pratense, T. hybridum*). Phyllody of flowers is the major symptoms in these hosts. The pedicels and calyx lobes become elongated and leaf like in advanced stage of infection. The ovary may proliferate and become leaf like. Fruits remain small, hard and green.

TRANSMISSION

The disease is transmitted experimentally by grafting and by dodder. Several species of leafhoppers (*Aphrodes bicincta, Macrosteles fascifrons, M. viridigraceus, M. cristatus, Anoscopus albifrons, Scaphytopius acutus, Paraphlepsius irroratus, pseudotettix subfusculus, Eucelis linealata, E. incises*) are known to transmit green petal disease in a persistent manner. *A. bicincta* is the most efficient vector which transmit green petal pathogen from and to strawberry (Chiykowski, 1962; Thompson, 1968). Other leafhopper species transmit green petal to clover (*Trifoliate* spp.) but not back to strawberry. The disease is also transmitted by several species of dodder, *Cuscuta campestris, C. subinclusa, C. arvensis, C. europea,* and *C. gronovii* from strawberry to strawberry and other hosts like periwinkle (*Catharanthus roseus*) and clover. The symptoms of green petal disease of strawberries in Britain resemble those caused by the aster yellows in strawberries in California. Symptoms of the aster yellows type were also found in red and white clover (*Trifolium pratense* and *T. repens*), plantain (*Plantago major*) and mayweed (*Anthemis cotula*) (Frazier and Posnette,1956).

THE PATHOGEN

Phytoplasmas are associated with Strawberry green petals disease. The phytoplasma belongs to the subgroup I-C of which type strain is clover phyllody phytoplasma (Honetslegrova *et al.*, 1996; Sinha and Paliwal, 1969, 1970).

DIAGNOSIS AND MANAGEMENT

The causal phytoplasma can be detected by its transmission to periwinkle and celery. The symptoms in periwinkle appear as chlorosis of leaves, leaf reduction and phyllody. The pathogen is symptomless in celery but this host can be used as differential host. The pathogen can also be detected by electron microscopy, serologically by ELISA and also by PCR. The management of the disease is possible by minimizing vector population (Thompson *et al.*, 1973), Use of antibiotics (Oxytetracycline, tetracycline and doxycycline) as root dips, foliage dips or spray. However, symptoms will mask after the treatment with antibiotics but may reappear (Sinha and Peterson, 1972). Heat therapy of affected plant and use of pathogen free planting material are also used as management practices.

STRAWBERRY LETHAL DECLINE

Strawberry lethal decline is probably related to X-disease of stone fruits (Frazier and Jensen, 1970). The incidence of the disease was upto 25% in some cultivars of strawberry (Schwartze and Frazier, 1970).

SYMPTOMS

The symptoms of the disease appear in young leaves as dwarfing of infected plants and upward cupping of young leaves. Older leaves are chlorotic, roll upward and lower surface become red or purple. In advance cases, the symptomatic leaves die and only a few infected leaves remain on the plant. Young plants coming out from infected ones are chlorotic and stunted. Infected daughter plants are dead within weeks in the nursery. The roots of infected plants become brown and black. Fruits do not mature on infected plants (Schwartze *et al.*, 1987). Most of the strawberry cultivars are susceptible to lethal decline infection. *Fragaria vesca* is a indicator host of lethal decline and develops symptoms as described above.

TRANSMISSION

Leathal decline agent is transmitted by leafhopper, *Aphrodes bicincta* from strawberry to strawberry (Converse, 1987). However, leafhopper, *Colladonus geminates* transmitted peach yellow leaf roll agent (Western X disease) to strawberry (*F. virginiana* cv. UC-10 and Hood) causing symptoms similar to lethal decline (Frazier and Jensen, 1970). The disease is also transmitted by dodder and petiole insert leaflet grafting (Braun and Keplinger, 1962).

THE PATHOGEN

A phytoplasma similar to western X phytoplasma of peach and cherry is associated with the disease. The phytoplasma has been characterized as '*Candidatus* Phytoplasma australiense' (Liffting *et al.*, 2007).

DIAGNOSIS AND MANAGEMENT

The diagnosis is done by leaflet grafting to *F. vesca* and observing typical symptoms of the disease. The phytoplasma can be detected in PCR using peach X specific primers. Rouguing of infected plants and use of disease free planting material are recommended for the management of the disease.

STRAWBERRY YELLOWS PHYTOPLASMA DISEASE

Strawberry yellows disease was initially reported as lethal yellows from Australia (Stubbs, 1968). The symptoms of yellows disease in strawberry appear as general stunting and yellowing of leaves. Young leaves are shorter with reduced petioles. Leaf margins become chlorotic or yellow. Older leaves show bronze or purple colour and rolled upward. There is no flower or fruit production. Affected plants may die within a few weeks after infection. *Fragaria vesca* seedlings and clone 'UC-1' are used as indicator hosts and develop similar symptoms upon graft inoculation (Shanmuganathan and Garrett, 1976). The disease is transmitted by leaf grafting and also by dodder (*Cuscuta campestris*). The leafhoppers vector of the disease are *Orosius* species (Gryll, 1979).

A phytoplasma is associated with yellows disease which was designated '*Candidatus* phytoplasma Australiense'. Phylogenetic analysis of 16S rRNA gene sequences indicated that Strawberry Yellows phytoplasma, '*Candidatus* Phytoplasma australiense', '*Candidatus* Phytoplasma asteris', stolbur phytoplasma and '*Candidatus* Phytoplasma japonicum' shared a common ancestor, but were mutually distinct. The results of 16S rRNA indicated that strawberry yellows phytoplasma is a representative of a novel taxon '*Candidatus* phytoplasma fragariae' (Valiunas *et al.*, 2006). The disease can be diagnosed by leaf grafting on indicator plants. The associated phytoplasmas can be viewed in electron microscope. Specific detection is possible in PCR using yellows-specific primers (Valiunas *et al.*, 2006). Roguing of infected plants, minimizing vector population, use of disease free planting material and treatment with antibiotics are recommended for management of the disease.

STRAWBERRY WITCHES BROOM AND MULTIPLIER DISEASES

Strawberry multiplier diseases (SMD) are also known as bushy plant, spindly plant diseases, and multiplier plant. Witches broom on strawberry has been reported from North America, the Netherlands and Japan. Multiplier disease (MD) is present in USA (Demaree and Marcus, 1951; Sehgal and Boone, 1963; Zeller, 1927). Both Witches' broom and multiplier diseases are of minor importance.

SYMPTOMS

Under natural conditions, both diseases affect *Fragaria* species. Infected plants by witches broom are dwarfed, bushy in appearance, sterile, have multibranched crowns and erect spindly petioles with small leaves. Fruits are rarely formed. The plants infected with SMD are dwarfed and show crown proliferation, cupping of leaves on short and eract petioles. Crown proliferation and short rulers are also characteristic. Flowers are reduced in number and size (Converse, 1987).

TRANSMISSION

Witches broom disease (WBD) is transmitted by leaf and stolon grafts and also by dodder, *Cuscuta subinclosa* to *Fragaria vesca* 'EMC' and to Strawberry (Frazier, 1974a; Demaree and Marcus, 1951; Sehgal and Boone, 1963). SMD is transmitted only by grafting but leafhopper vectors (*Macrosteles orientalis, M. fascifrons*) have been reported from Japan.

HOSTRANGE

Hostrange of strawberry witches' broom is restricted to *Fragaria vesca* var. samperflorens ser cv. 'Alpine and *F. chiloensis*. The leaf graft inoculated plants of these species show crown proliferation with several small leaves on thin petioles (Frazier, 1974a). Similar symptoms were observed when leaf grafts were made from SMD to several strawberry cultivars and *F. vesca* clones 'EMC', UC-1, *F. vesca* vars. Alba, Bracteata, *F. cheloensis* and *F. orientalis* (Sehgal and Boone, 1963).

THE PATHOGEN

Strawberry witches' broom disease is now known to be caused by a phytoplasma (Huhtanen and Converse, 1971). SMD is also a Phytoplasma disease (Heimann *et al.*, 1992) and both the phytoplasmas can infect the similar hosts. The taxonomic position of witches broom phytoplasma is: Bacteria: Tenericutes: Mollicutes: Phytoplasmas.

DIAGNOSIS AND MANAGEMENT

The disease can be diagnosed with symptoms in field and by transmission tests on indicator hosts in laboratory. *F. virginiana* clone UC 12 is best indicator host for witches broom and *F. chiloensis* for SMD. The associated phytoplasmas can be detected by electron microscopy, structures resembling phytoplasmas can be seen in the phloem of infected petioles of strawberry. Tissues may be assayed for phytoplasmas by PCR analysis using a pair of universal primers that amplify a 1.4-kbp phytoplasma 16S rDNA fragment. Roguing of infected plants and use of certified planting material are recommended as control measures.

ASTER YELLOWS IN STRAWBERRY

Aster yellows, in general, occurs in North America, Europe, Japan and Russia. The disease is also known as Eastern aster yellows, Western aster yellows occurring on eastern and western parts of USA. Aster yellows disease naturally infects strawberry and was transmitted experimentally to strawberry from aster by leafhopper vector, *Macrosteles fascifrons* in USA (Frazier and Thomas, 1953).

SYMPTOMS

Aster yellows has very wide host range infecting about 300 plant species in about 50 families. In strawberry, the symptoms appear as smaller petioles of young leaves. The young leaves are smaller in size, chlorotic and normally cupped. These cupped leaves turn red, later brown and flat. The flowers become partly or fully sterile with small green petals. The flowers become leaf like, proliferated and phylloid. Natural infection is not found in wild strawberry (*F. vesca*) but it can be infected experimentally (Frazier and Thomas, 1953).

TRANSMISSION

The pathogen is transmitted by dodder, *Cuscuta campestris* from strawberry to periwinkle and back to strawberry (Fulton, 1957). From periwinkle, the disease is transmitted by stolon grafts and excised leaf graft to strawberry (Braun and Keplinger, 1962). About 27 species of leafhopper are known to transmit western strain of aster yellows. *Macrosteles fascifrons, Colladonus montanus* and *C. geminatus* can transmit the disease experimentally but do not prefer *Fragaria* as a host. The eastern strain is transmitted only by *M. fascifrones* (Chiykowski, 1987).

THE PATHOGEN

The disease is caused by a phytoplasma (Maramorosch *et al.* 1968). RFLP analysis of the R16(I)F1/R1 amplified fragments and sequence analysis revealed that strawberry plants were infected with phytoplasmas belonging to three different sub-groups of the Aster yellows group: 16SrI-A, -B and –R (Cieslinska, 2011). Phylogenetic analysis of 16S rRNA gene sequences indicated that StrawY phytoplasma, '*Candidatus* Phytoplasma australiense', '*Candidatus* Phytoplasma asteris', stolbur phytoplasma and '*Candidatus* Phytoplasma japonicum' shared a common ancestor, but were mutually distinct. Studies suggest that StrawY phytoplasma is a representative of a novel taxon, '*Candidatus* Phytoplasma fragariae' (Valiunas *et al.*, 2006).

DIAGNOSIS AND MANAGEMENT

Pleomorphic bodies of phytoplasma can be seen in sievetubes of phloem tissues by viewing ultrathin sections in electron microscope. The disease can also be detected by experimental transmission to periwinkle through dodder. The phytoplasma infection is also detected by nested polymerase chain reaction (PCR). RFLP analysis using MseI, HhaI and HpaII restriction endonucleases are used to identify the causal agents of the disease (Cieslinska, 2011). Phytoplasma free planting material must be used to manage this disease.

STRAWBERRY YELLOWS CAUSED BY RICKETTSIA-LIKE ORGANISMS

Both phytoplasma and Rickettsia like organisms (RLOs) have been reported with yellows symptoms in strawberry (Streten *et al.*, 2005). RLOs could be distinguished from phytoplasma in thin sections of diseased tissue when viewed in electron microscope (Greber and Gowanlock, 1979). The RLOs are restricted to Xylem and Phloem vessels of infected plant. These organisms have round to elongated in morphology and measure about 0.4 μm in diameter and upto 3 μm in length. The organisms have rigid cell wall and inner cell membrane. Ribosome-like fibrillar structures are found in RLO cells. The symptoms of the two diseases overlapped, hence identification of individual diseases in the field is difficult. Young leaves of plants infected by this disease show interveinal chlorosis and tend to bend back along the midvein. The leaves become brittle and can be tear easily. Older leaves become bronze or purple. At higher temperatures, petioles and stolons become necrotic. Affected plants may show wilting or die leaving a cluster of chlorotic and infected runners. Indicator plants *F. vesca* cv. 'UC-4' produce similar symptoms when grown around 30°C. The disease is transmitted by leaf grafts and also by dodder. The disease is also transmitted by *Austroagalliia torrid*, the vector of clover rugose leaf curl disease (Gryll, 1979).

The disease can be identified by leaf grafting on indicator host or by dodder transmission. The pathogen is also detected in electron microscope by observing thin sections of diseased tissue. Polymerase chain reaction (PCR) is a reliable method to detect RLOs and phytoplasma from individual infection or in mixed infections (Streten *et al.*, 2005). The management is similar as has been described with yellows disease caused by phytoplasma. However, treatment of RLO infected plants with penicillin drenches (1000 ppm sodium penicillate) at 5 days interval for 8 weeks is effective to cure plants from infection (Greber, 1987).

B. VIRUS DISEASES OF RUBUS (RASPBERRY AND BLACKBERRY)

There are several species of *Rubus.* Plants under this genus are known as the blackberry, red raspberry, and the dewberry. Raspberries such as *Rubus arcticus* (Arctic Raspberry), *R. crataegifolius* (Korean Raspberry), *R. idaeus* (European Red Raspberry), *R. leucodermis* (Whitebark or Western Raspberry, native: Blue Raspberry), *R. occidentalis* (Black Raspberry), *R. odoratus* (Flowering Raspberry), *R. phoenicolasius* (Wine Raspberry or Wineberry), *R. strigosus* (American Red Raspberry) (syn. *R. idaeus* var. *strigosus*) are commonly used for eating and other purposes. Blackberries and raspberries are perennial plants and both are brambles and grow on canes.

Blackberry also comes under the genus *Rubus* and taxonomically known as *Rubus macropetalus, Rubus loganobaccus,* and *Rubus ursinus*. Erect blackberry bushes that are recognized as native genera are: *Rubus frondosus, Rubus argutus,* and *Rubus allegheniensis.* Blackberries are rich in anti-oxidants and vitamins along with being a good source of the minerals potassium, phosphorus, iron, and calcium.

Rubus species are infected with large number of viruses - Apple mosaic virus, Arabis mosaic virus, Blackberry chlorotic ringspot virus, Bramble yellow mosaic virus, Cherry leaf roll virus, Cherry rasp leaf virus, Cucumber mosaic virus, Raspberry bushy dwarf virus, Raspberry ringspot virus, Strawberry latent ringspot virus, Tobacco ringspot virus, Tobacco streak virus, Tomato black ring virus, Tomato ringspot virus, Wineberry latent virus. Some of these viruses are aphid transmitted (Raspberry mosaic, Rubus yellow net, Black raspberry necrosis, raspberry leaf mottle and leaf spot, raspberry leaf curl, cucumber mosaic virus in raspberry, Thimbleberry ringspot, vein chlorosis), one is leafhopper borne disease (Rubus stunt), and a few are nematode-borne viruses (raspberry yellow dwarf, ringspot, cherry leaf roll virus in Rubus, tomato ringspot virus,tobacco ringspot). Other viruses known to infect *Rubus* spp. are pollen borne and in some cases vector is not known. A new *Emaravirus* was observed in blackberry plants affected with yellow vein disease showing leaf mottling, chlorotic ringspot and curved midrib (Hassan *et al.*, 2012). The most common methods for management of raspberry viruses are: Use certified planting material, Plant resistant or immune cultivars, if available, avoid pollen borne viruses by planting in large fields, and control insect vectors. A putative new Flexivirus tentatively named as *Rubus canadensis virus-1* closely related to members of the genus *Foveavirus* in family *Betaflexiviridae* was isolated from blackberry in United States (Sabanadzovic *et al.*, 2012).

BLACKBERRY CALICO VIRUS

The disease is found in North America and possibly in Europe. The symptoms of the disease are: yellow spots, blotchy chlorosis, rings or line pattern on leaves of affected plants. Symptoms normally appear in hot weather. The associated virus infects wild and cultivated *Rubus ursinus*, blackberry x raspberry hybrids such as boysenberry and loganberry. The virus is graft and sap transmitted (Converse, 1987). A carlavirus with flexuous rod shaped particles of about 15x627 nm in length is associated with calico disease (Jones *et al.*, 1990). The virus can be detected by serological techniques like ELISA. The disease can be managed by thermotherapy for 35 days at 37°C.

RASPBERRY RINGSPOT DISEASE

Raspberry ringspot disease was first described by Cadman,1956 and Harrison, 1958. This virus disease is widespread in the major red raspberry-producing areas of several countries in Europe, Canada, Russia and Turkey (Murant, 1970a).

SYMPTOMS

Raspberry ring spot virus (RRSV) causes diseases in strawberry and raspberry. It causes spoon leaf symptoms in red currant and also cause rasp leaf symptoms in cherry in association of Prunus necrotic ringspot virus. RRSV infects plant species in more than 14 families. The virus causes mottle and death to *Fragaria vesca* (Strawberry). The most characteristic symptom of the disease is the production of small, crumbly berries that fall apart when touched. Infected plants of raspberry are usually patchily distributed in crops suggesting nematode transmission. Young, expanding leaves of susceptible infected plants develop a conspicuous, yellowish-green, ringspot or oak-leaf pattern; chlorotic blotches and a net-like chlorosis along the smaller veins. Leaves may be smaller, slightly wavy, distorted with deep sinuses. Secondary symptoms are enations or rasp leaves and short internodes. Some necrosis and dieback may also occur (Murant, 1970b).

TRANSMISSION

The virus is transmitted by grafting. It is mechanically transmissible to a number of herbaceous plants and is also transmitted through seed in at least six plant species including strawberry and raspberry (Lister and Murant, 1967). Weeds provide an important means of survival of RRSV inoculum in soils (Murant and Lister, 1967, 1987). RRSV is transmitted by two species of the nematode genus *Longidorus* (Harrison, 1964). Scottish and Dutch strains are transmitted most efficiently by *L. elongatus* (Taylor, 1962), while the English form is transmitted by *L. macrosoma* (Debrot, 1964). Both larvae and adults of *L. elongatus* can transmit the virus, but the virus does not pass to its progeny, nor is the virus retained when the nematode moults (Taylor, 1962; Harrison, 1964; Taylor and Murant, 1969). *Chenopodium quinoa* and *Spinacia oleracea* (spinach) are convenient 'bait plants' in nematode transmission.

HOST RANGE

Raspberry ringspot virus (RRSV) occurs naturally in many species of wild and cultivated dicotyledonous and monocotyledonous plants such as *Apium graveolens, Robinia pseudoacacia, Rosa* ssp., *Euonymus europaeus, Aesculus carnea, Rubus idaeus, Prunus avium, P. domestica, Ribes grossularia, R. nigrum, R. rubrum* and *Fragaria* spp. A number of other hosts have been infected experimentally. The symptoms on important herbaceous hosts developed upon inoculation are described below: (Lister, 1964; Schmelzer, 1969; Murant, 1974). Symptoms on some of the inoculated hosts are given below:

- *Chenopodium amaranticolor:* Chlorotic or necrotic local lesions on inoculated leaves. It is the most reliable local lesion host.
- *C. quinoa*: Chlorotic or necrotic local lesions followed by systemic chlorotic mottle or necrosis symptoms.
- *Nicotiana rustica:* Local chlorotic or necrotic spots or rings; systemic rings, spots and line-patterns with variable amounts of necrosis.
- *N. tabacum cv:* White Burley (tobacco). Chlorotic local lesions; scattered systemic chlorotic spots and rings.
- *Phaseolus vulgaris cv:* Prince (French bean). Dark brown necrotic local lesions, systemic chlorotic mottle, necrosis and distortion at higher temperature.
- *Petunia hybrida:* Local chlorotic lesions; systemic veinal chlorosis or line-pattern

THE PATHOGEN

A RNA-containing virus with isometric particles about 30 nm in diameter is associated with the disease (Harrison and Nixon, 1960). Raspberry ring spot virus is an ICTV approved species of the genus *Nepovirus* in the family *Comoviridae* (Murant 1970a). The virus particles are of the same size but sediment as three components, apparently empty protein shells (T) and two kinds of nucleoprotein with different amounts of RNA (M and B). The virus genome is bipartite.The nucleic acid of the virus is single-stranded RNA. The genome comprises two essential RNA molecules, RNA-1 and RNA-2, with M. Wt 2.4×10^6 and 1.4×10^6 respectively (Murant *et al.*, 1972). Virulence is determined by both RNA species (Harrison *et al.*, 1972, 1974; Harrison and Hanada, 1976; Hanada and Harrison, 1977). The coat protein has a M. Wt of 54,000 daltons (Mayo *et al.*, 1971). The complete genome is 11338 nucleotides long. The RNA-1 and RNA-2 have been fully sequenced and complete sequences are 7496 nucleotides and 3842 nucleotides long respectively. Raspberry ringspot virus resembles in many properties to other nepoviruses (arabis mosaic, grapevine fanleaf, strawberry latent ringspot, tobacco ringspot, tomato black ring and tomato ringspot), but is unrelated to them serologically (Murant *et al.*, 1968). Serologically two strains have been described: Scottish strain (RRSV-S) and English strain (RRSV-E).

DIAGNOSIS AND MANAGEMENT

Prunus avium 'Bing', *P. persica* 'GF305', *Chenopodium amaranticolor, C. quinoa, Spinacia oleracea* are good indicator hosts and can be used for disease diagnosis. The virus can also be detected by serological tests like ELISA. The disease is managed by certification schemes, use of immune cultivars (Murant, 1970, 1974) or by treatment of soil with fumigant nematicides (Murant and Taylor, 1965; Taylor and Murant, 1968; Trudgill and Alphey, 1976). Crop rotation can help to minimize the disease incidence (Taylor, 1967; Thomas, 1969).

RUBUS YELLOW-NET

This disease was originally isolated from Himalayan blackberry and was found to be a part of raspberry mosaic disease complex (Stace-Smith, 1956). Rubus yellow net virus (RYNV) infects *Rubus* species and cultivars worldwide and is an essential component of raspberry vein-banding mosaic, a virus disease complex that causes serious decline in plant vigour and productivity. The virus spreads in Eurasia and North America.

SYMPTOMS

The virus has restricted natural hostrange in genus Rubus. *Rubus strigosus* (red raspberry), *R. procerus* (Himalayan blackberry) and *R. occidentalis* (black raspberry) are susceptible to virus infection and symptoms in these hosts appear as chlorotic netting of leaf veins (Stace-smith, 1955a). The virus is found in mixed infections with other aphid-borne viruses such as black raspberry necrosis, raspberry leaf spot and others.

TRANSMISSION

Rubus yellow net virus is transmitted by grafting and by an aphid vector, *Amphorophora agathonica* in America. In Europe, the vector is another species, *A. idaei*. The mode of transmission is of semi-persistent type (Stace-Smith, 1955).

HOST RANGE

Experimentally rubus yellow net virus (RYNV) infected 12 of 34 plant species upon graft inoculation.The susceptible plant species are : *Fragaria vesca, Rubus idaeus, Rubus macraei, Rubus occidentalis, Rubus procerus, Rubus strigosus, Rubus albescens* and *R. henryi* (Stace-Smith and Mellor, 1957; Stace-Smith and Jones, 1978; Jones and Badenoch, 1982). *Rubus idaeus, R. occidentalis, R. procerus* and *R. macraei* develop vein netting symptoms but symptoms in *Fragaria vesca* appear as down curling of leaves, necrosis of petiole, wilting and death of infected plants. *R. macraei* was recommended as an improved indicator plant (Jones, 1991).

THE PATHOGEN

Rubus yellow net virus particles are bacilliform measuring 80-150 nm long and 25-31 nm wide in ultrathin sections (Jones and Roberts, 1976). Virus is a tentative member of the genus *Badnavirus* in the family *Caulimoviridae*. The genome of the virus is double-stranded DNA. The virus codes for an RNA-dependent DNA polymerase. Sequence analysis suggests that RYNV is possibly a distinct species in the genus *Badnavirus* and is most closely related to Gooseberry vein banding associated virus (GVBAV) and Spiraea yellow leaf spot virus, the other two badnaviruses (Jones *et al.*, 2006).

DIAGNOSIS AND MANAGEMENT

The virus can be detected by inoculating indicator host species. The virus can also be detected in electron microscopy and in PCR (Jones *et al.*, 2006). The virus can specifically detected in blackberry by TaqMan® quantitative reverse transcription PCR (qRT-PCR) (Ho and Tzanetakis, 2012). Most effective method to control the disease is the use of vector-resistant cultivars. Heat treatment followed by tissue culture was also recommended for control of RYNV (Stace-Smith and Jones, 1978).

BLACK RASPBERRY NECROSIS

Black raspberry necrosis virus (BRNV) was first reported in *Rubus idaeus* from Canada (Stace-Smith, 1955a). Now this virus is known to occur in several countries of Europe, North America, Australia and New Zealand. In Great Britain, upto 100 % infection was recorded in cv. Malling Jewel in the first fruiting year (Jones, 1979).

SYMPTOMS

The virus remains symptomless in several *Rubus* species but in some hosts like *R. occidentalis* and *R. henryi,* the symptoms of stem necrosis appear and persist. The causal virus induces necrosis and mosaic in black raspberry. Some cultivars may show veinal chlorotic mottle or line pattern during early growth. However in some *Rubus* species symptoms disappear soon after infection (Stace-Smith, 1955a; Jones and Murant, 1972; Jones and Wood, 1979; Converse and Bartlett, 1979; Stace-Smith and Jones, 1987).

TRANSMISSION

The virus is transmitted by grafting from *Rubus* to *Rubus* and by mechanical inoculations from *Rubus* spp. to herbaceous hosts. The vectors of the virus are several aphid spp., *Amphorophora agathonica, A idaei, A rubicumberlandii, Illinoia rubicola.* Virus is transmitted in a semi-persistent manner and lost by the vector when it moults.

HOST RANGE

Natural infection of the disease is restricted to genus *Rubus* and has been observed in *Rubus idaeus, R. fruticosus, R. procerus, R. ursinus, R. lasiocarpus, R. phoenicolasius, R. loganobaccus.* The plant species in the family Amaranthaceae, Chenopodiaceae, Rosaceae and Solanaceae are susceptible (Jones and Murant, 1972; Murant *et al.,* 1976). The following species were susceptible to experimental virus inoculation: *Chenopodium amaranticolor, Chenopodium murale, Chenopodium quinoa, Gomphrena globosa, Nicotiana debneyi, Petunia x hybrida, Rubus fruticosus, Rubus henryi, Rubus idaeus, Rubus lasiocarpus, Rubus loganobaccus, Rubus occidentalis, Rubus phoenicolasius, Rubus procerus, Rubus ursinus, Spinacia oleracea.* In *Rubus, Rubus occidentalis, R. henryi* are good systemic assay hosts showing apical tip necrosis and leaf mottle. For diagnosis, local lesion hosts, *Chenopodium amaranticolor, C. quinoa, Gomphrena globosa, Nicotiana debneyi, Petunia × hybrida, Spinacia oleracea* can be used.

THE PATHOGEN

The virus was initially reported as '52V virus' from Europe (Jones and Murant, 1972) but now equated with black raspberry necrosis virus (Jones and Roberts, 1977). The virions are isometric measuring 25-30 nm in diameter. Virus preparation contains few virions (Murant *et al.,* 1976). The genome is monopartite. The genome organization was identical to that of *Strawberry mottle virus,* a member of the genus *Sadwavirus.* Phylogenetic analysis indicates that BRNV belongs to the genus *Sadwavirus* (Halgren *et al.,* 2007). The viral genome encodes structural proteins.

DIAGNOSIS AND MANAGEMENT

The virus can be detected by grafting to *Rubus* indicator hosts or by mechanical inoculation to herbaceous local lesion hosts. The virus can also be detected by RT-PCR (reverse-transcriptase polymerase chain reaction) (Halgren *et al.*, 2007). The best management of the virus is to develop and grow vector resistant cultivars (Jones and Jennings, 1986). The virus can be eradicated from *Rubus* spp. by heat treatment at 34-37°C for 1–4 weeks (Converse, 1963).

RASPBERRY LEAF MOTTLE AND LEAF SPOT

Leaf spot mosaic disease was first described in Great Britain (Harris, 1933). The symptoms of the disease were put in two categories: mosaic-1 and mosaic-2 (Harris, 1933). Now this disease is known to occur in several countries in Europe, Australia, New Zealand and North America (Cadman, 1951; Converse, 1981; Jones and Wood, 1979). Raspberry leaf mottle virus (RLMV) and raspberry leafspot virus (RLSV) are normally found together but can also occur singly (Jones and Jennings, 1980).

SYMPTOMS

Most of the infected red raspberry cultivars remain symptomless but there is degeneration of vigor in these cultivars (Jones, 1981). Most of the *Rubus* species and cultivars are susceptible to RLMV and RLSV infection either as latent or with conspicuous symptoms. Natural infection of these viruses has been observed in both cultivated and wild European raspberries. Infected plants of red raspberry (*Rubus idaeus*) show angular chlorotic yellow spots of about 1–2 mm are scattered over the leaves of infected primocanes. Some spots are merged together covering larger surface of a leaf. Leaves and fruiting canes become small and distorted. Infected plants become stunted and often die within 2–3 years following infection (Jones, 1987). In blackberry plants, symptoms develop as faint chlorotic mottle but at cooler temperature (Jones and Jennings, 1980). Other *Rubus* species, *R. henryi, R. occidantalis, R. molaccanus, R. saxatalis* are susceptible to infection and develop symptoms such as apical necrosis, mosaic and tip necrosis after 4-8 weeks of graft inoculation. *R. phoenicolasius* did not develop any symptom (Cadman, 1951).

TRANSMISSION

Both Raspberry leaf mottle virus (RLMV) and raspberry leafspot virus (RLSV) are graft and aphid transmitted. The vector is an aphid, *Amphorophora idaei* (Cadman, 1951; Jones and Jennings, 1980). The virus transmission by the vector is probably semi-persistent type (Converse *et al.*, 1970).

THE PATHOGEN

No information is available on particle morphology of RLMV and RLSV. However, a few isometric virus-like particles measuring 30 nm were obtained in partially purified preparations from RLMV infected raspberry plants (Jones, 1976a). Electron microscopy of ultra thin sections of leaves of symptomless 'Himalaya Giant' blackberry and of the virus indicator species, *Rubus macraei*, showing severe leaf curl symptoms following graft inoculation with scions from this blackberry material, detected highly flexuous virus-like particles with an unusual 'beaded' structure (Jones *et al.*, 2004).

DIAGNOSIS AND MANAGEMENT

The RLMV and RLSV can be detected in susceptible hosts by graft or vector transmission (Jones and Jennings, 1980). The commonly used virus indicators, *R. henryi* and *R. occidentalis* develop severe tip necrosis followed by mosaic when infected. The mother plants free from viruses can be obtained by heat treatment at 32–37^{0}C for 10–20 days (Chambers, 1961). Heat treatment couple with meristem tip culture is the best procedure to eliminate these viruses (Baumann, 1982). Control of vector and breeding of virus or vector resistant varieties are the approaches recommended for management.

RASPBERRY MOSAIC

Raspberry mosaic disease (RMD) is worldwide in distribution especially in red, black or purple raspberries. The mosaic is a complex disease involving symptoms caused by several viruses such as black raspberry mosaic virus, Rubus yellow net virus, raspberry leaf mottle virus, raspberry leaf spot virus and a few more. The mosaic disease to be briefly described here is a mixed infection of Raspberry leaf mottle virus (RLMV) and raspberry leafspot virus (RLSV). These two viruses have been described in earlier chapter. The yield of red raspberries due to mosaic disease is reduced from 11% to 14% in British Columbia (Freeman and Stace-Smith, 1970), 28% in Scotland (Wood and Anderson, 1959).

SYMPTOMS

Symptoms of mosaic disease in black raspberry appear during cool weather as mottle and blistering of leaves. At higher temperature, plants remain symptomless. Affected plants develop shoot tip necrosis, canes become short appearing as rosette, fruits are small and seedy. In red respberry, symptoms similar to black raspberry appear but affected plants show decreased vigor and pollen production. In blackberry, plants show more tolerance than red raspberry and may show less vigor and fruit yield and quality may reduce. On highly susceptible varieties, leaves become puckered with large, dark green blisters surrounded by yellowish or yellowish-green tissue.

TRANSMISSION

Mosaic disease of raspberry is graft and aphid transmitted. The vectors belong to the genus *Amphorophora* (*A. idaei* spp. idaei in Europe and *A. agathonica* in North America). The transmission by vectors is of semi-persistent type (Murant, 1974; Cadman, 1951, 1954). The other aphid vectors of one or more viruses associated with RMD are: *A. sensoriata, A. rubicumberlandi, Aulacorthum solani, Macrosiphum euphorbiae, M. fragriae, Illinoia rubicola and Myzus ornatus* (Converse *et al.*, 1987).

HOSTRANGE

The natural hostrange of the virus is restricted to *Rubus* spp (Converse, 1981). The disease is experimentally transmitted to *Fragaria vesca* in addition to *Rubus* spp. Black raspberry necrosis virus, a component of RMD is transmitted mechanically to *Chenopodium quinoa* in which the virus induces chlorotic local lesions, mottle or necrosis.

THE PATHOGEN

Since the mosaic disease is associated by several viruses, it is difficult to assign a virus to the mosaic disease. The virus particles associated with individual viruses have been described earlier (Raspberry yellow-net virus-bacilliform 80-150 × 25-31 nm, Black raspberry necrosis virus-isometric 25-30 nm, Raspberry leaf mottle virus-possibly isometric, 30 nm, Rraspberry leafspot virus -no particle seen).

DIAGNOSIS AND MANAGEMENT

Diagnosis of mosaic disease is possible by grafting or aphid transmission to *Rubus* spp. The inoculated plants must be maintained at low temperature (18-25^{0}C) to get symptoms. Following methods are suggested to manage the disease: Use propagation material free from viruses of RMD complex through certification programme (Winter, 1929). Eliminate viruses from contaminated stock by heat treatment (Converse *et al.*, 1987) or tissue culture (Putz, 1971). Heat therapy and tissue culture can be combined for better results.Control aphid population by using insecticide or by breeding vector resistant varieties (Daubeny, 1982). Red raspberry cvs like Glen Clova, Norfolk Giant and Willamette support low population of aphids (Cadman and Fisken, 1958). RMD is now decreasing in several countries where proper management practices are employed (Jones, 1981).

RASPBERRY BUSHY DWARF VIRUS

Raspberry bushy dwarf virus (RBDV) is distributed worldwide where susceptible *Rubus* cultivars are grown. In susceptible raspberry cultivars, symptoms develop as yellowing of interveinal areas of leaves followed by rings or line patterns. In some cultivars fruits become crumbly. The causal virus is pollen and seed borne in *Rubus*. It is also transmitted by grafting and by mechanical inoculation to herbaceous hosts such as *Chenopodium amaranticolor, C. quinoa, C. murale, Phaseolus vulgaris* and *Nicotiana clevelendii* (Barnett and Murant, 1970; Barbara *et al.*, 1984). All these hosts develop local lesions upon inoculation except *N. clevelendii* where symptomless systemic infection develop. The virus particles are quasi-isometric about 33 nm in diameter with single stranded positive sense RNA as its genome. The virus can be detected by sap inoculation to *Chenopodium quinoa* which develops local chlorotic or necrotic lesions followed by systemic infection. ELISA is the best procedure for detection of RBDV. The virus can be eradicated by thermotherapy at 36°C followed by meristem culture.

RASPBERRY LEAF CURL

Raspberry leaf curl disease is wide spread in *Rubus* in USA and Canada (Stace-Smith and Converse, 1987) and perhaps the first virus disease reported in *Rubus* from North America (Rankin and Hockey, 1922). The yield of infected raspberries can be reduced up to 70 percent. Two forms of the leaf curl agent as alpha and beta are found (Bennett, 1927). The alpha form infects red and purple raspberry whereas the beta form infects black raspberry in addition to red and purple. The fruit quality is reduced due to virus infection (Bolton, 1970).

SYMPTOMS

Infected raspberry plants remain stunted and contain excessive branching. Leaves on infected plants are uniformly small, dark green, crinkled, and tightly curled downward and inward. When diseased shoots first appear, they are pale yellowish-green, but they soon turn dark green, become stiff and brittle. Black raspberry plants may degenerate and die after two or three years of infection. The characteristic symptoms are leaf curling, shoot proliferation, stunting, with small and crumbly fruits. In red raspberry (*R. idaeus* and *R. idaeus* var. *strigosus*), leaves of affected plants are curled and slightly yellow. Short fruit laterals, proliferation of shoot and stunting and large number of new canes branched at the leaf axils, small and crumbly fruits (Stace-Smith, 1962). In purple raspberry (*R. neglectus*), symptoms are comparatively milder and recover spontaneously.

TRANSMISSION

The raspberry leaf curl virus is transmitted by grafting and by the small raspberry aphid, *Aphis rubicola* and *A. idaei* in a persistent manner (Stace-Smith and Converse, 1987, Stace-Smith, 1962). *Aphis rubicola* transmits both alpha and beta strains.

HOST RANGE

Rubus spp. such as *R. allegheniensis, R. idaeus, R. idaeus* var. *strigosus, R. neglectus, R. occidentalis* and *R. phoenicolasius* are the natural hosts but other *Rubus* spp. may also be infected (Stace-Smith and Converse, 1987). The symptoms in these hosts develop as downward leaf curling, leaf yellowing, proliferation of shoots and stunting. Diagnostic host species are : *Rubus idaeus, R. albescens, R. henryi, Fragaria vesca* which upon inoculation will show downward curling of leaves and leaf yellowing. Several species of *Rubus* and *Fragaria vesca* (Alpine strawberry) have been found to be susceptible under experimental conditions. Symptoms on wineberry (*R. phoenicolasius* and *R. henryi*) developed almost similar to as has been described in red raspberry. The leaf curl agent seems to be confined to phloem tissue (Bennett, 1927).

THE PATHOGEN

Raspberry leaf curl virus possibly belongs to Luteoviruses (Matthews, 1982). Virus capsid is not enveloped, round with icosahedral symmetry. The genome is not segmented and contains a single molecule of linear positive-sense, single-stranded RNA. However, Electron microscopy of ultrathin sections of leaves of symptomless 'Himalaya Giant' blackberry revealed highly flexuous virus-like particles with an unusual 'beaded' structure. This virus, provisionally named *Hawaiian rubus leaf curl virus* (HRLCV), symptomlessly infected a wide range of Rubus species and cultivars (Jones *et al.*, 2004). However, the virus associated with leaf curl disease is yet to be characterized.

DIAGNOSIS AND MANAGEMENT

The virus can be detected by grafting on *Rubus idaeus* cultivars. Aphid transmission to the sensitive indicator species *Rubus phoenicolasius* induces pronounced leaf curl symptoms within 10 days. Natural spread of the virus takes place by aphid vectors and the virus can also be spread through the distribution of contaminated planting material. Therefore, management of insect vector by application of insecticides, roguing infected plants and use of virus-free certified propagation material or use of resistant varieties are important to contain the leaf curl disease (Stace-Smith, 1962).

RASPBERRY STREAK

Raspberry streak disease was reported from North America (Converse, 1972; Brunt and Stace-Smith, 1976). The causal virus attacks only black raspberries. This virus was considered to be a distinct virus (Lister and Converse, 1972) or a strain of tobacco streak ilarvirus (TSV) (Jones and Mayo, 1975; Brunt and Stace-Smith, 1976). Many other strains of TSV are known such as strawberry necrotic shock strain, which also infects *Rubus* (Frazier, 1966) and is closely related to it serologically (Stenger *et al.*, 1987). A virus serologically related to TSV in raspberry is also present in Italy and Australia (Poggi-Pollini and Giunchedi, 1987; Jones and Wood, 1979; Guy *et al.*, 1982).

During warm weather, symptoms of streak disease appear as several blue to purplish or grey dots and narrow water-soaked lines or streak on or under surface on the lower parts of young canes. Terminal leaves on infected new canes are often hooked or recurved, twisted or rolled and darker green than normal. Leaves show yellowing along the veins or mottling on the lower side. Fruits on infected plants are reduced in size, dull, seedy, crumbly, and lacking in flavor. Normally tobacco streak virus symptoms are somewhat deformed leaves blotched with yellow areas and poor fruit production. Though TSV infects a wide range of crop and weed species probably worldwide, it does not infect *Rubus* in Europe. TSV occurs in many variants that differ serologically and in natural and experimental host range (Fulton, 1985). Tobacco streak virus is common but symptomless in field-run 'Munger' and 'Boysenberry'.

The virus is transmitted by grafting. No vector is known for raspberry streak virus but the spread in field to healthy plants adjoining to infected plants has been observed (Converse, 1979). The virus is pollen-borne (Converse and Lister, 1969). TSV is seed-borne in several natural and experimental hosts, including *Rubus* (Converse and Lister, 1969; Brunt and Stace-Smith, 1976). TSV is readily transmitted by mechanical inoculation to several herbaceous species (Fulton, 1985). TSV is transmitted by thrips vector; *Frankliniella occidentalis* and *Thrips tabaci*; (Sdoodee and Teakle, 1987). TSV has quasi-isometric particles with properties typical of ilarviruses (Fulton, 1985). The virus classification and its genome have already been discussed in earlier chapters.

Detection is possible by mechanical inoculation from *Rubus* sap to herbaceous test plants. (Converse, 1978). Sap must be extracted in pH 8-9 buffer that contains an antioxidant because TSV particles are very labile (Fulton, 1985). Detection can be done by serological tests like ELISA and other (Jones and Mayo, 1975; Fulton, 1985). The best way to control the disease is by growing plants immune or resistant to infection. Healthy planting material should be used; EPPO developed (OEPP/EPPO, 1994) a virus-free certification scheme for *Rubus*.

RASPBERRY YELLOW SPOT VIRUS

Raspberry yellow spot virus (RYSV) has been reported from Poland and Scotland on red raspberry. Symptoms of yellow spots appear on leaves of infected plant. In severe cases whole leaf become yellow. Affected leaves are comparatively smaller and malformed. The disease is transmitted by grafting and aphid vector, *Amphorophora idaei* in a semipersistent manner. No information is available on aetiological agent. The virus can be eradicated by thermotherapy (Basak, 1974).

RASPBERRY LEAF BLOTCH VIRUS

Raspberry leaf blotch virus has been reported from, United Kingdom, Serbia and Finland. Affected raspberry plants showed conspicuous yellow blotches and abnormal development of leaf hairs in the corresponding areas of the abaxial side of leaves (Bi *et al.*, 2012). The virus was shown to be transmitted by a mite, *Phyllocoptes gracilis* always infesting raspberry plants (MacFarlane *et al.*, 2012). A new, segmented, negative-strand RNA virus with morphological similarities and sequence similarities to the genus *Emaravirus* was associated with the disease. This virus was tentatively named as raspberry leaf blotch virus (McGavin *et al.*, 2012). The virus has five RNAs, each potentially encodes a single protein on the complementary strand. Other members of this genus are European mountain ash ringspot associated virus, fig mosaic virus and rose rosette virus, all are transmitted by mite.

MOSAIC DISEASE OF HIMALAYAN RASPBERRY

Himalayan raspberry (*Rubus ellipticus* Smith) is a widely distributed weed in Shimla hills, Himachal Pradesh, India. A very high incidence of mosaic disease was recorded on this host (Azad and Sehgal, 1958; Dhingra and Niazi, 1972). Symptoms of the disease are yellow or light green mosaic mottle and oak leaf pattern on leaves of affected plants. Symptoms of line pattern and complete chlorosis and smalling of leaves were also observed in some plants (Fig. 50). The incidence of the disease is above 50% in most of the localities (Azad and Sehgal, 1958). The disease is transmitted by inarch grafting as well as bottle-graft technique to *R. ellipticus, R. macilentus*. The transmission of the disease to healthy *R. ellipticus* was also obtained by dodder (*Cuscuta reflexa* Roxb.) (Azad and Sehgal, 1958).

Fig. 50: Mosaic disease of Himalayan Raspberry

RINGSPOT DISEASE OF HIMALAYAN RASPBERRY

A ringspot disease in Himalayan raspberry is widely prevalent in Shimla hills of Himachal Pradesh in India. In naturally infected plants, the symptoms were observed as irregular chlorotic rings on leaves (Fig. 51). The plants carrying ringspot symptoms also show symptoms of mosaic, yellow net vein, oak leaf and vein clearing. Affected plants were stunted (Dhingra and Niazi, 1972). The disease is transmitted to healthy *Rubus* plants by inarch grafting. Inoculated plants show symptoms of ringspot within 4-6 months after grafting. The ringspot disease appears to be soil transmitted probably by nematodes (Dhingra and Niazi, 1972). However, it needs confirmation.

Fig. 51: Ringspot disease of Himalayan Raspberry

CUCUMBER MOSAIC VIRUS INFECTING RASPBERRY

Cucumber mosaic virus (CMV) was reported on *Rubus idaeus* cv Lloid George in Scotland (Harrisson, 1958).

SYMPTOMS AND HOST RANGE

Cucumber mosaic virus has a very wide hostrange worldwide (Francki *et al.*, 1979). CMV may be lethal to some *Rubus* species like *R. phoenicolasius.* In other *Rubus* species, CMV produces mild symptoms or remains symptomless (Jones, 1976b). In susceptible species, symptoms appear as severe chlorotic blotching, mottling on leaves, ringspots, smalling of leaves and line pattern. Experimentally infected plants of *R. occidentalis* showed mild mosaic of leaves (Harrison, 1958). From infected *Rubus* spp.,

CMV infects following hosts experimentally: *Chenopodium amaranticolor, C. quinoa, C. murale, Phaseolus vulgaris* are local lesion hosts and *Cucumis sativus, Nicotiana clevelandii, N. glutinosa, N. tabacum* cv. White Burley and Xanthi-nc are systemic hosts showing mosaic symptoms.

TRANSMISSION

Cucumber mosaic virus is transmitted by about 60 aphid spp. (Kennedy *et al.*, 1962). However, *Rubus* isolates are transmitted by *Amphorophora idaei, Macrosiphum euphorbiae* and *Myzus persicae* in a non-persistant manner (Jones, 1987). The virus is readily transmitted by mechanical inoculations to above mentioned herbaceous hosts. Graft transmission is possible in *R. phoenicolasius* but not in others. It is seed transmitted in many hosts but not in *Rubus.*

THE PATHOGEN

Cucumber mosaic virus (CMV) was isolated from American wineberry (*Rubus phoenicolasius*) plants with leaves showing yellow blotching and line pattern symptoms (Jones, 2007). Virions are round about 30 nm in diameter with icosahedral symmetry. The capsid consists of 32 capsomers. Virus preparations contain more than one particle component. Virus is the type of the genus *Cucumovirus* in the family *Bromoviridae*. The genome is tripartite with linear positive-sense, single-stranded RNA. Minor species of non-genomic nucleic acid are also found in virions. The Mr of the genome constitutes 18% of the virion by weight. Proteins constitute about 82% of the particle weight. The encapsidated nucleic acid is mainly of genomic origin, but virions may also contain subgenomic RNA and satellite RNA (in some isolates, virions may contain mRNA derived from genomic RNA-3). The complete genome is 8621 nucleotides long. RNA-1 and RNA-2 are fully sequenced and complete sequence is 3389 and 3035 nucleotides long respectively. RNA-4 is subgenomic and has been sequenced. The multipartite genome is divided among more than one type of particle and the segments are distributed between 3 different types of particles. The largest particles contain each one molecule of RNA-1 (sedimenting component B). The medium sized particles contain each one molecule of RNA-2 (sedimenting component M). The smallest particles contain one molecule each of RNA-3 and RNA-4 (sedimenting component T). The viral genome encodes structural proteins and non-structural proteins. Virions consist of 1 structural protein(s) located in the capsid. Capsid protein is involved in viral encapsidation and involved in protein movement (ICTVdB Management, 2006).

DIAGNOSIS AND MANAGEMENT

CMV in *Rubus* spp. can be detected by mechanical inoculation to herbaceous hosts. The virus can also be detected by serological tests like agar gel double diffusion test (Jones, 1976b). It is possible to detect the virus by other serological tests like ELISA and others. PCR can also be used for virus detection using CMV specific primers. Growing virus free plants are useful to contain the disease.

RASPBERRY VEIN CHLOROSIS

Raspberry vein chlorosis disease was first described from Scotland, U.K. (Cadman, 1952). This disease is also known as vein chlorosis mosaic and raspberry chlorotic net. Now the disease is probably distributed worldwide wherever *Rubus* is grown and has been reported from Canada, Europe, and New Zealand (Cadman and Stace-Smith, 1970).

SYMPTOMS

In raspberry cultivars, main symptoms of the disease are chlorosis of minor leaf veins resulting yellow net vein pattern on the whole leaf or a part of the leaf. In some cultivars leaf deformity and epinasty are common symptoms (Jordovic, 1963). In some cultivar, the virus remains symptomless. *Fragaria vesca* and *Rubus idaeus* can be infected experimentally. Most of the red raspberry varieties are susceptible either naturally or the virus was transmitted experimentally to them by grafting (Jones *et al.*, 1987). *Rubus idaeus* cvs. Lloyd, Malling delight, Norfolk Giant, Washington and *Fragaria vesca* are good indicator hosts (Jones *et al.*, 1977).

TRANSMISSION

The vein chlorosis virus is transmitted by grafting and also by aphid vector, *Aphis idaei* in a persistent manner (Jordovic, 1963). In aphids, the virus particles were observed in muscle, brain and salivary gland tissue. The virus multiplies in the vector.

HOSTRANGE

In addition to *R. idaeus* the virus was transmitted to loganberry (*R.*

loganobaccus) and *Fragaria vesca* (Stace-Smith, 1961). *R. idaeus* cvs. Lloyd George, Malling Delight, Norfolk Giant and Washington are susceptible. In these hosts, chlorosis of the minor veins develops in small patches in young expanding leaves upon inoculation. In cv. Malling Delight, epinasty and stunting of young expanding leaves may also develop. Mild, moderate and severe strains of vein chlorosis in red raspberry have been reported (Cadman, 1952).

THE PATHOGEN

Two types of bacilliform virus particles are associated with the disease (a) particles rounded at both ends and measured as *c.* 442-560 x 73-91 nm (Stace-Smith and Lo, 1973) and as *c.* 430-500 x 65-80 nm (Jones *et al.*, 1974) and (b) shorter particles rounded at one end. The nucleocapsid shows some cross-banding with a periodicity of *c.* 4-5 nm (Stace-Smith and Lo, 1973; Jones *et al.*, 1974). The particles were restricted to the cytoplasm and occurred singly or in bundles. Large bacilliform particles occurred in the cytoplasm and perinuclear space of a small proportion of xylem parenchyma cells (Jones *et al.*, 1974). The virus is a tentative member of the genus *Nucleorhabdovirus* in the family *Rhabdoviridae*; order *Mononegavirales.* The genome of the virus is a single stranded RNA and the viral genome encodes structural proteins and non-structural proteins.

DIAGNOSIS AND MANAGEMENT

The diagnosis of the disease is done by transmission to virus-free raspberry clones through aphid vector. The virus can also be detected by thin sections electron microscopy of infected plant tissue and also the viruliferous aphids (Murant and Roberts, 1981). The disease can be managed effectively by the use of certified planting material and control of aphid vector. The plants of raspberry cv Rubin were made free from the heat-stable raspberry vein chlorosis virus (RVCV) by meristem culture (Sobczykiewicz, 1986; Baumann, 1982). Immune cultivars like *R. idaeus* (cvs. Cuthbert, Latham and Viking), *R. occidentalis* and *R. henryi* must be used (Cadman, 1952; Stace-Smith, 1961).

RUBUS STUNT

Rubus stunt is a serious disease causing severe losses in several countries in Europe and United States of America. Heavy damage to the crop is done especially in places where the disease becomes epidemic (Van der Meer, 1987). The disease is also known as witches' broom disease of Rubus.

SYMPTOMS

The major symptoms of the stunt disease are: virescence, phyllody and other severe flower and fruit abnormalties. Stunted growth, short internodes, few spindly canes, proliferation of shoots from the base and leaf reddening in early autumn were also common on infected plants. Affected plants show excessive lateral branches together with phyllody and flower malformation followed by death of affected plants. The causal pathogen may infect strawberry experimentally causing symptoms like witches' broom, phyllody of flowers, growth reduction and death of some affected plants. The pathogen infects cultivars of raspberry and blackberry and their hybrids.

TRANSMISSION

The disease is transmitted by grafting and also by leafhopper vectors, *Macropsis fuscula* (Van der Meer, 1987). Leafhopper, *Allygus mayri* also transmitted Rubus stunt agent to celery (Jenser *et al.*, 1981) and *Euscelis plebeja* transmitted to *Chrysanthemum carinatum* (Lehman, 1973).

THE PATHOGEN

Phytoplasmas were found by electron microscopy in the sieve tubes of raspberry plants showing symptoms of *Rubus* stunt disease (Murant and Roberts, 1971). *Rubus* stunt phytoplasma, a member of the elm yellows phytoplasma group, is widespread in *Rubus* in continental Europe (Vindimian *et al.*, 2004). The *Rubus* stunt phytoplasma 16S rRNA gene was sufficiently differentiated to represent a novel putative taxon: '*Candidatus* Phytoplasma rubi' (Sylvie Malembic-Maher *et al.*, 2011).

DIAGNOSIS AND MANAGEMENT

The Rubus stunt phytoplasma can be detected by the symptoms in the field and by graft transmission to raspberry cvs. 'Radboud' and 'Malling Landmark' or 'thornless evergreen' blackberry in Laboratory. Phytoplasma can also be detected by thin section electron microscopy. PCR detection of Rubus stunt phytoplasma is possible by selection of primers from 16S rRNA gene (Davies, 1998). The disease can be managed by controlling insect vectors. Hot water treatment of dormant root cuttings or rooted shoot at 45^{0}C for 2-3 h can cure stunt phytoplasma

(Thung, 1952). Infected blackberry plants showed remission of the symptoms by aureomycin therapy (Murant and Roberts, 1971).

BOYSENBERRY DECLINE

Decline disease of Boysenberry (*Rubus ursinus*) has been reported from New Zealand (Wood, 1991). The flowers on infected plants become large with crinkled and large distorted sepals. The stamens become shorter but stigma and styles enlarged. Shoot growth proliferates and develop small down-curled leaves. Fruits remain small and hard. The causal pathogen may also infect loganberry and blackberry upon graft inoculation. The disease is transmitted by graft. A leafhopper, *Ribautiana tenerrima* has been suspected the vector of this disease. The disease is possibly caused by a phytoplasma. The disease can be identified by grafting on boysenberry. No management practices are known. A phytoplasma, *Candidatus* Phytoplasma australianse has been reported to be associated with boysenberry decline diseases (Liefting *et al.*, 2011).

THIMBLEBERRY RINGSPOT

Ringspot disease in Thimbleberry (*Rubus parviflorus*) was first observed in British Columbia, Canada (Stace-Smith 1958). The infected plants show irregular leaf ring spots and oak leaf pattern and net-like vein chlorosis on leaves. Canes are stunting in plants showing severe symptoms. The virus is transmitted by grafting and by aphid vector, *Amphorophora parviflorus* and *Macrosiphum (Oestlundia) davidsonii, M.(O.) maxima*. The causal virus was not transmitted by mechanical inoculation; by contact between plants; by seed and by pollen but was transmitted by grafting. The following plant species are susceptible to experimental virus inoculation: *Rubus henryi, Rubus idaeus, Rubus occidentalis, Rubus parviflorus*. These species show symptoms of ringspot and oak leaf pattern, faint chorotic markings or spots and diffuse mottle upon inoculation. *Rubus parviflorus* and *Rubus occidentalis* are used as indicator hosts.

In thin section electron microscopy of infected tissue, the particles were isometric, round and exhibits icosahedral symmetry measuring 25 nm, and confined to single rows enclosed in tubules (Stac-Smith, 1987). The capsomer arrangement was clearly visible. Thimbleberry ringspot virus is unclassified and has not been assigned to a genus and family. The viral genome encodes structural proteins.

The virus can be detected by visual examination of plants. Other methods of detection is transmission to indicator hosts or by thin section electron microscopy. The virus is confined to wild hosts in nature and information on its spread to other cultivated *Rubus* spp. is not available. Hence, no control measures were recommended.

BRAMBLE YELLOW MOSAIC

Bramble yellow mosaic virus has been reported from South Africa. The virus causes yellow mosaic and line pattern in *Rubus rigidus*. The virus is transmitted by grafting and by mechanical inoculation to some herbaceous hosts like *Chenopodium murale, C. quinoa, Gomphrena globosa,* and *Nicotiana tabacum* (ICTVdB, 2006). The causal virus is also seed born. Inoculated *C. murale* plants develop large local chlorotic or yellow lesions followed by necrotic and systemic chlorotic or yellow rings. The virus is of the genus P*otyvirus* in family *Potyviridae* with flexuous particles of about 15 × 730 nm are associated with the disease (Engelbrecht, 1987). The genome is monopartite. The genome is positive sense single stranded RNA. The virus can be detected by graft inoculation on *R. Henryi* or by sap inoculation to *C. murale*.

WINEBERRY LATENT

The latent virus on wineberry was originally reported from USA but also detected in Scotland on wineberry (*Rubus phoenicolasius*). The symptoms of the disease resembles with blackberry calico disease (Converse, 1987). The virus is transmitted by grafting and by mechanical inoculation to herbaceous hosts. Virus particles are flexuous filamentous about 12 x 620 nm. The virus showed several similarities with blackberry calico Carlavirus (Jones *et al.*, 1990). The virus produces necrotic local lesions in *Chenopodium quinoa* and *C. amaranticolor* and these hosts are used for virus diagnosis. Detection can also be done by electron microscopy. There is no information on therapy (Jones, 1987).

RED RASPBERRY LATENT VIRUS

In Red raspberry (*Rubus idaeus*) severe crumbly fruit symptoms were observed in North America. This disease was provisionally named as Raspberry latent virus. Studies revealed that two viruses, Rice ragged stunt (RRSV), a ten segmented dsRNA *Oryzavirus* that belongs to family *Reoviridae* alongwith a member of *Closteroviridae* associated with raspberry leaf mottle virus (RLMV) were isolated from affected plants (Quito *et al.*, 2010).

APPLE MOSAIC VIRUS INFECTING *RUBUS*

Apple mosaic virus disease is wide spread in Europe and North America in *Rubus* spp., *R. idaeus, R. occidentalis* and *R. ursinus* (Converse and Casper, 1975; Baumann *et al.*, 1982). Most infected *Rubus* plants in North America are symptomless but some *R. idaeus* plants in Germany showed conspicuous yellow spots and/or line patterns (Baumann *et al.*, 1982). Apple mosaic virus (ApMV) is transmitted by seed and may be through pollen. No vector of the virus has been established though thrips are known to be associated with pollen transmission. In addition to *Rubus*, apple mosaic virus (ApMV) is common in woody perennials like *Malus*, *Prunus*, *Rosa* and *Humulus*. An ilarvirus is associated with the disease. As most plants are symptomless, testing by mechanical inoculation to herbaceous indicator plants like *Chenopodium quinoa* is important for disease diagnosis (EPPO/CABI, 1996b). Upon inoculation *C. quinoa* plants develop chlorotic or necrotic lesions followed by systemic necrosis, rings and line pattern. ApMV can be detected by serological tests (Fulton, 1972; Casper, 1973). Healthy planting material should be used and the virus could be eliminated by meristem-tip culture (Theiler-Hedtrich and Baumann, 1989).

C. VIRUS AND VIRUS-LIKE DISEASES OF BLUEBERRY AND CRANBERRY

Blueberry is grown in North America, South America, Europe and Asia. Following species in the genus *Vaccinium* are known :*Vaccinium angustifolium, V. boreale, V. caesariense, V. corymbosum, V. darrowii, V. darrowii, V. elliottii, V. formosum, V. fuscatum, V. hirsutum, V. operium, V. myrtilloides, V. pallidum, V. simulatum, V. tenellum,* and *V. virgatum.* There are subgenera in V*accinium* like *Cyanococcus* (true blueberries), *Oxicoccus* (cranberry) and *Euvaccinium* (Bilberries and Whortleberries). Blueberries are used in a variety of consumer goods such as jellies, jams, pies, muffins, snack foods, and cereals. Several viruses and virus like diseases have been reported on *Vaccinium*. The important ones will be discussed in this chapter.

BLUEBERRY SHOESTRING

Shoestring disease of blueberry is restricted in its distribution in United States of America and was originally described on highbush blueberry, *Vaccinium corymbosum* (Hutchinson, 1950; Stretch and Hilborn, 1970). It is a highly damaging virus disease and annual losses are estimated several million dollars in Michigan state only (Ramsdell, 1987).

SYMPTOMS

The most diagnostic symptoms of the disease on highbush blueberry appear as elongated reddish streak of about ¼ to 1 inch on 1yr old stem. Leaves of some infected plants become strap-like from where the name of the disease has been derived. Some leaves of infected bushes become malformed and twisted. A few leaves may show red vein banding or streaks along the veins or oak-leaf pattern. Some petals of flowers will also show red streaks. The fruits on infected plants develop reddish purple colour instead of blue colour of normal berries. After a few years, berry production drops dramatically (Ramsdell, 1987, 1995).

TRANSMISSION

The virus is transmitted by chip budding and whip grafting and also by mechanical inoculation. In nature, the virus is spread by aphid vector, *Illinoia pepperi* (*Masonaphis pepperi*) and the virus is transmitted in a semi-persistent manner (Lensey *et al.*, 1978). This aphid spends its entire life on blueberry plants.

HOST RANGE

Shoe string disease has been observed in highbush cvs. Blueray, Burlington, Coville, earliblue, Jersey, June, Rancocas, Rubel and Weymouth. Under experimental conditions, *Vaccinium angustifolium* and *V. corymbosum* are susceptible. These hosts show symptoms of malformation, oak leaf pattern on leaves, shoestring leaves, red streaks on stems and pink blossoms (Ramsdell, 1970a).

THE PATHOGEN

The virions are isometric, not enveloped measuring 27 nm in diameter with round profile and without a conspicuous capsomere arrangement (Lensey *et al.*, 1978). The virus belongs to genus *Sobemovirus*. Genome of the virus is a linear and single stranded RNA. Virions contain 20 % nucleic acid and 80% protein. Genome is unipartite (Ramsdell, 1979b). Coat protein is single with M 30000. Genome was sequenced, but only an estimate is available, complete sequence is 4200 nucleotides long (Ramsdell, 1979b). The viral genome encodes structural proteins and non-structural proteins. The virus does not show serological relationships to 33 viruses with isometric virions (Ramsdell, 1979a).

DIAGNOSIS AND MANAGEMENT

The virus can be detected by grafting to susceptible host like *V.*

corymbosum which will show characteristic symptoms of the disease. The virus can be detected in electron microscopy tests and by serological methods like ELISA (Lesney and Ramsdell,1976). The disease can be managed by planting disease-free plants. Infected bushes must be removed and should replant with healthy stock. The blueberry aphid must be kept under control by using well-timed aphidicide sprays (Morimoto and Ramsdell, 1985; Morimoto *et al.*,1985). Blueberry cultivars Bluecrop and Atlantic have good field resistance and are planted where disease pressure is more.

BLUEBERRY LEAF MOTTLE

Blueberry leaf mottle disease was first observed in Michigan, USA in 1977 (Martelli *et al.*, 1977). It appears to be restricted in USA and Bulgaria. The affected plants become unproductive and unhealthy causing great loss to the growers.

SYMPTOMS

Affected plants of cv. Rubel show dieback, stunted growth and deformation in new growth. Leaves on affected plants are mottled sometime showing chlorotic roughly circular 'windows'. In case of severe infection, malformation of leaves including curling and strapping is common. Blueberry leaf mottle virus (BBLMV) has a narrow host range (Ramsdell and Stace-Smith, 1979).

TRANSMISSION

Blueberry leaf mottle virus is transmitted by grafting and pollen similar to other Nepoviruses such as tomato ringspot and cherry leaf roll viruses (Harrison and Murant, 1977). Pollen contains a high level of virus. Natural transmission of BBLMV is by honey bees (*Apis mellifera*) (Boylan-Pett *et al.*, 1991). The virus is easily sap transmitted by mechanical inoculations to herbaceous hosts.

HOST RANGE

Vaccinium corymbosum cv. Rubel and *V. corymbosum* cv. Jersey are natural hosts. Affected plants show poor growth, stunting, dieback and malformed mottled leaves. The following species were susceptible to experimental virus infection: *Chenopodium album, C. amaranticolor, C. capitatum, C. foetidum, C. murale, C. quinoa, Cucumis sativus, Cyamopsis tetragonoloba, Gomphrena globosa, Lycopersicon esculentum, Nicotiana benthamiana, N. clevelandii, N. debneyi, N. glutinosa, N. rustica, N. sylvestris, N. tabacum, Phaseolus vulgaris* and *Vaccinium corymbosum*. Experimentally infected hosts show symptoms of local lesions, mottle and tip necrosis. The following hosts are being used as assay or indicator hosts (Ramsdell and Stace-Smith 1979).

- *Chenopodium amaranticolor, C. quinoa and C. album* : Chlorotic local lesions, mottle and apical death of inoculated plants.
- *Cucumis sativus*: Chlorotic local lesions on inoculated cotyledons.
- *Nicotiana clevelandii*: Pinpoint local lesions on noninoculated leaves.

THE PATHOGEN

The virus associated with the disease is a putative member of N*epoviruses* in family *Comoviridae*. BBLMV has isometric particles measuring 28 nm (Ramsdell and Stace-Smith, 1981, 1983). The genome is segmented; bipartite, segements are distributed among 2 particle types of linear, positive-sense, single-stranded RNA. The complete genome is 14600 nucleotides long. The multipartite genome is divided among different particles, each segment encapsidated separately and the segments are distributed between 2 different types of particles. The largest particles contain each one molecule of RNA-1 (sedimenting component B). The medium sized particles contain each one molecule of RNA-2 (sedimenting component M). The smallest particles contain no nucleic acid and are found in the top (sedimenting component T) band after sedimentation (Ramsdell and Stace-Smith, 1981). The viral genome encodes structural proteins and non-structural proteins.

DIAGNOSIS AND MANAGEMENT

The virus can be detected by transmission on indicator hosts. Serological tests like agar gel double diffusion and ELISA can be used for detection of this virus. For management of the disease, regular inspection of field and rouging of infected plants are important (Childress and Ramsdell, 1987).

BLUEBERRY NECROTIC RINGSPOT

Blueberry ringspot disease is known to occur in USA (Converse and Ramsdell, 1982). The associated virus is related to tobacco ringspot virus. The disease is economically important as the infected bushes of susceptible varieties are killed outright.

SYMPTOMS

Leaves of Infected plants of *Vaccinium corymbosum* are deformed, somewhat thickened, become chlorotic and show necrotic spots followed by stem dieback and stunting. The tissue of necrotic spots on leaves fell down giving a shothole or tatterd effect (Ramsdell, 1987). Fruit yield and quality is reduced. In some susceptible cultivars like Concord and Stanley, affected plants show short internodes and small strap-like leaves.

TRANSMISSION

Blueberry necrotic ringspot virus (BBNRSV) is transmitted by bud or graft to susceptible cultivars like 'Pemberton'. It is also transmitted by nematode, *Xiphinema americanum* (Stace-Smith, 1985). The virus is transmitted mechanically to herbaceous hosts. Seed transmission has been reported in cucumber and soybean and probably in several other hosts.

HOSTRANGE

The causal agent of raspberry necrotic ringspot disease is tobacco ringspot virus. The virus has a very wide hostrange both in herbaceous and woody plants (Stace-Smith, 1970). The important susceptible hosts are: soybeans (*Glycine max*), *Nicotiana tabacum, N. rustica, N. clevelandii, Vaccinium spp., especially V. corymbosum, Petunia hybrida, Datura stramonium, Phaseolus vulgaris, Cucumis sativus, Chenopodium amaranticolor* and *C. quinoa*. Hosts like *Chenopodium quinoa, Cucumis sativus* cv. National Pickling and *N. tabacum* cv. Burley are used as diagnostic and indicator hosts. The necrotic or chlorotic lesions or ringspots develop on inoculated leaves of indicator hosts followed by systemic infection.

THE PATHOGEN

Virus particles are isometric measuring 28 nm in diameter. The virus is a member of nepovirus group (Harrison and Murant, 1977). It is multicomponent virus. In sucrose density gradient, virus particles sediment in three components, two components are nucleoproteins containing different amount of RNA (2.4 and 1.4 × 106 Da). One component is empty protein shells. Some isolates are associated with a satellite RNA (Stace-Smith, 1985). Other nepoviruses infecting raspberry are peach rosette mosaic and tomato ringspot virus. Both the viruses are transmitted by *X. americanum* and by mechanical inoculations to herbaceous hosts.

DIAGNOSIS AND MANAGEMENT

The virus can be detected by mechanical transmission to herbaceous indicator plants, *Chenopodium quinoa*, *Cucumis sativus*, and *Nicotiana tabacum* (Stace-Smith, 1985). The virus is also detected in ELISA (Converse and Ramsdell, 1982). To manage the virus, following recommendations are given: use of virus-tested certified stock, rogue out infected plants before planting blueberries, test soil in spring for dagger nematodes and do not plant until soil has been fumigated.

BLUEBERRY SHOCK VIRUS

Blueberry shock virus occurs only in United States. Naturally, the virus infects *Vaccinium corymbosum* causing symptoms of leaf and flower necrosis during flowering season. Young leaves on affected plants show wilting, black veins and black streaks on petiole. Plants remain infected for a few years and then become symptomless and produce normal fruits. Virus is transmitted by grafting and mechanical inoculations (Martin and Bristow, 1995). No vector has been established as yet. It is pollen borne in blueberry. The virus is mechanically transmitted to *Nicotiana benthamiana, N. clevelandii, N. sylvestris, N. tabacum* cvs Samsun and Havana-425 causing symptoms of necrotic local lesions and systemic necrosis. In mechanically inoculated *Vaccinium corymbosum* plants symptoms appear as necrosis of leaves and flowers. *Nicotiana clevelandii* can be used as assay host.

Virus particles are isometric to quasi-isometric measuring 27 nm in diameter. Virus preparations contain more than one particle component (MacDonald *et al.*, 1991). The virus is serologically related to apple mosaic and prunus necrotic ringspot viruses. The virions are found in cytoplasm.

Blueberry necrotic shock virus is an ICTV approved species of the genus *Ilarvirus* in family *Bromoviridae*. The genome is segmented; tripartite, segements are distributed among 3 particle types of different size; consists of four segments of linear positive-sense, single-stranded RNA. Minor species of non-genomic nucleic acid are also found in virions. The multipartite genome segments are distributed between 3 different types of particles. The largest particles contain each one molecule of RNA-1, the medium sized particles contain each one molecule of RNA-2, and the smallest particles contain one molecule each of RNA-3 and RNA-4. The complete genome is estimated to be 7717 nucleotides long. Capsid protein is involved in viral encapsidation and involved in protein movement (MacDonald *et al.,* 1991). The virus can be detected by mechanical inoculation to herbaceous hosts. To manage the virus, use virus-tested certified stock and rogue out infected plants.

BLUEBERRY MOSAIC

Blueberry mosaic disease was first reported by Varney, 1957 from New Jersey. The disease is found in USA and Canada in plantings of cultivated blueberry. The disease has been observed in blueberry cvs. Cabot, Concord, Earliblue, Pioneer, Rubel and Stanley. Blueberry mosaic virus causes light green, brilliant yellow, and sometimes pink and white patchy mottling on leaves. Symptoms may be irregularly distributed on an infected plant (Ramsdell and Stretch, 1987). Fruits on affected plants are reduced in quantity and quality and may ripe late. The associated virus is transmitted by grafts (Raniere, 1960), sap and dodder and infects *Vaccinium corymbosum* and *V. vacillans*. No insect vector is known. Recently putative causal agent has been fully characterized (Thekke-Veetil *et al.,* 2012). All symptomatic plants collected from Arkansas, Michigan, New Jersey, Kentucky and Oregon were infected with a new member of the negative-sense, single-stranded RNA genus *Ophiovirus.* Symptoms on blueberry cvs. are itself diagnostic and removal of infected plants is the only measure known to contain the disease.

BLUEBERRY SCORCH VIRUS

Blueberry scorch virus (BBScV) disease was reported from USA on highbush blueberry (*Vaccinium corymbosum*). In some cultivars symptoms appear as flower necrosis and marginal leaf chlorosis while some cultivars react with complete necrosis of flowers and leaves followed by death of infected bushes. The disease is transmitted by grafting and by aphid vector, *Fimbriaphis fimbriata*. A flexuous rod shaped *Carlavirus*, about 610-700 nm in length is associated with this disease (Martin and Bristow, 1988, 1995). The virus has a positive sense, single stranded RNA as its genome.

BLUEBERRY RED RINGSPOT

Blueberry red ringspot disease was first described from United States on high bush blueberry (*Vaccinium corymbosum* Dunal) (Hutchinson, 1950, Hutchinson and Varney, 1954). The disease is economically important in some parts of USA (Kim *et al.*, 1981) as the crop loss was estimated 25% in Michigan. The symptoms of the disease are characterized by red spots, rings, and oak-leaf patterns that appear on older leaves in late June or July. Similar spots can be found on epidermis of canes older than one year. The berries are pock-marked and unattractive. Some varieties may have light blotching on the fruit. 'Bluecrop' is resistant; 'Blueray', 'Bluetta', 'Burlington', 'Coville', 'Darrow', 'Earliblue,'and 'Rubel' are susceptible. Host range is restricted to *Vaccinium corymbosum* and *V. australe*. The vector of the disease is suspected to be a mealybug, (*Dismococcus* sp.) (Ramsdell *et al.*, 1987). Red ringspot is an isometric virus measuring 42-46 nm and is a member of genus *Soymovirus* in family *Caulimoviridae* (Kim *et al.*, 1981; Ramsdell, 1995). A similar virus has also been reported from cranberry (*V. macrocarpon*) causing ringspot symptoms in leaves and fruits (Stretch, 1987). Testing for this virus is done in ELISA and PCR. The disease is managed by using disease-free plants and rouging infected plants from the field.

CRANBERRY RINGSPOT

A ringsport disease was reported on cranberry (*Vaccinium macrocarpon*) in USA (Stretch, 1987). Fruits of infected plants were malformed with pale or white rings when ripe. Ringspot symptoms also developed on leaves. There is some premature terminal bud development in the fall and increased lateral bud development in the following spring. Keeping quality of affected fruits is greatly reduced. The disease was systemic. Experimental transmission has not yet been accomplished. Ringspot disease in cranberry is similar to blueberry red ringspot disease. Electron microscope studies of affected cranberry show inclusion bodies similar to those found in blueberry tissue infected with red ringspot virus. These two diseases may be caused by the same virus probably a Caulimovirus-a dsDNA virus (Boone, 1994).

PHYTOPLASMA DISEASES OF BLUEBERRY AND CRANBERRY

Stunt and witches' broom-two phytoplasma diseases have been recorded on blueberry (*Vaccinium myrtillus*) and false blossom has been reported on cranberry (*Vaccinium macrocarpon*) (Thung, 1952; Gorcio and Dale, 1982; Chen, 1971).

STUNT DISEASE OF BLUEBERRY

Stunt is an important disease of blueberry throughout the United States and Eastern Canada. Stunt infected blueberry plants are weak and stunted. Leaves may be spoon shaped with chlorotic margins. Leaves and internodes are short giving a bushy appearance to infected plants. Numerous thin branches develop on infected plants. Leaves develop yellow colour along the edges which turn red during fall and are mottled and cupped downward. Affected plants develop a few small, hard and tasteless berries which ripe late and remains hanging on the plant for longer time. Most of the cultivars of highbush blueberries are susceptible to stunt infection (Van der Meer, 1987).

Stunt agent naturally infects *Vaccinium vacillans, V. atrococcum, V. stamineum* and *V. mirtilloides*. Experimentally, the disease is transmitted by grafting from blueberry to *V. amoenum, V. altomontanum, V. elloitti* and *V. ashei* (Hutchinson *et al.*, 1960, Dale and Mainland, 1981). It spreads by cuttings taken from stunt-infected plants. The pathogen is also transmitted from blueberry to periwinkle (*Catharanthus roseus*) through dodder (*Cuscuta compestris* and *C. subinclusa*). Leaves of infected periwinkle show pale green mottling with green veins. Leaves and flowers are reduced in size. Stunt disease is also transmitted from diseased to healthy plants by the sharp-nosed leafhopper (*Scaphytopius magdalensis*) (Maramorosch, 1955) and other closely related leafhoppers, *S. magdalensis, S. acutus* and *S. frontalis* (Jenser *et al., 1981;* Tozzi *et al*, 1993).

Pleomorphic phytoplasma bodies measuring from 160-700 nm have been observed in sievetubes of phloem tissue of infected plants (Chen, 1971; Murant and Roberts, 1971; Hartman *et al.*, 1972). The pathogen can be detected by indexing on cultivar 'Cabot', by dodder transmission to periwinkle or by colour test like DAPI or in electron microscope. For management of the disease, grow disease-free plants or resistant varieties like Rancocas, remove diseased plants and spray insecticides to kill vectors.

WITCHES' BROOM DISEASE

Witches broom disease of blueberry has been reported on *Vaccinium myrtillus* in several countries of Europe causing losses upto 15 % (Blattny and Blattny, 1970). The affected plants develop several thin and erect branches with a dense and bushy growth. The leaves become narrow with erect position and drop later in autumn. Some leaves may show reddening and infected plants do not flower (Bos, 1960). Similar symptoms have also been observed in *V. uliginosum, V. vitisidaea* and *V. occicoccus*. Two types of symptoms have been described, type-A and B. Type A symptoms are severe than B type. Interestingly, both types cross protect each other.

The disease in *V. myrtillus* is transmitted by graft (Bos, 1960) and may be by dodder to periwinkle. This disease is also transmitted by the leafhopper vector, *Idiodonus cruentatus* to *V. myrtillus*. The phytoplasma bodies have been found associated with the disease (Blattny and Vana, 1974; de Leeuw, 1975). On the basis of RFLP and nucleotide sequence analyses of 16S rDNA, the phytoplasma (strain VAC-L) was classified in the group 16SrIII (X-disease phytoplasma group), subgroup F (Valiunas *et al.*, 2004).

The disease can be detected by transmission through grafting and leafhopper vector. It can also be detected in thin sections of diseased tissue in electron microscope. Phytoplasmas are now detected by Immunofluorescence (Ahlawat, 2010) and also by PCR technology. Eradication of infected plants and control of leafhopper vector are recommended for management.

CRANBERRY FALSE BLOSSOM

Cranberry false blossom disease was first described from USA (Shear, 1908). The disease is known to occur in American cranberry (*Vaccinium macrocarpon*) and European cranberry (*Vaccinium oxicoccus*). False blossom is readily recognized at bloom because the flowers of infected plants are upright in position. Calyx become enlarged, petals become short with red and green streaks. Stamen and pistils become abnormal and the flowers become sterile. Several axillary buds sprouted and give a bushy or witches broom appearance. Terminal flower buds enlarged (Stretch, 1987).

The disease is graft and dodder (*Cuscuta esculentum*) transmitted from cranberry to periwinkle, tomato and other 28 plant species in 10 different families. The disease is also transmitted by the blunt-nose leafhopper, *Scleroacus vaccinii* = *Eucelis striatulus* (Dobroscky, 1929, 1931).

Cranberry false blossom disease is caused by a phytoplasma (Chen, 1971). The phytoplama pleomorphic bodies measuring 80-300 nm are found in sieve tubes of phloem tissue of infected plants but not in healthy. The disease can be identified by symptoms at the time of blooming. Dodder transmission to periwinkle is also a reliable diagnostic test. Electron microscopy of thin sections from disease tissues is advantageous but negative result does not confirm the absence of pathogen. Serological tests like immunofluorescence (Ahlawat, 2010) may be useful to diagnose phytoplasmas. The disease is managed by controlling leafhopper vector through insecticides or by growing vector or pathogen-resistant cultivars. Heat therapy at 42-43°C for 8 days was also suggested.

D. VIRUS DISEASES OF RIBES: GOOSEBERRY, BLACK CURRANT AND RED CURRANT

Ribes genus contains about 150 species of shrubs in two distinct groups, the currants and the gooseberries in the family *Grossulariaceae*. Some authorities separate the gooseberries as the genus *Grossularia*. Gooseberries are grape-like fruits that can be eaten fresh when ripe, or used for cooking. Red currants actually come in three main colors, red, pink, and white. They are sweet and sour flavored, good for cooking and used fresh in salads or topping ice cream and other desserts. Ribes crops are affected by several virus diseases like 'vein banding', 'black current reversion', 'black current yellows', 'gooseberry mosaic', 'red current interveinal white mosaic' and 'yellow leaf spot'. Ribes crops are also infected by viruses of other plants such as cucumber mosaic virus, arabis mosaic virus, raspberry ringspot virus, strawberry latent ringspot virus, tobacco ringspot virus and tobacco rattle virus.

D-1.VIRUS DISEASES OF GOOSEBERRY

GOOSEBERRY VEIN BANDING

Gooseberry vein banding disease (GVBD) affects *Ribes* species and cultivars worldwide. This disease also known as gooseberry mosaic and is found in Europe, North America and New Zealand. After black currant reversion disease, it is the most serious virus-like disorder in these crops. Most of the cultivars of gooseberry, black current and red current appear to be susceptible to vein banding virus.

SYMPTOMS

Leaves on GVBD affected gooseberry plants become pale yellow with translucent areas along the main veins. Leaves of infected black and red currents show yellow banding pattern along the main veins. The symptoms are conspicuous in early spring growth but harder to see in mid-summer. Back cross derivatives of *R. sanguineum* X gooseberry of East Malling Research Station selection numbers, 1385/81 and 1385/90 are good indicator hosts (Knight, 1981).

TRANSMISSION

Vein banding disease is transmitted by chip budding or patch grafting to gooseberry and black currents. The disease is also transmitted by several aphid vectors in semi-persistent manner. The aphid vectors are: *Aphis grossulariae, A. schneideri, Hyperomyzus pallidus, H. lactucae, Nasonovia ribisnigri, Cryptomyzus ribis* and *Myzus persicae* (Adams and Posnette, 1987).

THE PATHOGEN

Gooseberry vein-bandind virus (GVBV) is a *Pararetrovirus* of the genus *Badnavirus* in family *Caulimoviridae* (Xu *et al.*, 2011). The virus particles are 120 × 30 nm in size and contained dsDNA as its genome. Phylogenetic analyses suggest that GVBV is most closely related to Spiraea yellow leaf spot virus (Jones *et al.*, 2001), Dioscorea bacilliform virus and Taro bacilliform virus (Xu *et al.*, 2011).

DIAGNOSIS AND MANAGEMENT

The virus can be detected by graft or aphid vector transmission to indicator hosts. The virus is also detected in PCR with primers specifically designed for it. These primers were also used in PCR to assay for GVBV in a range of *Ribes* germplasm affected with GVBD, with other unrelated virus-like diseases and viruses found in *Ribes* (Jones *et al.*, 2001). The disease can be managed by the production of virus-free planting material through meristem culture (Jones and Vine, 1968; Hedtrich and Feucht, 1981).

GOOSEBERRY DETERIORATION

Gooseberry deterioration disease is caused by raspberry ringspot virus which has been described earlier in this chapter. This disease has been reported from the Netherlands (Houtman, 1951, Van der Meer, 1965). The symptoms of the disease are: mosaic on leaves, small and malformed berries which often ripe late in the season. The disease has been recorded on gooseberry cvs. 'Whitesmith' and 'Whynham's Industry' under natural condition. Experimentally, the virus has been transmitted mechanically to *Chenopodium quinoa* and *C. amaranticolor*. These hosts develop necrotic lesions on inoculated leaves. *C. quinoa* may also show systemic necrosis of inoculated plants. Similar to raspberry ringspot virus, several other herbaceous plant spp. are the host of the virus. The virus is transmitted by nematode, *Longidorus elongatus*.

A virus with RNA genome having isometric particles about 30 nm in diameter is associated with the disease (Harrison and Nixon, 1960). Virus is an ICTV approved species of the genus *Nepovirus* in the family *Comoviridae*. The virus can be detected by sap inoculation to *C. quinoa* and other hosts of raspberry ringspot virus. There is no information on management of the disease (Van der Meer, 1987).

OTHER DISEASES OF GOOSEBERRY

Gooseberry is known to be infected with arabis mosaic virus (AMV) reported from East Germany (Kleinhempel, 1970, 1972). Gooseberry plants did not show any symptoms upon infection with AMV but black current plants show chlorotic blotches, mosaic and ringspot symptoms. *Chenopodium amaranticolor* and *C. quinoa* are used as indicator hosts showing local lesion on inoculated leaves followed by mosaic and stunted. The disease is transmitted by nematode, *Xiphinema diversicaudatum* and by sap inoculation. AMV particles are isometric measuring 28 nm in diameter.

A mosaic and leaf malformation disease on gooseberry has also been reported from Poland and England respectively (Basak and Maskiewiez, 1980; Thresh, 1970). Symptoms of mosaic on gooseberry cv. 'Lady Delamare' develop as bright yellow mottle and vein yellowing (Basak and Maskiewiez, 1980). The disease is transmitted by grafting and infects black current species and hybrids. The malformation disease reported from England is not graft transmitted (Adams, 1987).

D-2. VIRUS DISEASES OF CAPE GOOSEBERRY (*Physalis peruviana*)

CAPE GOOSEBERRY MOSAIC

A mosaic disease of cape gooseberry (*Physalis peruviana*) is of common occurrence in India (Capoor and Sharma, 1965; Nariani and Pathanian, 1953). This disease is caused by tobacco mosaic virus (TMV) (Capoor and Sharma, 1965) and cucumber mosaic virus (CMV) (Nariani and Pathanian, 1953). The diagnosis of the disease caused by individual virus in the field is difficult as the symptoms produced by both the viruses are same and mixed infection is also common. The symptoms of the disease are mosaic mottling, puckering, dark green patches or blisters on leaves (Fig. 52). Symptoms caused by TMV are premature defoliation of leaves and drop of unripe fruits. Whereas, in plants infected by CMV show reduced leaf size and malformed and distorted leaves. Both these viruses are mechanically transmitted. CMV has aphid vectors, *Myzus persicae* and *Aphis gossypii*. The host reactions, physical properties, transmission and serological reactions confirmed it to be caused by a virulent strain of cucumber mosaic Cucumovirus (Gupta and Singh, 1996). Both viruses can be detected in electron microscopy by their morphology (TMV-rode shaped and CMV-spherical particles).

Fig. 52: Mosaic symptoms on cape gooseberry due to Cucumber mosaic virus

LEAF CURL OF CAPE GOOSEBERRY

The major symptoms of the leaf curl disease appear as curling, puckring, smalling of leaves. Leaves show dark green enations on veins on underside. The disease is not sap transmissible but can be easily transmitted by whitefly, *Bemisia tabaci* (Nariani and Pathanian, 1953).

A TOSPOVIRUS INFECTING CAPE GOOSEBERRY IN BRAZIL

A virus was isolated from a commercial crop of Cape gooseberry (*Physalis peruviana*) in October 2009 in Santa Maria, State of Rio Grande do Sul, Brazil. The incidence of the disease was observed upto 100%. Affected plants showed symptoms of stunting, mosaic, necrosis and foliar distortion. Spherical particles measuring 80-100 nm in diameter were observed by electron microscopy, giving the initial evidence that a tospovirus might be involved in this disease. Mechanical inoculation on healthy indicator plants: *Petunia hybrida* and *Chenopodium amaranticolor* reacted with local lesions; *Datura stramonium, Solanum gilo, S. lycopersicum, S. nigrum, Physalis angulata, P. floridana* and *P. peruviana* displayed systemic symptoms (Eiras *et al.*, 2012). A tospovirus infecting gooseberry has also been reported from South Africa (Da Graca *et al.*, 1985)

E. VIRUS DISEASES OF BLACK CURRANT (*RIBES NIGRUM L.*)

BLACK CURRANT REVERSION DISEASE

Black currant reversion disease (BCRD) is the most important virus disease in *Ribes spp*. BCRD occurs in most of the countries where black currant is grown. It is wide spread in European countries. The disease adversely affects the crop yield (Cropley *et al.*, 1964). The disease affects black, red and white currants. Two strains have been identified in black currant reversion pathogen (i) the common European strain and (ii) a virulent strain present in Scandinavia, eastern and central Europe, and countries of the former Soviet Union.

SYMPTOMS

The symptoms of reversion disease in black currant appear as smalling and chlorosis of leaves with chlorotic vein pattern on inoculated plants. Leaves on infected bushes are flatter with smaller basal sinus and decrease serrations. The hairiness of sepals decrease, and brightly coloured buds become grey with downy appearance (Adams and Thresh, 1987). Flowers are severely malformed. In red currant, symptoms are almost similar but with faint flowers and leaf symptoms. Natural infection of BCRD has been recorded in several *Ribes* spp. and hybrids such as *R. bracteosum, R. rubrum* var. *Pubescens, R. alpinum, R. spicatum, R. glutinosum* × *R. nigrum* etc. The black currant cultivars 'Baldwin' and 'Ojebyn'are used as indicator hosts. These cultivars show clear leaf symptoms after one year of inoculation.

TRANSMISSION

Reversion disease is transmitted by chip or patch grafting. The vector of the disease is a gall mite, *Cecidophyopsis ribis*. Since the mite remains in the bud, the transmission by budding must be avoided (Thresh, 1965). In a single gall there may be upto 35000 mites, which may crawl and infect large number of plants. Interestingly, reversion infected plants are more susceptible to infestation by mite than healthy plants (Thresh, 1964a).

THE PATHOGEN

Black currant reversion has been reported to be associated by Phytoplasmas (Zirka *et al.* 1977), Bacteria (Silvere and Remeikis, 1973) and potato virus Y (Jakob, 1976). However, recent studies have shown that the *Black currant reversion associated virus* (BCRAV) belongs to the genus *Nepovirus*, has isometric particles *c.* 28 nm in diameter that contain a single major polypeptide of *c.* 55 KDa and two polyadenylated ssRNA species of 7700 nt and 6400 nt. Some particle preparations also contain a satellite ssRNA species of 1432 nt (Jones, 2000; Lemmetty *et al.*, 1998; Pøibylová *et al.*, 2008). The RNA1 of black currant reversion associated nepovirus (BCRAV) is 7711 nucleotides (nt) long, excluding the poly-A tail, and contains one long open reading frame (ORF) which is translated into a polyprotein of 2094 amino acids (Pacot-Hiriart *et al.*, 2001).

DIAGNOSIS AND MANAGEMENT

Black currant reversion virus can be detected either by symptoms on the plants or by chip or patch grafting to indicator hosts mentioned above. The presence of BCRV is also detected by the RT-PCR (Lemmetty *et al.*, 2001; Pyibylova *et al.*, 2001). The virus can be managed by adopting strict quarantine measures, use of virus-free planting material and control of mites by acaricides. Mites can also be eliminated from cuttings by warm water treatment (Thresh,1964b). Virus and gall mite resistant varieties should be used and breed commercial varieties from resistant sources already known. For example, some cultivars derived from *R. nigrum* var. sibiricum are resistant to virus infection as well as resistant to gall mites (Potapov and Grinenko, 1968; Anderson, 1971).

GREEN MOTTLE DISEASE OF BLACK CURRANT

Most of the black currant cultivars are susceptible to green mottle infection. Infected bushes of black currant are stunted resulting reduction in crop yield. In black currant, the leaves of inoculated plants become pale green which may be along main veins, sometimes giving a 'watermark' effect which is visible against transmitted light. This disease is caused by cucumber mosaic virus (CMV) and is found in some countries of Europe (Adams and Thresh, 1987). CMV also causes mosaic disease of red currant and an arc mosaic of golden currants (*Ribes aureum*) in Germany. CMV has been described in detail under 'cucumber mosaic virus infecting raspberry.

The virus is sap transmissible from *Ribes* to herbaceous hosts. *Chenopodium amaranticolor* and *C. quinoa* are good local lesion hosts. The virus is also transmitted by grafting. CMV is transmitted by more than 60 aphid species in a non-persistent manner. In black currant, CMV is transmitted by *Myzus persicae* and five other species which colonize on *Ribes* crop (*Aphis grossulariae, A. schneideri, Cryptomyzus ribis, Hyperomyzus lactucae, Nasonovia ribisnigiri*) (Thresh, 1970). The virus in *Ribes* can be detected by sap inoculation to *Chenopodium* spp., by ELISA and PCR. The elimination of infected bushes from nursery stocks are the recommended management practice.

VEIN CLEARING AND VEIN NET DISEASE OF BLACK CURRANT

Vein banding and vein net disease is rare in black current in Europe. The disease is caused by gooseberry vein banding virus (Adams and thresh, 1987). Leaves of affected black currant plants develop vein clearing and yellow banding and distortion in some cases. The causal agent is transmitted by grafting and by aphids, *Nasonovia ribisnigri, Hypermyzus lactucae, Myzus persicae* and *Cryptomyzus ribis.* The disease is associated with badnavirus-like particles similar to gooseberry vein banding virus which is also associated with a badnavirus-like particles (Jones *et al.,* 2001). The virus has been described under gooseberry vein banding. The virus can be identified based on its conspicuous symptoms in infected bushes and can be managed by the use of healthy plantings and control of aphid vectors.

YELLOW MOTTLE OF BLACK CURRANT

Chlorotic mottling and ringspot symptoms in leaves of blackcurrant (*Ribes nigrum*), cv. Baldwin are the main symptoms of the disease. The disease is associated with Arabis mosaic Nepovirus and severe crop loss was associated with infection by this Nepovirus infecting black currant (Jones and McGavin, 1996). The vector of the disease is a nematode, *Xiphinema diversicaudatum*.The disease is graft and sap transmitted to herbaceous hosts. Black currant cv. Amos Black is a suitable indicator for virus diagnosis by graft and *C. amaranticolor* and *C. quinoa* by sap inoculation. *Chenopodium amaranticolor* and *C. quinoa* develop local and systemic symptoms similar to strawberry latent ringspot virus.Virus can also be detected by ELISA (Adams and Thresh, 1987). The site where virus and vector are present should be avoided for planting new bushes. Healthy planting material should be used.

BLACK CURRANT YELLOWS

Black currant yellows disease was reported from an English nursery in 1952 and spread widely in several plantations (Posnette, 1952). The growth of bushes affected with black-currant yellows virus are reduced. The total crops from the Wellington and Baldwin bushes were decreased by 50 and 33% respectively (Cropley *et al.*, 1964). The symptoms of the disease appear as chlorotic flecks and olive-green mosaic on leaves of infected plants. The virus also causes transitory vein-pattern symptoms and decreased the number and size of fruits. The disease is transmitted by grafting and by aphids, *Hyperomyzus lactucae, H. pallidus* and *Nasonovia ribisnigri*. There is no information on causal pathogen. Diagnosis of the virus is done on the symptom caused in black currant and also by graft and vector transmission. Use of healthy planting material is important to manage the disease.

INFECTIOUS VARIEGATION OF BLACK CURRANT

Infectious variegation disease has been reported from some countries in Europe. The symptoms of the disease appear as chrome or pale yellow mosaic, vein banding and vein net pattern. The host range is restricted to black currant only. The black currant cultivars 'daniel's september' and 'Laxton's Nigger' produce conspicuous symptoms of the disease. The disease appears to be graft transmitted only and there is no information on the causal agent. No other information is available on the disease (Adams and Thresh, 1987).

D-2 VIRUS AND VIRUS-LIKE DISEASES OF RED CURRANT

RED CURRANT VEIN BANDING

Red currant vein banding disease may be present wherever red currant is grown. It has been shown to be present in some countries of Europe. The growth of seedlings is reduced upto 50%. This is one of the most important diseases of red currant because of its wide prevalence and less conspicuous symptoms in field plants which are easily overlooked. The symptoms of the disease appear as vein clearing and vein banding of leaf veins. Infected plants may be stunted. Black currant and seedlings of red currant cv. Jonkheer van tets are used as indicator hosts which may show shock reaction. The disease is transmitted by patch grafting and also by several species of aphids, *Aphis schneideri, A. triglochinis, Cryptomyzus galeopsides subsp. Citrinus, C. ribis, Hypermyzus lactucae, Nasonovia ribisnigri* in non-persistent manner. The disease can be detected based on its symptoms on indicator host upon inoculation by grafting or by vector. Use of healthy planting material is the known management strategy. The disease has been described in detail by Van der Meer (1987).

GREEN MOTTLE OF RED CURRANT

Green mottle disease in red current is found in the Netherlands and England (Van der Meer, 1987; Thresh, 1967). This disease of red currant is caused by a strain of cucumber mosaic virus (CMV). Cucumber mosaic virus has a wide host range covering several genera. *Ribes* seems to be infected only occasionally. In *Ribes* CMV causes green mottle, line pattern, yellowing, and malformation of leaves. Affected bushes may show dieback symptoms. The virus is transmitted by grafting, and by sap inoculation to herbaceous hosts. It is transmitted by several aphid species like *Myzus persicae* and others in a nonpersistent manner. After mechanical inoculation *Chenopodium quinoa* develops necrotic lesions on inoculated leaves within 3-4 days and *Nicotiana tabacum* cv. White Burley, *N. glutinosa, N. rustica* and cucumber develop mild systemic mottle and severe mosaic symptoms respectively.

The disease is caused by cucumber mosaic virus (see cucumber mosaic virus infecting raspberry for detail). Diagnosis of CMV in red currant can be done by mechanical inoculation to herbaceous hosts described above. The virus can be detected by serological tests like agar gel double diffusion test (Maat, 1966) and ELISA. PCR can also be used for virus detection using CMV specific primers. The virus can be eradicated by thermotherapy at 36°C for three weeks followed by apical tip propagation. Growing virus free plants are useful to contain the disease.

SPOON LEAF OF RED CURRANT

Spoon leaf disease of red currant is caused by a strain of raspberry ringspot virus. The disease has been reported from the Netherlands and Germany. The virus can cause a crop reduction upto 32% in some cultivars (Van der Meer, 1987). The leaves on affected plants curl down or up giving a spoon shaped appearance from which the disease derives its name. Infected plants may also show bright yellow mosaic and ringspot on leaves. Experimentally, gooseberry and black currant are susceptible. The virus is transmitted by grafting and by a nematode vector, *Longidorus elongatus*. It is also transmitted by sap inoculation to *Chenopodium quinoa* and *C. amaranticolor* which develop local lesions and systemic necrosis upon inoculation. Tomato and petunia are also susceptible causing bright systemic yellowing in inoculated plants.

The disease is caused by raspberry ringspot virus (RRSV) which is an isometric virus with about 28 nm in diameter. The genome consists of two species of single stranded RNA with Mol. wt. of 2.8 and 1.4 x 10^6 daltons. The virus can be detected by sap inoculation to herbaceous hosts and by serological testing. The virus can be managed by using healthy planting material and to avoid planting in field where virus or nematode vector are present.

STRAWBERRY LATENT RINGSPOT VIRUS AND ARABIS MOSAIC VIRUS INFECTING RED CURRANT

Arabis mosaic virus (AMV) and Strawberry latent ringspot virus (SLRV) have been isolated from red currant bushes in some countries of Europe

(Van der Meer, 1987). The affected plants by these viruses either did not show any symptoms or may show bright yellow symptoms on leaves. These viruses also infect black currant. *Chenopodium amaranticolor* and *C. quinoa* infected by sap inoculation show faint chlorotic lesions, systemic mosaic, stunting and death of apex. The vector of these viruses is a nematode, *Xiphinema diversicaudatum*. AMV and SLRV from red currant resemble to those infecting strawberry. AMV has isometric particles measuring 28 nm in diameter and SLRV particles are 30 nm in diameter (Murant, 1987). These viruses can be detected by sap inoculation to *Chenopodium amaranticolor* and *C. quinoa* which develop specific symptoms upon sap inoculation. The disease can be managed in a similar way as mentioned for spoon leaf disease of red currant.

TOMATO RINGSPOT VIRUS INFECTING RED CURRANT

Tomato ringspot virus infecting some cultivars of red currant has been reported from USA and Yugoslavia (Williams *et al.*, 1987). Infected plants show chlorosis, spotting and vein banding symptoms at low temperature (21^{0}C). At higher temperature (27^{0}C), only chlorotic spots and rings appear on young leaves. The disease is transmitted by grafting and also by sap inoculation to following herbaceous hosts: *Chenopodium amaranticolor, C. quinoa, Cucumis sativus, Phaseolus vulgaris, Nicotiana tabacum, Petunia hybrida, Lycopersicon esculentum* and *Vigna unguiculata*. Symptoms in these hosts appear as chlorotic lesions/spots, systemic chlorosis or mottle and tip necrosis. *Chenopodium amaranticolor* and *C. quinoa* can be used as indicator or assay hosts. Tomato ringspot virus is transmitted by a nematode vector, *Xiphinema americanum* and infection to red currant is probably spread by this vector (Williams *et al.*, 1987).

Tomato ringspot virus is a member of Nepovirus group (Harrison and Murant, 1977). The virus particles are isometric measuring 28 nm in diameter with single stranded RNA as its genome. Details of tomato ringspot virus have already been described earlier. The virus can be detected by sap inoculation to herbaceous hosts and also by serological tests like agar gel test and ELISA. Indexed and virus-free planting material should be used to contain the disease.

RED CURRANT LEAF PATTERN

Leaf pattern disease is caused by tobacco rattle virus (TRV). The disease has been reported from the Netherlands and Germany. The symptoms of the disease in red currant and *Ribes sanguineum* appear as line pattern or oak leaf pattern and yellow to light green mosaic (Van der Meer, 1987). Tobacco rattle virus is transmitted by nematodes of the genera *Trichodorus* and *Paratrichodorus* (Harrison, 1970). It is also seed transmitted. Experimentally, the virus was sap transmitted to herbaceous hosts, *Chenopodium quinoa* and *Nicotiana tabacum*. These hosts react with local lesions on inoculated leaves (Van der Meer, 1987). TRV particles are long and small rods measuring 190 nm and 45-115 nm respectively with a width of 23 nm (Robinson and Harrison, 1989). Long particles contain genetic information for the production of infectious RNA and small particles have genetic information for the production of coat protein (Harrison, 1970).

The virus can be detected by sap inoculation to herbaceous hosts and also by Electron microscopy as serological tests may not be reliable. For management of the disease, use of healthy propagation stock and avoidance of plantings in fields known to have virus and vector earlier.

YELLOW LEAF SPOT OF RED CURRANT

The yellow leaf spot disease is known to occur in some European countries and has been described as 'Aucuba mosaic' or 'mosaic' (Van der Meer, 1987). The symptoms of the disease appear as small chlorotic or yellow spots distributed over the leaf. The leaf margins become light yellow and infected plants become stunted. Experimentally inoculated plants may show pale yellow patches or bands along the main veins. The disease is transmitted by grafting. No information is available on its causal agent. Graft inoculation to sensitive red currant cultivars "Laxton No.1" or "Fay's prolific" can detect the pathogen. Selection and propagation of healthy planting material give desired control.

INTERVEINAL WHITE MOSAIC OF RED CURRANT

White mosaic disease in red currant has been described from the Netherlands and Great Britain (Van der Meer, 1987). The characteristic symptoms of the disease appear in the angles between main and secondary veins of the leaves as light yellow or white patches. Subsequently, leaves show bright yellow to white margin. The disease is graft transmitted and also by aphids in a non-persistent manner. It is also transmitted by seed in some hosts and by mechanical inoculation from 'Fay's Prolific' to the following hosts:

- *Callistephus chinensis* (China aster): Mosaic and chlorosis of leaves.
- *Nicotiana rustica*: Local and systemic yellowing and small gray rings on leaves develop at low temperature. At higher temperature, no symptoms develop.
- *Chenopodium quinoa*: Necrotic local lesions and systemic mosaic.

The disease has been associated with alfalfa mosaic virus (ALMV). It has bacilliform particles of different lengths. All particles are approximately 16nm wide. However, the association of ALMV with interveinal white mosaic in The Netherlands is yet to be confirmed. The virus can be detected by characteristic symptoms in red currant and also by sap inoculation to *C. quinoa*. Selection and propagation of healthy planting material has been recommended for its control.

FULL BLOSSOM OF RED CURRANT

This disease has been reported from Czechoslovakia (Rukus, 1978). Flower malformation is the typical symptom of the disease. Stamens are absent, styles may be several in a flower, ovaries are half-inferior or superior (in healthy flower it is inferior). Petals and sepals are enlarged and become leaf-like. The berries become misshaped and appear cauliflower-like (Van der Meer, 1987). The disease is transmitted by patch grafting. The symptoms of the disease are suggestive that the causal organism may be a phytoplasma but no aetiological agent has been established as yet. The disease can be diagnosed by flower malformation symptoms on red currant or by graft transmission to sensitive cultivars like 'Houghton Castle'. Use of certified virus-free planting material is recommended for management of the disease.

LEAF MALFORMATION OF RED CURRANT

Malformation of leaves in red currant has been reported from some countries in Europe. In affected red currant, leaves are more rounded than normal with reduced number of lobes and marginal serration. In severe cases of infection leaves are rugose, distorted and asymmetrical. There is no information on modes of natural spread and the causal pathogen associated with the disease. The disease can be diagnosed by visual symptoms. Graft inoculation to sansitive black currant cultivars is another method of diagnosis. Use of disease-free stock for propagation is the best control of the disease (Van der Meer, 1987).

BUDLESSNESS OF RED CURRANT

Budlessness disease probably occurs wherever sensitive cultivars like 'Fay's Prolific' are grown. It has been reported from several countries in Europe. The symptoms of the disease are: stoping of shoot extension growth during early summer, dark green and rounded leaves with reduced dentation, failure of axillary bud development, reduced size of buds which develop weak laterals, failure to develop terminal and axillary buds which may also die. Poor bud formation is directly related to the yield loss in infected bushes. There is no information on the method of transmission and the agent associated with the disease. The diagnostic symptoms are the only method of disease diagnosis and use of cuttings from healthy plants is the known management practice at present (Van der Meer, 1987).

PHYTOPLASMA DISEASES OF BLACK CURRANT AND RED CURRANT

A disease with redness and downward curling of leaves in affected shrubs has been reported from Canada in black currant (*Ribes nigrum*) and red currant (*Ribes rubrum*). Studies indicated that these two plant species were infected with phytoplasma members of group 16SrX (apple proliferation), subgroup 16SrX-A (Arocha *et al.*, 2012).

REFERENCES

Adams, A.N. 1987. Gooseberry mosaic and leaf malformation of gooseberry. p. 132. Pacot-Hiriart, C., Latvala, S. and Lehto, K. 2001. Nucleotide sequence of black currant reversion associated nepovirus RNA-1. *Virus Research,* 79: 145–152.

Adams, A.N. and Posnette, A.F. 1987. Gooseberry vein banding. *In*: Virus Diseases of Small Fruits (Ed). Converse, R.H. USDA Agric. Handbook No. 631: 129–130.

Adams, A.N. and Thresh, J.M. 1987. Infectious variegation. *In*: Virus Diseases of Small Fruits (*Ed*.). Converse, R.H. USDA Agric. Handbook, 631: 141–142.

Adams, A.N. and Thresh, J.M. 1987. Reversion of black currant. *In*: Virus Diseases of Small Fruits (*Ed*). Converse. R.H. USDA Agric. Handbook, 631: 133–136.

Adams, A.N. and Thresh, J.M. 1987. Yellow mottle of black currant. *In*: Virus Diseases of Small Fruits (*Ed*). Converse, R.H. USDA Agric. Handbook, 631: 139–140.

Adams, A.N. and Thresh, J.M.1987. Vein clearing and vein net disease of Black Currant. *In*: *Virus Diseases of Small Fruits* (*Ed*). Converse, R.H. USDA Agric. Handbook, 631: 137–138.

Ahlawat, Y.S. 2010. *In*: Diagnosis of plant viruses and allied pathogens. Ed, Y.S. Ahlawat. Studium Press (India) Pvt. Ltd., New Delhi, pp. 73–74.

Allen, W.R., Davidson, T.R. and Briscoe, M.R. 1970. Properties of a strain of strawberry latent ringspot virus isolated from sweet cherry growing on Ontario. *Phytopathology,* 60: 1262.

Allen, W.R. and Dias, H.F. 1977. Properties of the single protein and two nucleic acids of tomato ringspot virus. *Can. J. Bot.,* 55: 1028–1037.

Anderson, M.M. 1971. Resistance to gall mite in the Eucoreosma section of *Ribes. Euphytica* 20: 422–426.

Arocha, Y., Agrawal, V., Scott, J., Bertaccini, A. and Michelutti, R. 2012. Detection and identification of a phytoplasma affecting black currant and red currant shrubs. *In*: Proc. 22nd Interl. Conf. Virus and other graft transmissible diseases of fruit crops. Held in Rome June 3-8, 2012. p. 91.

Azad, R.N. and Sehgal, O.P. 1958. Some virus diseases of temperate fruits. *Indian Phytopath.* 11: 159–164.

Barbara, D.J., Jones, A.T., Henderson, S.J., Wilson, S.C. and Knight, V.H. 1984. Isolates of raspberry bushy dwarf virus differing in Rubus host range. *Ann. Appl. Biol.*, 105: 49–54.

Barnett, O.W. and Murant, A.F. 1970. Host range, properties and purification of raspberry bushy dwarf virus. *Ann. Appl. Biol.*, 65: 435–449.

Baumann, G. 1982. Elimination of a heat-stable raspberry virus by combining heat treatment and meristem culture. *Acta Hort.* (ISHS), 129: 11–12.

Baumann, G. 1981. Moglichkeiten und grenzen der virus-freimachung von himbeeren durch gewebekultur. *Obstau*, 6: 88–89.

Baumann, G., Casper, R. and Converse, R.H. 1982. The occurrence of apple mosaic virus in red and black raspberry and in blackberry cultivars. *Acta Hort.*, 129: 13–20.

Basak, W. 1974. Yellow spot-a virus disease of raspberry. *Bull. Acad. Pol. Sci. Cl.V. Ser. Sci. Biol.*, 22: 47–51.

Basak, W. and Maskiewiez, J. 1980. Gooseberry mosaic. *Acta Hort.*, 95: 49–52.

Belli, G., Fortusini, A., and Bianco, P.A. 1986. Etiology of peach willow leaf rosette. *Acta Hort.*, 193: 63–65.

Bennett, C.W. 1927. Virus diseases of raspberries. *Michigan Agricultural Experiment Station Technical Bulletin* 80: 38.

Bi, Y., Artola, K., Kurokura, T., Hytonen, T. and Valkonen, J.P.T. 2012. First report of leaf blotch virus in raspberry in Finland. *Plant Dis.*, 96: 1231.

Blattny, C. and Blattny, C. Jr. 1970. Blueberry witches broom, *In*: Virus diseases of small fruits and grapevines. Frazier N.W. (ed) University of California, Div. Agric. Sci. Berkeley, pp. 177–189.

Blattny C. and Vana V.l. 1974. Mycoplasmalike organisms in *Vaccinium myrtillus* L. infected with blueberry witches broom. *Biologia Plantarum*, 16: 476–478.

Blattny, C. and Janeckova, M. 1980. Apricot bare twig and unfruitfulness. *Acta Hort.*, 94: 383–390.

Bolton, A.T. 1964. Virus diseases of strawberries in Eastern Canada. *Canadian Plant Dis. Survey*, 44: 83–90.

Bolton, A.T. 1967. The inactivation of vein banding and latent C viruses in strawberries by heat treatment. *Canadian J. Plant Sci.*, 47: 375–380.

Bolton, A.T. 1970. Strawberry leafroll. *In*: Frazier, N.W. (*Ed.*) Virus diseases of small fruits and grapevines. Univ. California, Div. Agric. Sci. Berkeley, p. 290.

Bolton, A.T. 1970. Spread of raspberry leaf curl virus. *Canadian J. Plant Sci.*, 50: 667–671.

Boone, D.M. 1994. Ringspot. *In*: Compendium of blueberry and cranberry diseases. *Eds.* Caruso, F.L. and Ramsdell, D.C. APS Press, Paul, MN.

Bos L., 1960. A witches broom virus disease of *Vaccinium myrtillus* in the Netherlands. *Tijdschrift Plantezieten*, 66: 259–263.

Boylan-Pett, W., Ramsdell, D.C., Hoopingarner, R.A. and Hancock, J.F. 1991. Honeybee foraging behaviour, in-hive survival of infections, pollen-borne blueberry leaf mottle virus and transmission of the virus in highbush blueberry. *Phytopathology*, 81: 1407–1412.

Braun, A.J. and Keplinger, J.A. 1962. Transmission of viruses to strawberry by excised leaf grafts from plants other than strawberry. *Phytopathology*, 52: 726.

Brown, D.J.F. 1985. The transmission of two strains of strawberry latent ringspot virus by populations of *Xiphinema diversicaudatum*. *Nematologia Mediterranea*, 13: 217–223.

Brown, D.J.F. 1989. Virus transmitted by nematodes. *OEPP/EPPO Bull.* 19: 453–461.

Brunt, A.A. and Stace-Smith, R. 1976. The occurrence of the black raspberry latent strain of tobacco streak virus in wild and cultivated *Rubus* species in British Columbia. *Acta Hort.*, 66: 71–76.

Brunt, Crabtree, Dallwitz, Gibbs, Watson and Zurcher, (eds.) 1997. 'Plant viruses online: Descriptions and lists from the VIDE Database. Version: 9th April 1997.' URL http://biology.anu.edu.au/Groups/MES/vide/, 1996.

Cadman, C.H. 1951. Studies in Rubus virus diseases I. A latent virus of Norfolk giant raspberry. *Ann. appl. Biol.*, 38: 801–811.

Cadman, C.H. 1952. Studies on Rubus virus diseases. II. Three types of vein chlorosis of raspberries. *Ann. appl. Biol.*, 39: 61–68.

Cadman, R.H. 1954. Studies on Rubus virus diseases. IV. Aphid transmission of raspberry leaf mottle virus. *Annl. Appl. Biol.*, 41: 207–214.

Cadman, R.H. and Fisken, A.G. 1958. Susceptibility of raspberry varieties to infection by aphid borne viruses. *J. Hort. Sci.*, 32: 13–20.

Cadman, C.H. 1956. Studies on the etiology and mode of spread of Scottish raspberry leaf curl disease. *J. Hort. Sci.* 31: 111.

Cadman, C.H. and Stace-Smith, R. 1970. *In:* Virus Diseases of Small Fruits and Grapevines, University of California Press, p. 119.

Capoor, S.P. and Sharma D.C. 1965. Studies on cape gooseberry mosaic disease. *Indian Phytopath.*, 18: 58–65.

Casper, R. 1973. Serological properties of Prunus necrotic ringspot and apple mosaic virus isolates from rose. *Phytopathology,* 63: 238–240.

Chambers, J. 1961. The production and maintenance of virus-free raspberry plants. *J. Hort. Sci.*, 36: 48–54.

Chen, T.A. 1971. Mycoplasma-like organisms in sievetube elements of plants infected with blueberry stunt and cranberry false blossom. *Phytopathology,* 61: 233–236.

Childress, A.M. and Ramsdell, D.C. 1987. Bee mediated transmission of Blueberry leaf mottle virus via infected pollen in high bush blueberry. *Phytopathology,* 77: 167–172.

Chiykowski, L.N. 1962. Clover phyllody and strawberry green petal diseases, caused by the same virus in eastern Canada. *Canadian J. Botany,* 40: 1615–1617.

Chiykowski, L.N. 1987. Aster yellows in strawberry. *In*: Virus diseases of small fruits. *Ed*. Converse, R.H. USDA Agriculture Hand Book No. 631. pp. 31–34.

Cieslinska, M. 2011. Detection and characterization of phytoplasmas associated with diseases of *Rubus* spp. in Poland. *J. Plant. Path.*, 93: 51–56.

Converse, R.H. 1963. Influence of heat-labile components of the raspberry mosaic complex on growth and yield of red raspberries. *Phytopathology,* 53: 1251–1254.

Converse, R.H. 1972. Tobacco streak virus in black raspberry. *Phytopathology,* 62: 1001–1004.

Converse, R.H. 1978. Uneven distribution of tobacco streak virus in Santiam blackberry before and after heat therapy. *Phytopathology,* 68: 241–244.

Converse, R.H. 1979. Transmission of tobacco streak virus in *Rubus*. *Acta Hort.*, 95: 53–61.

Converse, R.H. 1981. Infection of cultivated strawberries by tomato ringspot virus. *Phytopathology,* 71: 1149–1152.

Converse, R.H. 1981. Virus tested Rubus cultivars in the USDA research collection. *Fruit varieties Journal,* 35: 141–145.

Converse, R.H. 1987. Virus diseases of small fruits. *Ed*. Converse, R.H. United States Department of Agriculture. Agriculture Handbook No. 631, Illus. p. 277.

Converse, R.H. 1987. Strawberry witches' broom and multiplier diseases. *In*: *Virus diseases of small fruits*. United States Department of Agriculture. Agriculture Handbook No. 631. pp. 66–68.

Converse, R.H. 1987. Blackberry calico. *In*: Virus diseases of small fruits. *Ed*. Converse, R.H. USDA Agricultuse handbook No. 631.

Converse, R.H. and Lister, R.M. 1969. The occurrence and some properties of black raspberry latent virus. *Phytopathology*, 59: 325–333.

Converse, R.H. and Casper, R. 1975. A member of the apple mosaic virus group Occurring in Rubus. *Acta Hort.*, 44: 63.

Converse, R.H. and Bartlett, A.B. 1979. Occurrence of viruses in some wild *Rubus* and *Rosa* species in Oregon. *Plant Dis. Reptr.*, 63: 441–444.

Converse, R.H. and Ramsdell, D.C. 1982. Occurrence of tobacco and tomato ringspot viruses and of dagger and other nematodes associated with cultivated highbush blueberries in Oregon. *Plant Dis.*, 66: 710–712.

Converse, R.H. and Stace-Smith, R. 1987. Tomato ringspot virus in strawberry. *In*: Virus diseases of small fruits. R.H. Converse. Editor. USDA Agriculture Handbook No. 631. pp. 52–55.

Converse, R.H. and Aliniazee, M.T. 1988. Effect of insecticides applied in the field on incidence of aphid borne viruses in cultivated strawberry. *Plant Dis.*, 72: 127–129.

Converse, R.H. and Volk, E. 1990. Some effects of Pallidosis disease on strawberry growth under greenhouse conditions. *Plant Dis.*, 74: 814–816.

Converse, R.H., Stace-Smith, R. and Cadman, C.H. 1970. Raspberry mosaic. *In*: Virus diseases of small fruits and grapevines. Frazier, N.W. (*Ed*) University of California, Div. Agric. Sci. Berkeley, p. 290.

Converse, R.H., Martin, R.R. and Spiegel, S. 1987. Strawberry mild yellow-edge. *In*: *Virusdiseases of small fruits*. R.H. Converse. Editor. USDA Agriculture Handbook No. 631. pp. 25–29.

Converse, R.H., Stace-Smith, R. and Jones, A.T. 1987. Raspberry mosaic. *In*: *Virus Diseases of Small Fruits*; ed. RH Converse, USDA Hdbk No. 631. pp. 168–175.

Cornelissen, B.J.C., Janssen, H., Zuidema, D. and Bol, J.F. 1984. Complete nucleotide sequence of tobacco streak virus RNA 3. *Nucleic Acids Research*, 12: 2427–2437.

Costa, A.S. and da Lima Neto, V. 1976. Transmiao do virus da necrose branca do fumo por *Frankniella* sp. *Fitopatologi*, 11: 11.

Cropley, R., Posnette, A.F. and Thresh, J.M. 1964. The effect of black currant yellows virus and a strain of reversion virus on yield. *Annl. Appl. Biol.*, 54: 177–182.

Da Graça, J.V., Trench, T.N. and Martin, M.M. 1985. Tomato spotted wilt virus in commercial Cape gooseberry (*Physalis peruviana*) in Transkei. *Plant Pathology*, 34: 451–453.

Dale W.T. and Hendy, A.F. 1967. Control of Arabis mosaic virus in strawberry with nematicides. *Plant Pathology*, 16: 187–191.

Dale, J.l. and Mainland, C.M. 1981. Stunt disease in rabbit eye blueberry. *Hort. Sci.*, 16: 313–314.

Danet, J.L., Fimbeau, S., Salar, P., Faure, C., Pommier, J.J., Turquet, M., Marais, A. and Foissac. 2012. Strawberry marginal chlorosis (SMC): A disease of various bacterial and viral etiologies in France. *In*: Proc. 22nd Interl. Conf. Virus and other graft transmissible diseases of fruit crops. Held in Rome June 3-8, 2012. p. 61.

Daubeny, H.A. 1982. Breeding for insect, mite and nematode resistance. *In*: Methods in fruit breeding. *Eds*. Moore, J.N. and Janick, J. Purdue Univ. Press, Lafayette, Ind., p. 650.

Davies, D.L. 1998. The occurrence of two distinct phytoplasma isolates associated with *Rubus* species in the U.K. *Acta Hort.*, 471: 63–65.

Debrot, E.A. 1964. Studies on a strain of raspberry ringspot virus occurring in England. *Ann. Appl. Biol.*, 54: 183–191.

de Leeuw G.T.N. 1975. Presence of mycoplasma-like organisms in the phloem of witches' broom diseased *Vacinium myrtillus* plants in the Netherlands. *Phytopath. Z.*, 83: 91–94.

Demaree, J.P. and Marcus, C.P. 1951. Virus diseases of strawberries in the United States with special reference to distribution, indexing and insect vectors in the East. *Plant Dis. Reptr.*, 35: 527–537.

Dhingra, K.L. and Niazi, F.R. 1972. Transmission studies of raspberry ring spot virus. *Indian Phytopath.*, 26: 265–268.

Dobroscky, I.D. 1929. Cranberry false blossom spreads by a leafhopper. *Science* 70: 635.

Dobroscky, I.D. 1931. Studies on cranberry false blossom disease and its insect vector. *Boyce Thompson Institute Contributions*, 3: 59–83.

Edwardson, J.R. and Christie, R.G. 1991. Handbook of viruses infecting legumes, CRC press, Boca Raton, Univ. Florida, USA. p. 293.

Eiras, M., Costa, I.F.D., Chaves, A.L.R., Colariccio, A., Harakava, R., Tanaka, F.A.O., Garcêz, R.M. and Silva, L.A. 2012. First report of a tospovirus in a commercial crop of Cape gooseberry in Brazil. *New Disease Reports*, 25: 25.

Eliseeva, Z.N., Kulikova, M.T. and Slobozhanina, I.S. 1974. Virus diseases of strawberries in the Amla Ata District. *Rev. Plant Pathol.*, 53: 546.

Engelbrecht, D.J. 1987. Black raspberry necrosis. *In*: *Virus Diseases of Small Fruits*; ed. RH Converse, USDA Hdbk No. 631. pp. 178–183.

EPPO/CABI. 1996a. Strawberry crinkle cytorhabdovirus. *In*: *Quarantine pests for Europe*. 2nd edition (*Eds.* by Smith, I.M., McNamara, D.G., Scott, P.R. and Holderness, M.). CABInternational, Wallingford, UK.

Ficcioli, G. 1969. Caratteristiche sintomatologiche e principali proprieta fisiche di un virus isolato da radici di fragola (*Fragaria vesca* L.) edella sua variante instabile. *Phytopathologia Mediterranea*, 8: 162–170.

Ficcioli, G. 1970. Necrotic rosette: a new virus infecting strawberry.] Extract of the Proceedings of 4th National Strawberry Conf., Cese, Italy, May 21-23, 1970 (In Italian)

Finlay, J.R. and Teakle, D.S. 1969. The effect of pH on the particle stability of phosphotungstate stained tobacco necrosis virus. *J. gen. Virol.*, 5: 93.

Francki, R.I.B., Mossop, D.W. and Hatta, T. 1979. Cucumber mosaic virus. *CMI/AAB Descr. Pl. Viruses* No. 213, Commonwealth Agricultural Bureaux, Slough.

Francki, R.I.B., Milne R.G. and Hatta, T. 1985. Atlas of plant viruses. Vol. II, CRC Press, Boca Raton, Florida, USA. pp. 23–38.

Frazier, N.W. 1955. Strawberry vein banding virus. *Phytopathology*, 45: 307–312.

Frazier, N.W. 1966. Natural and experimental infection of *Rubus* with strawberry necrotic shock virus. *Phytopathology*, 56: 568–569.

Frazier, N.W. 1974a. Detection of graft transmissible diseases in strawberry by a modified leaf grafting technique. *Plant Dis. Reptr.* 58: 203–207.

Frazier, N.W. 1974b. Six new strawberry indicator clones evaluated for the detection and diagnosis of twelve graft transmissible diseases. *Plant Dis. Reptr.*, 58: 28–31.

Frazier, N.W. 1975. Strawberry mild yellow-edge in North Carolina. *Plant Dis. Reptr.* 59: 41–42.

Frazier, N.W. 1975. Possible transmission of strawberry pallidosis by the leafhopper *Coelidia olitoria. Plant Dis. Rep.*, 59: 40–41.

Frazier, N.W. 1987. Strawberry leaf roll. *In: Virus diseases of small fruits.* R.H. Converse, Ed.,USDA Agriculture Handbook No. 631. pp. 61–62.

Frazier, N.W. 1987. Strawberry feather-leaf. *In: Virus diseases of small fruits.* Ed. R.H. Converse, USDA Agriculture Handbook No. 631. pp. 62–63.

Frazier, N.W. 1987. Strawberry pseudo mild yellow edge. *In: Virus diseases of small fruits. Eds.* Converse, R.H. USDA Agriculture Handbook No. 631. pp. 63–64.

Frazier, N.W. and Thomas, H.E. 1953. Strawberry a host of western aster yellows virus. *Plant Dis. Reptr.*, 37: 272–275.

Frazier, N.W. and Posnette, A.F. 1956. Leafhopper transmission of a clover virus causing green petal disease in strawberry. *Nature,* 177: 1040–1041.

Frazier, N.W. and Posnette, A.F. 1957. Transmission and hostrange studies of strawberry green petal virus. *Ann. Appl, Biol.*, 45: 580–588 .

Frazier, N.W. and Stubbs, L.L. 1969. Pallidosis-A new virus disease of strawberry. *Plant Dis. Rep.*, 53: 524–526.

Frazier, N.W. and Jensen, D.D. 1970. Strawberry, an experimental host of peach western X disease. *Phytopathology,* 60: 1527–1528.

Frazier, N.W. and Mellor, F.C. 1970a. *In: Virus Diseases of Small Fruits and Grapevines.* Ed. Frazier, N.W. Univ. Calif., Div. Agric. Sci., Berkeley. pp. 18–23.

Frazier, N.W. and Mellor, F.C. 1970b. Tomato ringspot virus in strawberry. *In: Virus diseases of small fruits and grapevines.* Frazier, N.W. (ed). Univ. California, Div. Agric. Sci. Berkeley. pp. 43–45.

Frazier, N.W., Rarwood, C.E. and Gold, A.H. 1961. Yellow-bud virus endemic along California coast. *Plant Dis. Reptr.,* 45: 649–651.

Frazier, N.W., Jorjensen, P.S., Thomas, H.E. and Johnson Jr. H.A. 1962. Necrotic shock – a virus disease of strawberries. *Plant Dis. Reptr.*, 46: 547–550.

Frazier, N.W., Sylvester, E.S. and Richardson, J. 1987. Strawberry crinkle. *In: Virus diseases of small fruits. Ed.* Converse, R.H. *USDA Agriculture Handbook* No. 631.,

Freeman, J.A. and Stace-Smith, R. 1970. Effect of raspberry mosaic viruses on yield and growth of red raspberries. *Can. J. Pl. Sci.*, 50: 521–527.

Fritsch, C., Koenig, I., Murant, A.F., Raschke, J.H., and Mayo, M.A. 1984. Comparisons among satellite RNA species from five isolates of tomato black ring virus and one isolate of myrobalan latent ringspot virus. *J. Gen. Virol.*, 65: 289–294.

Fritzsche, R. and Schmidt, H.B. 1963. *Xiphinema paraelongatum* Altherr und *Xiphinema n.*sp., zwei Vektoren der Arabis-Mosaik Virus. *Naturwis-senschaften* 50: 163.

Fry, P.R. and Wood, G.A. 1973. Further viruses of Prunus in New Zealand. *N.Z.J. Agric. Res.* 16: 131–142.

Fulton, J.P. 1957. Aster yellows virus affecting strawberries in Arkansa. *Plant Dis. Reptr.*, 41: 521–523.

Fulton, R.W. 1948. Hosts of tobacco streak virus. *Phytopathology* 38: 421.

Fulton, R.W. 1971. Tobacco streak virus. Commonwealth Mycological Institute and the Association of Applied Biologists. *Description of plant viruses* No. 44, p. 4.

Fulton, R.W. 1972. Apple mosaic virus. *CMI/AAB Descriptions of Plant Viruses* No. 83. Association of Applied Biologists, Wellesbourne, UK.

Fulton, R.W. 1985. Tobacco streak virus. *CMI/AAB Descriptions of Plant Viruses* No. 307. Association of Applied Biologists, Wellesbourne, UK.

Gorcio, M.J. and Dale, J.L. 1982. Investigation of blueberry stunt disease. *Plant Dis.*, 66: 244–245.

Greber, R.S. 1979. Virus diseases of Queensland strawberries and the epidemiological effects of strawberry runner approval scheme. *Queensland J. Agric. and Animal Sci.*, 36: 93–103.

Greber, R.S. 1987. Strawberry rickettsia yellows and mycoplsma yellows. *In*: Virus diseases of small fruits. Converse, R.H. United States Department of Agriculture. Agriculture Handbook No. 631, p. 277.

Greber, R.S. and Gowanlock, D.H. 1979. Rickettsia-like and mycoplasma-like Organisms associated with two yellows-type diseases of strawberries in Queensland. *Australian J. Agric. Sci.*, 30: 1101–1109.

Gryll, N.E. 1979. Leafhopper vectors and the plant disease agents they transmit in Australia. *In*: Maramorosch, K. and Harris, k. (*Eds*.), Leafhopper vectors of plant disease agents Academic press New York, pp. 179–212.

Gupta, S.P. and Singh, B.R. 1996. Severe mosaic of cape gooseberry due to cucumber mosaic virus. *Indian J. Virol.*, 12: 155–156.

Guy, G.L., Sampson, P.Y., McGechan, J. and Stace-Smith, R. 1982. Occurrence of viruses in *Rubus* cultivars and species in Australia. *Acta Hort.*, 29: 31–39.

Hadidi, A., Montasser, M.S., Levy, L., Goth, R.W., Converse, R.H., Madkour, M.A. and Skrzeckowski, L.J. 1993. Detection of potato leaf roll and strawberry mild yellow-edge luteovirus by reverse transcription-polymerase chain reaction amplification. *Plant Dis.*, 77: 595–601.

Halgren, A., Tzanetakis, I. E., and Martin, R. R. 2007. Identification, characterization, and detection of *Black raspberry necrosis virus*. *Phytopathology*, 97: 44–50.

Hanada, K. and Harrison, B.D. 1977. Effects of virus genotype and temperature on seed transmission of nepoviruses. *Ann. Appl. Biol.* 85: 79–92.

Hanson, C.M. and Campbell, R.N. 1979. Strawberry latent ringspot virus from 'Plain' parsley in California. *Plant Dis. Reptr.*, 63: 142–146.

Harris, R.V. 1933. Mosaic disease of raspberry in Great Britain. I. symptoms and varietal susceptibility. *J. pomology and Hort. Sci.*, 11: 237–255.

Harrison, B.D. 1958. Cucumber mosaic virus in raspberry. *Plant Pathology*, 7: 109-111.

Harrison, B.D. 1958. Further studies on raspberry ringspot and tomato black ring, soil borne viruses that affect raspberry. *Ann. Appl. Biol.*, 46: 571–584.

Harrison, B.D. 1964. Specific nematode vectors for serologically distinctive forms of raspberry ringspot and tomato black ring viruses. *Virology*, 22: 544–550.

Harrison, B.D. 1970. Tobacco rattle virus. CMI / Association of applied biologists. Description of plant viruses No.12, p. 4.

Harrison, B.D. and Nixon, H.L. 1960. Purification and electron microscopy of three soil borne plant viruses.*Virology*, 12: 104–117.

Harrison, B.D. and Hanada, K. 1976. Competitiveness between genotypes of raspberry ringspot virus is mainly determined by RNA-1. *J. Gen. Virol.*, 31: 455-457.

Harrison, B.D. and Murant, A.F. 1977. Nepovirus group. CMI/AAB description of plant viruses No.185. Association of Appl. Biologists, Kew Surrey, England,4P.

Harrison, B.D., Mowat, W.P. and Taylor, C.E. 1961. Transmission of a strain of tomato black ring virus by *Longidorus elongatus* (Nematoda). *Virology*, 14: 480–485.

Harrison, B.D., Murant, A.F. and Mayo, M.A. 1972. Two properties of raspberry ringspot virus determined by its smaller RNA. *J. Gen. Virol.*, 17: 137–141

Harrison, B.D., Murant, A.F., Mayo, M.A. and Roberts, I.M. 1974. Distribution of determinants for symptom production, host range and nematode transmissibility between the two RNA components of raspberry ringspot virus. *J. Gen. Virol.*, 17: 137–141.

Harrison, R.V. and Cadman, C.H. 1969. Role of dagger nematode (*Xiphinema* sp.) in outbreak of plant diseases caused by arabis mosaic virus. *Nature, Lond.*, 184: 1624–1626.

Hartman, J.X., Hooper, G.R. and Bath, J.E. 1972. Occurrence and nature of Mycoplasma-like organisms in stunt disease of Michigan high bush blueberry. *Michigan Academician*, 4: 461–467.

Hassan, M., Keller, K.E., Martin, R.R., Sabanadzovic, S. and Tzanetakis, I.E. 2012. A new *Emaravirus* detected in Blackberries affected by yellow vein disease. *In*: Abs. 22nd Interl. Conf. Virus and other graft transmissible diseases of fruit crops. Held in Rome June 3-8, 2012. p. 189.

Hedtrich, C.M. and Feucht, W. 1981. Die bedeutung der Meristem-Vermehrung-der derzeitige Stand beim Beerrenobst. *Obstbau*, 6: 82–86.

Heimann, M.F., Spear, R.N. and Jeffers, S.N. 1992. Mycoplasma-like organisms associated with strawberry multiplier disease. *Phytopathology*, 82: 1133.

Henriques, M. and Schlegel, D.E. 1975. Studies on the nature of strawberry pallidosis disease. *Proc. American Phytopathol. Soc.*, 2: 114.

Hepp, R.F. and Martin, R.R. 1991. Occurrence of strawberry mild yellow-edge associated virus in wild *Fragaria chiloensis* in South America. *Acta Hort.*, 308: 57–60.

Ho, T. and Tzanetakis, I.E. 2012. Sensitive detection of three berry fruit viruses by TaqMan®QPCR. *In*: Abs. 22nd Interl. Conf. Virus and other graft transmissible diseases of fruit crops. Held in Rome June 3-8, 2012. p. 190.

Hokanson, S.C., Martin, R.R., Heflebower, Jr., R.F., Rouse, R. and Maas, J. 2000. Survey of strawberry viruses occurring in commercial plantings in the state of Maryland, USA. *Adv. Strawberry Res.*, 18: 25–32.

Hollings, M. 1965. Some properties of celery yellow vein, a virus serologically related to tomato black ring virus. *Annl. appl. Biol.*, 55: 459–470.

Honetslegrova, J.F., Vibio, M. and Bertaccini, A. 1996. Electron microscopy and molecular identification of phytoplasma associated with strawberry green petals in the Czech Republic. *European J. Plant Pathol.*, 102: 831–835.

Houtman, G. 1951. Het verlopen van kruisbessenstruiken. [deterioration of gooseberries]. *Fruitteelt*, 41: 720–721.

Huhtanen, D.P. and Converse, R.H. 1971. Temporary symptom remission of strawberry June yellows and witches' broom by use of oxytetracycline. *Phytopathology*, 61: 1137–1139.

Hunter, B.G., Richardson, J., Dietzgen, R.G., Karu, A., Sylvester, E.S., Jackson, A.O. and Morris, T.J. 1990. Purification and characterization of strawberry crinkle virus. *Phytopathology* 80: 282–287.

Hutchinson, M.T. 1950. Can you recognize the symptoms of stunt disease. Proc. 19th Ann. Blueberry Open house 19: 9–11.

Hutchinson, M.T. and Varney, E.H. 1954. Ringspot: A virus disease of cultivated blueberry. *Plant Dis. Reptr.*, 38: 260–263.

Hutchinson , M.T., Goheen, A.C. and Varney, E.H. 1960. Wild sources of blueberry stunt virus in New Jersey. *Phytopathology*, 50: 308–315.

ICTVdB Management.2006. 00.010.0.04.001. Cucumber mosaic virus. *In*: *ICTVdB – The Universal Virus Database*, version 4. Büchen-Osmond, C. (*Ed.*), ColumbiaUniversity, New York, USA.

ICTVdB Management. 2006. 00.057.0.81.009. Bramble yellow mosaic virus. *In*: *ICTVdB – The Universal Virus Database*, version 4. Büchen-Osmond, C. (*Ed*), Columbia University, New York, USA

Jakob, H. 1976. Investigations on symptomatology, transmission, etiology and host specificity of black currant reversion virus. *Acta Hort.*, 66: 99–104.

Jelkmann, W., Maiss, E., Breyel, M.E. and Casper, R. 1988. Production and use of cDNA clones from arabis mosaic virus. *Ann. Appl. Biol.*, 113: 483–491.

Jelkmann, W., Martin, R.R., Lesemann, D.E., Vetten, H.J. and Skelton, F. 1990. A new potexvirus associated with strawberry mild yellow edge disease. *J. Gen. Virol.*, 71: 1251–1258.

Jelkmann, W., Maiss, E. and Martin, R.R. 1991. The nucleotide sequence and genome organization of strawberry mild yellow edge associated potexvirus (SMYEAV). *J. Gen. Virol.*, 73: 475–479.

Jenser, G.A., Hegab, A. and Kollanyi, L. 1981. Vectors of Rubus stunt in Hungary. *In:* Proc. 9th Conf. of the Czechoslovak plant Virologists. pp. 73–75.

Jha, A. 1961. Arabis mosaic virus in strawberry. *J. Hort. Sci.*, 36: 219–227.

Jha, A. and Posnette, A.F. 1959. Transmission of a virus to strawberry plants by a Nematode (*Xiphinema* sp.). *Nature, Lond.*, 36: 962.

Johnson, Jr. H.A., Converse, R.H., Amorao, A., Espejo, J.I., Frazier, N.W. 1983. Seed transmission of tobacco streak virus in strawberry. *Plant Dis.*, 68: 390–392.

Jones, A.T. 1976a. Purification of aphid borne viruses directly from raspberry. *In*: *Annual Report* 1975 of Scottish Hort. Res. Inst. pp. 73–74.

Jones, A.T. 1976b. An isolate of cucumber mosaic virus from *Rubus phoenicolasius* Maxim. *Plant Pathology*, 25: 137–140

Jones, A.T. 1979. Further studies on the effect of resistance to *Amphorophora idaei* in raspberry (*Rubus idaeus*) on the spread of aphid borne viruses. *Annl. Appl. Biol.*, 92: 119–123.

Jones, A.T. 1981. Recent research on virus diseases of red raspberry – the viruses, their effects and control. *Scottish Hort. Res. Inst. Assoc. Bull.* 19: 15–25

Jones, A.T. 1981. Performance of aphid resistant cultivars. *In*: *Scottish Hort. Res. Inst.Ann. Report*. pp. 98–99.

Jones, A.T. 1987. Raspberry leaf mottle and raspberry leaf spot. *In*: *Virus Diseases of Small Fruits; Ed.* Converse, R.H. U.S. Dep. Agric. Hdbk No. 631. pp. 183–187.

Jones, A.T. 1987. Cucumber mosaic virus in raspberry. *In*: *Virus diseases of small fruits.*Ed. Converse, R.H. *USDA* Agriculture Handbook No. 631.

Jones, A.T. 1987. Wineberry latent virus. *In*: *Virus Diseases of Small Fruits.*Ed. Converse, R.H. USDA Agric. Handbook 631. pp. 239–242.

Jones, A.T. 1991. Rubus host range of rubus yellow net virus and its involvement with other aphid-borne latent viruses in inducing raspberry vein banding mosaic disease. *Annl. Appl. Biol.*, 118: 331–338.

Jones, A.T. 2000. Black currant reversion disease – the probabale causal agent, eriophyid mite vectors, epidemiology and prospects of control. *Virus Research*, 71: 71–84.

Jones, A.T. 2007. An Isolate of Cucumber Mosaic Virus from *Rubus phoenicolasius* Maxim. *Plant Pathology*, 25: 137–140.

Jones, A.T. and Murant, A.F. 1972. Some properties of a mechanically transmissible Virus wide spread in raspberry (*Rubus idaeus*). *Pl. Path.*, 21: 166–170.

Jones, A.T. and Mayo, M.A. 1975. Further properties of black raspberry latent virus, and evidence for its relationship to tobacco streak virus. *Ann. Appl. Biol.*, 79: 297–306.

Jones. A.T. and Roberts. I.M. 1976. Ultrastrucrural changes and small bacilliform Particles associated with infection by rubus yellow net virus. *Ann. Appl. Biol.*, 84: 305–310.

Jones, A.T. and Roberts, I.M. 1977. Hostrange, symptomatology and ultrastructural effects of a British isolate of black raspberry necrosis virus. *Ann. Appl. Biol.*, 86: 381–387.

Jones, A.T. and Wood, G.A. 1979. The virus status of raspberries (*Rubus idaeus* L.) in New Zealand. *New Zealand J. Agric. Res.*, 22: 173–182.

Jones, A.T. and Jennings, D.L. 1980. Genetic control of the reaction of raspberry to Black raspberry necrosis, raspberry leaf mottle and raspberry leaf spot viruses. *Annl. Appl. Biol.,* 96: 59–65.

Jones, A.T. and Badenoch, J. 1982. Recent studies on virus and virus-like diseases of rubus in Scotland. *Acta Hort.* (ISHS), 129: 49–58

Jones, A.T. and McGavin, W.J. 1996. A damaging outbreak of arabis mosaic virus in black currant, the occurrence of other nepoviruses in Ribes species, and the demonstration that alfalfa mosaic virus is the cause of interveinal white mosaic in black currant. *Annl. Appl. Biol.,* 129: 47–55.

Jones, A.T., Roberts, I.M. and Murant, A.F. 1974. Association of different kinds of bacilliform particle with vein chlorosis and mosaic diseases of raspberry (*Rubus idaeus). Ann. Appl. Biol.,* 77: 283–288.

Jones, A.T., Murant, A.F. and Stace-Smith, R. 1977. Raspberry vein chlorosis virus. *CMI/AAB Descr. Pl. Viruses* No. 174, p. 4.

Jones, A.T., Murant, A.F. and Stace-Smith, R. 1987. *In*: *Virus Diseases of Small Fruits.*Ed. RH Converse. USDA Agric. Handbook No. 631. pp. 194–197.

Jones, A.T., Mitchell, M.J., McGavin, W.J. and Roberts, I.M. 1990. Further properties of wineberry latent virus and evidence for its possible involvement in Calico disease. *Ann. Appl. Biol.,* 117: 571–581.

Jones, A.T., McGravin, W.J., Geering, A.D.W. and Lockhart, B.E.L. 2001. A new badnavirus in *Ribes* species, its detection by PCR, and its close association with Gooseberry Vein Banding Disease. *Plant Dis.,* 85: 417–422.

Jones, A.T., Roberts, I.M. and McGavin, W.J. 2004. Unusual filamentous virus-like particles in symptomless blackberry associated with severe leaf curling in the virus indicator *Rubus macraei,* and anomalous data using a degenerate badnavirus primer in PCR of *Rubus* species. *Annl. Appl. Biol.,* 144: 87–94.

Jones, A.T., Megavin, W.J., Geering, A.D.W. and Lockhart, B.E.L. 2006. Identification of Rubus yellow net virus as a distinct badnavirus and its detection by PCR in *Rubus* species and in aphids. *Annl. Appl. Biol.,* 141: 1–10.

Jones, O.P. and Vine, S.J. 1968. The culture of gooseberry shoot tips for eliminating virus. *J. Hort. Sci.,* 43: 289–292.

Jordovic, M. 1963. Proucrvanja ekonomski najznacajnijih maline u Jugoslaviji. *Arhiv za Poljoprivredne Nauke,* 16: 3–28.

Kaiser, W.J., Wyatt, S.D. and Pesho, G.R. 1982. Natural hosts and vectors of tobacco streak virus in eastern Washington. *Phytopathology,* 72: 1508.

Kassanis, B. 1967. Tobacco necrosis virus and its satellite virus. *Nature* 214: 178.

Kennedy, J.S., Day, M.F. and Eastop, V.F. 1962. A conspectus of aphids as vectors of plant viruses. Commonwealth Institute of Entomology (London). p. 114.

Kim, K.S., Ramsdell, D.C., Gillett, J.M. and Fulton, J.P. 1981. Virions and ultrastructural changes associated with blueberry red ringspot disease. *Phytopathology,* 71: 673–678.

Kitajima, E.W., Betti, J.A. and Costa, A. S.1971. Isometric virus-like particles in leaf tissue of *Fragaria vesca* L. infected with strawberry mottle virus. *Ciencia e Cultura,* 23: 649–655.

Kitajima, E.W., Betti, J.A. and Costa, A.S. 1973. Strawberry vein banding virus, a member of the cauliflower mosaic group. *J. gen. Virol.,* 20: 117–119.

Kleinhempel, H. 1970.Verbreitung und schadwirkung von virosen an johannis-und stachelbeere. *Archive fur Gartenbau,* 18: 319–325.

Kleinhempel, H. 1972. Diagnose und Bekampfung von virusinfektionen an johannis-und stachelbeere. *Archive fur Gartenbau,* 19: 234–236.

Knight, V.H. 1981. Gooseberry vein banding virus. East Malling Research Station report, p. 114.

Komuro, Y., Iwaki, M. and Makino, T. 1973. [Occurrence of tobacco necrosis virus in roots of strawberry collected in the Watari district of Miyagi prefecture]. *Annal Phytopathol. Soc. Japan* 39: 134.

Kondakova, V. and Schusterster, G. 2008. Elimination of strawberry mottle virus and strawberry crinkle virus from isolated apices of three strawberry varieties by the addition of 2,4-Dioxohexahydro-1,3,5-triazine (5-Azadihydrouracil) to the Nutrient Medium. *J. Plant Pathol.*, 132: 84–86.

Krezal, H. 1979. Transmission of the strawberry mild yellow edge and crinkle viruses by the strawberry aphid *Chaetosiphon fragaefolii*. *Acta Hort.*, 95: 23–30.

Kulshrestha, S., Hallan, V., Raikhy, G., Adekunle, O.K., Ventra,N., Haq Q.M.R. and Zaidi, A.A. 2005. Reverse transcription polymerase chain reaction-based detection of Arabis mosaic virus and Strawberry latent ringspot virus in vector nematodes. *Curr. Sci.*, 89: 1759–1762.

Lamberti, F., Roca, F., Landriscina, S. and Ciancio, A. 1986. Seasonal transmissibility of strawberry latent ringspot virus by *Xiphinema diversicaudatum*. *Nematologia Mediterranea*, 14: 173–179.

Lehman, W. 1973. *Euscelis plebejus* (Fallen) and *Macrosteles crestatus* Ribaut als Ubertragervon pflanzenkrankheiten von vermuteter Mycoplasma-Atiologie. *Archiv Fur Phytopathologie und Pflanzenschutz*, 9: 363–370.

Lemmetty, A,. Susi, P., Latvala, S. and Lehto, K. 1998. Detection of putative causal Agent of blackcurrant reversion disease. *Acta Hort.*, 471: 93–98.

Lemmetty, A., Latvala-Kilby, S. and Lehto, K. 2001. Comparison of different isolates of Blackcurrant reversion virus. *Acta Hort.*, 551: 45–49.

Lesney, M.S. and Ramsdell, D.C. 1976. Purification and some properties of blueberry shoestring virus. *Acta Hort.* (ISHS), 66: 105–109

Lesney, M.S., Ramsdell, D.C. and Sun, M. 1978. Etiology of blueberry shoestring disease and some properties of the causal virus. *Phytopathology*, 68: 295–300.

Liefting, L.W., Veerakone, S., Clover, G.R.G. and Ward, L.I. 2011. An update on phytoplasma diseases in New Zealand. *Bull. Insectology*, 64: 93–94.

Liefting, L.W., Beever, R.E, Andersen, M.T. and Clover, G.R.G. 2007. Phytoplasma diseases in New Zealand. *Bull. Insectology*, 60: 165–166

Lister, R.M. 1964. Strawberry latent ringspot: A new nematode-borne virus. *Ann. Appl. Biol.*, 54: 167–176.

Lister, R.M. and Murant, A.F. 1967. Seed transmission of nematode borne viruses. *Ann. Appl. Biol.*, 59: 49–62.

Lister, R.M. and Converse, R.H. 1972. Black raspberry latent virus. *CMI/AAB Descriptions of Plant Viruses* No. 106. Association of Applied Biologists, Wellesbourne, UK.

Lister,R.M., Ghabrial, S.A. and Saksena, K.N. 1972. Evidance that particle size heterogeity is the cause of centrifugal heterogeneity in tobacco streak virus. *Virology*, 49: 290–299.

Maat, D.Z. 1966. Jaarverslag 1965 (Annual report). Institut fur plantenziektenkundig Onderzoek, pp. 80–81.

Mahmoudpour, A. 2004. Diagnosis and quantification of strawberry vein banding virus Using molecular approaches. *Acta Hort.* (ASHS), 656: 69–74

Maramorosch, K.1955. Transmission of blueberry stunt virus by *Scaphytopius magddalensis*. *J. Econo. Entomol.*, 48: 106.

Maramorosch, K., Shikata, E. and Granados, R.R. 1968. Mycoplasma like bodies in leafhoppers and diseased plants . *Phytopathology*, 58: 886.

Martin, R.R. and Converse, R.H. 1985. Purification, properties and serology of strawberry mild yellow-edge virus. *Phytopathologische Zeitschrift,* 114: 21–30.

Martin, R.R. and Bristow, P.R. 1988. A carlavirus associated with blueberry scorch disease. *Phytopathology,* 78: 1636–1640

Martin, R.R.,and Bristow, P.R. 1995. Blueberry shock. *In: Compendium of Blueberry and cranberry diseases. Eds.* Caruso, F.L. and Ramsdell, D.C. APS Press, St. Paul, MN.p. 87.

Martin R.R. and Tzanetakis, I.E. 2006. Characterisation and recent advances in detection of strawberry viruses. *Plant Dis.,* 90: 384–396.

Matthews, R.E.F. 1982. Classification and nomenclature of viruses. Fourth Report of the International Committee on Taxonomy of Viruses. *Intervirology,* 17: 140-141.

Mayo, M.A., Murant, A.F. and Harrison, B.D. 1971. New evidence on the structure of nepoviruses. *J. gen. Virol.,* 12: 175–178.

Mayo, M.A., Murant, A.F., Harrison, B.D. and Goold, R.A. 1974. Two proteins and two RNA species in particles of strawberry latent ring spot virus. *J. Gen. Virol.,* 24: 29–31.

Mayo, M.A, Barker, H. and Robinson, D.J. 1982. Satellite RNA in particles of strawberry latent ringspot virus. *J. gen. Virol.,* 63: 417–423.

MacDonald, S.G., Martin, R.R. and Bristow, P.R. 1991. Characterization of an ilarvirus associated with a necrotic shock reaction in blueberry. *Phytopathology,* 81: 210-214.

McGavin, W.J., Mitchell, C., Cock, P.J., Wright, K.M., and MacFarlane, S.A. 2012. Raspberry leaf blotch, a putative new member of the genus Emaravirus, encodes a genomic RNA. *J. Gen. Virol.,* 93: 430–437.

McGrew, J.R. 1970. Strawberry feather-leaf, *In*: N.W. Frazier Editor, virus diseases of small fruits and grapevines. Univ. California, Div. Agric. Sci. Berkeley,. pp. 46–47

McGrew, J.R. 1980. Meristem culture for production of virus-free strawberries. *In*: Proc. Conf. Nursery Production of fruit plants through tissue culture. Application and feasibility. USDA, SEA, Agricultural Res. Results. AAR-NE- pp. 80–85

McGrew, J.W. 1987. Strawberry latent C. *In*: *Virus diseases of small fruits. Ed.* Converse, R.H. USDA Agriculture Handbook No. 631. pp. 29–30.

MacFarlane, S., McGavin, W., Mitchell, C. and Cock, P. 2012. Raspberry leaf blotch virus, a new *Emaravirus* infecting raspberry. *In:* Proc. 22nd Interl. Conf. Virus and other graft transmissible disease. p. 63.

Martelli, G.P., Gallitelli, D., Abracheva, P., Savino, V. and Quacquarelli, A. 1977. Some properties of grapevine Bulgarian latent virus. *Ann. Appl. Biol.,* 85: 51–58.

Mellor, F.C. 1957. Studies on Rubus virus diseases in British Columbia. IV.Transmission of raspberry mosaic viruses to *Fragaria vesca* L. *Can. J. Bot.,* 35: 287–290.

Mellor, F. C. and Forbes, A. R., 1960. Studies of virus diseases of strawberries in British Columbia. *Can. J. Botany,* 38: 343–352.

Mellor, F.C. and Stace-Smith, R. 1963. Reaction of strawberry to a ringspot virus from raspberry. *Canadian J. Botany,* 41: 865–870.

Mellor, F.C. and Krezel, H. 1987. Strawberry mottle. *In*: *Virus diseases of small fruits.*(ed) R.H. Converse. USDA, Agriculture Handbook No. 631. pp. 10–16.

Miller, P.W. and Belkengren, R.O. 1962. Obtaining virus-free strawberries by excising and culturing of apical meristem of plants infected with the yellow edge, crinkling and vein banding viruses and certain other virus complexes. *Phytopathology,* 52: 743–744.

Miller, P. W. and Frazier, N. W., 1970. Strawberry vein banding virus. *In*: Virus diseases of small fruits and grapevines. Frazier, N.W. *Ed*. University of California, Division of Agricultural Sciences, Berkeley. pp. 8–10.

Morimoto, K.M. and Ramsdell, D.C. 1985. Aphid vector population dynamics and movement related to field transmission of shoe sting virus. *Phytopathology,* 75: 1217–1222.

Morimoto K.M., Ramsdell D.C., Gillett J.M. and Chaney W.G. 1985. Acquisition and transmission of Blueberry shoestring virus by its aphid vector *Illinoia pepperi. Phytopathology 75*: 709–712.

Morris, T.J., Mullin, R.H., Schlegel, D.E., Cole, A. and Alosi, M.C. 1980. Isolation of a caulimovirus from strawberry tissue infected with strawberry vein banding virus. *Phytopathology,* 70: 156.

Mullin, R.H., Frazier, N.W., and Schlegel, D.E. 1975. Meristem culture of *Fragaria chiloensis* infected with pallidosis. *Pl. Dis. Reptr,* 59: 268

Mullin, R.H., Smith, S.H., Frazier, N.W., Schlegel, D.E. and McCall, S.R. 1974. Meristem culture frees strawberries of mild yellow edge, pallidosis, and mottle diseases. *Phytopathology* 64: 1425–1429.

Murant, A.F. 1970. Arabis mosaic virus. *CMI/AAB Descriptions of Plant Viruses* No. 16. Association of Applied Biologists, Wellesbourne, UK.

Murant, A.F. 1970a. Raspberry ringspot virus. *CMI/AAB Descriptions of Plant Viruses* No. 6. Association of Applied Biologists, Wellesbourne, UK.

Murant, A.F. 1970b. Soilborne viruses and diseases in *Rubus*. *In*: Virus diseases of small fruits and grapevines, a handbook (*Ed*. by Frazier, N.W.), University of California, Berkeley, California, USA. pp. 132–136.

Murant, A.F.1974. Strawberry latent ringspot virus. *CMI/AAB Descr. Pl. Viruses* No. 126, p. 4. .

Murant, A.F.1974. Viruses infecting raspberry in Scotland. *Scottish Hort. Res. Inst. Association, Bull.,* 9: 37–43.

Murant, A.F. 1976. Strawberry latent ringspot virus. *CMI/AAB Descriptions of Plant Viruses* No. 126. Association of Applied Biologists, Wellesbourne, UK.

Murant, A.F. 1981. Nepoviruses. *In*: Handbook of plant virus infections and Comparative diagnosis (Kurstal, E. *Ed*.). Elsevier, Amsterdam, The Netherlands. pp. 197–238.

Murant, A.F. 1987. Raspberry ringspot and associated diseases of *Rubus* caused by raspberry ringspot and tomato black ring viruses. *In*: *Virus diseases of small fruits.* Ed. Converse, R.H., *USDA* Agric. Hdbk No. 631. pp. 211–220.

Murant, A.F. 1987. Raspberry bushy dwarf. *In*: *Virus Diseases of Small Fruits.* Ed. RH Converse, USDA Agric. Hdbk No. 631. pp. 229–234.

Murant, A.F. 1987. Raspberry yellow dwarf and associated diseases of Rubus caused by Arabis mosaic and Strawberry latent ringspot viruses. *In*: *Virus Diseases of Small Fruits* (ed). R.H .Converse. USDA Agric. Hdbk. 631. pp. 204–211.

Murant A. F. and Lister R. M. 1967. Seed transmission in the ecology of nematode-borne viruses. *Ann Appl. Biol.,* 59: 63–76.

Murant, A.F. and Lister, R.M. 1987. European nepoviruses in strawberry. *In*: *Virus diseases of small fruits.* Ed. Converse, R.H., USDA Agric. Hdbk No. 631. pp. 46–51.

Murant, A.F.and Taylor, C.E. 1965. Treatment of soil with chemicals to prevent transmission of tomato blackring and raspberry ringspot viruses by *Longidorus elongatus* (de Man). *Ann. Appl. Biol.,* 55: 227–237.

Murant, A.F. and Roberts, I.M.1971. Mycoplasma-like bodies associated with Rubus Stunt disease. *Ann. Appl. Biol.,* 67: 389–393.

Murant, A.F. and Roberts, I.M. 1981. Particles of raspberry vein chlorosis virus in the aphid vector, *Aphis idaei*. *Acta Hort.* (ISHS), 94: 103–106

Murant, A.F., Taylor, C.E. and Chambers, J. 1968. Properties, relationships and transmission of a strain of raspberry ringspot virus infecting raspberry cultivars immune to the common Scottish strain. *Ann. Appl. Biol.*, 61: 175–186.

Murant, A.F., Mayo, M.A., Harrison, B.D. and Goold, R.A. 1972. Properties of virus and RNA components of Raspberry ringspot virus. *J. gen. Virol.*, 16: 327–338.

Murant, A.D., Jones, A.T. and Roberts, I.M. 1976. Recent research on '52V virus of Raspberry. *Acta Hort.*, 66: 39–46.

Nariani, T.K. and Pathanian, P.S. 1953. *Physalis peruviana* L., a new host of tobacco leaf curl virus. *Indian Phytopath.* 6: 143–145.

Németh, M. 1986. Virus, mycoplasma and rickettsia diseases of fruit trees. Martinus Nijhoff Publishers, The Netherlands, p. 840.

Nourrisseau, J.G., Lansac, M., and Garnier, M. 1993. Marginal chlorosis, a New Disease of Strawberries Associated with a Bacterium-like Organism. *Plant Dis.* 77: 1055–1059.

OEPP/EPPO. 1994. Certification scheme No.11. Pathogen tested strawberry OEPP/ EPPO Bull. 24: 875–889.

Pacot-Hiriart, C., Latvala-Kilby, S. and Lehto, K. 2001. Nucleotide sequence of black currant reversion associated nepovirus RNA1. *Virus Res.*, 79: 145–152.

Pelet, F. 1989. Small fruit viruses. *Revue Suisse de Viticulture, Arboriculture, Horticulture* 21: 113–116.

Petrzik, K., Benes, V., Mráz, I., Honetslegrová-Fránová, J., Ansorge,W. and Spak, J. 1998. Strawberry vein banding virus – definitive member of genus *Caulimovirus*. *Virus Genes*, 16: 303–305.

Poggi-Pollini, C. and Giunchedi, L. 1987. Preliminary research on decline of raspberry cv. Lampone di Peveragno in Piemonte. *Phytopathologia Mediterranea*, 26: 132–136

Posnette, A.F. 1952. New virus diseases of Ribes. East Malling Research Stn. Report for pp. 133–135.

Posnette, A.F. 1953. Green petal, a new virus disease of strawberries. *Plant Pathology*, 2 : 17

Posnette, A.F. and Cropley, R. 1958. Heat treatment for the inactivation of strawberry viruses. *J. Hort. Sci.*, 33: 282–288.

Potapov, S. and Grinenko, A. 1968. [Reversion in black currant and problems of breeding] *Moskva*, 13: 77–80.

Pyibylova, J., Špak, J., Kubelková, D. and Lemmetty, A. 2008. The presence of Blackcurrant reversion virus in red and white currants with symptoms of full blossom disease. *Acta Hort.*, 656: 131–136.

Pøibylová, J., Spak, J., Petrzik,D., Kubelkova, D. and Spakova,V. 2008. Sequence comparison and transmission of black currant reversion virus isolates in black, red and white currants with black currant reversion disease and full blossom disease symptoms. *European J. Plant Path.*, 121: 67–75.

Putz, C. 1971. Obtention de framboisiers (var. Bois Blank) sans virus par la technique Descultures de meristemes. *Annl. De Phytopathologie*, 3: 493–501.

Quito, D., Jelkmann, W., Alt, S., Leible, S. and Martin, R.R. 2010. A new member of the family *Reoviridae* may contribute to severe crumbly fruit in red raspberry, *Rubus idaeus* 'Meeker'. *In*: Proc. 21st Interl. Conf. Virus and other graft transmissible diseases of fruit crops. Julius-Kuhn-Archiv 427: 18–23.

Ragozzino, A. and Alioto, D. 1992. Cherry rasp leaf etiology in Campania. *Acta Hort.*, 309: 115–118.

Ramiah, M., Bhat., A.I., Jain, R.K., Pant, R.P., Ahlawat., Y.S., Prabhakar, K. and Varma, A. 2001. Partial characterisation of an isometric virus causing sunflower necrosis disease. *Indian Phytopath.* 54: 246–250.

Ramsdell, D.C. 1979*a*. Blueberry shoe sting virus.*CMI/AAB Descr. Pl. Viruses* No. 204, p. 4.

Ramsdell, D.C. 1979*b*. Physical and chemical properties of blueberry shoesting virus. *Phytopathology,* 69: 1087–1091.

Ramsdell, D.C. 1987. Blueberry shoesting. *In*: *Virus Diseases of Small Fruits; Ed.* Converse, R.H. U.S. Dep.agric. Hdbk No. 631.

Ramsdell, D.C. 1987. Necrotic ringspot of blueberry. *In*: *Virus diseases of small fruits,* USDA Agriculture Handbook No. 631. pp. 114–116.

Ramsdell, D.C. 1995. Shoestring. *In: Compendium of Blueberry and Cranberry diseases.* Caruso, F.L. and Ramsdell, D.C. *Eds.* APS Press, St. Paul, MN. pp. 50–51.

Ramsdell, D.C. 1995. Red ringspot. *In: Compendium of Blueberry and cranberry diseases.* Eds. Caruso, F.L. and Ramsdell, D.C. APS Press, St. Paul, MN. p. 87.

Ramsdell, D.C. and Stace-Smith, R.1979. Blueberry leaf mottle, a new disease of high bush blueberry. *Acta Hort.,* 95: 37–45.

Ramsdell, D.C. and Stace-Smith, R. 1981. Physical and chemical properties of the particles and Ribonucleic acid of bluberry leaf mottle virus. *Phytopathology,* 71: 468–472.

Ramsdell, D.C. and Stace-Smith, R. 1983. Blueberry leaf mottle virus. CMI/AAB Description of plant viruses No. 267. Association of applied biologists, Wallesbourne, UK.

Ramsdell, D.C. and Stretch, A.W. 1987. Blueberry mosaic. *In*: *Virus diseases of small fruits.* USDA Agriculture Handbook No. 631. pp. 119–120.

Ramsdell, D.C., Kim, K.S. and Fulton, J.P. 1987. Red ringspot of Blueberry. *In*: *Virus diseases of small fruits.* USDA Agriculture Handbook No. 631. pp. 121–123

Randles, J.W. and Rattigen, J.P. 1995. *Luteovirus* genus. *Archives of Virology,Supplement,* 10: 379–383.

Randles, J.W., Harrison, B.D., Murant, A.F. and Mayo, M.A. 1977. Packaging and biological activity of the two essential RNA species of tomato black ring virus. *J. Gen. Virol.,* 36: 187–193.

Raniere, L.C.1960. Responses of cultivated highbush blueberry varieties to the known blueberry viruses. *Proc. 28th Ann. Blueberry Open house,* 28: 18–20

Rankin, W.H. and Hockey, J.F. 1922. Mosaic and leaf curl (yellows) of the cultivated red raspberries. *Phytopathology* 12: 253–264.

Ratti, C., D'Alonzo, M., Babini, A.R., Pisi, A. and Rubies, A.C. 2006. Real-time RT-PCR (TaqmanÂ®) protocol for *Strawberry mottle virus* detection and quantification. *In*: XX International Symposium on Virus and Virus-like Diseases of Temperate Crops Antalya (Turkey), 22-26 May 2006. p. 148.

Richardson, J., Frazier, N.W. and Sylvester, E.S. 1972. Rhabdovirus-like particles associated with strawberry crinkle virus. *Phytopathology,* 62: 491–492.

Richter, J. and Proll, E. 1970. Purification and fractionation of Strawberry latent ringspot virus *Acta phytopath. Acad. Sci. Hung.,* 5: 151–166.

Roberts, I.M. and Brown, D.J.F. 1980. Detection of six nepoviruses in their nematode vectors by immunosorbent electron microscopy. *Ann. Appl. Biol.,* 96: 187–192.

Robinson, D.J. and Harrison, B.D. 1989. Tobacco rattle virus. AAB description of plant viruses No. 346. Association of Applied Biologists, Wellesbourne.

Roossinck, M.J., Bujarski, J.J., Ding, S.W., Hajimorad, R., Hanada, K., Scott, S. and Tousignant, M. 2000. *In*: *Virus Taxonomy. Seventh Report of the International Committee on Taxonomy of Viruses, Eds* Regenmortel, M.H.V. *et al,* pp. 11–62

Rüdel, M.1985. Grapevine damage induced by particular virus-vector combinations. *Phytopathologia Mediterranea,* 24: 183–185.

Rukus, D. 1978. The incidence of currant full blossom in the CSIR. *In* Proc. VII Czechoslovak Plant Prot. Conf. Nitra, 6-8th Sept., 1978. p. 132.

Rydén, K. 1972. *Pelargonium* ringspot - a virus disease caused by tomato ringspot virus in Sweden. *Phytopathologische Zeitschrift* 73: 178–182.

Sabanadzovic, S., Ghanem- Sabanadzovic, N.A. and Tzanetakis, I.E. 2012. In Abs. 22nd Interl. Conf. Virus and other graft transmissible diseases of fruit crops. Held in Rome, p. 188.

Scarborough, B.A. and Smith, S.H. 1977. Effects of tobacco and tomato ringspot viruses on the reproductive tissues of Pelargonium x hortorum. *Phytopathology,* 67: 292–297.

Schmelzer, K. 1963. [Investigations on viruses of ornamental and wild woody plants. 2nd part. Viroses of *Forsythia, Lonicera, Ligustrum* and *Laburnum*]. *Phytopathologische Zeitschrift* 46: 105–138.

Schmelzer, K. 1969. [SLRV from *Euonymus, Robinia* and *Aesculus*]. *Phytopathologische Zeitschrift,* 66: 1–24.

Schöniger, G. 1958. Erdbeervirosen in Deutschland. III. Das Erdbeer-Nekrosevirus: Fin weiteres nicht-persistentes virus. *Phytopathologische Zeitschrift,* 32: 325-332.

Schwartze, C.D. and frazier, N.W. 1970. Lethal decline of strawberry, *In*: Frazier, N.W. *Ed.*, virus diseases of small fruits and grapevines, Univ. California, Div. Agric. Sci. Berkeley, 290. pp. 50–51,

Schwartze, C.D., Frazier, N.W. and Converse, R.H. 1987. *In*: Virus diseases of small fruits . Ed. Converse, R.H., USDA Agriculture Handbook No. 631: pp. 38–41.

Scott,S.W., Zimmerman, M.T. and Ge, X. 1998. The sequence of RNA 1 and RNA 2 of Tobacco streak virus: Additional evidence for the inclusion of alfalfa Mosaic virus in the genus Ilarvirus. *Archives of Virology,* 143: 1187–1198.

Scotto La Massese, C., Marenaud, C. and Dunez, J. 1973. Analyse d'un phenomena de degenerescence du pecher dans la vallee de l'Eyrieux. *C.R. séances Acad. Agric. Fr.* 59: 327–339.

Sdoodee, R. and Teakle, D.S. 1987. Transmission of tobacco streak virus by *Thrips tabaci*: A new method of plant virus transmission. *Plant Path.,* 36: 377.

Sehgal, O.P. and Boone,D.M. 1963. Multiplier disease of strawberry in Wisconsin. *Plant Dis. Reptr.,* 47: 46–48

Shanmuganathan, N. and Garrett, R.G. 1976. Little leaf of strawberry. *Australian Plant Pathology Soc. News letter,* 5(suppl.): 83.

Shear, C.L. 1908. Cranberry disease in wisconsin. Wisconsin State Cranberry growers *Association Ann. Rep.,* 21: 17–21.

Silvere, A.P. and Remeikis, M.A. 1973. [Formation and release of mesosome-like microbodies in the sporogenesis of the endophytic current reversion bacillus]. *Eesti NSV Teaduste Acadeemia Toimetised Bioloogia,* 22: 274–277.

Sinha, R.C. and Paliwal, Y.C. 1969. Association, development and growth cycle of mycoplasma-like organisms in plants affected with clover phyllody. *Virology,* 39: 759–767.

Sinha, R.C. and Paliwal, Y.C. 1970. Localization of a mycoplasma-like organism in tissues of a leafhopper vector carrying clover phyllody agent. *Virology* 40: 664–672.

Sinha, R.C. and Peterson, E.A. 1972. Uptake and persistence of oxytetracycline in aster plants and vector leafhoppers in relation to inhibition of clover phyllody agent. *Phytopathology,* 62: 377–383.

Smith, K.M. and Markham, R. 1944. Two new viruses affecting tobacco and other plants. *Phytopathology,* 34: 324–329.

Smith, K.M. 1946. Tomato black ring : A new virus disease. *Parasitology,* 37: 126–130.

Smith, K.M..1972. *A textbook of plant virus diseases* (edition 3), Longman, London, UK. pp. 541–544.

Sobczykiewicz, D. 1986. Elimination of heat stable raspberry vein chlorosis virus by meristem culture. *Acta Hort.* (ISHS), 186: 47–50

Spiegel,S. and Martin, R.R. 1998. Virus and virus-like diseases. *In*: Compendium of strawberry diseases.J.L.Maas ed. APS, St. Paul, MN. pp. 62–72.

Spiegel, S., Cohen, J. and Converse, R.H. 1986. Detection of strawberry mild yellow-edge virus by serologically specific electron microscopy. *Acta Hort.,* 86: 95.

Spiegel, S., Tam, Y., Martin, R.R. and Ter Borg, M. 1995. Uneven distribution of tobacco streak virus in strawberry plantlets grown *in vitro*. *Acta Hort. (ISHS)* 385: 122–125.

Stace-Smith, R. 1955. Studies on Rubus virus diseases in British Columbia. 1. Rubus yellow- net. *Can J. Bot.,* 33: 267–274.

Stace-Smith, R. 1955. Studies on Rubus virus diseases in British Columbia II. Black raspberry necrosis. *Canadian J. Bot.,* 33: 314–322.

Stace-Smith, R.1956. Studies on Rubus virus diseases in British Columbia. 111. Separation of components of raspberry mosaic. *Can J. Bot.,* 34: 435–442.

Stace-Smith, R. 1958. Studies on Rubus virus diseases in British Columbia. V. Thimbleberry ringspot. *Can. J. Bot.,* 36: 385–388.

Stace-Smith R. 1961. Studies on Rubus virus diseases in British Columbia. VII. Raspberry vein chlorosis. *Canadian J. Bot.,* 39: 559–565.

Stace-Smith, R. 1962. Studies on Rubus virus diseases in British Columbia. IX. Ringspot disease of red raspberry. *Can. J. Bot.,* 40: 905–912.

Stace-Smith, R.1962. Studies on Rubus virus diseases in British Columbia. VIII. Raspberry leaf curl. *Canadian J. Bot.,* 40: 651–657.

Stace-Smith, R. 1970. Tomato ringspot virus. Description of plant viruses No. 17. Commonwealth Mycological Institute and the association of applied biologists, Kew, Surrey, England.

Stace-Smith, R. 1984. Tomato ringspot nepovirus. *CMI/AAB Descriptions of Plant Viruses* No. 290. Association of Applied Biologists, Wellesbourne, UK.

Stace-Smith, R. 1985. Tobacco ringspot virus. *CMI/AAB Descriptions of Plant Viruses* No. 309 (No. 17 revised). Association of Applied Biologists, Wellesbourne, UK.

Stace-Smith, R. 1987. Thimbleberry ringspot. *In*: *Virus Diseases of Small Fruits.* Ed. R.H. Converse, USDA Agric. Hdbk No.631. pp. 192-194.

Stace-Smith, R. and Mellor, F.C. 1957. Studies on Rubus virus diseases in British Columbia. IV. Transmission of Raspberry mosaic virus to *Fragaria vesca* L. *Canadian J. Bot.,* 35: 287–290.

Stace-Smith, R. and Frazier, N.W. 1971. Tobacco streak virus isolated from strawberry infected with necrotic shock. *Phytopathology,* 61: 757.

Stace-Smith, R. and Lo, E. 1973. Morphology of bacilliform particles associated with raspberry vein chlorosis virus. *Canadian J. Bot.,* 51: 1343–1345

Stace-Smith, R. and Jones, A.T. 1978. Rubus yellow net virus. *CMI/AAB Descr. Pl. Viruses* No. 188, Commonwealth Agricultural Bureaux, Slough.

Stace-Smith, R. and Ramsdell, D. 1987. Nepoviruses of the Americas. *In*: Current topics in vector research, Vol. 3 (Harris, R.V. *Ed.*) Springer-Verlag, New York USA pp. 131–166.

Stace-Smith, R. and Jones, A.T. 1987. Black raspberry necrosis. *In*: *Virus Diseases of Small Fruits.* Ed. R.H. Converse, USDA Agric. Hdbk No. 631. pp. 178–183.

Stace-Smith, R. and Converse, R.H. 1987. Raspberry leaf curl. *In*: *Virus diseases of small Fruits.* Ed. Converse, R.H. *USDA Agriculture Handbook* No. 631. pp. 187-190.

Steinkellner, H., Himmler, G., Laimer, M., Mattanovich, D., Bisztray, G. and Katinger, H. 1989. Construction of cDNA of arabis mosaic virus and its use for diagnosis. *Mitteilungen Klosterneuburg, Rebe und Wein, Obstbau und Früchteverwertung,* 39: 242–246.

Stenger, D.C., Mullins, R.H. and Morris, T.J. 1987. Characterization and detection of the strawberry necrotic shock isolate of tobacco streak virus. *Phytopathology* 77: 1330–1337.

Stenger, D.C., Mullin, R.H. and Morris, T.J. 1988. Isolation, molecular cloning and detection of strawberry vein banding virus DNA. *Phytopathology,* 78: 154–159.

Stretch, A.W. 1987. Ringspot of cranberry. *In*: *Virus diseases of small fruits. Ed.* Converse, R.H. USDA Agriculture Handbook No. 631. pp. 123–124.

Stretch, A.W. 1987. Cranberry false blossom. PP. 110-111. In: *Virus diseases of small fruits. Ed.* Converse, R.H. USDA Agriculture Handbook No. 631.

Stretch, A.W. and Hilborn. 1970. *In*: *Virus Diseases of Small Fruits and Grapevines,* University of California Press. CA. p. 186.

Streten, C., Waite, G.K., Herrington, M.E., Hutton, D.G., Persley, D.M. and Gibb, K.S. 2005. Rickettsia-like-organisms and phytoplasmas associated with diseases in Australian strawberries. *Australian Plant Pathology,* 34: 157–164.

Stubbs, L.L. 1968. Lethal yellows disease of strawberry. *Australian Plant Dis. Recorder,* 20: 33–34.

Sylvester, E.S., Frazier, N.W. and Richardson, J. 1976. Strawberry crinkle virus. *CMI/AAB Descriptions of Plant Viruses* No. 163. Association of Applied Biologists, Wellesbourne, UK.

Sylvie Malembic-Maher, Pascal Salar, Luisa Filippin, Patricia Carle, Elisa Angelini and Xavier Foissac. 2011. Genetic diversity of European phytoplasmas of the 16SrV taxonomic group and proposal of '*Candidatus* Phytoplasma rubi. Interl. *J. Syst.Evolu. Microbiol,* 61: 2129–2134.

Taylor, C.E. 1962. Transmission of Raspberry ringspot virus by *Longidorud elongatus* (de Man) (Nematoda Dorylaimidae). *Virology,* 17: 493–494 .

Taylor, C.E. 1967. The multiplication of *Longidorus elongatus* (de Man) on different host plants with reference to virus transmission. *Ann. appl. Biol.,* 59: 275–281

Taylor, C.E. and Murant, A.F. 1968. Chemical control of raspberry ringspot and tomato black ring viruses in strawberry. *Plant Pathology,* 17: 171–178.

Taylor, C.E. and Murant, A.F. 1969. Transmission of strains of raspberry ringspot and tomato black ring viruses by *Longidorus elongatus* (de Man). *Ann. Appl. Biol.,* 64: 43–48.

Teliz, D., Grogan R.G. and Lownsberry, G.F. 1966. Transmission of tomato ringspot, peach yellow-bud mosaic, and grape yellow-bud mosaic, and grape yellow-vein viruses by *Xiphinema americanum. Phytopathology,* 56: 658–663.

Theiler-Hedtrich, R. and Baumann, G. 1989. Elimination of apple mosaic virus from infected red raspberry (*Rubus idaeus*) by tissue culture. *J. Phytopathol.,* 127: 193–199.

Thekke-Veetil, T., Martin, R.R. and Tzanetakis, I.E. 2012. Univ. Arkansas, Fayetteville, AR, USA, T.-V.T. and T.I.E.) and USDA-ARS, Corvallis, OR, USA (M.R.R.) Unpublished data.

Thomas, P.R. 1969. Crop and weed plants compared as hosts of viruliferous *Longidorus elongatus* (de Man). *Pl. Path.* 18: 23–28

Thompson, I.J. and McKenry, M. 1975. Chemical control of nematode vectors of plant viruses. *In*: Lamberti, F., Taylor, C.E. and Seinhorst, J.W. *Eds,* Nematode vectors of plant viruses. NATO adv. Study Inst. Series A: Life Sciences, Vol.2, Plenum Press, London, p. 460.

Thompson, L.S. 1968. Strawberry green petal virus transmission in the field. *Plant Dis.Reptr.*, 52: 393–394.

Thompson, L.S., Cutcliffe, J.A., Gourley, C.O. and Murray, R.A. 1973. Evaluation of several insecticides for control of strawberry green petal disease. *Canadian Plant Disease Survey*, 53: 16–18

Thompson J.R. and Jelkman, W. 2003. The detection and variation of strawberry mottle virus. *Plant Dis.*, 87: 385–390.

Thompson, J. R., Leone, G., Lindner, J.L., Jelkmann, W. and Schoen, C.D. 2002. "Characterization and complete nucleotide sequence of Strawberry mottle virus: A tentative member of a new family of bipartite plant picorna-like viruses". *J. Gen. Virol.* 83(Pt 1): 229–239.

Thompson JR., Wetzel, S., Klerks, M.M., VaÅ¡ková, D., Schoen, C.D., Å pak, J. and Jelkmann, W. 2003b. Multiplex RT-PCR detection of four aphid-borne strawberry viruses in *Fragaria* spp. in combination with a plant mRNA specific internal control. *J. Virological Methods,* 111: 85–93.

Thresh, J.M. 1964a. Increased susceptibility to the mite vector caused by infection with black currant reversion virus. *Nature,* London 202: 1028.

Thresh, J.M. 1964b. Warm water treatments to eliminate the gall mite from black currant cuttings. *In*: East malling research station report for. pp. 131–132.

Thresh, J.M. 1965. The effect of gall mite on the leaves and buds of black currant bushes. *Plant Pathol.*, 14: 26–30.

Thresh, J.M. 1967. Virus diseases of red currant. *In*: East Malling Res. Stn. Report, pp. 146–152..

Thresh, J.M.1970. Virus and virus-like diseases of gooseberry and current. *In*: *Virus diseases of small fruits and grapevine.* Ed. N.W. Frazier, Univ. California, Div. Agric. Sci, Berkeley, pp. 75–104.

Thung, T.H. 1952. Observations on stunt in Raspberry and wild blackberry (in Dutch). *Tijdschrift Over Planteziekten,* 58: 255–259.

Ting Wei, Guangjin Lu and Gerard Clover.2008. Novel approaches to mitigate primer interaction and eliminate inhibitors inmultiplex PCR, demonstrated using an assay for detection of three strawberry viruses. *J Virol Methods,* 151: 132–139.

Tozzi, D.C.M., Ramsdell, D.C., Taboada, O., Lee, I.M. and Davis, R.E.1993. Epidemiological studies on the stunt disease of highbush blueberry. *Ann. Appl. Biol.*, 123: 579–599

Trudgill, D.L.and Alphey, T.J.W. 1976. Chemical control of the vector nematode *Longidorus elongatus* and of *Pratylenchus crenatus* in raspberry plantations. *Plant Pathology,* 25: 15–20.

Tzanetakis, I.E. and Martin, R.R. 2004. Complete nucleotide sequence of a strawberry isolate of Beet pseudo yellows virus. *Virus Genes,* 28: 239–246.

Tzanetakis, I.E. and Martin, R.R. 2007. Strawberry chlorotic fleck: Identification and characterization of a novel Closterovirus associated with the disease. *Virus Res.*, 124: 88–94.

Tzanetakis, I.E. and Martin, R.R. 2008. Strawberry chlorotic fleck disease may be caused by a novel *closterovirus. Acta Hort.* (ISHS), 780: 21–26

Tzanetakis, I.E., Wintermantel, W.M. and Martin, R.R. 2003. First report of *Beet pseudo yellows virus* in strawberry: A second crinivirus able to cause pallidosis disease. *Plant Dis.*, 87: 1398.

Uyemoto, J.K. 1975. A severe outbreak of virus-induced grapevine decline in cascade grapes in New York. *Plant Dis. Reptr.,* 59: 98–101.

Valiunas, D.,Alminaitel, A., Jomantienel, R., Davis, R.E. and Maas, J.L. 2004. Possible cause of european blueberry disease is related to North American milkweed yellows phytoplasma. *J. Plant Pathology,* 86: 135–140.

Valiunas, D., Staniulis, J. and Davis R.E. 2006. '*Candidatus* phytoplasma fragariae' a novel phytoplasma taxon in yellows diseased strawberry, *Fragaria ananassa. Internl. J. Syst. and Evolut. Microbiology,* 56: 277–281

Van der Meer, F.A. 1965. Virusziekten van bessen. (Virus diseases of currents) Jaarverslag 1964 (annual report). Institut fur Plantenziektenkundig Onderzock, pp. 85–87.

Van der Meer, F.A. 1987. Gooseberry deterioration. *In*: *Virus Diseases of Small Fruits.* Ed. R.H .Converse. USDA Agric. Handbook 631 p. 131.

Van der Meer, F.A. 1987. Leaf malformation of red currant. *In*: *Virus Diseases of Small Fruits. Ed.* Converse. R.H. USDA Agric. Handbook pp. 158–159.

Van der Meer, F.A. 1987. Budlessness of red currant. *In*: *Virus Diseases of Small Fruits* (*Ed*). Converse, R.H. USDA Agric. Handbook 631. p. 160.

Van der Meer, F.A. 1987. Leaf pattern of red current. *In*: *Virus Diseases of Small Fruits* (*Ed*). Converse. R.H. USDA Agric. Handbook 631. pp. 150–152.

Van der Meer, F.A.1987. Interveinal white mosaic. *In*: *Virus Diseases of Small Fruits.* Ed. Converse, R.H. USDA Agric. Handbook 631. pp. 153–155.

Van der Meer, F.A. 1987. Full Blossom of red currant. *In*: *Virus Diseases of Small Fruits. Ed.* R.H. Converse. USDA Agric. Handbook 631. pp. 155–157.

Van der Meer, F.A. 1987. Green mottle of red currant. *In*: *Virus Diseases of Small Fruits.* Ed. Converse, R.H . USDA Agric. Handbook No.631. pp. 145–146.

Van der Meer, F.A. 1987. Red current vein banding. *In*: *Virus diseases of small fruits.* Ed. Converse, R.H. USDA Handbook No. 631. pp. 143–145.

Van der Meer, F.A. 1987. Spoon leaf of red currant. *In*: *Virus Diseases of Small Fruits.* Ed. Converse. R.H. USDA Agric. Handbook 631. pp. 145-146

Van der Meer, F.A. 1987. Infection of red currant with Arabis mosaic and Strawberry latent ringspot viruses. *In*: *Virus Diseases of Small Fruits.* (*Ed*). Converse. R.H . USDA Agric. Handbook 631. pp. 148–149.

Varney, E.H. 1957. Mosaic and shoestring virus diseases of cultivated blueberry in New Jersey. *Phytopathology,* 47: 307–309.

Vindimian, M.E., Grassi, A., Ciccotti, A., Pollini, C.P. and, Terlizzi, F. 2004. Epidemiological studies on Rubus stunt (RS) in blackberry orchards located near Trento(Italy). ISHS *Acta Hort.* 656 : X International Symposium on small fruit virus diseases.

Ward, L.I., Quinn,B.D., Tang, J., Wei, T. and Clover, G.R.G. 2008. Strawberry mottle Virus detected on strawberry in New Zealand. *New Disease Reports,* 17: 13.

Williams, H.E., Holdemann, Q.L. and Casper, R.1987. Tomato ringspot virus in red currant.. *In*: *Virus Diseases of Small Fruits. Ed.* Converse, R.H. USDA Agric. Handbook 631. pp. 149–150

Winter, J.D. 1929. Raspberry mosaic. *J. Economic Entomology,* 22: 486–490

Wood, C.A. and Anderson, M.M. 1959. Raspberry cultivation in light of recent research. *Commercial Grower* 3323: 338–340.

Wood, G.A. 1991. Three graft transmissible diseases and a variegation disorder of small fruit in New Zealand. *N.Z.J. Crop Hort. Sci.,* 19: 313–323

Xu, D., Mock, R., Kinard, G. and Li, R. 2011. Molecular analysis of the complete genomic sequences of four isolates of gooseberry vein banding associated virus. *Virus Genes,* 43: 130–137.

Yang, I. L., Deng, T.C. and Chen, M.J. 1986. Sap transmissible viruses associated with grapevine yellow mottle disease in Taiwan. *J. Agricultural Research China,* 35: 504–510.

Yoshikawa, N. and Inouye, T. 1986. Purification, characterization and serology of strawberry pseudo mild yellow edge virus. *Ann. Phytopathol. Soc. Japan,* 52: 643–652.

Yoshikawa, N. and Inouye, T. 1988. Strawberry viruses occurring in Japan. *Acta Hort.,* 236: 59–67.

Yoshikawa, N. and Converse, R.H. 1990. Strawberry pallidosis disease: Distinctive dsRNA species associated with latent infection. *Phytopathology,* 80: 543.

Yoshikawa, N. and Converse, R.H. 1991. Purification and some properties of strawberry mottle virus. *Ann. Appl. Biol.,* 118: 565–576.

Yoshikawa, N., Ohki, S.T., Kobatake, H., Osaki, T. and Inouye, T. 1984. Luteovirus-like Particles in phloem tissue of strawberry mild yellow edge virus infected plants. *Ann. Phytopathol. Soc. Japan,* 50: 659–663.

Yoshikawa, N., Inouye, T. and Converse, R.H. 1986. Two types of rhabdovirus in strawberry. *Ann. Phytopathol. Soc. Japan,* 52: 437–444.

Yoshikawa, N., Poolpol, P. and Inouye, T. 1986. Use of dot immunobinding assay for rapid detection of strawberry pseudo mild yellow edge virus. *Ann. Phytopathol. Soc. Japan,* 52: 728–731.

Zeller, S.M. 1927. Preliminary studies on witches' broom of strawberry. *Phytopathology,* 17: 329–335.

Zirka, T.I., Skrypal, I.G., and Alekseenko, I.P. 1977. [Possible etiological factors of black currant reversion. DDR 152: 101–105.

Annexure-1
ORIGIN OF FRUITS

Most of the fruit plants have their origin in North America, Mediterranean region, Australia and Asia. Origin of important fruit plants in these regions has been shown below:

Fruits of North American origin

Canada, Mexico, and the United States are home to a large number of fruits. Some of these fruits are either grown commercially on a global scale (grapes, cranberries, and blueberries) or they are still eaten locally:

Fruit plant	*Botanical name*	*Family*
American Mayapple	*Podophyllum peltatum*	Berberidaceae
American plum	*Prunus americana*	Rosaceae
American persimmon	*Diospyros virginiana*	Ebenaceae
Beach Plum	*Prunus maritima*	Rosaceae
Black cherry	*Prunus serotina*	Rosaceae
Black raspberry	*Rubus occidentalis* or *Rubus leucodermis*	Rosaceae
Blueberry	*Vaccinium*, sect. *Cyanococcus*	Ericaceae
Buffaloberry	*Shepherdia argentea*	Elaeagnaceae
Chokecherry	*Prunus virginiana*	Rosaceae
Cocoplum	*Chrysobalanus icaco*	Chrysobalanaceae
Cranberry	*Vaccinium oxycoccus*	Ericaceae
False-mastic	*Mastichodendron foetidissimum*	Sapotaceae
Florida Strangler Fig	*Ficus aurea*	Moraceae
Ground Plum	*Astragalus caryocarpus*	Fabaceae
Mayhaw	*Crataegus aestivalis*	Rosaceae
Huckleberry	*Vaccinium parvifolium*	Ericaceae
Maypop	*Passiflora incarnata*	Passifloraceae
Pawpaw	*Asimina triloba*	Annonaceae
Prickly pear	*Opuntia* spp	Cactaceae
Red mulberry	*Morus rubra*	Moraceae
Pigeon plum	*Coccoloba diversifolia*	Polygonaceae
Salal berry	*Gaultheria shallon*	Ericaceae
Salmonberry	*Rubus spectabilis*	Rosaceae
Saskatoonberry	*Amelanchier alnifolia*	Rosaceae
Saw Palmetto	*Serenoa repens*	Arecaceae
Southern crabapple	*Malus angustifolia*	Rosaceae
Strawberry	*Fragaria × ananassa*	Rosoideae
Thimbleberry	*Rubus parviflorus*	Rosaceae
Toyon	*Heteromeles arbutifolia*	Rosaceae
American grape	*Vitis labrusca*	Vitaceae

Table (*Contd...*)

Table (*Contd...*)

Fruits of Mediterranean Origin

Fruit plant	*Botanical name*	*Family*
Black mulberry	*Morus nigra*	Moraceae
Cornelian cherry	*Cornus mas*	Cornaceae
Date palm	*Phoenix dactylifera*	Arecaceae
Fig	*Ficus* spp	Moraceae
Grape, called raisin, sultana, or currant when it is dried.	*Vitis* spp	Vitaceae
Jujube	*Ziziphus zizyphus*	Rhamnaceae
Olive	*Olea europea*	Oleaceae
Pomegranate	*Punica granatum*	Punicaceae
Sycamore fig	*Ficus sycomorus*	Moraceae

Fruits of Asian origin

Fruit plant	*Botanical name*	*Family*
Citrus fruits	*Citrus* spp.	Rutaceae
Arhat	*Siraitia grosvenorii*	Cucurbitaceae
Coconut	*Cocos nucifera*	Arecaceae
Che Also called Cudrania, Chinese Mulberry, Cudrang, Mandarin Melon Berry, Silkworm Thorn, zhe	*Cudrania tricuspidata*	Moraceae
Durian	*Durio* spp	Malvaceae
Goumi	*Elaeagnus multiflora ovata*	Elaeagnaceae
Hardy Kiwi	*Actinidia arguta*	Actinidiaceae
Kiwifruit or Chinese gooseberry	*Actinidia* spp.	Actinidiaceae
Mock Strawberry or Indian Strawberry	*Potentilla indica*	Rosaceae
Lanzones	*Lansium domesticum*	Meliaceae
Lapsi	*Choerospondias axillaris*	Anacardiaceae
Longan	*Dimocarpus longan*	Sapindaceae
Lychee	*Litchi chinensis*	Sapindaceae
Mangosteen	*Garcinia mangostana*	Clusiaceae
Nungu	*Borassus flabellifer*	Arecaceae
Peach	*Prunus persica*	Rosaceae
Persimmon	*Diospyros kaki*	Ebenaceae
Rambutan	*Nephelium lappaceum*	Sapindaceae
Rhubarb	*Rheum rhaponticum*	Polygonaceae
Sageretia	*Sageretia theezans*	Rhamnaceae

Table (*Contd...*)

Table (*Contd...*)

Fruits of Australian origin

Fruit plant	*Botanical name*	*Family*
Atherton Raspberry	*Rubus probus*	Rosaceae
Black Apple	*Planchonella australis*	Sapotaceae
Blue tongue	*Melastoma affine*	Melastomataceae
Bolwarra	*Eupomatia laurina*	Eupomatiaceae
Burdekin Plum	*Pleiogynium timorense*	Anacardiaceae
Broad-leaf Bramble	*Rubus hillii*	Rosaceae
Cedar Bay cherry	*Eugenia carissoides*	Myrtaceae
Cluster fig	*Ficus racemosa*	Moraceae
Common apple-berry	*Billardiera scandens*	Pittosporaceae
Conkerberry	*Carissa lanceolata*	Apocynaceae
Davidson's plum	*Davidsonia* spp.	Cunoniaceae
Desert fig	*Ficus platypoda*	Moraceae
Desert lime	*Citrus glauca*	Rutaceae
Doubah	*Marsdenia australis*	Apocynaceae
Emu Apple	*Owenia acidula*	Meliaceae
Fibrous Satinash	*Syzygium fibrosum*	Myrtaceae
Finger Lime	*Citrus australasica*	Rutaceae
Illawarra Plum	*Podocarpus elatus*	Podocarpaceae
Little gooseberry	*Buachanania arborescens*	Anacardiaceae
Kakadu lime	*Citrus gracilis*	Rutaceae
Kutjera	*Solanum centrale*	Solanaceae
Kakadu plum	*Terminalia ferdinandiana*	Combretaceae
Karkalla	*Carpobrotus rossii*	Aizoaceae
Lady apple	*Syzygium suborbiculare*	Myrtaceae
Lemon aspen	*Acronychia acidula*	Rutaceae
Midyim	*Austromyrtus dulcis*	Myrtaceae
Mountain pepper	*Tasmannia* spp	Winteraceae
Muntries	*Kunzea pomifera*	Myrtaceae
Native Cherry	*Exocarpus cupressiformis*	Santalaceae
Native currant	*Acrotriche depressa*	Ericaceae
Native gooseberry	*Physalis minima*	Solanaceae
Pigface	*Carpobrotus glaucescens*	Aizoaceae
Pink-flowered Native Raspberry	*Rubus parvifolius*	Rosaceae
Purple apple-berry	*Billarderia longiflora*	Pittosporaceae
Quandong	*Santalum acuminatum*	Elaeocarpaceae
Riberry	*Syzygium luehmannii*	Myrtaceae
Rose-leaf Bramble	*Rubus rosifolius*	Rosaceae
Rose myrtle	*Archirhodomyrtus beckleri*	Myrtaceae
Sandpaper Fig	*Ficus coronata*	Moraceae
Small-leaf tamarind	*Diploglottis campbellii*	Sapindaceae

Table (*Contd...*)

Table (*Contd...*)

Fruit plant	*Botanical name*	*Family*
Snow berry	*Gaultheria hispida*	Ericaceae
Sweet apple-berry	*Billarderia cymosa*	Pittosporaceae
Tanjong	*Mimusops elengi*	Sapindaceae
White aspen	*Acronychia oblongifolia*	Rutaceae
Wild orange	*Capparis mitchellii*	Capparaceae
Wongi	*Manilkara kaukii*	Sapotaceae
Yellow plum	*Ximenia americana*	Olacaceae
Zig Zag Vine	*Melodurum leichhardtii*	Annonaceae

SUBJECT INDEX

B

D

E

F

G

H

I

J

K

L

M

N

O

P

Q

R

S

T

Y

Z